World Survey of Climatology Volume 14

CLIMATES OF THE POLAR REGIONS

World Survey of Climatology

World Survey of Climatology Volume 14

Climates of the Polar Regions

edited by S. ORVIG

Department of Meteorology
McGill University
Montreal, Que. (Canada)

ELSEVIER PUBLISHING COMPANY Amsterdam-London-New York 1970

ELSEVIER PUBLISHING COMPANY
335 Jan van Galenstraat
P.O. Box 211, Amsterdam, The Netherlands

ELSEVIER PUBLISHING COMPANY LTD.
Barking, Essex, England

AMERICAN ELSEVIER PUBLISHING COMPANY, INC.
52 Vanderbilt Avenue
New York, New York 10017

Library of Congress Card Number: 79–103355

ISBN 0-444-40828-2

With 91 illustrations and 193 tables

Printed in The Netherlands

World Survey of Climatology

Editor in Chief: H. E. LANDSBERG

Volume 1 General Climatology, 1
Editor: H. FLOHN

Volume 2 General Climatology, 2
Editor: H. FLOHN

Volume 3 General Climatology, 3
Editor: H. FLOHN

Volume 4 Climate of the Free Atmosphere
Editor: D. F. REX

Volume 5 Climates of Northern and Western Europe
Editor: C. C. WALLÉN

Volume 6 Climates of Central and Southern Europe
Editor: C. C. WALLÉN

Volume 7 Climates of the Soviet Union
Editor: P. E. LYDOLPH

Volume 8 Climates of Northern and Eastern Asia
Editor: H. ARAKAWA

Volume 9 Climates of Southern and Western Asia
Editor: H. ARAKAWA

Volume 10 Climates of Africa
Editor: J. F. GRIFFITHS

Volume 11 Climates of North America
Editor: R. A. BRYSON

Volume 12 Climates of Central and South America
Editor: W. SCHWERDTFEGER

Volume 13 Climates of Australia and New Zealand
Editor: J. GENTILLI

Volume 14 Climates of the Polar Regions
Editor: S. ORVIG

Volume 15 Climates of the Oceans
Editor: H. THOMSEN

List of Contributors to this Volume:

S. Orvig
Department of Meteorology
McGill University
Montreal, Que. (Canada)

P. Putnins
Environmental Science Services Administration
Rockville, Md. (U.S.A.)

W. Schwerdtfeger
Department of Meteorology
University of Wisconsin
Madison, Wisc. (U.S.A.)

E. Vowinckel
Department of Meteorology
McGill University
Montreal, Que. (Canada)

Contents

Chapter 3. THE CLIMATE OF THE NORTH POLAR BASIN
by E. VOWINCKEL and S. ORVIG

Chapter 4. THE CLIMATE OF THE ANTARCTIC
by W. SCHWERDTFEGER

Chapter 1

Introduction

S. ORVIG

Climatology is usually defined as being the description of average atmospheric conditions. Generally, it has been agreed to use a thirty-year period for the average value of the various climatic elements. The calculation of such mean values can serve three purposes:
(*1*) recording the history of climatic events;
(*2*) predicting future climatic averages from the past;
(*3*) testing and verifying theories of physical meteorology and general circulation with real values.

The first of these points can hardly be regarded as a scientific endeavour, although it is important, and it will be disregarded in the following.

In the early days of scientific climatology it was assumed that the second purpose could be fulfilled. However, investigations into climatic change and the longer period of observational data, as well as accumulated weather maps, have shown that climate is by no means stable. Therefore, point (*2*) cannot be fulfilled strictly. In most areas of the world the fluctuations of climate are of little importance, at least for practical purposes. The level of the major climatic elements is neither raised nor lowered sufficiently to affect human beings or life significantly. Climatic fluctuations may produce profound changes only in areas where one of the main elements becomes marginal, for instance at the fringe of deserts or near the poles. Slight changes in the energy balance, especially in the north, may cause large scale variations in ice and snow cover with resulting fundamental climatic changes.

If one considers that the climatic fluctuations during the past half century were more marked in the far north than in other parts of the world, then it becomes apparent that the 30-year period for climatic normals has little meaning in the Arctic, and predictions from such a data period will be doubtful.

The final of the three purposes, that of testing and verifying theories with real data, becomes very important in all regional climatologies. However, for the polar regions it would seem to be the major guide, if not the sole guideline, for a climatology. Some of the discussion in the present volume is, accordingly, somewhat different from conventional climatology. There is, particularly for the North Polar Basin, less emphasis on distributions and frequencies and rather more consideration of the governing physical processes which find their expression in the energy balance. This method has as a consequence that minor climatological elements become highly significant, although they may not usually be considered or even measured in detail. The climatological observing and recording networks are conservative and not always adapted to the requirements of such climatology. As a result, the data coverage for such elements is very meagre. One should consider the values resulting from using such data as being approximations and definitely not as certain as the distributions of the major climatic elements.

From these considerations it becomes apparent that a 30-year period of data is not useful for the prediction of future climatic conditions in high latitudes, and such a series of observations is not required for the purpose of testing theories of physical meteorology and atmospheric circulation. This is helpful in the present connection, because the polar regions have not been sampled extensively in time or space.

The deficiency of having only a short period of observations, however, is not of great consequence. Much more important is the fact that the data available refer to quite different periods from one station to another. In view of the great variability of weather and climate in the polar regions, the errors introduced by using such data, obtained at different times, are likely to be large and unpredictable. At present this difficulty cannot be overcome and should be kept in mind.

Another problem is caused by the paucity of observing stations in the polar regions. A similar number of stations in a continental area would make it impossible to prepare a climatology. However, the uniformity of the ice, snow and water surfaces makes each station much more representative of a larger area than would be the case at a continental site. The details of any analysis of climatic elements in the polar regions, as far as they exist, remain somewhat uncertain. However, it is quite safe to state that the large scale distributions are reasonably well known, as are the general atmospheric conditions and processes. Over the North Polar Basin this is indicated also by the fact that the results now obtained, using much more plentiful data, essentially correspond to those reported in his pioneering works by H. U. Sverdrup in the 1930's.

Chapter 2

The Climate of Greenland

P. PUTNINS

Introduction

Greenland is a huge island, extending from the 60th parallel to approximately 83°N, almost entirely covered by an ice sheet whose length, along the 44°W meridian, is about 2,400 km. The greatest width of the ice sheet, at 77°N, is about 1,100 km.

The area of Greenland, including the islands, is 2,186,000 km². The main ice cap covers an area of 1,726,000 km² or approximately 4/5 of the island. The area of isolated peripheral glaciers is 76,000 km², and the area of the ice-free land is only 383,000 km².

In its central part, Greenland's ice cap lies at 3,000 m or more above sea level and 1,504,200 km², or 87%, of the ice cap is above 1,220 m. The highest part of the ice sheet, i.e., that portion between 3,050 and 3,230 m in elevation, has an area of 114,200 km² or 6.6% of the total area of the cap.

The volume of the ice mass is huge. Almost 1/3 of its area lies below sea level. The mean thickness of the ice cap is 1,515 m. Thus, the volume of ice on Greenland is about 2,600,000 km³ with a water equivalent of 2,350,000 km³. If this water were distributed over the world's oceans, the sea level would be raised 6.5 m (see BADER, 1961).

Greenland, located partly in the predominant paths of low pressure systems, must certainly influence the patterns of the general atmospheric circulation over a sizeable portion of the Northern Hemisphere. Due to its height, orographic influences it exerts are particularly pronounced: shallow lows from the west are usually retarded and move northward along its west coast. Under some circumstances, a low approaching the southern tip of Greenland can be split into two parts. One of the resulting lows then moves off toward the northeast or east, and the other moves northward along the west coast. The upper portions of lows and frontal disturbances cross the ice cap but, because of the almost total lack of observations from the interior, details concerning the three-dimensional structure of these disturbances while crossing the ice cap are not yet well-known.

The bottom parts of lows occlude orographically or move off in directions other than the initial one, i.e., they avoid the huge ice massif. Apparently the meridional exchange of air masses is very strong in the Greenland area. Interdiurnal temperature changes as large as 30°C or more have been observed on the ice cap. A large interdiurnal temperature change on the ice cap is frequently observed together with the destruction, or building, of a surface inversion and thus may not be particularly representative of air mass exchange. Still, it is known that temperature variations in the lower part of the troposphere are large over Greenland and in its vicinity. Except in summer, when the greatest changes in temperature are recorded over the Canadian Archipelago, extreme departures from the normal values are observed over Greenland or near its coast, e.g., between south Greenland and Labrador and between Iceland and the east coast of Greenland. In general, the same holds for heights of the 500-mbar level, i.e., centres of greatest positive and negative departures from normal are located near Greenland or over the island. This strong meridional air mass exchange is also noted in the upper part of the troposphere. Air mass changes in this part of the troposphere can also be detected by the variations of the height of the tropopause.

One extreme case was observed in February, 1950, during P. E. Victor's expedition. On this occasion, the height of the tropopause changed from 5,100 m to 11,390 m in 28 h, a change of 6,290 m! This increase in the height of the tropopause occurred as a result of a strong invasion of warm air throughout the entire troposphere.

One of the most pronounced phenomena on the ice cap is the almost permanent surface inversion. This inversion, created by strong radiational effects, exerts considerable influence on the circulation. Cold air sinks from the higher elevations of the centre of the ice cap toward the periphery and, deflected to the right due to the Coriolis effect, appears as a clockwise circulation near the coast and creates the impression of the "glacial anticyclone" postulated by Hobbs. This outflow of cold air produces conditions favorable for a vertical circulation near the coast (Sandström's rule and extreme baroclinicity). These conditions, strengthened by the flow of warm air from the south along Davis Strait, contribute considerably to the initiation of cyclogenesis near the west coast of Greenland.

The most peculiar, and perhaps problematical, phenomenon is the presence of open water in Smith Sound (northwest of Thule, on the northwest coast of Greenland) from December through March. During this season, nearly all of Davis Strait is ice covered. The open water areas in Smith Sound apparently decrease the stability of the air in this region and very likely contribute to the cyclogenesis frequently observed.

The complexity of the climatic problem in Greenland is also illustrated by the presence of ice free regions. These regions occupy strips of the coast as much as 160 km wide, south of 73°N in the west and between 70° and 78°N on the east coast, and along the entire north coast in latitudes 80°–82°. These parts are inhabited by various polar animals. In the northeast these include the musk-ox which, during the winter or polar night when there is very little snow, feeds on grass which has been frozen and preserved. The ice free strips are predominantly plateaux of 1,000 m elevation (Öpik, 1957).

Due to the great meridional length of Greenland and the high elevation of the ice cap, it is to be expected that the island should exhibit types of climate which differ quite substantially, such as those of the coast and those of the ice cap. Besides, the northern part of Greenland is located in the polar area while the southern end of the island is in that part of the Atlantic where the drift of subtropical air masses toward the pole takes place and thus shows a considerable positive anomaly. However, the inland ice together with the cold sea and the largest ice current in the world exert sufficient influence to largely wipe out the differences. Generally, the climate in the north is purely polar, whereas in the south it represents, as Petersen (1928) states, a curious transition stage between the cold climate of the east coast of North America at the same latitude (Labrador), and the mild oceanic climate of northwestern Europe.

However, in spite of the general differences between the climates of the northern and southern parts of Greenland, the strong meridional exchanges of air masses along the coasts erase the latitudinal differences in some areas.

Weather dynamics in the Greenland area

Greenland's role in the general atmospheric circulation

Greenland, extending over somewhat more than 23° of latitude and 50° of longitude and rising to more than 3,000 m above sea level in the central part, may well be expected to influence the atmospheric circulation in general. For example, a departure of the upper wind from geostrophic conditions results in the formation of waves which are propagated with rather great velocities. After a time, the pressure field adjusts itself to come into balance with the velocity field (OBUKHOV, 1949). Since the elevated ice cap generally disturbs the wind field by acting as a barrier, and also by the outflow of cold air from the ice sheet, ageostrophic conditions are to be expected. Thus, the influence of Greenland should be manifested in the forms of certain perturbations in the dynamic field.
Musaelian (as quoted in VOROBJEVA, 1962) computed, for the Northern Hemisphere, the pressure disturbance caused by orographic obstructions (mountain chains) to zonal flow. He showed that Greenland, along with the Rocky Mountains and the Urals, is one of the obstacles which is the cause of some of the stronger disturbances. The presence of Greenland should result in the generation of a wave system consisting of two waves, each with a wave length of approximately 7,000 km. This solution is valid only if constant zonal flow is assumed upstream of the obstruction. Considering only Greenland, this assumption must be questioned when one takes note of the presence of a circulation centre over the northeastern part of the Canadian Archipelago and the fact that the annual displacement of the centre is small. Due to the accompanying configurations of pressure patterns, meridional air mass exchange plays a significant role in the Greenland area.
The mean circulation pattern, as shown by average sea-level pressure maps for January, April, July and October, has a ridge over Greenland but no closed isobars. At 700 mbar, however, only the mean map for April shows a ridge over the island (NAMIAS, 1958).
Since disturbances form and develop periodically in baroclinic areas, i.e., over water areas adjacent to cold continents and warm oceans in winter, such disturbances have a tendency to develop at both coasts of Greenland. Here, cyclonic curvature of the air trajectories in the lower troposphere (700 mbar) reinforces the northward advection of warm air as well as the formation of cut-off cold masses when other conditions are favourable. Blocking activity is generally most pronounced in the Arctic in April and May. These tendencies may be due in part to increased pressure over the compact ice cover as is frequently the case over the Gulf of Bothnia and the Gulf of Finland in spring.
Actually, differences between synoptic scale phenomena of the Arctic and those of temperate latitude are probably not as great as formerly supposed. One of the more significant differences appears to be that the Polar Basin is either a sort of transit area or a sink for cyclones and anticyclones which develop elsewhere (NAMIAS, 1958). The entrance of lows into the Polar Basin is frequently favoured by blocking highs and by the blocking influence of Greenland itself. Greenland is a permanent hindrance to lows approaching from the west, which are therefore diverted northwards. Quite frequently the lows deepen and undergo cyclogenesis. The quite permanent outflow of cold air from

the ice cap, the strong zonal temperature gradient, and the advection of warm air in the lower levels are conditions for this type of development.

The atmospheric circulation is also dependent on the temperature distribution. Thickness maps for the 500–1,000 mbar layer show, for December through February, a tongue of cold air over the northeastern part of the Canadian Archipelago. In July–August, the cold air is still over the Canadian Archipelago, but it covers the greater part of Greenland as well (TVERSKOI, 1962).

Observational data from the Arctic stations supplemented with data from the central Arctic (drifting stations) show the following pressure distribution (GAIGEROV, 1962): In January, high pressure covers the ice cap (>1,015 mbar), and a pronounced trough extends from Davis Strait to Baffin Bay. In April, the pressure over the ice cap is somewhat higher (1,020 mbar or more), a low pressure centre is situated over Iceland and a trough extends to Davis Strait and Baffin Bay from the south. In July, high pressure dominates the ice cap. The northeastern part of the Canadian Archipelago, including Davis Strait, shows a flat low pressure distribution which is even more pronounced than the Icelandic low. In October there is high pressure over the ice cap and a low pressure centre is located over Davis Strait. This low extends north and northwestward and is more pronounced than the weak trough which stretches eastward toward Iceland.

The influence of Greenland on the general atmospheric circulation is very different from that of Antarctica. The Antarctic continent, completely covered by a permanent ice sheet, is centred at the pole. Greenland, however, is partly situated in the zone of general west–east drift.

The participation of Greenland in the general circulation is, therefore, more pronounced than is that of Antarctica. The role of Greenland as a barrier in the zone of the westerlies also causes a furthering of the meridional exchange in this area, providing preferred sites for the outbreak of cold air southward, as well as for the injection of warm subtropical air deep into the Arctic Basin. Thus Greenland, as opposed to Antarctica, is itself directly involved in the process of the general circulation.

As the intensity of the atmospheric circulation also depends on the horizontal and vertical distribution of temperature, it is interesting to see how this parameter is distributed in the Greenland area.

ARNASON and VUORELA (1955) give a two-parameter model of the normal temperature distribution, in the 500–1,000 mbar layer, such that the normal temperature height curve is approximated by a straight line which was adjusted so that the 700–1,000 mbar and 500–700 mbar thicknesses were correctly given by the model.

According to this definition, the "representative" 1,000 mbar temperature, τ_0, cannot deviate too much from the mean 700–1,000 mbar thickness. The computed values for January show a maximum of the τ_0 gradient along the southern coast of Greenland and across the North American continent. The maximum coincides, approximately, with the 50th parallel. By July, the maximum of the τ_0 gradient has shifted eastward from Greenland, but now another maximum appears on the northern shore of the island. These zones with maximum τ_0 gradient are the sites of maximum low level baroclinicity. In January, the "adjusted" lapse-rate κ (τ_0 and κ are uniquely determined from the 700–1,000 mbar and 500–700 mbar thickness normals) shows the strongest changes in the region of the Canadian Archipelago and in the Kap Tobin area of eastern Greenland; in July, a strong change of κ-values is present in the vicinity of the north shore of Greenland. Generally, a large gradient of κ-values appears in winter between Iceland and Canada, and in summer between the central part of Greenland and the North Pole. The area of large lapse-rates centred over Iceland can be explained by the frequent outbreaks of cold continental air from Canada and Greenland. The long trajectory over water increases the instability.

One of the most important factors in the climatic features of Greenland is the Icelandic low together with the trough extending northward from this low. If the semi-permanent Icelandic low is regarded as the result of the "capture" of occluding lows on their way eastward and northward, the contribution of the trough to sustaining the Icelandic low is not minimized.

The following sequence can be visualized: at the east coast of Greenland there is apparently a nearly continuous outflow of arctic air from the Polar Basin. Every cold air invasion is accompanied by pressure falls aloft, thus creating a semi-permanent trough, which strengthens the Icelandic low. This trough also contributes, potentially, to cyclogenesis at the east coast, north of the Icelandic low, which can occur when a pressure fall area crosses the ice cap. How it occurs depends on the circumstances. It could be that warm air crossing the ice cap from the west meets cold arctic air at the east coast, resulting in a development in the usual way. Sometimes a low disappears at the west coast to reappear some time later on the east coast. A pressure fall area moving eastward across the ice cap will apparently strengthen the trough also in those cases when development does not take place.

TABLE I

VALUES OF THE RATIO $\bar{u}/V_r$

Station	Winter						Summer					
	1,000 mbar	850 mbar	700 mbar	500 mbar	300 mbar	200 mbar	1,000 mbar	850 mbar	700 mbar	500 mbar	300 mbar	200 mbar
Eureka	1.00	0.90	0.81	0.93	0.30	0.76	0.99	0.21	0.67	0.93	0.94	0.97
Isachsen	0.49	0.45	0.38	0.60	0.92	0.97	0.91	0.94	0.98	0.99	0.99	0.97
Resolute	0.73	0.12	0.23	0.38	0.69	0.77	0.26	0.44	0.58	0.73	0.82	0.76
Alert	0.98	0.96	0.19	1.00	0.53	0.86	0.02	0.91	0.12	0.95	0.62	0.91
Thule	0.97	0.75	0.11	0.22	0.38	0.52	0.81	0.73	0.20	0.62	0.92	1.00
Egedesminde	1.00	0.18	0.15	0.32	0.53	0.63	0.91	0.23	0.22	0.45	0.93	0.93
Narssarssuaq	0.77	0.88	0.84	0.78	0.67	0.79	0.99	0.95	0.58	0.95	0.96	0.96
Angmagssalik	0.31	0.93	0.94	0.70	0.89	0.87	0.39	0.88	0.95	0.89	0.99	0.99
Kap Tobin	0.65	0.34	1.00	0.97	0.97	0.99	1.00	0.22	0.97	0.99	0.96	0.99
Danmarkshavn	0.75	0.84	0.97	0.99	–	–	0.64	0.81	0.84	0.93	0.95	0.99
Goose Bay	0.94	0.88	1.00	0.97	0.95	0.96	0.93	0.99	1.00	1.00	1.00	1.00
Ship "B"	0.92	1.00	0.95	0.88	0.84	0.85	0.63	0.93	0.98	0.99	0.98	0.98
Ship "A"	0.87	0.64	0.62	0.77	0.85	0.89	0.33	0.24	0.65	0.98	1.00	1.00
Keflavik	0.93	0.04	0.60	0.81	0.88	0.90	0.77	0.21	0.49	0.91	0.93	0.95
Jan Mayen	0.20	0.15	0.96	0.99	1.00	0.91	0.65	0.02	0.97	0.62	0.94	0.85
Ship "C"	0.96	0.90	0.91	0.90	0.92	0.94	0.90	0.96	0.98	0.99	1.00	1.00
Ship "I"	–	0.82	0.88	0.92	0.96	0.98	–	0.93	0.95	0.98	0.99	1.00
Ship "M"	0.81	0.92	0.96	1.00	0.99	0.98	0.83	0.68	0.78	0.82	0.87	0.89

There has been the recurring suggestion that the atmospheric circulation in the Greenland area includes a substantial meridional component. Table I shows the ratio $\bar{u}/V_r$ for the stations in the Greenland area for a number of pressure levels in winter (December through March) and in summer (June through August) (PUTNINS et al., 1959b). Wind observations were summarized for the approximate period 1949–1958. Here $\bar{u}$ is the mean zonal wind component and V_r is the resultant wind. $\bar{u}/V_r = 0$ represents a pure meridional wind and $\bar{u}/V_r = 1$ corresponds to a pure zonal wind, provided $V_r \neq 0$. In case $\bar{u} = \bar{v}$, $\bar{u}/V_r = 0.707$. This table only indicates the predominance of zonal or meridional wind components, e.g., in winter at Egedesminde the surface wind is pure zonal while at the 850, 700 and 500 mbar levels, the meridional component is predominant. Further, by taking into account the signs of the wind components, we come to the following interpretation.

Winter

On the east coast at 1,000 mbar, a meridional component from the north is dominant at Angmagssalik, Kap Tobin, and Jan Mayen. At 850 mbar a dominantly meridional component from the south is present at Ship "A" and Keflavik, while Kap Tobin and Jan Mayen mean winds still have a dominant component from the north. At 700 mbar, Ship "A" and Keflavik retain a slightly larger component from the south. Above this level, winds on the east side are very uniform such that they have a predominantly zonal component from the west.

In the west, the situation is considerably more interesting. Only Isachsen shows a primarily meridional wind from the north at 1,000 mbar, while the Greenland coastal stations and also Eureka and Resolute, have a wind with primary component from the east. At 850 mbar, Resolute has joined Isachsen, and Egedesminde now is primarily meridional from the south. At 700 mbar Alert, Isachsen and Resolute now have a relatively large north component, while Thule and also Egedesminde exhibit a relatively large component from the south. The situation at 500 mbar differs little from that at 700 mbar. The surprise appears at 300 mbar where now only Isachsen exhibits a primarily zonal component from the west. Resolute's major component is from the north. Alert, Eureka, Thule, Egedesminde are all mostly southerly. Narssarssuaq may also be included in this latter group, though it has only a slightly smaller westerly component. At 200 mbar Thule and Egedesminde still have a relatively large meridional component, while all remaining stations exhibit a relatively large component from the west.

Summer

Winds at 1,000 mbar show very little organization, as might be expected. At 850 mbar there is still alternation of the dominant component from station to station. At 700 mbar there is now some organization: Alert, Eureka, Thule, Egedesminde, and Narssarssuaq (all on the west side) have a relatively large component from the south. To this group may be added Keflavik, while a northerly component prevails at Resolute and at Ship "A". All other stations are zonal westerly. By 500 mbar the zonal character of the flow is well established in this season, with only Thule and Egedesminde showing a relatively large southerly component, and Jan Mayen a relatively large northerly component. At 300 mbar only Alert has a larger component from the south and even here the west component is nearly as large. At 200 mbar there is no exception to a predominantly zonal flow from the west and the magnitude of the ratio is such that in every case the meridional component is very small (see also SCHALLERT, 1960, 1964a). Since the east and west parts of Greenland are generally governed by different circulation regimes, an attempt was made to investigate the zonal indices (Z.I.) for two regions, west and east of the southern part of Greenland. For comparison, the regions south of Greenland, west and east of the 50th and 40th west meridians were also examined (PUTNINS, 1963, 1965, 1966).

Here we shall discuss the zonal indices in the following areas: (*1*) around the southern part of Greenland, i.e., 60°–70°N and 50°–90°W and 40°W–0°; (*2*) south of Greenland, i.e., 50°–60°N and 50°–90°W and 40°W–0°.

Zonal indices were computed from monthly values taken from 500 mbar contour maps. Table II presents the mean monthly values of Z.I. for the years 1949–1961/62 for the indicated areas, and Fig.1 and 2 present these values graphically. In the zone 60°–70°N (Table II and Fig.1), differences are most pronounced from December through April. During other months, the course of Z.I. is quite similar. The maximum in the area east of Greenland (40°W–0°) is strongly pronounced in winter months with the peak in March and the minimum in summer. In the area west of Greenland (50°–90°W), there are two maxima—the principal one occurs in October and a secondary appears during February–April. The principal minimum occurs in summer, as it does in the area east of Greenland, and the secondary comes during December–January.

The annual variation of the Z.I. was analyzed harmonically. Table III shows the amplitude of the first three harmonics expressed as per cent of the variance, the magnitudes of the phase angles and the times of maxima.

In the zone comprising the southern part of Greenland (60°–70°N) the semi-annual variation of Z.I. is predominant in the western part of the zone. The time of the first maximum of the semi-annual variation is approximately April 6 and that of the annual variation about December 6. In the eastern part of the zone (40°W–0°), the annual variation is dominant—the amplitude of the annual variation is nearly four times that of the semi-annual variation. The times of the maxima are: January 6, March 16 and March 6. Fig.1 also shows the calculated annual course of Z.I. as given by the sum of the first three harmonics. The agreement is quite good. The difference in the behaviour of the annual variation of the Z.I. in the western and eastern parts of the zone from 60°–70°N can partly be explained by the annual movement of the cold circulation centre (PUTNINS et al., 1959b).

TABLE II

ZONAL INDEX GPM (1949–1961, 1962)
(After PUTNINS, 1963)

Latitude	Jan.	Feb.	Mar.	Apr.	May	June	July	Aug.	Sept.	Oct.	Nov.	Dec.	Longitude
60°–70°N	56.64	88.64	85.48	93.16	64.66	40.78	44.58	51.12	83.66	108.02	99.44	73.88	50°–90°W
	103.78	116.24	124.32	118.34	59.02	50.22	45.90	48.58	86.46	95.42	103.68	104.00	40°W–0°
50°–60°N	167.38	151.86	86.42	105.76	117.80	128.08	131.76	147.34	175.86	193.60	163.58	148.18	50°–90°W
	197.08	156.02	107.52	138.04	100.44	125.40	149.68	148.58	158.30	181.78	166.04	203.36	40°W–0°

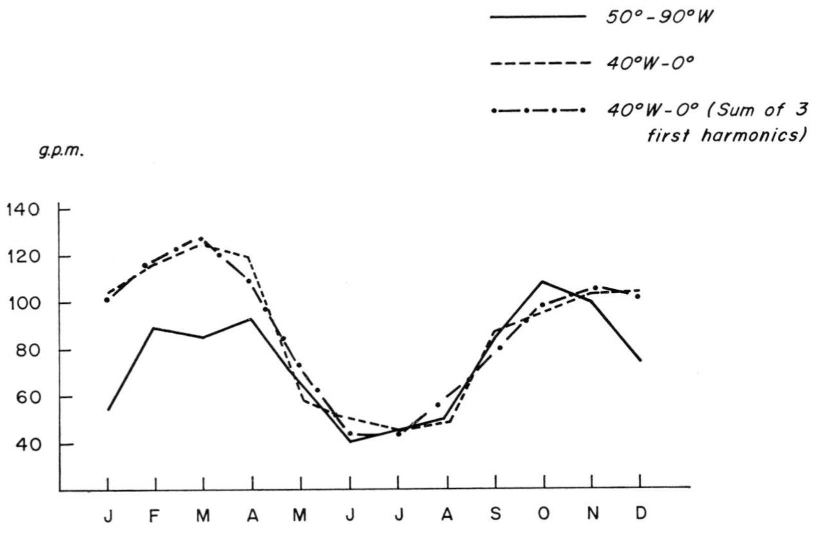

Fig.1. Zonal index 60°–70°N (1949–1961, 1962).

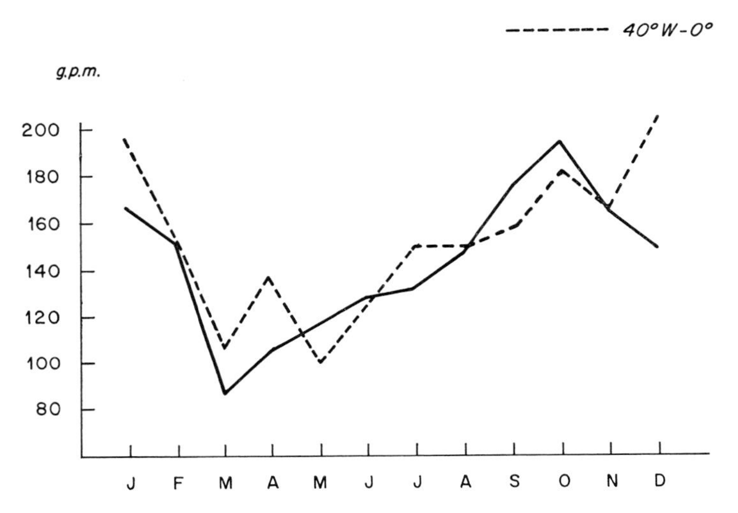

Fig.2. Zonal index 50°–60°N (1949–1961, 1962).

Whether the difference in the annual variation of Z.I. in the areas west and east of Greenland is explained entirely by the movement of the circulation centre in the area north of the Canadian Archipelago, or whether Greenland itself exerts an influence, remains a question. For this reason, the variation of Z.I. in the zone from 50° to 60°N was also investigated (Tables II, III and Fig.2).

In this southern zone, the largest differences in Z.I. appear in December and January. Although the courses of Z.I. in eastern and western parts of the zone are generally parallel during the remaining ten months, differences still appear, especially in April and May. In the area west of Greenland the maximum occurs in October, as it does in the area north of it. A pronounced minimum occurs in March. East of Greenland (40°–0°) the maximum appears in December–January and the minimum in March and May. The Z.I. increases in April.

TABLE III

AMPLITUDES OF THE FIRST THREE HARMONICS EXPRESSED IN PER CENT OF THE VARIANCE (a, b, c), MAGNITUDES OF THE PHASE ANGLES (φ_a, φ_b, φ_c) AND TIME OF MAXIMUM (t_a, t_b, t_c) IN TENTHS OF THE MONTH
(After PUTNINS, 1963)

Areas		Amplitude			Phase (degrees)			Time of maximum (tenth of month)		
		a	b	c	φ_a	φ_b	φ_c	t_a	t_b	t_c
60°–70°N	50°–90°W	28	56	03	114.2	259.4	308.2	11.2	3.2	1.6
60°–70°N	40°–0°W	68	18	01	83.3	299.1	249.9	0.2	2.5	2.2
50°–60°N	50°–90°W	69	00	17	170.6	78.0	48.7	9.3	0.2	0.5
50°–60°N	40°–0°W	66	07	02	144.4	105.5	129.4	10.0	5.7	3.6

Harmonic analysis shows that the annual component of the variation of Z.I. is dominant in both sections of the zone. The times of maxima for the western part are: October 10 for the annual variation, January 6 for the first maximum of the semi-annual variation, and about January 15 for the first maximum of the four-monthly component. In the region east of Greenland, the times of maxima are: November 1 for the first harmonic, June 21 for the first maximum of the second harmonic, and about April 18 for the first maximum of the third harmonic.

Apparently there should be a relationship between the zonal index and the meridional temperature gradient. The correlation between these quantities is negative because temperature changes are not uniformly distributed about the pole, but rather in the form of more or less confined outbreaks of cold arctic air from the north and tongues of warm air injected from the south.

The correlation between the temperature gradient at 500 mbar, as measured by the temperature difference between Portland, Maine and Alert, N.W.T., and the zonal indices for the area bounded by 40° and 50°N and by 90° and 40°W, during the period from December 1960 to February 1961, was a maximum (−0.26, significant at 5% level) with 24-h lag. With a two-day lag, the correlation was still −0.20, but dropped strongly with a lag of three days. More interesting, perhaps, are the correlations between (*1*) the temperature differences between Kap Tobin and Nord (both on the east coast of Greenland); and (*2*) the zonal index in the area 60°–70°N 40°W–0°, for June–August, 1960. The correlation, using simultaneous values, was 0.04. The ratio increased strongly with two days lag (zonal index values two days behind the temperature gradient), then decreased very slowly as the lag was increased to ten days (−0.21) and dropped strongly with lags greater than ten days.

It is understandable that the correlation between temperature gradients and zonal indices becomes larger with increasing time lag. Since the atmospheric circulation depends directly on the temperature distribution, we should expect a correlation between these parameters. We can also expect the circulation to adjust itself to the temperature distribution only after a lag of time. Of course, it is a little surprising to find a case with a good correlation being sustained for all lags from 2 to 10 days (PUTNINS, 1963).

The correlations cited here were computed on a day to day basis, i.e., with no consideration for specific weather situations. The correlation between these parameters may be expected to be greater when specific and similar weather situations are taken into account and also when the data are taken by seasons. They can then possibly contribute to a better understanding of Greenland's influence on the atmospheric circulation.

Surface pressure changes, e.g., three-hourly pressure tendencies, can be regarded as an indication of atmospheric activity. These changes can be very large on the coast of Greenland. For example, Thule recorded a pressure drop of more than 15 mbar in a 3-h period in December, 1957. It is evident that such changes can be observed only at the coast where very pronounced air mass changes occur, and that such extreme changes will take place seldom.

Three-hourly pressure changes were computed for Station Centrale (only 10 months, September 1949 through June 1950, were available) and for Eismitte (August 1930–July 1931). These data were compared with Brønlunds Fjord, on the northeast coast of Greenland (82°10'N 30°30'W) during August 1948–July 1950 (Putnins et al., 1959b). Because Brønlunds Fjord belongs to another climatic regime, one cannot expect to find a close similarity between the pressure variation at this station and that at the stations on the ice cap. As the periods of record are short and different, and since weather conditions in the Arctic are quite changeable, one cannot expect the daily variation of pressure tendencies, by months, to be similar for the ice cap stations.

Generally, the daily variations of three-hourly pressure tendencies by months, at the two ice cap stations, show a certain similarity. For example in January, Station Centrale shows an increase of barometric tendencies between 03h00 and 15h00 as does Eismitte between 00h00 and 15h00. In February, Station Centrale had increasing tendencies between 09h00 and 21h00 and tendencies also increased at Eismitte between 00h00 and 18h00. In June, Station Centrale had increasing barometric tendencies between 00h00 and 15h00 and so did Eismitte from 00h00–12h00. In November, tendencies increased from 00h00–09h00 at Station Centrale and from 00h00–06h00 at Eismitte. At Brønlunds Fjord a daily variation of barometric tendencies is barely recognizable.

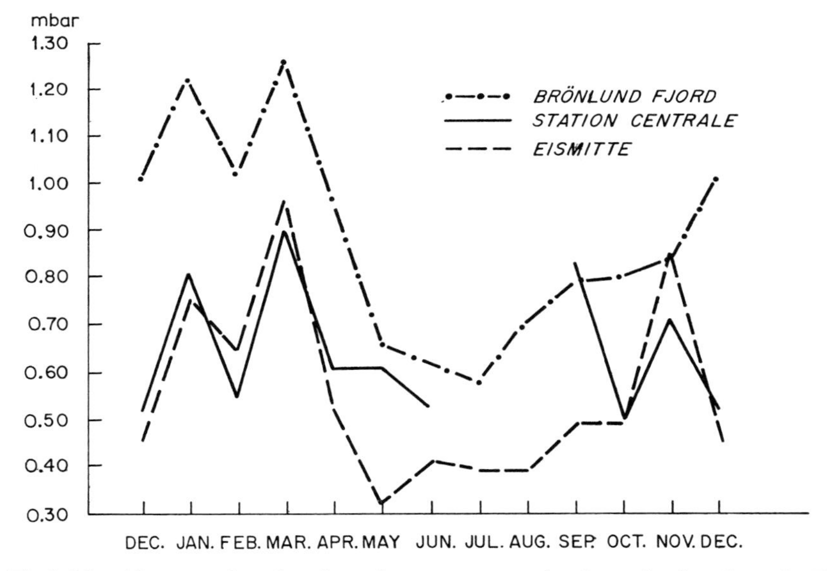

Fig.3. Monthly mean three-hourly surface pressure tendencies at Station Centrale, Eismitte and Brønlunds Fjord.

In the daily course of barometric tendencies for the year, all three stations show a slight increase during the afternoon (12h00–15h00 or 15h00–18h00). Station Centrale shows an increase in tendencies from 00h00 to 15h00 and at Eismitte it increases from 00h00 to 18h00. Brønlunds Fjord does not show a daily variation in tendencies, even for the year (PUTNINS et al., 1959b).

The annual variation of the three-hourly pressure tendency, by months, is quite interesting. All three stations show much larger tendencies in winter than in summer (Fig.3). A rather strong increase in the mean pressure tendency from November to December at Brønlunds Fjord, indicates that November still belongs to the transition season. Brønlunds Fjord, besides having larger tendencies, also has a relatively smaller number of occasions with no pressure change than do the stations in the central part of the ice cap (PUTNINS et al., 1959b). The ice cap stations show a pronounced maximum tendency in November. December, however, shows a secondary minimum on the ice cap as does February at all three stations.

Another factor to be considered in the problem of the general circulation is the distribution of the 24-h pressure fall centres in the Greenland area. Daily weather maps for a five-year period (1953–1957) were examined and the frequency of surface pressure centres was reduced to a standard area of 10^5 km^2. The same technique was applied to the frequency of 500 mbar minimum centres (PUTNINS et al., 1959a). The annual maps are presented here.

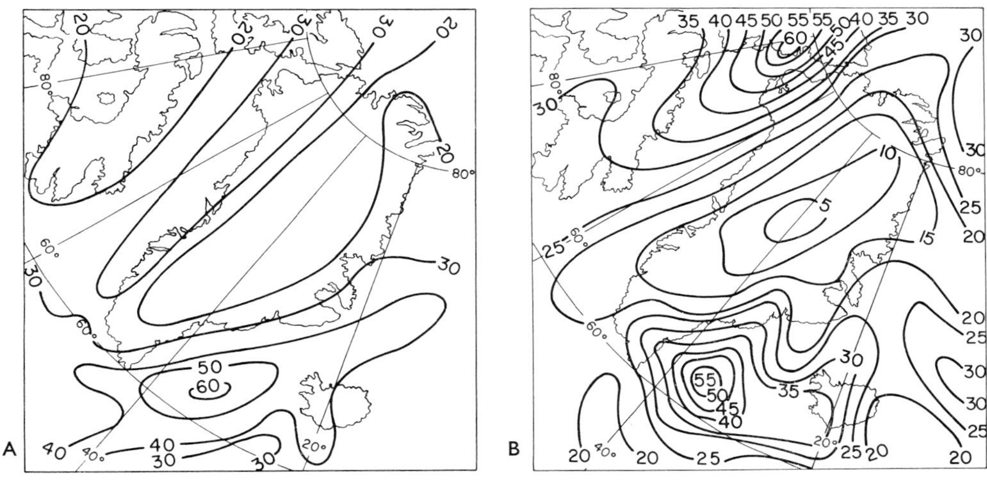

Fig.4A. Frequency of falling pressure surface, annual. B. Frequency of height minima, 500 mbar, annual.

The map for centres of falling pressure, Fig.4A, shows a minimum frequency over the interior of Greenland. This map illustrates the situation quite well for all months of the year. Each month saw areas over the ice cap with zero frequency. A relatively strong maximum appears at a centre located roughly midway between Kap Farvel and Iceland. This is the result of a centre which appears fairly regularly in or near this position month by month. A secondary maximum appears along the west coast of Greenland. Monthly maps emphasize the fact that this is a meteorologically interesting region. In January, a maximum appears in the vicinity of Thule. In April, a maximum is present in Davis Strait. In May and June, the pattern along the west coast is quite weak and in

July a minimum is centred in the vicinity of Disko Island. The balance of the year shows a somewhat erratic return to mid-winter conditions.

The map representing the frequency of 500 mbar minima (Fig.4B) is quite similar to the one showing annual frequencies of centres of falling pressure. There is a minimum frequency of lows over the interior of the island. Each month of the 4-year period showed several regions with zero frequency in the centre of the island. The maximum in the frequency of low centres over the northern part of Davis Strait is even more pronounced than the maximum located southeast of Greenland. Again the situation along the west coast is the most variable and interesting. A maximum of frequencies appears near Thule in January, March, April, July, October, November and December. In June there is a minimum near Thule and, to a lesser extent, this also occurs in September (PUTNINS et al., 1959a).

Also of interest is the frequency of occurrence of the "cold pole" in the Greenland area. MAEDE (1956) investigated the movement of the "cold pole" in the Arctic for a period of seven years, 1949–1955. In winter, the locations of the "cold pole" are concentrated north of 55°N and between 50° and 100°W; in summer they are much more scattered and also appear over Greenland and the adjacent seas. In winter, and especially in January and February, the "cold pole" is located over the Canadian Archipelago. The centre is located in the vicinity of 70°N 85°W, for 65–70% of the time. The annual course of the "cold pole" in the Greenland sector differs from that in other sectors of the Arctic. The annual frequency of occurrence of the "cold pole" in the Greenland sector is about 14%. In winter it is only about 8% and increases to 24% in summer (MAEDE, 1956).

One of the phenomena which is quite pronounced in Arctic regions is the so-called advance of cold air aloft. In the thirties and forties it was assumed that, in the neighbourhood of the Polar Basin and consequently near the polar front, the advance of cold air aloft should be a quite regular occurrence (PUTNINS, 1942). As the problem is immensely important to an understanding of atmospheric dynamics in the Arctic, interdiurnal changes of thickness of several atmospheric layers were computed from ascents made during Victor's expedition. For comparison, the same was done for a number of other arctic stations (PUTNINS et al., 1959a).

Since surface temperatures from the ice cap could not be used due to the effects of radiation, the thickness of the 500–700 mbar layer was taken as a surface layer and changes in thickness of this layer were compared with changes in thickness of the 300–500 mbar layer. Such a comparison fails to count some cases when the advance of cold air aloft takes place, because the advance frequently starts well below the 500 mbar level (PUTNINS, 1942). A decrease in the mean virtual temperature of the 300–500 mbar layer, during a 24-h period, which was numerically greater than a similar change in the 500–700 mbar layer, was assumed to indicate the advance of cold air aloft. During the two years of the expedition, there were only 51 successive ascents taken at 24-h intervals. In fifteen of these cases (about 30%), cooling aloft preceded the cooling in the lower layer.

For comparison, the ascents at Nord, Alert, Thule, Danmarkshavn, Kap Tobin, Jan Mayen, Angmagssalik, Keflavik, Ship "A", Ship "B", Narssarssuaq, Frobisher, Egedesminde and Arctic Bay were evaluated for January and July 1956. On an average, cooling aloft preceded cooling at the surface in 37% of all cases in January and in 41% of all cases in July. Ascents at the U.S.S.R. drifting station (approximate location 76.5°–

79.5°N 180°–160°E) were also evaluated for the period April–September 1950. In 130 of 256 cases (51%) a 24-h decrease in temperature aloft preceded the cooling at the surface (PUTNINS et al., 1959a).
The implications of these results are far-reaching. Cooling taking place first in middle and high layers of the troposphere results in a decrease in stability which, in turn, can release large scale vertical mass displacements. These strong vertical movements can extend into the upper part of the troposphere and can also affect the tropopause and consequently the lower stratosphere. We thus obtain a series of mutually interacting oscillations in the troposphere and stratosphere (PUTNINS, 1942).
Thickness changes can be regarded as a relative measure of the intensity of the vertical circulation. Since the thickness also represents the specific volume of a layer, it is possible to interpret the circulation acceleration (dC/dt) in terms of thicknesses. Further, a change in thickness is equivalent to the change in circulation acceleration—a change in energy which is not usually compensated in the atmosphere. In this way, the thicknesses of several layers in an area, as well as the thickness changes, can reveal not only the sites with high or low virtual temperatures but also the sites with more or less pronounced vertical circulation.
January and July, 1956 were investigated. Table IV and V show thicknesses and thickness changes at a number of Greenland stations.
Comparing January and July of 1956 shows that temperatures are much more uniform in July, especially in the lower layer. The temperature range is generally larger in the western part of the area. Also, the variability and the interdiurnal change of thicknesses (or temperatures) were much larger in January.
The values for January reveal some interesting features. In the lower layer (700/1,000 mbar), the lowest mean temperatures were observed at Thule; 10° lower than at Alert and Arctic Bay. Similar temperatures were recorded at Nord and Danmarkshavn. In this layer, interdiurnal changes in thickness are a maximum in the Alert–Arctic Bay area. In the Egedesminde–Narssarssuaq region, the maximum appears in the 500–700 mbar layer. In the 300–500 mbar layer, the maximum shifts to the Arctic Bay–Frobisher region of the Canadian Archipelago. There were no marked areas of maximum interdiurnal change in the eastern part of the Greenland area and only in the southern part is there an indication of an increase of interdiurnal thickness changes.
In July, 1956 (Table V) the interdiurnal change of thickness increased from the 500–700 mbar to the 300–500 mbar layer in both parts of the area. The interdiurnal thickness changes in the 300–500 mbar layer increased toward the south in the western part, but decreased in the eastern part of the area. In January, the interdiurnal change of temperature was more pronounced in lower levels. In July the interdiurnal change of temperature was approximately the same in the three layers considered.
This pilot study also indicates that, in the Greenland area, the meridional advection of air masses is quite pronounced though there probably are areas with greater cooling and more intense vertical exchange. Apparently the level of most intense vertical movement varies from site to site (PUTNINS et al., 1959a).
Another question is: how well are the upper air elements correlated? R. Berggren, S. Petterssen and R. Murrey showed that the correlation between the 500 mbar temperature and the 300–500 mbar thickness is generally very good.
A number of 91 ascents was examined for the winter of 1950/1951 (December through

TABLE IV

GREENLAND AREA, THICKNESSES (m), JANUARY 1956 (03h00Z)
(After PUTNINS et al., 1959a)

Stations	700–1,000 mbar							500–700 mbar							300–500 mbar						
	mean	var.	max. dev.	interdiurn. change				mean	var.	max. dev.	interdiurn. change				mean	var.	max. dev.	interdiurn. change			
				mean	$n+$	$n-$	greatest change				mean	$n+$	$n-$	greatest change				mean	$n+$	$n-$	greates change
Western part:																					
Alert	2,673	87	231	43	16	14	−101	2,323	36	185	30	15	14	137	3,299	43	102	33	16	13	−119
Thule	2,577	25	106	24	16	14	78	2,340	27	56	26	17	13	−92	3,311	36	80	34	14	16	105
Arctic Bay	2,687	92	219	50	16	14	−207	2,363	57	134	32	13	13	125	3,339	41	105	46	12	13	186
Egedesminde	2,621	33	101	21	8	10	70	2,355	40	87	46	8	11	111	3,359	53	96	30	9	10	92
Frobisher	2,634	72	175	34	17	12	−134	2,402	59	145	30	14	13	−71	3,382	46	155	46	7	11	186
Narssarssuaq	2,712	34	110	38	16	14	−112	2,447	49	182	43	14	16	−113	3,454	38	118	32	12	14	−146
Ship "B"	2,766	48	127	37	14	13	−102	2,493	59	183	34	13	17	−146	3,490	43	117	29	14	16	−129
Eastern part:																					
Nord	2,567	41	109	29	12	11	102	2,346	26	55	30	8	14	90	3,298	44	108	29	11	10	64
Danmarkshavn	2,618	44	93	29	13	10	64	2,352	39	151	24	8	13	62	3,299	38	105	30	8	13	−72
Jan Mayen	2,709	41	171	27	10	16	72	2,395	37	186	23	13	12	−141	3,339	61	220	42	14	11	−183
Kap Tobin	2,655	50	123	26	13	15	90	2,364	48	188	28	14	14	101	3,323	41	105	27	13	13	66
Angmagssalik	2,690	34	86	35	8	10	107	2,403	66	150	36	7	10	−83	3,385	47	150	37	10	7	−191
Keflavik	2,738	45	111	40	14	15	96	2,417	41	147	35	16	13	85	3,407	49	183	35	12	17	−104
Ship "A"	2,750	30	77	28	14	14	−86	2,455	43	116	48	16	12	−133	3,459	52	178	43	10	16	−118

TABLE V

GREENLAND AREA, THICKNESSES (m), JULY 1956 (03h00Z)
(After PUTNINS et al., 1959a)

Stations	700–1,000 mbar							500–700 mbar							300–500 mbar						
	mean	var.	max. dev.	interdiurn. change				mean	var.	max. dev.	interdiurn. change				mean	var.	max. dev.	interdiurn. change			
				mean	$n+$	$n-$	greatest change				mean	$n+$	$n-$	greatest change				mean	$n+$	$n-$	greatest change
Western part:																					
Alert	2,860	29	70	26	16	14	66	2,546	29	57	14	16	13	40	3,533	38	114	19	13	17	73
Thule	2,853	25	67	14	14	15	46	2,544	28	80	13	11	16	49	3,534	32	75	19	16	14	65
Arctic Bay	2,866	44	105	22	15	14	−69	2,550	31	89	26	14	13	−57	3,557	31	68	26	14	13	−79
Egedesminde	2,869	24	55	21	14	14	−74	2,557	26	58	20	12	12	−53	3,551	43	122	26	10	13	−49
Frobisher	2,892	22	63	16	13	17	57	2,575	29	78	18	12	13	−55	3,576	39	110	27	10	16	80
Narssarssuaq	2,882	17	53	17	12	18	60	2,574	23	58	14	14	14	95	3,584	36	87	23	16	14	70
Ship "B"	2,897	24	61	19	14	14	−64	2,586	28	64	16	13	15	55	3,598	36	73	21	10	16	81
Eastern part:																					
Nord	2,856	21	60	22	11	16	−71	2,546	19	45	19	13	15	57	3,540	27	73	30	12	16	88
Danmarkshavn	2,862	23	76	24	14	15	115	2,554	22	52	23	19	11	−63	3,556	30	88	32	12	14	68
Jan Mayen	2,866	16	43	18	15	15	−45	2,571	19	40	19	17	13	46	3,583	26	79	28	16	11	±70
Kap Tobin	2,866	59	150	17	14	15	−90	2,558	16	42	17	15	15	52	3,543	26	94	35	15	13	165
Angmagssalik	2,858	16	40	18	14	15	−63	2,564	22	55	18	14	12	−47	3,560	32	112	26	11	13	−71
Keflavik	2,872	18	54	15	10	19	48	2,553	17	45	17	13	15	39	3,553	25	72	20	14	14	81
Ship "A"	2,873	15	43	14	15	15	−44	2,573	52	148	17	15	15	35	3,576	31	113	26	12	14	−99

March). On these occasions, soundings were made simultaneously at Station Centrale (on the ice cap), Egedesminde (on the west coast) and Kap Tobin (on the east coast) (ROSENDAL, 1961). The correlation between 500 mbar temperature and 300–500 mbar thickness is given in Table VI for each of the three stations.

TABLE VI

CORRELATION BETWEEN 500 MBAR TEMPERATURE AND 300–500 MBAR THICKNESS

Station	*r*	Standard error
Egedesminde	0.60	0.068
Station Centrale	0.79	0.040
Kap Tobin	0.94	0.013

The decrease in the correlation coefficient from east to west seems to result from the fact that the tropopause height (near the 300 mbar surface) also decreases from east to west as one approaches the mean circulation centre. One also finds larger interdiurnal changes in temperature (especially in the column below the 400–500 mbar level over Egedesminde). At 700 mbar, in winter, the standard deviation of the mean temperature is 7.4°C at Egedesminde compared with 6.0° at Kap Tobin. At 500 mbar, the standard deviation is 6.4° at Egedesminde and 5.6° at Kap Tobin.

500 mbar heights, 300 mbar heights, 500–700 mbar thicknesses and 300–500 mbar thicknesses were correlated for the station pairs: Station Centrale–Egedesminde and Station Centrale–Kap Tobin. The results are shown in Table VII.

TABLE VII

CORRELATION OF HEIGHT AND THICKNESS FOR TWO STATION PAIRS

Element	Station Centrale–Egedesminde		Station Centrale–Kap Tobin	
	r	standard error	*r*	standard error
H_{500}	0.88	0.024	0.78	0.041
H_{300}	0.84	0.031	0.73	0.049
$h_{500/700}$	0.75	0.046	0.57	0.071
$h_{300/500}$	0.68	0.056	0.59	0.068

The slightly better correlations between Station Centrale and Egedesminde may be due in part to the shorter horizontal distance between the stations (160 km less than between Station Centrale and Kap Tobin) and also by the fact that Egedesminde and Station Centrale are, more frequently than the second pair, in the same atmospheric circulation pattern.

Greenland, as a large massif mostly covered by ice, develops very low temperatures near the surface. The temperature differences between the surface layer on the ice cap and the 700 mbar level in the atmosphere around Greenland are of interest in themselves and also because they can indicate the location of regions with more or less intense circulation around and over the ice cap.

All available radio ascents from stations around the ice cap, for the period of Victor's expedition (September 1949–July 1951) were used to compute the mean seasonal 700 mbar temperatures (PUTNINS et al., 1959a).

TABLE VIII

HORIZONTAL TEMPERATURE GRADIENTS[1] (°C/100 km)

Stations	Winter		Spring		Summer	Autumn	
	1949/1950	1950/1951	1950	1951	1950	1949	1950
Station Centrale–Thule	−1.05	−1.13	−0.79	−1.40	*−0.75*	−0.97	−1.07
Station Centrale–Egedesminde		−2.72		−2.91	−1.44		−2.64
Station Centrale–Søndre Strømfjord	−2.76					−2.34	
Station Centrale–Narssarssuaq	−1.58	−1.93	−1.76	−1.94	−0.94	−1.53	−1.62
Station Centrale–Scoresbysund	−2.44	*−2.94*	−2.29	−2.29	−1.53	−2.18	−2.54

[1] Maximum and minimum values are in italics.

Table VIII shows the horizontal temperature gradients, (in degrees centigrade per 100 km). On an annual basis, the mean horizontal temperature gradient was greater in the west–east direction than in the north–south direction. Thus, some of the annual mean gradients for 1950 were: Station Centrale–Thule, −0.88°C/100 km; Station Centrale–Narssarssuaq, −1.48°C/100 km; while Station Centrale–Scoresbysund was −2.15°C/100 km. This implies that the air flow over the surface of the ice cap, in the vicinity of the coastal regions, is predominantly meridional rather than zonal.

The largest monthly mean horizontal temperature gradient was obtained for March 1950 between Station Centrale and Søndre–Strømfjord: −3.44°C/100 km. In April 1950 and 1951 the gradients between Station Centrale–Søndre Strømfjord and Station Centrale–Egedesminde were −3.43°C/100 km. The smallest gradient occurred between Station Centrale and Thule: −0.54°C/100 km during January 1950.

This negative temperature gradient between the surface layer over the ice cap and the free atmosphere near the edge of the ice cap should be different with elevation and, very probably, even change sign.

Although the 500 mbar level is likely to be outside the influence of the Greenland ice cap, the distribution of horizontal (isobaric) temperature gradients at this level is still of interest. The temperature gradients were computed for January 1950 and for July, 1951. The temperature gradients between the ice cap (Station Centrale) and coastal stations were changeable in sign, indicating quite strong activity.

TABLE IX

HORIZONTAL TEMPERATURE GRADIENTS AT 700 AND 500 MBAR LEVEL
(After PUTNINS et al., 1959a)

	Alert–Thule				Thule–Arctic Bay				Egedesminde–Frobisher				Narssars.–Ship "B"				Jan Mayen–Kap Tobin		Keflavik–Angmagssalik				Ship "A"–Narssars.	
	700 mbar		500 mbar		700 mbar		500 mbar		700 mbar		500 mbar		700 mbar		500 mbar		700 mbar	500 mbar	700 mbar		500 mbar		700 mbar	500 mbar
	03h00	15h00	03h00	15h00	03h00	15h00	03h00	15h00	03h00	15h00	03h00	15h00	03h00	15h00	03h00	15h00	03h00	03h00	03h00	15h00	03h00	15h00	03h00	03h00
January 1956:																								
Mean (°C/10³ km)	−1.57	−0.60	−2.12	−1.46	−4.11	−4.08	−5.22	−5.20	−1.64	−1.76	−1.72	−2.26	−8.54	−8.52	−6.30	−6.31	3.69	3.40	2.56	4.68	1.70	2.80	2.05	0.42
$n+$	14	14	15	12	10	10	10	9	10	9	7	6	2	3	2	3	17	19	15	17	13	14	16	16
$n-$	16	15	14	16	21	19	20	19	13	13	16	14	27	25	28	26	8	5	9	6	9	8	13	14
Mean, abs. values	6.36	6.05	5.83	5.20	7.40	7.84	7.37	7.27	5.82	5.18	3.63	4.38	9.31	9.34	7.68	6.81	6.93	4.62	5.72	8.43	7.23	7.60	8.25	9.02
Persistence (%)	25	10	36	28	56	52	71	72	28	34	48	52	92	91	82	92	53	73	45	56	24	37	25	5
No. of changes	9	6	4	6	5	7	12	7	10	8	10	7	3	4	4	4	7	10	8	10	8	4	8	6
July 1956:																								
Mean (°C/10³ km)	−0.19	0.00	0.80	1.38	−0.16	−2.12	−1.47	−3.64	−1.52	−2.34	−2.23	−0.37	−1.61	−0.68	−1.45	−1.30	1.76	5.23	−0.60	−1.39	−2.53	−3.08	−0.78	0.23
$n+$	13	17	18	21	13	6	9	6	11	10	9	11	14	15	11	11	17	28	12	8	8	5	17	14
$n-$	16	14	13	8	17	23	20	21	17	16	16	12	17	14	17	19	14	3	16	18	18	21	14	17
Mean, abs. values	2.45	2.47	2.44	2.64	2.85	4.37	3.29	5.80	3.42	4.17	4.86	2.51	3.46	4.02	3.16	3.56	6.19	7.32	3.44	4.45	4.28	4.04	3.50	4.76
Persistence (%)	8	0	33	52	6	48	45	63	44	56	46	15	47	17	46	36	29	79	17	31	59	76	23	5
No. of changes	10	15	11	8	17	8	13	8	11	8	13	9	5	10	8	10	13	4	12	10	6	4	10	13

For January and July, 1956, isobaric temperature gradients were calculated for pairs of stations in the area around Greenland. Table IX shows the mean gradients, the number of positive and of negative gradients, persistence and the number of changes.
Larger temperature gradients occurred in January. The strongest persistence occurred in January in the area south of Greenland (Narssarssuaq–Ship "B"). Persistence was also quite strong at the 500 mbar level in the Thule–Arctic Bay area. Relatively weak persistence occurred in the Keflavik–Angmagssalik (Icelandic low) and Egedesminde–Frobisher area. In July the persistence of the isobaric temperature gradients was generally smaller. It must be noted, however, that this table is based on only two months of data. As the temperature distribution in the Arctic varies greatly from year to year, wide variations in the distribution of the temperature gradient must also be expected. These preliminary results suggest the following:
(*1*) In general, the temperature gradient over the ice cap is directed toward the west in accordance with the location of the cold centre over the northeastern part of Canada.
(*2*) A comparison of the direction of the temperature gradient in the layer near the surface of the ice cap with that in the free atmosphere indicates that the air flow should be more disturbed over the western part of the ice cap and along the west coast than over the eastern part of the island. Thus, pressure fluctuations along the west coast should be more pronounced. Air flow aloft over the southern corner of Greenland should be especially disturbed. Due to the presence of an extended ice cap, 2,500 m high, in close proximity of the coast, the temperature gradient is always directed from the coast toward the ice sheet. At higher elevations, the gradient is frequently reversed. This results in strongly disturbed air flow aloft together with the divergence and convergence of air masses. These conditions should contribute to the sudden deepening of lows in this region.
For comparison, some mean isobaric gradients are given in Table X.

TABLE X

MEAN TEMPERATURE GRADIENTS (°C/1000 km) FOR WINTER (DECEMBER–MARCH) AND SUMMER (JUNE–AUGUST) FOR THE PERIOD, GENERALLY, 1949–1958[1]

Stations	Winter			Summer		
	700 mbar	500 mbar	300 mbar	700 mbar	500 mbar	300 mbar
Alert–Thule	−0.44	−0.74	−1.04	−1.79	−1.48	−1.04
Thule–Resolute	+1.56	+1.17	−0.65	+0.13	−0.13	−0.78
Narssarssuaq–Ship "B"	−2.34	−3.93	−2.55	−4.71	−4.32	−2.34
Jan Mayen–Kap Tobin	+2.23	+0.81	+1.01	+0.81	+0.61	+1.82
Keflavik–Angmagssalik	+4.84	+3.60	+2.35	+1.52	+1.11	+1.38

[1] Differences are taken from north to south and from east to west.

Development of disturbances in the Greenland area and the problem of disturbances crossing the ice cap

KEEGAN (1958) analyzed data from fifteen winter months during 1952–1957 and determined the frequency of occurrence of lows and highs. The frequency of occurrence of the centres was reduced to an area of 100,000 sq. miles. Two maxima were present in the Greenland area—one over the southeastern part of Greenland with a centre (>3.6%) east of the southern tip of the island, and the other located off Baffin Bay (>2.4%). The magnitude of this maximum was the same as another located northwest of the northern part of Norway. The northern part of the Greenland ice cap, north of 70°N, showed a minimum (>0.4%) which was similar to a minimum over Alaska and eastern Siberia. The frequency of occurrence of highs was a maximum over western Canada (>2.8%) with a secondary maximum over the Greenland ice cap (>2.4%). Another, slightly lesser maximum (>2.0%), was present over eastern Siberia.

Lows over northeastern Greenland and the eastern part of the Canadian Archipelago moved with lower speeds than those over eastern Greenland and the Atlantic.

GAIGEROV (1962) found that low centres seldom appear over Greenland though the peripheries of lows frequently cover a substantial part of the island.

The frequency of cyclones is a maximum over the southern end of Greenland and also in Baffin Bay during January, April, July and October. The maxima are particularly well marked in January, April and October. This shows that some Atlantic lows, and some which are diverted into more northern tracks in this region, are almost entirely blocked by the Greenland ice cap. There are cases where low centres are known to have crossed the Greenland ice cap in every direction, but this occurs relatively infrequently. Such crossings are usually associated with strong tropospheric jet streams. The most frequently observed track for lows in the North Atlantic frontal zone is along Denmark Strait. In such cases a regeneration of the low usually takes place due to an influx of cold air from Greenland.

A concentration of tracks of lows in Baffin Bay is the result of low centres moving northward along the west coast of the island from the southern tip of Greenland.

A ridge which appears over Greenland on the 700 mbar contour maps is possibly the result of a shallow layer of cold air. Due to this layer, pressure at the surface of the ice cap will be higher than the pressure, at the same elevation, in the free atmosphere around Greenland.

The strongest cyclonic activity in the central Arctic is usually observed near the end of the summer season when the frontal zones are in this most northern position, and also in winter when the meridional exchange is quite intense. In general, a zonal circulation is dominant in the Arctic in summer while a meridional circulation is dominant in winter. With a meridional circulation, strong temperature contrasts develop and intense advection of warm air takes place along with the displacement of frontal zones far to the north, and the development of cyclonic whirls in the troposphere.

At sea level in the central Arctic, cyclonic and anticyclonic circulations have the same probability of occurrence. In mid-troposphere, however, a cyclonic circulation prevails. The frequency of lows is much higher in summer and in autumn. Anticyclones prevail at the beginning of winter and in spring. Anticyclones, and to some extent cyclones, show little movement in spring and summer, but in winter they move rapidly.

Many of the lows extend to great heights in the central Arctic. Cyclogenesis seldom occurs here but rather regeneration, deepening, and dissolving of lows takes place. In the majority of cases fronts in the central Arctic also reach into the higher troposphere. High reaching lows and highs are more frequent in summer than in the remaining seasons. With a good zonal circulation, the development and movement of lows occur over the adjacent seas, but the central Arctic is dominated by slow moving depressions which extend to great heights (GAIGEROV, 1962).

Observational data indicate that the basic field of the arctic stratosphere in summer consists of a circumpolar anticyclone, as opposed to the mean cyclonic field of the troposphere. In winter, a cyclonic circulation dominates both the troposphere and the stratosphere. Also in winter, the mean planetary troughs, oriented toward East Asia and toward eastern North America, are well marked in the stratosphere and in mid-troposphere. East of these troughs, over the northern parts of the Atlantic and Pacific Oceans, warm stratospheric highs are formed (GAIGEROV, 1962).

Concerning the problem of a "glacial anticyclone", one should not forget that the dense air over the ice cap—cooled by radiation in winter—results in relatively high pressure over the ice cap compared to pressure at the same level in the surrounding area. This is almost a permanent winter feature. The layer of cold air, however, is so shallow that it can be destroyed very easily. This is shown by the large temperature variations observed over the ice cap. For this reason, it is difficult to agree with Hobbs' concept of a permanent anticyclone.

Basically, VAN EVERDINGEN (1926) does not accept the existence of a permanent anticyclone over extended ice plateaus such as found in Greenland and Antarctica. In the case of a stationary anticyclone, disturbances occurring at the boundary can intensify the pressure gradient but decrease the pressure in the anticyclone itself. Observations made in the Antarctic in 1911 (Scott and Amundsen expeditions) showed large mean pressure differences in October and December. Van Everdingen concludes that this could be explained by a shifting of the site of the maximum in the whole continent. Apparently, half of the Antarctic was under the influence of lows at this time.

Van Everdingen's major objection is that, under the assumption of the existence of a stationary glacial anticyclone, the Greenland ice sheet could not be maintained. In the Antarctic, for example, of the 30 mm of precipitable water in the lowest 4 km thick layer of the atmosphere, only 6 mm would reach the surface as compared with the 40 mm per year which are required to maintain the ice sheet. (Recent results indicate that 14–19 cm water equivalent precipitation is necessary to maintain Antarctic ice balance.) For Greenland, De Quervain computed an annual precipitation amount of 400 mm as being necessary to maintain the ice sheet. The larger amount is required in Greenland because of the greater slope of the edge of the ice (VAN EVERDINGEN, 1926).

The examination of data from A. Wegener's expedition at Eismitte led to the conclusion that there was no evidence of a virtually permanent anticyclone over the Greenland ice cap. However, there is consistent evidence from all parts of Greenland that the weather over the island is controlled by alternating lows and highs. The entire ice cap is nourished by precipitation from rising maritime air (MATTHES, 1946).

The German weather service in Hamburg found that 53% of the lows experienced in the Greenland area were of the type which split at the southern tip of the island. Of these, 7 or 8% showed the resulting western lows to be the stronger. The eastern lows were stronger in 30% of the cases; the western portion was stronger at first but declined slowly in 13% of the cases, while the eastern portion developed; in the remaining 3% the two resulting lows were nearly equal in intensity (JENSEN, 1961).

According to KOPP and HOLZAPFEL (1939), the observations at Eismitte did not prove the existence of any pronounced lows centered on the ice cap. However, Eismitte was frequently influenced by lows located near the coasts, and at times its weather was influenced by two lows, one being on the east and the other on the west coast. Apparently the layer of cold air as well as the Greenland massif itself, functioned as an obstruction for these lows. Kopp and Holzapfel assume that the Greenland anticyclone does not exist. The high frequency of "bad" weather over the ice cap can perhaps better be attributed to radiation fog rather than the influence of lows.

The analysis of weather situations during 1930/1931 showed that weather developments during the winter season were generally as follows: lows usually approached the southern tip of Greenland from a southerly direction. The speed of the approaching lows frequently increased. When a low reached the coast, it almost always split; the western centre then began to dissolve while the eastern part moved toward the northeast with renewed energy. Frequently both parts of the split low dissolved at the west coast (Kopp and Holzapfel, 1939).

When a real high is located over Greenland, lows pass Greenland far to the south. This is in contrast to the storm tracks, which are very close to the coast, during the presence of a shallow glacial anticyclone. Such a local glacial anticyclone can be weakened by migratory lows to such an extent that the anticyclonic properties become barely recognizable.

Kopp and Holzapfel argue that lows are not likely to cross the ice cap because the pressure variation observed at Eismitte was much smaller than at Weststation and at Oststation, and also that the sequence of pressure rise and fall is not such as should be expected from a low crossing from west to east.

Considering the magnitude of the pressure variations at the coasts and over the central part of the ice cap, it is indeed difficult to assume that the entire low system should cross the ice cap. Only the upper portion of a low can cross the ice cap and for this reason pressure variations in central Greenland should be smaller than at the coast. The argument concerning the time sequence is more persuasive however: in case pressure falls or rises occur almost simultaneously at the coasts and on the ice cap, it is difficult to explain a "crossing". An almost simultaneous fall of pressure at the west coast and over the central part of the ice cap can be explained by the strong advection of warm air from the south, but the east coast could hardly be subjected to the same behaviour of the warm air.

Sometimes a low appears simultaneously at the east coast, perhaps due to a pressure fall area moving in from some direction other than the west.

With reference to lows crossing the ice cap, Georgi (1933) points out the differences in wind directions at the ice cap station of Watkins' expedition and at Eismitte, i.e., at Watkins' base and ice cap stations, strong winds from the south were recorded followed by north and northwest gales in the rear of a depression. On the other hand, northerly gales were entirely missing at Eismitte, located only 440 km north of Watkins' Station. This was also supported by the appearance of the snow surface, i.e., members of Wegener's expedition, arriving in mid-summer of 1930, encountered sastrugi only from the southeast and south.

A careful investigation of some weather situations (January 27–February 1, and February 16–21, 1931) by Georgi (1933) indicated that the influence of lows did not extend to Eismitte. During those periods when the weather at Watkins' Station was typical of that in the rear of a low, with northwesterly and northerly gales, Eismitte recorded only light winds from various directions. During these two periods, the tracks of the low centres were diverted by Greenland's continent in such a way that lows from Davis Strait were blocked off the west coast of the island, and lows approaching Greenland's south coast moved along the east coast in a northeasterly direction. These lows affected the Watkins station but were not perceptable in the central part of the inland ice; there, warm air masses from the depression in Davis Strait were dominant (GEORGI, 1933). It must be pointed out, however, that these cases deal with lows moving along the Greenland coasts, and not with lows crossing any part of the ice cap.

DORSEY (1945) suggests the following for lows from Davis Strait passing over the southern part of Greenland: the upper part of the low crosses the ice cap and initiates a new disturbance on the east coast. Meanwhile, the old low stagnates and fills near the western coast. DORSEY's (1945) investigation indicates that few fronts make a complete west–east transit of the ice cap north of 67°N, the location of a major "saddle" in the crest of the ice cap.

The influence of fronts on the ice cap is, however, demonstrated by the fact that the cold air layer can be destroyed at any time and place by the inflow of warm air from adjacent seas. Warm and moist oceanic air penetrates inland even to the innermost parts of the ice cap in all seasons, but especially in winter (LOEWE, 1936). Loewe assumes that the forces controlling conditions on the ice cap are seated in the free atmosphere. At Watkins' Ice Cap Station, pressure oscillations were as great as at Angmagssalik on the coast. At Eismitte, pressure conditions are no more stable than in central Europe with its prevailingly cyclonic weather. Also, pressure oscillations on the ice cap are considerably larger in winter than in summer. Very possibly, the katabatic winds are "steered" by outside pressure changes and may, at times, be strengthened to gale force by an increased pressure gradient, or completely stopped at other times by a reversal of the gradient (LOEWE, 1936).

Lows around and near Greenland

The influence of Greenland on the general circulation was investigated by WALDEN (1959) with particular emphasis on the behavior of lows moving toward Greenland from the west and southwest.

Walden examined the tracks of all lows which approached Greenland and which somehow influenced weather conditions on the island from January 1948 to December 1957. He finally defined 52 weather patterns. However, he found that the 52 types could be collected into seven groups depending on the prevailing directions of the storm tracks. Only the four groups most important in the circulation were evaluated statistically:

(*1*) A low moving from a direction between southwest and west-northwest, toward the central part of Greenland (includes 12 of the 52 types).

(*2*) A low approaching south Greenland from southwest to west (13 types).

(*3*) A low moving into Davis Strait from the south (7 types).

(*4*) A low moving toward southern Greenland from south to southeast (7 types).

629 cases were included in these four groups. Group (*2*) was the most frequent, with 139 (lows extending to great heights and crossing the southern corner of Greenland alone account for 15% of all cases). Next in order of frequency was group (*1*), with the most frequent sub-group again consisting of lows extending to great heights which

crossed the ice cap. 26% of all cases were vertically extended lows which crossed a portion of the ice cap.

Walden assumes that vertically extended lows cross the ice cap quite frequently. The crossings occur in varying directions and with varying speeds. Crossings from west to east and from northwest to southeast could often be forecast from the previous movement of the centres. Frequently the original low will split; one part may remain at the west coast while the other part crosses the ice cap. In some cases when a low appears and deepens aloft over the east coast, only a trough will appear aloft over the west coast. The tendency for cyclogenesis to take place at the east coast could be explained as an increase of vorticity due to stretching and shrinking of the air column (PETTERSSEN, 1956, p. 226), provided there a sustained zonal flow. The latter, however, is doubtful.

Walden indicates that it is generally possible to detect the appearance of a low on the east coast one to three days after it arrives on the west coast. In case a high is situated aloft over Greenland, neither lows nor isallobaric centres cross the ice cap.

An interesting development sometimes takes place when a low aloft (500 mbar) crosses the ice cap: a centre aloft crossed the ice cap from northwest to southeast while the surface low moved along the west coast toward Thule. A new low centre appeared on the east coast about 15 h before the upper centre reached the east coast. This surface low was almost stationary initially, but it moved eastward rapidly after it became associated with the low centre aloft. WALDEN (1959) assumes the cyclogenesis to have been associated with a front over the ice cap.

It sometimes happens that a low, with a core which extends to considerable height, approaches from the northwest, but after reaching the west coast the upper centre moves toward the south-southeast and disappears. Sometime later one or more low pressure centres appear at the east coast. Walden assumes a marked divergence of the streamlines aloft to be a necessary condition for this development.

The following pattern is observed frequently: a low approaches the central part of Greenland from the southwest. After the centre arrives at the west coast, a new development occurs near the southern tip of the island, at times as much as 1,000 km south of the primary centre. The new centre moves off toward the east or northeast. The cyclogenesis takes place along the eastward pulsating front. Divergence aloft, over the southern part of Greenland, is apparently a prerequisite for this type of development also (WALDEN, 1959).

Group (*2*), comprised of lows approaching the southern part of Greenland from southwest to west, represents the most frequent dynamic development. The role of the southern part of the island is often that of a so-called "switching" influence on the lows (GEORGI, 1933; DORSEY, 1945). Lows frequently split in the vicinity of Kap Farvel with one part moving along Davis Strait while the other moves off toward Iceland. This poses the question: how much do migratory lows, moving toward Greenland, contribute to the Icelandic low as a quasi-permanent structure? It seems very likely that the Icelandic low is indeed nourished very largely by these lows.

Generally, when a low approaches the southern corner of Greenland from the southwest, one of several developments can occur. One such development is a splitting of the initial low with one part moving toward the northeast. This portion can become retrograde near the southeast coast. The frontal parts of these lows, which progress along the frontal zone as eastward moving secondary lows, follow the direction of the streamlines aloft. When a low moves northward along the west coast the "lee-effect" may be important. The limited frontal zone along the west coast probably plays a role also. Evidently, this frontal zone is created as the boundary between cold air from Baffin Bay and warm air of Atlantic origin close to the shore. Since a ridge is located over Greenland in these cases, the splitting of the lows is, as WALDEN (1959) also points out, a "real" phenomenon.

In some cases, the northern part of the split low first moves along the west coast but later moves eastward and crosses the ice cap. The crossing occurs north of the 70th parallel.

Sometimes the eastern part of a split low becomes stationary in the Irminger Sea, while a new cyclogenesis occurs in the vicinity of Scoresbysund. The fact that the eastern part of the split low frequently becomes stationary in the vicinity of the Irminger Sea can possibly be explained by the influence of the Greenland Massif, i.e., the "lee-effect" at the east coast during a westerly current, combined with relatively high water temperatures. Frequently cyclogenesis occurs independently over the Irminger Sea as a "lee-low" when northwesterly and westerly winds are observed at the west coast of south Greenland.

Lows which are deep, in vertical extent, easily cross the Greenland Massif. However, the speeds of these lows differ from case to case. Frequently their speed decreases after the crossing. At times these lows have a tendency to divide into two centres. The surface low frequently deepens over the Irminger Sea. In their early stages, the types included in group (*2*) are quite similar and can be distinguished only by variations in their later development.

SCHNEIDER (1930) describes a case when a trough intensified at the west coast of Greenland while a high pressure system was located over the ice cap. Due to the strong pressure gradient along the west coast, a southeast foehn appeared. The foehn intruded into the system of northerly winds along the coast and in Davis Strait and finally played the role of the warm sector in the development, which finally resulted in a closed low. Following the occlusion, northerly winds were again dominant.

WALDEN (1959) included in group (*3*) those weather developments which result from the northward movement of lows along Davis Strait. The low centres are some distance offshore. The following developments can occur:

(*a*) A low moves northward with no visible secondary effect on the pressure field over Greenland, or the low is "captured" by a stationary low located west of the southern corner of Greenland. This usually occurs when there is a ridge over Greenland with a uniform southerly current.

(*b*) The most frequent type in this group is characterized by the development of a secondary low at the southern tip of Greenland. The secondary generally moves in a northeast direction. This development is apparently very similar to that of the well-known Skagerrak low. In this case, the 500 mbar contour maps usually show a south-southwest flow over Davis Strait and southwest winds in the vicinity of Kap Farvel. The new development is unlikely if southerly winds are dominant above South Greenland.

(*c*) Depending on the upper air conditions over Greenland, a seeming development may occur along the east coast. These developments could possibly be interpreted as being caused by a part of a low having crossed the ice cap.

In the less common group (*4*) are included cases when a low moves from the south to southeast sector toward the southern corner of Greenland. Some of these lows become stationary at the east coast of the island, some move along the east coast, and some move along the west coast and later cross the ice cap. Also, some move along the east coast but a part of the low crosses the southern part of Greenland in a northwesterly direction. The most frequent type in group (*4*) is identified with a low moving from the southeast and crossing the southern corner without being divided. If the low does divide, one part crosses southern Greenland while the second part moves along the east coast.
Another group comprises cases where a low approaches Greenland from the west, northwest or southwest. Included in the group are cases when a low crosses the northern part of Greenland in an easterly or southeasterly direction. In some cases the low approaching the northern part of Greenland from the west becomes stationary. Occasionally, cyclogenesis occurs at the east coast on the frontal system of a low moving toward northern Greenland from the southwest.
Very frequently cyclogenesis occurs in the Kap Farvel–Godhavn region in the form of "lee-lows" due to strong easterly or northeasterly flow (WALDEN, 1959).
A final group includes a variety of cases of cyclogenesis, stationary lows, etc. Cyclogenesis can occur in the frontal zone of a low centred a considerable distance away. Real cut-offs result in spontaneous new development after crossing the ice cap.
Sometimes there is a regeneration of a stationary low in the Irminger Sea due to a pressure fall field moving over the ice cap. Occasionally, a stationary low in the southern part of Davis Strait causes a secondary development in the western part of the Irminger Sea. Quite frequently, a deep stationary low in the western part of the Irminger Sea will start to move eastward or even to cross the ice cap.
The availability of more complete data from the ice cap, including ascents and more accurate weather maps for the period of Victor's expedition (September 1949–August 1951), prompted an examination of these data for indications of disturbances crossing the ice cap (PUTNINS et al., 1959b).
It is generally very difficult to detect when a low crosses the ice cap. Only the upper parts of vertically extended lows can cross the ice cap under favorable conditions. On the other hand, lows moving along the coasts can frequently influence even the central part of the ice cap. This influence is manifested in the form of moving disturbances. These disturbances exhibit the same properties as surface fronts: changes of pressure and temperature (very strong changes of temperature at times), wind shifts, cloudiness, and precipitation.
Both parts of a low, which split in the vicinity of Kap Farvel, can influence the central part of the ice cap if the low deepens along the coast. Occasionally, a retrograde and deepening low can also influence the central part of the ice cap.
An interesting example of a low crossing the ice cap was analyzed and discussed by WALDEN (1958). The following is a summary of his investigation.
Wind and pressure observations indicated that a low pressure area was passing north of Station Centrale during April 12–14, 1950. The station was in the warm sector from 00h00Z until shortly after 15h00Z on April 13. The passage of the warm front was not very distinct, but snowfall became heavier and the temperature increased 20°C in 24 h. By 06h00Z on April 14, the temperature had decreased 28°C to −51°. The following synoptic development took place. A low pressure area located in the northern part

of Hudson Bay on April 10, deepened slightly and moved to the west coast of north Greenland, to the vicinity of Upernavik. At 18h00Z on April 12, the low had a distinct cold front. Observations from the west coast also indicated the presence of a weak warm front, though its exact location could not be determined. On the morning of April 13, the low pressure area could not be detected over Greenland, probably due to the lack of stations. The observations from Station Centrale suggested that a low was passing north of 71°N 41°W on April 13. At 06h00Z on April 13, pressure at the east coast began to fall more rapidly than before, and a low pressure area developed rapidly at Scoresbysund and moved in a southerly direction until the morning of April 14. Upper air data from the morning of April 13 indicated that southwesterly winds prevailed everywhere over Greenland to a height of 5,000–5,500 m. Winds in this layer veered into the northwest only from 15h00Z on April 13 to 03h00Z on April 14. During this time, a rather weak upper trough passed Station Centrale. From upper air charts it could be concluded that no independent upper low had crossed Greenland at 500 mbar during this period. It was not until 03h00Z on April 14 that a small low developed about 5,000 m above the surface low at Scoresbysund and deepened during the following days.

The warm air advanced on Greenland on the western flank of a high reaching anticyclone which approached from the west. After passing Kap Farvel on April 12, the high was strengthened, especially in the upper layers, by warm air advancing from the south. The 500 mbar map for 03h00Z, April 13, shows a well defined upper ridge and a tongue of warm air over south Greenland. It is certain that the tongue of warm air, which was found at 3,000 m west of southern Greenland on the previous day, contained air which was later brought to the region of Station Centrale by southwesterly winds. This provided a situation where the warm air available in middle layers could easily be drawn into the low pressure area. It could be assumed that the deepening of the originally weak disturbance could be traced to this cause at least in part.

In northeastern Greenland, Peary Land and as far south as 79°N, the passages of fronts are apparently manifested quite differently. Perhaps this is due to the fact that since ground elevation is low, there is no appreciable diminution of frontal activity.

That frontal disturbances affect northeastern Greenland was shown by TRANS (1955). For example, toward the end of July and the beginning of August, 1949, the ice cap was dominated by a high pressure system. A large and intense low was centred north of Scandinavia. During this time some frontal disturbances crossed the northeastern and eastern parts of Greenland. These disturbances exhibited all the indications and properties of fronts. The frontal passages were indicated in the temperature and relative humidity from Zackenberg (74°28′N 20°38′W) and Jørgen Brønlunds Fjord (82°10′N 30°30′ W). On this occasion the low itself did not pass the area, but frontal disturbances did cross the eastern part of Greenland.

Trans also cites a case when a frontal disturbance from the northwest crossed the northern part of Greenland (August 16–18, 1949). A flat pressure distribution dominated Greenland. An extensive high pressure system was located west of Greenland. Observations from an airplane indicated the passage of fronts from the northwest. These observations were fully supported by the records from Thule, Jørgen Brønlunds Fjord, Danmarkshavn, and other coastal stations. According to the weather maps, these frontal disturbances consisted of a series of cold air "pushes". The disturbances also resulted in cyclogenesis along the east coast.

In general, investigations of the period when Victor's expedition was in the field, indicate that the passage of disturbances across the ice cap is more than an occasional occurrence. Yet it is easier to detect a frontal crossing than a crossing by a cyclonic centre. This is mainly due to the existence of only one station on the ice cap during this time. However, only the upper parts of lows can cross the ice cap, and at these levels the surface properties of lows are partly distorted. In some cases, a regeneration of the low takes place after it reaches the east coast, provided favourable conditions exist, such as the appearance of an upper trough, the inflow of cold air from the ice cap or Arctic Ocean, the advection of surface warm air from the south, etc."

Cyclogenesis and stationary fronts

In general, weather conditions over the east and west coasts and over the ice cap are all different. Conditions are relatively similar only when a comparatively intense anticyclone dominates the island (BAUMANN, 1933).
The following very illuminating description of weather conditions over the ice cap, extracted from LOEWE (1936), proves the frequent influence of disturbances.
"The records of Mid-Ice (Eismitte) show that unfavourable conditions are more frequent during summer even in the interior than had been supposed before. With the barometer falling over Davis Strait, Baffin Bay, Denmark Strait or the Greenland Sea, the katabatic wind is strengthened by the pressure gradient from the ice cap towards the coast. Under such conditions heavy foehn storms develop in the narrow fjords, whilst at an altitude of ten thousand feet (3,000 m) a uniform sheet of altostratus and stratus moves slowly towards the interior of the ice cap. Lance-shaped lenticular clouds stand with brilliant white edges motionless above the border of the ice cap ... When the upper air tendencies gain predominance on the ice cap, visibility decreases and small snowflakes fall, mostly in minute compact forms; in the outer parts of the ice cap these are succeeded by small droplets of rain and frozen raindrops. Up to 6,000 ft. (2,000 m) even heavy rainfall may be met. At such times the katabatic wind subsides and veers to south ... Sky and ground are wrapped into a milky white cover ..."
"The results of earlier sledge journeys had led to the conception of a big 'central area' with generally very slight winds or complete calm; but this zone is non-existent or at least its size is very restricted. Even with the lowest temperatures there was at Mid-Ice (Eismitte) never a complete calm as in Siberia, but always a sensible wind, making the low temperature very trying... But of more than a thousand observations at Mid-Ice none showed a calm; only 2% of all observations gave velocities less than 2.5 miles/h (1.1 m/sec), and only one observation in eight gave a velocity less than 4.5 miles (2 m/sec). Even with a temperature of nearly −80°F (−60°C) a wind of 12 miles/h (>5 m/sec) and a slight amount of driven snow have been experienced. . ."
"In the interior the change to overcast weather and precipitation is, in winter as well as in summer, announced by a display of cirrostratus and altostratus in the western sky. The ground temperature remains low. The wind is slight from southeast to east; but wonderful cirrus displays often show a more southerly wind in the higher atmosphere. Near the ground the first sign of a weather change is a slight gustiness of the katabatic wind. Then the wind veers to the south, its strength increases, and the cloud cover comes down to the ground. Gradually the gusty wind reaches its maximum with a southerly to

south-southeasterly direction. The resultant wind direction for all winds exceeding 25 miles/h (11 m/sec) is at Mid-Ice (Eismitte) south-southeast, whilst the mean direction of all winds is east-southeast; there 7% of all winds exceed 20 miles/h (9 m/sec). Temperatures increase considerably and, with strong cyclonic winds, temperatures at Mid-Ice even in mid-winter may exceed 0°F, 45°F (25°C) above the respective monthly mean. Here, on sixteen occasions between September and April, temperatures increased more than 35°F (20°C) during 24 hours, always in connection with increasing wind force" (LOEWE, 1936).

The following are some general features related to the development of disturbances at the coast.

When a low with an occluded front approaches Greenland from the south, southwest or southeast, new disturbances frequently develop on the occlusion east of Greenland. If the lows are nourished by cold air from east Greenland, they are capable of deepening quite rapidly (RODEWALD, 1955b).

The generation of so-called "lee-lows" at the southeast coast of Greenland is usually associated with a strengthening high pressure system located over Labrador. With warm air above 3,000 m the high moves from the west-northwest over the southern part of Greenland at high speed. Usually such a "lee-low" appears only as a westward extension of the Icelandic low (RODEWALD, 1955b).

An area off the west coast, near Egedesminde, is the site of frequent cyclogenesis, which can sometimes be initiated by the horizontal temperature gradient aloft, given other conditions favourable for development such as the advection of warm air at the surface. For example, cyclogenesis took place on the west coast during March 15–16, 1950 (PUTNINS, 1961, 1965). Temperature differences between the west and east coasts were quite pronounced prior to cyclogenesis and particularly on March 14. The temperature difference persisted to higher elevations.

A strong temperature gradient should cause an increase of the winds aloft. As the increase of temperature followed the northeast–southwest direction, northwesterly winds should increase (or southeasterly winds should decrease). This should result in mass divergence aloft which should be reflected as a decrease of surface pressure.

Although temperatures generally decreased at the west coast on March 15, temperature differences between the west and east coasts remained large. Comparing Egedesminde with Kap Tobin (15h00), temperatures differed 18°C at 850 mbar, 16°C at 800 mbar, 17°C at 750 mbar, 15°C at 700 mbar, and 8°C still at 500 mbar. During March 15, the pressure dropped on the west coast: <1 mbar at Ivigtut, 5 mbar at Jakobshavn, nearly 9 mbar at Umanak, and 12.7 mbar between 10h00 and 24h00 at Upernavik. A 10-mbar decrease was observed at Station Centrale.

The weather map for 00h00 March 16, showed that a low was now established west of Egedesminde. During March 16, pressures decreased slightly, south of 70°N on the west coast (about 4 mbar) but increased about 3–5 mbar north of 70°N. At Egedesminde, the temperature decreased substantially at all levels between 900 and 280 mbar so that a super-adiabatic lapse rate was established near the surface. As the temperature dropped over Egedesminde, the temperature gradient between Egedesminde and Kap Tobin decreased also, as compared with that on March 15. On March 17, the low persisted but in a greatly weakened state.

The climate of the ice cap

Ice

Seismic measurements of ice thickness have been made at many sites around Station Centrale, in the central part of the ice cap, as well as at locations to the south, northeast and west (HOLTZSCHERER and DE ROBIN, 1954). These observations showed that the inland ice is in the shape of a lenticular mass bounded on the east and west by mountain chains, the eastern chain being the higher. The average thickness of the ice, along a west–east profile, is about 2,300 m. The greatest measured thickness is 3,300 m at a site about 100 km northeast of Station Centrale, in an area corresponding to the central dome of the inland ice.

Farther north, at 79°N, the west–east profile is similar to that across the central part of the ice cap but since the altitude of the ground surface is less, the ice thicknesses are smaller, with a median of about 1,500 m.

The north–south profile shows that the rock surface descends to sea level north of 67°N, but south of this parallel the surface rises again and remains about 1,000 m above sea level. In this region the flat-shaped basin changes to a more irregular mountainous zone.

An attempt was made to calculate the volume of the inland ice using data from profile measurements. This gave a volume of $2.7 \cdot 10^6$ km^3 $\pm 5\%$ and a mean thickness of 1,600 m for the entire ice mass (HOLTZSCHERER and DE ROBIN, 1954). The seismic investigations showed that the altitude of the bed rock, over considerable portions of the area covered by the central part of the ice cap, is near sea level so that the thickness of the ice is as much as 3,400 m (HAMILTON, 1956).

Air pressure

Actual pressures at the surface of the ice cap are not as important as their annual and daily variations. Observations from the ice cap at Eismitte showed that the influence of temperature is particularly strong: the minimum occurs in February, the maximum in July; the pressure decreases from September to October and increases markedly from March to April. These features correspond to the temperature changes in the layer between the ice cap and sea level (LOEWE, 1935).

At Weststation (1,000 m) a minimum in December and a maximum in May corresponded to the annual variation at coastal stations. However, the amplitude was greater than at coastal stations because the winter minimum and the summer maximum are intensified due to the temperature changes in the layer between Weststation and the coast (LOEWE, 1935).

The extent to which migratory lows influence the ice cap is shown by the aperiodic daily pressure variation. Pressure at Eismitte and Watkins' Station was reduced to sea level by multiplying observed pressures by 1.5 and 1.4 respectively (Table XI).

In winter the pressure variation is smaller on the ice cap than on the coast. However, the difference is less over the southern part of the ice cap. In summer there is no difference between the ice cap and the coast.

The mean daily range of pressure is shown in Table XII (LOEWE, 1936). At Watkins' Ice Cap Station, the greatest daily barometric range (midnight to midnight) in November

TABLE XI

APERIODIC PRESSURE AMPLITUDES (mbar)
(After LOEWE, 1935)

Station	Jan.	Feb.	Mar.	Apr.	May	June	July	Aug.	Sept.	Oct.	Nov.	Dec.	Year
Eismitte	8.0	7.7	8.1	5.9	3.6	4.5	4.0	4.8	4.1	4.4	9.2	4.4	5.73
Danmarkshavn	10.5	10.1	9.2	7.1	4.5	3.5	3.9	4.7	6.8	7.7	7.6	8.0	6.94
Watkins' Ice Cap Station	10.0	10.1	10.0	(8.5)	–	–	–	–	–	8.1	11.1	6.8	–
Watkins' Base Station	10.3	11.1	10.8	8.7	5.1	4.4	(4.0)	(5.9)	5.3	7.5	10.8	8.1	7.70

TABLE XII

MEAN DAILY RANGE OF PRESSURE (mbar)
(After LOEWE, 1936)

Station	Spring	Summer	Autumn	Winter
Eismitte	5.9	4.4	5.9	6.7
Watkins' Station	9.2 (Mar.–Apr.)	–	9.6 (Oct.–Nov.)	9.0
Danmarkshavn	7.8	4.0	7.4	9.5

was 24 mbar. Ranges of 10–20 mbar were not uncommon in November and from January to April. At the Base Station, the greatest daily range was 34 mbar, observed on the same day as the 24 mbar range at the Ice Cap Station. The rises and falls of pressure did not usually exceed 2–3 mbar/h at the Base Station and 2 mbar/h at the Ice Cap Station (MIRRLESS, 1932).

To compare the ice cap data with those obtained from drifting arctic stations ("A" and "B"), the pressure maximum in the central Arctic appears at the end of winter (February–March), but the minimum extends over the period from August to November. A strong decrease in pressure occurs from October to November.

A comparison of the annual course of pressure on Greenland's ice cap with that in the central part of the Antarctic shows a certain similarity. According to four years of observations from the Amundsen-Scott Station, whose elevation is almost the same as that of Eismitte, the maximum occurs during the summer (December–January), while the minimum appears at the beginning of spring (September). A strong increase in pressure from October to November and a marked decrease from February to March correspond well with the temperature variation between the ice cap and the coast.

Wind

The wind on the ice cap is generally a katabatic one. At Weststation, 3/4 of all observations showed east and southeast winds and the same was recorded at Eismitte. At Watkins' Station, the predominant directions, north and northwest, were not as pronounced (LOEWE, 1935).

Since the predominant wind on the ice cap is katabatic, the wind and the temperature gradient should be related. Thus, during February and November when the strongest

winds were observed at Weststation, temperature differences between Weststation and Eismitte were also the largest. In the central part of the ice cap, however, this relationship is not as pronounced because the more sloping surfaces at the edge of the ice cap are more conducive to downflowing winds than are the more level regions at higher elevations. Also, the directions of the strong winds indicate that the edges of the ice sheet are more subject to katabatic flow than the centre of the ice cap.
Table XIII (LOEWE, 1935) shows that the persistence is greater the stronger the wind at Eismitte. On clear days, when the temperature in the central part of the ice cap is below average, the katabatic wind is better developed.

TABLE XIII

FREQUENCY OF WINDS (%)
(After LOEWE, 1935)

	N	NE	E	SE	S	SW	W	NW	Result. (deg.)	Persistence (%)
Weststation:										
≧ 10 m/sec (192 cases)	–	1	56	40	2	1	–	–	110	90
≧ 15 m/sec (38 cases)	–	–	62	35	–	3	–	–	108	90
Eismitte:										
> 8 m/sec (118 cases)	–	3	30	28	26	12	1	–	141	68
> 12 m/sec (24 cases)	–	–	23	27	42	8	–	–	152	79
clear days (248 cases)	2	22	59	14	2	0	0	1	87	83

An absence of strong winds from northerly directions indicates that the air in the rear of lows is too shallow to influence the central part of the ice cap.
The frequency of surface wind direction was stabulated for Station Centrale (Table XIV) for the period of Victor's expedition. This table shows a remarkable lack of diurnal variation in wind direction. Surface winds were observed most frequently in the 30°

TABLE XIV

PERCENTAGE FREQUENCY OF SURFACE WINDS AT STATION CENTRALE AS A FUNCTION OF WIND DIRECTION AND TIME
(After PUTNINS et al., 1959b)

Time	Direction:	North			East			South			West		
	Calm (%)	350–10	20–40	50–70	80–100	110–130	140–160	170–190	200–220	230–250	260–280	290–310	320–340
00h00	2.2	0.9	3.4	6.2	14.3	29.4	19.4	8.5	8.8	2.4	1.6	2.4	0.9
03h00	1.8	0.8	3.8	6.0	16.6	28.1	21.0	8.3	7.4	2.9	1.1	2.4	0.5
06h00	3.4	0.5	3.6	6.4	16.0	28.7	21.2	7.1	6.7	3.3	1.4	1.1	0.8
09h00	3.3	0.6	3.6	6.8	15.2	30.9	20.1	6.8	6.5	3.2	1.1	0.9	1.1
12h00	2.2	0.6	3.5	5.2	14.8	30.9	20.6	9.2	7.5	2.6	1.4	0.7	0.9
15h00	1.5	0.6	3.3	4.8	14.2	28.7	21.6	9.7	8.7	2.6	1.5	2.0	0.9
18h00	1.9	0.9	3.6	5.5	14.5	26.0	21.2	8.7	10.7	2.5	2.0	1.7	0.9
21h00	2.7	0.3	3.5	6.2	14.1	26.2	19.7	9.2	10.3	2.7	1.7	2.4	1.1

sector centred at 120°. This represents a wind blowing downslope but with a large component parallel to the contours of the surface topography.
As a rule, the vertical extent of the inland wind is small, generally not exceeding some 100 m.
Using three observations per day, Loewe obtained the daily variation of the wind speed shown in Table XV.

TABLE XV

DAILY VARIATIONS OF WIND SPEED (m/sec)
(After LOEWE, 1935)

Station		Jan.	Feb.	Mar.	Apr.	May	June	July	Aug.	Sept.	Oct.	Nov.	Dec.	Year
Weststation	(08h00)	8.6	5.0	6.1	6.0	4.7	5.7	4.8	4.6	6.4	8.2	8.7	5.5	6.2
	(14h00)	9.3	6.4	4.7	4.2	3.2	2.0	3.5	3.4	4.9	7.6	9.1	6.3	5.4
	(21h00)	8.9	6.9	5.7	5.2	4.6	3.6	4.0	3.8	6.0	7.6	8.5	6.7	6.0
Eismitte	(08h00)	4.5	3.7	5.6	5.0	4.1	4.1	4.5	3.6	4.8	4.4	4.6	6.0	4.6
	(14h00)	4.5	3.7	5.5	5.5	4.8	4.3	4.5	3.9	5.0	4.5	3.9	6.0	4.7
	(21h00)	5.0	4.0	5.6	5.4	3.7	3.3	3.6	3.5	4.7	4.3	3.9	6.2	4.4

At Weststation the daily variation of wind speed is evident as long as the temperature amplitude is relatively large (March–October). The maximum occurs in the morning and the minimum in the afternoon. The wind is strongest when the temperature difference between the coast and the central part of the ice cap has its maximum value but the speed is a minimum when temperature differences are small.
At Eismitte, the maximum wind speed occurs at noon from April through October; the minimum appears in the evening. A. Wegener explains the noon maximum as a consequence of the absence of the surface inversion during the time of highest temperature. At this time, the energy of faster moving air particles at higher elevations is being transported to the surface layers as a result of turbulence. Loewe assumes that this explanation is supported by a large temperature amplitude in the central part of the ice cap by the (probably) smaller thickness of the cold air layer, and by the smaller energy of the katabatic wind along the less tilted surface. LOEWE (1935) points out that there are indications that another wind regime with an evening maximum prevails at Eismitte and Weststation during the winter months.
The mean annual wind speed at Eismitte was 4.6 m/sec for the year (August 1930–July 1931); the largest monthly mean appeared in December (6.1 m/sec); the means for March and April were 5.7 and 5.3 m/sec respectively, and the lowest wind speed occurred in summer, 3.6 and 4.2 m/sec in August and July respectively. The maximum speeds recorded were 17 m/sec in January and 18 m/sec in March, 1931. Generally, the highest wind speeds are from the east-northeast to south (WEGENER, 1939).
Eismitte was located on the west slope of the inland ice, but Watkins' Ice Cap Station was in the wind regime prevailing on the east slope. Wind conditions at these two stations are therefore quite different, in spite of the relatively short distance (400 km) between them and of the fact that there is no topographical obstruction between them.

At Watkins' Base Station, light southeast winds prevailed in summer and north winds in winter. At Ice Cap Station, however, northerly winds were dominant during the entire expedition period, September, 1930–April, 1931 (MIRRLESS, 1932).
The frequency of different surface wind speeds at Weststation and Eismitte is shown in Table XVI (LOEWE, 1935).

TABLE XVI

PERCENTAGE FREQUENCY OF SURFACE WIND SPEEDS AT WESTSTATION AND EISMITTE
(After LOEWE, 1935)

Speed (m/sec):	0	≦ 2	2–4	4–6	6–8	8–10	10–12	12–14	> 14
Weststation	5	11	20	20	17	12	7	4	4
Eismitte	–	12	36	30	11	6	3	1	1

At Watkins' Base Station the percentage of calms was high all year, but particularly high in winter. At Ice Cap Station, calms were infrequent. There, the most frequent winds (73%) were from 1 to 4.4 m/sec and more than 23% were at speeds from 6.7 to 15.5 m/sec. Gales occurred at all times of the year at Base Station, 42 days with gale, many of them with great violence. Gales were much less frequent at the Ice Cap Station. During the period when observations were made at both stations, there were 33 days with gale at the Base Station. Only on seven days were gale winds observed at both stations. There were also two days with gale at the Ice Cap Station when such strong winds were not experienced at Base Station. Gales were recorded as follows (MIRRLESS, 1932, 1934):

	Jan.	Feb.	Mar.
Ice Cap Station	4	4	1
Base Station	11	11	6

Table XVII summarizes wind speed frequencies at a number of ice cap stations (HAYWOOD and HOLLEYMAN, 1961).

TABLE XVII

WIND SPEED FREQUENCIES AT DIFFERENT STATIONS
(After HAYWOOD and HOLLEYMAN, 1961)

Station	0–5 0–2.6	6–20 3.1–10.3	21–40 10.8–20.6	41–60 (knots) 21.1–30.9 (m/sec)
Eismitte	31	65	4	0
Station Centrale	23	70	6	1
Site 2	7	86	7	0
Northice	7	79	14	0

The character of winds on the inland ice is also revealed by the results of the Michigan University expedition. A winter station, located approximately 40 km inland from the edge of the ice cap at Mount Evans in the Holsteinborg District (elevation 350 m),

showed a predominance of southeast winds in the means for both the year and the winter half-year, even 400 m above the station.
According to the Swiss Greenland Expedition, 1912–1913, these winds over the inland ice, which are always present at the edge of the ice cap, do not usually reach the coast line. These winds reach the coast only when a depression approaches the coast and a critical pressure gradient is established (GEORGI, 1939).
Generally speaking, wind speeds over the ice cap are not strong, winds greater than 18 m/sec are of rare occurrence. With a single exception, the wind speeds are below 9 m/sec more than 90% of the time. Records from ice cap stations are relatively short, and they may not reflect the average occurrence of high winds. On the other hand, extremely high winds are not uncommon at the coastal stations. Speeds in excess of 60 m/sec have been recorded on the mountains above Thule. On the ice cap, however, it is possible that blowing snow together with winds of moderate speed may lead to a false impression of high speeds which are not substantiated by actual records.
Observations from different sites on the ice cap show generally that, in the northern part of the ice cap, winds of 3–9 m/sec are most frequent while in the southern part of the ice cap, winds of 0–2 m/sec are the most frequent (GERDEL, 1961).
The damp, cold and windy climate of the Greenland ice cap naturally influences the human body. Human reaction can be described, in part, by the so-called "windchill". According to MILLER (1956), the "windchill" is 2,200 kcal./m^2 h in interstorm periods—slightly higher than during storms, when it is about 2,000. Observations from two expeditions to the central part of the ice cap (Wegener's and Victor's) show that the frequency of interstorm periods, defined as a group of two or more days without snowfall, is found to be about the same—40 to 45% of the winter days fall into this category (MILLER, 1956).

Temperature

Monthly mean temperatures, as well as the mean and absolute extremes, on the ice cap during a number of expeditions are presented in Table XVIII.
Hourly temperature observations during Wegener's expedition and three-hourly observations during Victor's expedition were arranged in a frequency distribution with 2° steps. The comparison by years for these stations with almost identical sites is of considerable interest (Table XIX).
The variation of temperature from winter to winter is apparently quite considerable on the ice cap. For example, November 1930 was cooler than the same month in 1949 and in 1950. In 1930, nearly 50% of all observations were of temperatures between −60° and −46°C, while in 1949 and 1950, one half of the temperatures recorded were in the range from −58° to −32°C.
Each of the summer months of June, July and August, and the transition months of May and September showed a very similar temperature distribution in all three years. The year to year variation of temperature is slightly larger for the transition months, April and October.
Interdiurnal temperature changes may be used as one of the indications of the stability conditions at a station. Table XX shows the interdiurnal changes of temperature for the period of Victor's expedition. In addition to the interdiurnal temperature changes

TABLE XVIII

MEAN AND EXTREME TEMPERATURES AT ICE CAP STATIONS (°C)

Station	Year	Mean temp.	Mean max.	Mean min.	Extr. max	Extr. min.	Mean temp.	Mean max.	Mean min.
		January:					*February:*		
Watkins[1]	1931	−30.6	−26.7	−35.0	−6.7	−46.7	−36.1	−30.0	−42.2
Eismitte[2]	1931	−41.7	−36.0	−47.2	−20.8	−64.2	−47.2	−41.4	−53.0
Centrale[3]	1950	−31.0	−27.0	−35.0	−19.0	−50.0	−40.0	−36.0	−44.0
	1951	−42.0	−38.0	−45.0	−23.0	−55.0	−40.0	−35.0	−45.0
Northice[4]	1953	−40.0	−36.0	−44.0	−18.0	−57.0	−46.0	−42.0	−49.0
	1954	−42.0	−37.0	−46.0	−20.0	−66.0	−37.0	−34.0	−41.0
Site 2*	1954	−31.7	−27.2	−36.1	−14.4	−54.4	−30.6	−27.2	−33.9
	1955	−36.1	−32.2	−40.0	−21.1	−51.7	−31.1	−26.7	−35.0
	1956	−36.7	−33.9	−40.0	−23.9	−50.0	−35.6	−30.6	−40.0
	1957	−36.7	−32.8	−41.1	−18.3	−56.1	−31.7	−28.9	−35.0
		May:					*June:*		
Watkins	1931	–	–	–	–	–	–	–	–
Eismitte	1931	−20.3	−14.3	−28.8	−6.8	−45.3	−15.6	−10.1	−22.9
Centrale	1950	−20.0	−14.0	−26.0	−8.0	−46.0	−13.0	−8.0	−19.0
	1951	−19.0	−14.0	−24.0	−9.0	−43.0	−14.0	−9.0	−19.0
Northice	1953	−21.0	−17.0	−25.0	−11.0	−37.0	−12.0	−8.0	−16.0
	1954	−20.0	−16.0	−25.0	−6.0	−36.0	−12.0	−9.0	−17.0
Site 2	1953	–	–	–	–	–	–	–	–
	1954	−16.1	−11.7	−20.6	−2.2	−30.0	−8.9	−5.6	−12.8
	1955	−16.7	−15.0	−24.4	−6.1	−38.3	−8.3	−5.0	−12.8
	1956	−19.4	−15.6	−24.4	−5.6	−33.3	−10.0	−6.7	−13.9
	1957	−15.0	−11.7	−20.6	−6.1	−40.0	–	–	–
		September:					*October:*		
Watkins	1930	−16.7	−10.0	−23.3	−1.7	−32.8	−27.2	−23.3	−31.1
Eismitte	1930	−21.7	−16.6	−28.2	−8.8	−40.1	−35.8	−31.2	−40.3
Centrale	1949	−26.0	−20.0	−31.0	−9.0	−45.0	−27.0	−22.0	−32.0
	1950	−24.0	−20.0	−28.0	−15.0	−39.0	−33.0	−28.0	−38.0
Northice	1952	–	–	–	–	–	–	–	–
	1953	−26.0	−22.0	−30.0	−11.0	−45.0	−36.0	−33.0	−40.0
Site 2	1953	−20.0	−16.7	−23.3	−8.3	−39.4	−28.9	−26.7	−31.7
	1954	−25.6	−21.7	−30.6	−12.8	−42.8	−25.6	−21.7	−29.4
	1955	−18.9	−16.1	−23.9	−7.8	−30.0	−18.3	−15.6	−21.7
	1956	−21.7	−17.8	−26.7	−6.7	−37.2	−26.1	−23.9	−32.2

[1] 67°03′N 41°49′W; 2,440 m. [2] 70°53.8′N 40°42.1′W; 3,000 m. [3] 70°55′03″N 40°38′22″W; 2,993 m.
[4] 78°04′N 38°29′W; 2,343 m. * 76°59′42″N 56°04′30″W; 2,128 m

Extr. max.	Extr. min.	Mean temp.	Mean max.	Mean min.	Extr. max.	Extr. min.	Mean temp.	Mean max.	Mean min.	Extr. max.	Extr. min.
		March:					*April:*				
−21.1	−51.1	–	−24.4	−37.2	−13.3	−50.6	–	–	–	–	–
−23.2	−64.3	−39.4	−34.2	−45.5	−16.7	−64.8	−31.0	−25.6	−37.9	−13.0	−57.9
−22.0	−65.0	−35.0	−30.0	−40.0	−15.0	−62.0	−35.0	−28.0	−42.0	−13.0	−53.0
−15.0	−61.0	−41.0	−35.0	−46.0	−13.0	−58.0	−36.0	−29.0	−43.0	−14.0	−57.0
−19.0	−60.0	−44.0	−41.0	−48.0	−24.0	−61.0	−31.0	−27.0	−35.0	−17.0	−47.0
−22.0	−49.0	−41.0	−38.0	−45.0	−24.0	−56.0	−32.0	−28.0	−38.0	−14.0	−46.0
−12.2	−42.2	−33.9	−28.9	−38.9	−13.9	−49.4	−23.9	−17.2	−30.6	−6.1	−41.7
−10.6	−47.8	−34.4	−30.6	−38.9	−12.2	−47.2	−30.6	−26.1	−36.1	−15.0	−45.6
−11.1	−52.2	−35.0	−32.2	−38.9	−14.4	−50.6	−26.7	−22.8	−31.7	−16.1	−41.1
−18.3	−43.9	−35.6	−32.2	−40.0	−18.3	−52.8	−27.8	−23.9	−33.3	−13.9	−42.8
		July:					*August:*				
–	–	–	–	–	–	–	–	–	–	–	–
−3.5	−30.0	−11.2	−7.0	−17.6	−3.0	−28.5	−17.0	−12.1	−24.8	−6.0	−33.4
+1.0	−30.0	−13.0	−8.0	−18.0	−3.0	−28.0	−14.0	−10.0	−19.0	−4.4	−34.0
−3.0	−33.0	−13.0	−9.0	−17.0	−5.0	−26.0	−14.0	−9.0	−17.6	−4.0	−31.0
−1.0	−24.0	−10.0	−7.0	−13.0	−1.0	−23.0	–	–	–	–	–
−2.0	−27.0	−8.0	−6.0	−12.0	0.0	−18.0	−13.0	−10.0	−16.0	−4.0	−25.0
–	–	−6.1	−3.9	− 8.9	+1.1	−20.0	−8.3	−5.6	−10.6	−2.8	−22.2
−1.1	−20.0	−6.1	−3.9	−10.6	+1.7	−16.7	−7.2	−4.4	−11.7	+2.2	−20.6
+1.1	−17.8	−8.3	−5.6	−11.7	−0.6	−19.4	−12.2	−8.9	−16.1	−2.2	−26.1
+0.6	−24.4	−7.8	−4.4	−11.1	−0.6	−21.1	−8.9	−6.7	−11.7	−1.1	−19.4
–	–	–	–	–	–	–	–	–	–	–	–
		November:					*December:*				
−5.6	−44.4	−33.9	−29.4	−37.8	−13.9	−46.1	−31.1	−27.2	−35.0	−13.9	−48.9
−14.1	−55.6	−43.2	−36.6	−49.4	−19.1	−58.5	−38.5	−33.6	−42.6	−19.8	−56.0
−9.0	−46.0	−33.0	−28.0	−38.0	−18.0	−55.0	−41.0	−36.0	−46.0	−20.0	−64.0
−19.0	−49.0	−34.0	−29.0	−40.0	−11.0	−58.0	−36.0	−32.0	−40.0	−14.0	−59.0
–	–	−39.0	−35.0	−43.0	−24.0	−54.2	−40.0	−37.0	−43.0	−28.0	−52.0
−21.0	−49.0	−41.0	−37.0	−45.0	−20.0	−61.0	−44.0	−41.0	−48.0	−28.0	−59.0
−13.3	−40.0	−35.0	−32.2	−37.8	−15.0	−50.0	−41.7	−38.3	−45.0	−20.0	−55.0
−13.9	−43.3	−27.8	−23.3	−32.2	−13.9	−46.1	−38.3	−34.4	−42.2	−23.3	−53.3
−7.2	−33.3	−31.7	−28.3	−35.0	−12.8	−51.7	−36.7	−33.9	−39.4	−25.0	−48.3
−7.2	−43.9	−34.4	−31.1	−37.8	− 7.2	−51.7	−30.0	−27.2	−33.9	−15.0	−40.0

TABLE XIX

FREQUENCY DISTRIBUTION OF TEMPERATURES ON THE GREENLAND ICE CAP (%) ACCORDING TO OBSERVATIONS AT EISMITTE, AUGUST 1930–

Month	Year	Interval (°C): −65.9 −64.0	−63.9 −62.0	−61.9 −60.0	−59.9 −58.0	−57.9 −56.0	−55.9 −54.0	−53.9 −52.0	−51.9 −50.0	−49.9 −48.0	−47.9 −46.0	−45.9 −44.0	−43.9 −42.0	−41.9 −40.0	−39.9 −38.0	−37.9 −36.0	−35.9 −34.0
Jan.	1931	0.3	1.4	2.7	2.3	2.6	6.5	6.8	4.7	4.1	5.5	4.1	5.5	7.2	5.8	5.2	6.2
Jan.	1950								0.4	0.8	0.4	1.7	1.2	3.7	8.3	7.9	9.1
Jan.	1951						3.6	2.4	6.1	12.1	12.1	11.3	4.0	10.5	7.3	6.5	6.1
Feb.	1931	0.6	2.6	2.6	5.3	5.0	7.4	14.9	13.7	11.6	5.3	3.4	2.2	2.7	2.2	2.5	4.3
Feb.	1950	0.9	0.9	1.8	1.4	3.6	4.9	3.1	5.8	9.4	4.5	3.6	4.0	2.7	2.3	1.8	9.4
Feb.	1951			1.8	1.8	5.3	1.8	3.5	8.5	4.9	7.1	6.2	5.8	5.4	4.9	3.6	4.0
Mar.	1931	1.1	0.8	2.1	3.8	3.0	3.5	5.3	5.5	6.5	6.9	7.0	6.9	5.9	3.0	4.9	3.8
Mar.	1950		0.8	2.0	0.8	1.6	2.4	1.6	4.0	3.6	1.2	4.4	5.3	6.2	4.8	8.9	3.6
Mar.	1951				0.8	1.6	2.4	5.3	8.5	10.5	9.7	12.1	6.5	4.8	4.8	6.1	3.2
Apr.	1931					0.4	0.1	1.4	1.4	1.1	2.6	2.9	2.4	5.9	4.0	5.1	7.5
Apr.	1950							1.3	1.7	4.6	5.4	3.7	6.2	5.8	6.6	9.6	7.1
Apr.	1951					0.8	0.4		5.0	8.7	5.8	4.6	6.7	7.1	7.1	9.6	5.4
May	1931											0.5	0.3	0.3	0.1	0.4	2.4
May	1950										0.4	0.4		0.8	0.8	0.4	1.6
May	1951												0.4	0.4	0.4	0.4	0.4
June	1931																
June	1950																
June	1951																
July	1931																
July	1950																
July	1951																
Aug.	1931																
Aug.	1950																0.4
Aug.	1951																
Sept.	1930													0.1	0.4	1.0	1.9
Sept.	1949											2.4	5.3	3.8	4.3	5.3	3.4
Sept.	1950														2.1	2.5	3.7
Oct.	1930						3.6	6.1	6.6	5.5	1.5	2.7	4.1	4.2	5.8	5.6	4.2
Oct.	1949										0.4	0.8	1.6	6.1	2.8	5.2	5.2
Oct.	1950									1.6	2.4	3.2	4.0	7.7	8.1	8.9	8.9
Nov.	1930				0.4	5.8	7.8	10.1	8.3	9.0	6.5	5.8	5.4	7.2	5.3	5.0	3.1
Nov.	1949						0.4	1.3	1.3	2.5	1.7	3.8	6.2	4.6	10.0	5.4	7.5
Nov.	1950					2.5	5.4	2.9	4.2	6.7	5.0	2.5	4.6	5.0	1.7	3.4	3.4
Dec.	1930					0.3	2.9	5.0	3.5	5.5	9.1	8.3	7.4	6.7	8.1	7.8	10.1
Dec.	1949	0.4	2.0	1.6	0.8	2.8	5.2	1.6	7.7	6.1	4.0	7.7	8.9	8.5	4.0	4.8	6.1
Dec.	1950				3.2	1.6	2.0	2.8	2.4	3.6	1.2	3.6	6.9	6.9	5.7	6.5	4.4

* A. Wegener Expedition, one-hourly observations.
** P.E. Victor Expedition, three-hourly observations.

JULY 1931*, AND AT STATION CENTRALE, SEPTEMBER 1949–AUGUST 1951**

−33.9 −32.0	−31.9 −30.0	−29.9 −28.0	−27.9 −26.0	−25.9 −24.0	−23.9 −22.0	−21.9 −20.0	−19.9 −18.0	−17.9 −16.0	−15.9 −14.0	−13.9 −12.0	−11.9 −10.0	−9.9 −8.0	−7.9 −6.0	−5.9 −4.0	−3.9 −2.0	−1.9 ±0.0	+0.1 +2.0	+2.1 +4.0
5.7	3.5	6.5	6.3	2.2	2.6	1.0	0.4	0.6	0.3									
9.5	7.1	14.5	12.9	9.5	9.6	1.7	2.1											
5.2	5.2	2.0	2.0	2.0	1.6													
3.0	3.2	2.4	2.6	1.5	1.0													
9.0	5.8	7.6	4.9	6.3	4.9	1.4												
5.8	4.5	4.5	7.6	9.4	0.9		0.9	0.9	0.9									
2.7	1.8	2.1	2.0	2.4	2.9	10.8	2.4	2.9										
7.3	6.9	6.5	4.0	5.6	2.4	6.5	5.2	1.6	2.8									
4.5	4.0	1.6	0.4	1.6	0.8	2.0	2.0	3.6	1.2	2.0								
8.1	9.2	6.8	9.3	7.5	8.1	5.8	5.1	4.2	0.4	0.7								
9.6	7.5	8.8	4.6	7.1	2.9	2.5	0.8	0.4	2.1	1.7								
6.7	7.1	2.9	3.8	3.8	2.9	2.9	3.7	2.1	2.9									
2.2	3.8	4.3	6.2	6.4	8.5	11.7	10.2	11.7	13.2	12.2	5.1	0.4	0.1					
2.8	3.6	5.3	4.0	8.1	7.3	4.4	12.5	9.3	11.7	14.9	6.5	4.8	0.4					
2.4	1.6	3.6	4.4	6.4	8.9	8.9	9.3	12.5	19.0	8.1	9.7	3.2						
	0.3	0.7	2.5	5.1	4.6	7.8	7.0	11.8	16.7	19.3	18.2	5.1	0.8		0.1			
	0.4	0.4	2.9	4.2	5.0	6.7	5.8	5.8	7.5	7.5	12.1	13.8	10.0	8.3	3.3	3.3	1.3	1.7
0.4	0.4	0.8	2.1	5.0	4.6	4.1	6.7	7.1	6.7	13.8	15.4	18.8	8.3	2.9	2.9			
		0.3	0.6	1.2	2.6	2.7	2.9	2.4	6.0	12.7	25.9	20.7	16.9	3.5	1.5	0.1		
			1.7	1.3	4.1	6.6	5.0	7.0	15.6	11.9	11.5	14.8	15.2	4.1	1.2			
			0.8	2.0	2.4	3.2	6.1	5.2	6.1	15.7	26.2	19.4	10.1	2.8				
0.9	0.3	1.7	5.3	5.5	5.7	5.9	11.2	15.3	20.7	19.5	4.5	2.5	1.0					
0.4	1.2	2.4	1.6	1.2	2.0	3.6	8.5	8.1	12.1	16.5	19.4	12.9	6.9	2.4	0.4			
	0.9		1.8	2.7	2.7	5.3	4.5	4.5	9.8	19.6	22.3	12.5	5.4	7.1	0.9			
3.0	4.0	6.0	8.4	8.9	9.6	14.5	12.3	12.8	8.2	4.9	3.2	0.8						
1.9	8.7	8.2	3.9	7.2	6.7	5.7	6.3	5.3	6.7	10.6	3.8	0.5						
2.9	3.8	7.5	9.6	12.9	12.5	12.9	15.4	10.4	3.8									
8.2	5.3	5.3	5.5	9.4	3.2	0.9	4.3	5.2	2.8									
4.4	9.3	6.1	6.1	8.9	13.7	13.7	5.6	2.0	0.8	0.4	6.1	0.8						
9.6	8.9	6.0	6.5	8.5	4.4	9.7	1.6											
3.2	2.8	2.8	2.2	3.1	3.5	1.8	0.9											
6.7	12.5	6.2	4.1	6.2	5.4	12.5	1.7											
7.9	5.0	5.8	2.9	4.1	5.8	5.8	5.4	2.5	4.6	1.7	1.2							
8.5	6.7	1.7	2.6	3.1	2.3	0.4												
2.8	5.6	8.9	6.1	1.6	1.2	1.6												
10.5	11.7	5.3	6.5	5.2	2.4	3.6	2.8	0.4	0.4	0.4								

TABLE XX

INTERDIURNAL TEMPERATURE CHANGES (00h00Z–00h00Z). VICTOR EXPEDITION, SEPTEMBER 1949–JULY 1951, STATION CENTRALE (70°55′N 40°38′W, elev. 2,993 m)
(After PUTNINS et al., 1959b)

	Jan.	Feb.	Mar.	Apr.	May	June	July	Aug.	Sept.	Oct.	Nov.	Dec.	Year
n_0	0	0	0	2	1	0	1	0	1	1	0	0	6
n_-*	35	25	37	32	29	22	33	16	31	30	34	29	353
n_+*	27	31	25	26	32	28	28	15	23	31	26	33	325
Mean (°C)	5.4	7.7	7.5	6.8	6.3	3.7	3.5	5.6	6.3	7.1	8.4	7.2	6.3
Max. change (°C)	−22.6	+26.5	+30.0	±20.5	+20.3	+13.8	+11.4	+19.0	+28.1	+27.9	−29.9	−26.5	30.0
Mean +	6.8	6.9	9.2	7.9	6.6	3.8	3.8	5.9	7.4	7.2	9.1	6.7	6.8
Mean −	4.4	8.6	6.4	6.2	6.2	3.5	3.4	5.4	5.7	7.3	7.8	7.8	6.1

* $n_-/n_+ = 1.09$.

TABLE XXI

INTERDIURNAL TEMPERATURE CHANGES (00h00Z–00h00Z), AT STATION NORTHICE[1] (78°04′N 38°29′W, elevation 2,343 m)
(After PUTNINS et al., 1959b)

	Jan.	Feb.	Mar.	Apr.	May	June	July	Aug.	Sept.	Oct.	Nov.	Dec.	Year
n_0	5	0	4	6	9	8	10	4	1	2	2	5	56
n_-*	33	31	27	23	27	24	18	14	14	18	30	33	292
n_+*	22	25	31	31	26	28	18	13	15	11	27	24	271
Mean (°C)	7.7	5.7	4.0	3.9	3.1	3.0	5.1	2.4	5.9	5.2	6.6	5.0	4.8
Max. change (°C)	+33	−16	−18	+14	+10	+12	±6	−8	−16	+15	+16	+18	+33
Mean +	10.6	6.4	5.2	4.0	3.9	3.6	3.4	2.3	5.2	8.3	6.9	6.2	5.5
Mean −	7.0	5.1	5.3	4.9	3.2	3.4	2.8	3.2	7.1	4.3	6.1	4.9	4.8

[1] British North Greenland Expedition, 1952–1954.
* $n_-/n_+ = 1.08$.

TABLE XXII

INTERDIURNAL TEMPERATURE CHANGES (00h00Z–00h00Z), BRØNLUND FJORD, AUG. 1948–AUG. 1950 (82°10′N 30°30′W)
(After PUTNINS et al., 1959b)

	Jan.	Feb.	Mar.	Apr.	May	June	July	Aug.	Sept.	Oct.	Nov.	Dec.	Year
n_0	0	4	0	1	0	0	1	1	2	0	1	1	11
n_-*	37	27	29	29	29	25	32	40	36	40	29	33	386
n_+*	25	25	33	30	33	35	29	32	22	22	30	28	344
Mean (°C)	3.9	3.9	3.7	2.4	2.6	2.2	2.2	2.0	2.1	3.0	3.5	3.2	2.9
Max. change (°C)	+13.1	+12.4	+11.4	−7.2	+9.5	+9.4	+6.6	−10.0	−7.7	−8.0	+12.4	+8.1	13.1
Mean +	4.9	4.2	3.5	2.6	2.9	2.1	2.5	2.2	2.6	3.2	3.7	3.5	3.2
Mean −	3.3	4.3	3.9	2.3	2.3	2.4	2.0	2.0	1.9	2.9	3.4	3.1	2.8

* $n_-/n_+ = 1.12$.

(00h00Z–00h00Z), the number of occasions when an increase, n_+, or a decrease, n_-, of temperature was observed, maximum changes, and mean positive and negative changes are also shown. Extreme values were taken from eight observations per day.

Tables XXI and XXII give the interdiurnal temperature changes at Northice and Brønlunds Fjord, at the coast. At Northice, there was no ice-free land within 300 km of the station. The station at Brønlunds Fjord (82°N) was situated on the southern side of a 50 km long fjord; it was a little more than 100 km distant from the inland ice.

The mean values of the interdiurnal temperature changes in the central part of the ice cap (Station Centrale) are apparently among the largest observed on the earth's surface. The mean annual interdiurnal temperature change was 6.3°C. This exceeded, by 1.6°C, the annual value observed during Wegener's expedition. The highest mean values were observed in winter during both expedition periods, and the lowest values were obtained in summer. The highest values in interdiurnal temperature change were observed in November for both expeditions (mean 8.4° and 8.1°C. It is, however, premature to conclude that the strongest air mass exchange occurs in November, although physically this idea is appealing—an outbreak of cold arctic air and northward movement of maritime air masses still influenced by open waters.

Table XXI gives the value of interdiurnal temperature changes observed at station Northice. Again, the mean monthly values are relatively high although smaller than the mean values of the Victor expedition. The different period of observation (1952–1954) could possibly have an influence on the mean data. But Northice probably belongs to a different regime from that of Station Centrale. The maximum of interdiurnal temperature change was recorded in January (60 days) with a mean of 7.7°C. Notable is the secondary maximum in November (6.6°C) when the central part of the ice cap showed the maximum values.

Brønlunds Fjord, although it is the northernmost station (82°N), belongs to the regime of the coastal region. The influence of open water is, apparently, also strongly pronounced there. Table XXII shows the values of interdiurnal temperature changes at Brønlunds Fjord. The values are smaller throughout than at the central part and at the eastern part of the ice cap (Northice). The annual mean value is only 2.9°C against annual values of 6.3, 4.8 and 4.7°C on the ice cap. The maximum values are recorded in winter (January and February) as they are on the ice cap. Again, a secondary maximum appears in November, although much less pronounced. The monthly mean values of interdiurnal temperature change are more evenly distributed than at Weststation; the values in summer are at least twice as large as the summer values at Weststation.

For comparison, observational data were evaluated for an arctic site located far from a permanent ice massif, i.e., for a U.S.S.R. drifting ice station (1950–1951). The position of the station was approximately 76.5°N and ranged from 180–160°E. The means of the interdiurnal temperature changes are very similar to those at Wegener's Weststation. February showed the maximum change (5.4°C) with a secondary maximum (4.7°C) in November. This was much greater than in the coastal regions of Greenland. A strong decrease in interdiurnal temperature change occurred from March to April; this is similar to the changes at Brønlunds Fjord.

Generally, the number of days with "cooling" exceeds those with "warming", although the ratio of these two numbers is very close to one: 1.09 for the Victor expedition, 1.08 for Northice, 1.13 for Weststation, 1.12 for Brønlunds Fjord, and 1.11 for the

U.S.S.R. drifting station. The mean positive changes are a little higher over Greenland, both on the ice cap and on the coast; there is no difference, however, at the U.S.S.R. drifting station. The observed maximum interdiurnal temperature changes are predominantly positive although a very strong negative change of temperature (50°C in 78 h) was recorded during the Wegener expedition (LOEWE, 1935). The number of days with intense warming exceeds the number of days with extreme cooling over the central part of the ice cap: during the Wegener expedition there were 16 days with greater than 20°C warming and only 10 days with greater than 20°C cooling; the Victor expedition experienced 14 days of warming against 8 days of cooling of 20°C or more.

We see that the Greenland plateau is apparently the arena for extreme interchanges of air masses, although strong temperature changes are observed throughout the entire Arctic. The temperature changes over the Greenland ice cap seem to be greater than over the other parts of the Arctic. It is very possible that these strong temperature changes over the ice cap in winter, are the result both of intensified cyclonic activity and of occurrences within the cold surface layers. LOEWE (1935) notes the fact that at the Watkins' Station, located between the south coast of Greenland and Eismitte, smaller temperature fluctuations were observed than at Eismitte, in spite of the fact that Watkins' Station is located nearer to the predominant tracks of lows. The observation period at Watkins' Station was approximately the same as at Eismitte. The circumstance that the warming generally exceeds the cooling on the ice cap speaks, perhaps, in favour of predominant occurrences within the cold surface layer—warming due to the destruction of the surface inversion layer and additional warming due to warm air advection (PUTNINS et al., 1959b).

Temperatures very seldom rise to the freezing point in the central part of the ice cap. However, the occurrences of such temperatures are not absolutely excluded, e.g., during one month of Wegener's expedition temperatures above −3°C were observed on two occasions, and the mean for the warmest day was −6.2°C. In the southern part of the ice cap, positive temperatures occur at even greater elevations. Thus, +5°C was recorded at 63.5°N 44°W, at an elevation of 2,700 m (LOEWE, 1935).

At Eismitte, −10°C was exceeded on 57 days of the year the expedition was in the field; on one August day the minimum temperature was also above −10°C. On 44 days the minimum was not below −20°C. Loewe emphasizes that relatively high minima can occur, even in the winter half-year.

Pronounced evenness (Abstumpfung) of summer temperatures due to melting ice does not appear in the temperature means but, as could be expected, it does appear as an asymmetrical distribution of the daily means in the melting zone and in the maximum temperatures in the central part of Greenland (LOEWE, 1935).

The interaction of intense incoming and outgoing radiation produces large daily amplitudes in the ice cap temperatures. Table XXIII shows periodic and aperiodic daily temperature amplitudes at Eismitte (LOEWE, 1935).

The periodic amplitudes are largest in spring and summer and smallest in winter. The difference, aperiodic minus periodic (reduced variation) is greatest in winter with a maximum in November. The winter climate on the ice cap is the least stable of any season. The aperiodic temperature variations are smaller at the edge of the ice cap. Even at Watkins' Station, close to the tracks of lows, the variation is smaller than at Eismitte (LOEWE, 1935).

TABLE XXIII

PERIODIC AND APERIODIC DAILY TEMPERATURE AMPLITUDES (°C) AT EISMITTE
(After LOEWE, 1935)

	Jan.	Feb.	Mar.	Apr.	May	June	July	Aug.	Sept.	Oct.	Nov.	Dec.	Year
Period temp. amplitude	2.0	2.2	6.7	8.5	10.5	9.8	7.8	8.4	7.2	2.1	1.0	1.1	5.6
Aperiodic minus periodic	9.5	9.9	5.7	4.4	3.2	2.7	2.5	4.8	5.5	7.8	12.3	8.6	6.4

The evaluation of hourly temperatures at Northice during June–July, 1953, showed that the average diurnal temperature range during this period was 7.0°C. The maximum temperature occurred at 15 L.M.T. A similar analysis of the Britannia Sø thermograph for June 1953 showed a daily range of 3.7°C, only about half that at Northice. This may possibly be due to the influence of sea and on-shore winds, the melting of glaciers and ice in lakes, etc., in the neighbourhood. The time of the maximum was around 16 L.M.T. (HAMILTON, 1958).

A remarkable climatic feature of the inland ice are the large temperature amplitudes observed on clear days. While crossing the ice cap, A. Wegener found the following temperature variations on clear days in May and June:

Height (m)	below 2,000	2,000–2,500	2,500–2,800	>2,800
Temperature (°C)	8.0	8.7	11.0	15.5

The temperature variations on clear days, in the central part and at the edge of the ice cap are shown in Table XXIV (LOEWE, 1935).

TABLE XXIV

TEMPERATURE VARIATIONS (°C) ON CLEAR DAYS AT EISMITTE AND WESTSTATION
(After LOEWE, 1935)

	Mar.	Apr.–May	June	July–Aug.	Sept.	Oct.
Eismitte	9	16	16	14	14	5
Weststation	7	9	7	5	6	–

The maximum diurnal amplitudes occurred in the spring. Amplitudes on the ice cap are similar to those observed in a desert climate. These large amplitudes are explained by the high transparency of the air, the influence of the snow cover, and by the fact that here the cold air layer is very shallow. At the edges of the ice cap, however, the amplitudes are smaller. Presumably this is due to the convexity of the surface. In summer the influence of the melting (freezing) process is manifested in a depressing of the amplitudes (LOEWE, 1935).

Minimum temperatures recorded on the ice cap during several expeditions were: —70°C at Northice, —65°C at Eismitte and Station Centrale, and —53°C at Watkins' Station.

Apparently the low temperatures on the ice cap are also the result of the radiation effect: of 88 clear days at Eismitte, only eleven had temperatures exceeding the monthly mean. On the average, clear days were 9°C colder than the monthly means, during the winter half year. In summer, when the incoming radiation is extremely high, the temperature on clear days is generally still 4°C colder than the monthly mean temperatures, and 8°C colder than the means on cloudy days.
At Eismitte, the temperature fell below −50°C for the first time on October 10. In spring, the last occurrence was on April 12. However, temperatures about −50°C can occur in May. Table XXV shows the number of days, at Eismitte, with temperatures below −50°C. The number of days with temperatures below −60°C are shown in parentheses (LOEWE, 1935). At Watkins' Station, the temperature fell below −50°C on five days.

TABLE XXV

NUMBER OF DAYS AT EISMITTE WITH TEMPERATURES BELOW −50°C AND −60°C (in brackets)
(After LOEWE, 1935)

	Oct.	Nov.	Dec.	Jan.	Feb.	Mar.	Apr.	Year
Daily mean	5	8	3	9(1)	16(1)	5	1	47(2)
Daily max.	1	4	1	4	5	1	0	16
Daily min.	8	18	9	12(4)	22(5)	15(4)	3	87(13)

TABLE XXVI

MEAN INTERDIURNAL CHANGE IN TEMPERATURE (DIFFERENCES BETWEEN THE DAILY MEAN TEMPERATURES OF THE ADJACENT DAYS)
(After STEPANOVA, 1960)

Stations	Winter			Period of record
	mean	max+	max−	
Coastal stations				
Ivigtut	2.2	11.1	5.4	1953–1954
Godthaab	1.8	9.0	4.9	1953–1954
Upernavik	2.4	7.0	10.7	1953–1954
Thule	3.2	14.4	12.2	1948–1949
Brønlunds Fjord	2.7	8.5	10.0	1948–1949, 1949–1950
Scoresbysund	2.9	9.2	14.7	1930–1931
Angmagssalik	2.1	5.7	14.4	1930–1931, 1953–1954
Ice Plateau stations				
Northice	3.9	14.0	15.0	1952–1953, 1953–1954
Victor's Station Centrale	4.8	15.3	18.1	1949–1950
Eismitte	5.5	19.5	19.2	1930–1931
East Siberian cold pole region				
Verkhoyansk	3.6	13.6	14.7	1932–1933
Antarctica				
South Pole Station	3.6	22.0	17.0	June–Sept. 1958

The temperatures are relatively high at the edge of the ice cap. The minimum at West-station was —40°C and the temperature fell below —30°C on only 41 days. The minima at Borg Glacier (76°40'N 24°00'W): annual minimum —50°C with temperatures below —40°C occurring on 63 days, are explained by the stagnation of cold air in the fjord (LOEWE, 1935).

At the coast, the interdiurnal changes of temperature are much smaller than on the ice cap. Table XXVI (STEPANOVA, 1960) shows the mean interdiurnal changes of daily mean temperatures at Greenland coastal and ice cap stations, and also at stations in the Antarctic and in the east Siberian cold pole region (Verkhoyansk) during the winter (December through March).

Observations were made at three levels at Camp Century (77°10'N 61°08'W; elevation 1,925 m), on the slope of the ice cap east of Thule (see also SCHALLERT, 1962, 1963a, b). Table XXVII lists monthly mean temperatures.

TABLE XXVII

MONTHLY MEAN TEMPERATURES AT CAMP CENTURY (°C)

	7.5 cm	200 cm	Shelter
1962:			
June	−10.8	−10.8	−9.7
July	−7.1	−7.2	−5.6
Aug.	−11.9	−11.8	−10.6
Annual	−24.6	−24.3	−23.3
1963:			
June	−11.8	−11.9	−9.4
July	−8.6	−8.2	−6.1
Aug.	−9.9	−9.8	−7.8

It is seen that the monthly mean temperatures in the shelter are consistently higher than those at 7.5 cm in the summer months and, apparently, also during the entire year. There is little difference between the monthly mean temperatures at 7.5 cm and at 200 cm.

Table XXVIII shows the relationship between the numbers of cases with negative and positive differences. In the majority of cases (months) the number of negative differences between the temperature at 7.5 cm (12.5 cm) and the temperature in the shelter is larger than the number of positive differences: on sixteen occasions the negative differences were more frequent as opposed to seven cases when positive differences were more frequent. Camp Fistclench consistently showed a strong predominance of negative differences and only in August, at 21 L.S.T., were the number of negative and positive differences equal. The differences were usually less than 1°C and only on relatively few occasions did the differences exceed 1°C.

The temperature differences should partly depend on the cloudiness. Two days in July 1960 were chosen, and the temperature differences were computed for every hour (Table XXIX).

TABLE XXVIII

THE RELATIONSHIP ($n-/n+$) BETWEEN THE NUMBER OF CASES WITH NEGATIVE DIFFERENCES OF TEMPERATURES AT 7.5 cm (CENTURY), 12.5 cm (FISTCLENCH) AND THE TEMPERATURES IN SHELTER; ALSO THE VALUES OF MAXIMUM DIFFERENCES RECORDED

L.S.T.:	08h00		14h00		21h00	
1960	$n-/n+$	max. diff. (°C)	$n-/n+$	max. diff. (°C)	$n-/n+$	max. diff. (°C)
Century						
Mar.	0.20	1.7	0.60	2.8	0.42	−1.9
May[1]	3.00	−2.2	6.00	−1.4	0.50	−2.0
June	1.27	2.3	0.91	3.3	2.84	−2.3
July	1.90	1.2	0.87	−0.9	0.75	−2.9
Aug.	1.10	1.7	1.38	1.3	1.55	1.7
Fistclench						
June	12.50	−1.2	29.00	−1.6	1.90	−2.2
July	2.88	−1.5	1.80	1.6	2.00	−1.6
Aug.	1.38	−1.1	1.62	−1.5	1.00	1.7

[1] Only 9 days.

TABLE XXIX

HOURLY TEMPERATURE DIFFERENCES ON TWO DAYS IN JULY, 1960 (°C)

Station	n_-/n_+	Max. diff. (°C)	Mean cloudiness (tenths)
Century	3.40	−1.0	0.0
Century	0.09	+1.2	9.0
Fistclench	23.00	−1.2	0.0
Fistclench	0.37	+1.3	6.4

As expected, the negative temperature differences fully dominated the cloudless days The evaluation of data from Wegener's expedition showed further that in summer the snow is 1°C colder than the air at 21h00, and 2°C warmer than the air at 14h00. The daily range of the snow surface temperature is probably about 5°C greater than that of the air. Contact with the warming and cooling snow surface produces air temperatures with a range of 10°–15°C. Daily maximum and minimum temperatures lag behind the time of high and low sun by almost seven hours; the periods of most rapid air temperature rise and fall are centred at the hours of high and low sun respectively (MILLER, 1956).
The minimum temperatures recorded are not a peculiarity of the climate of the inland ice, but rather the mean temperature is—during three years of observation in the central part of the ice cap, the annual mean was −30°C. Here the annual variation of temperature (warmest minus coldest month) was 36.5°C. At Verkhoyansk the variation is 65.9°C.

The difference between the extremes (mean maximum minus mean minimum) is 62°C on the ice cap and 89°C at Verkhoyansk (GEORGI, 1953).
The annual temperature variation in the central part of the ice cap shows features imposed by radiation as well as by advection; the first is seen in the quick temperature increase from April to May, while the second appears in the damping of the mean temperature curve in winter. Apparently the advection processes are the cause of the "coreless" winters. The cause for the depression in the course of the temperature curve in summer is different; the temperature curve is "phase conditioned", i.e., it is linked to the ice phase of water, therefore the temperature near the surface is limited to a value near 0°C.
The temperature problem over the ice cap is a complex one for several reasons (GEORGI, 1953):
(*1*) Radiation processes: The outgoing radiation is particularly important during clear weather. Radiational cooling of the surface is transferred to the air by direct contact, molecular exchange, etc. The direct cooling of the air by the radiation process is of minor importance due to the small amount of water vapour in the air; consequently the back radiation is also small during clear weather. The warming of the air by incoming radiation is small in comparison with the outgoing radiation due partly to the high albedo of the inland ice and partly to the high cloudiness in summer.
(*2*) The small specific heat of the ice allows it to accumulate only a small amount of heat. The loose crude-crystalline structure of the ice surface has a density of only 0.1 g/cm^3 (sometimes 0.05 g/cm^3 or less), and the specific heat of the ice is only 0.5 cal./g degree. This results in a small heat capacity.
(*3*) The stabilizing influence of the ice sheet on air temperature because of its high albedo and its melting point of 0°C.
The central part of the ice cap (Eismitte, Station Centrale) is influenced more by weather occurrences from the west than from the east, in spite of the fact that these stations are located equidistant from both coasts. There are some similar features in the temperature regimes of the west coast and the central part of the ice cap.
In 1930/1931 the annual course of temperature at Eismitte was similar to that at the west coast, except that the maxima and minima appeared earlier at Eismitte. In 1950, the same delay of temperature increase from March to April occurred at Station Centrale, with an increase from April to May which also appeared on the west coast. However, small peaks which appeared at Station Centrale in January, 1950 and February, 1951 did not appear at west coast stations. In spite of general agreement between the temperature curves from the west coast and the central part of the ice cap, there are some discrepancies that may indicate that these two areas are governed by different temperature regimes.
In glacier areas with little or no melting, the amplitude of the annual temperature wave decreases to about 0.5°C at a depth of 8 m or so, and the snow temperature at this depth is very near the local annual mean air temperature.
DIAMOND (1958) used a mean lapse rate of 0.7°C/100 m, computed from the annual temperature at five coastal stations and the annual mean air temperatures obtained from snow profile studies at five inland stations at approximately the same latitude. This mean lapse rate was also used to compute the annual mean air temperature for the ice cap where air or ice temperatures were not available (Fig.5).

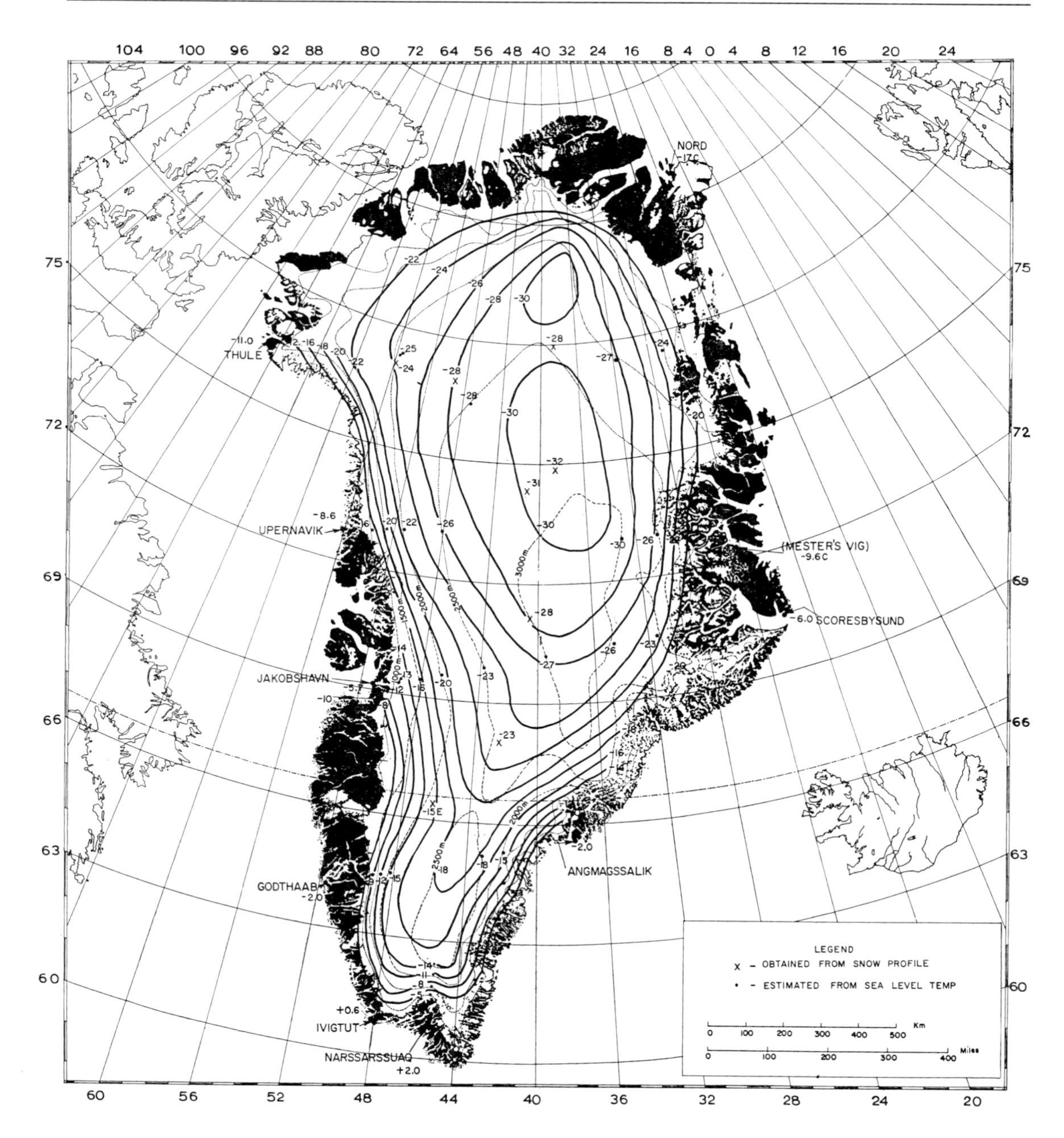

Fig.5. Annual mean air temperature (°C) on Greenland ice cap. (After DIAMOND, 1958.)

All available ascents from the Victor expedition (September 1949 through August 1951) were evaluated with respect to lapse rates (PUTNINS et al., 1959b).

The observation period was divided into three seasons: winter, summer, and transition periods. As the winter season on the ice cap, the months November through March were chosen. Thus we have nine months of observations in the winter season (no observations were made during March, 1951). The ascents were concentrated in the winter season, thus the number of ascents in the nine months was 130 as compared with 56 ascents during the nine months of the summer and transition periods. Of these 130 cases,

TABLE XXXII

PERCENTAGE FREQUENCY OF TEMPERATURE LAPSE RATES (°C/100 m) BETWEEN THE TOP OF THE INVERSION AND THE 400-MBAR LEVEL, STATION CENTRALE

Lapse rate:	0.10 to 0.19	0.20 to 0.29	0.30 to 0.39	0.40 to 0.49	0.50 to 0.59	0.60 to 0.69	0.70 to 0.79	0.80 to 0.89	
Winter period (Dec.–Mar.)									
no. of obs.	2	3	1	16	31	24	18	9	104
%	2	3	1	15	30	23	17	9	
Transition period (Sept.–Nov., Apr., May)									
no. of obs.	0	0	1	2	4	7	1	1	16
%	0	0	6	12	25	44	6	6	

TABLE XXXIII

GREENLAND AREA. MEAN TEMPERATURE LAPSE RATE (°C/100 m), 03h00Z
(After PUTNINS and CHOATE, 1960)

Station			Winter (Dec.–Mar.)*				Summer (June–Aug.)*			
			1	2	3	4	1	2	3	4
East of Greenland and Greenland east coast:										
Danmarkshavn, Greenland	76°46′N	18°46′W	−0.25	0.24	0.59	0.60	0.20	0.42	0.61	0.72
Jan Mayen	71°01′N	08°28′W	0.48	0.50	0.63	0.63	0.21	0.47	0.61	0.69
Kap Tobin, Greenland	70°25′N	21°58′W	−0.02	0.36	0.60	0.63	0.14	0.52	0.60	0.71
O. S. V. "M", Greenland	66°00′N	02°00′W	0.71	0.58	0.67	0.68	0.41	0.48	0.61	0.71
Angmagssalik	65°37′N	37°39′W	0.32	0.50	0.62	0.61	0.29	0.51	0.59	0.70
Keflavik, Iceland	63°57′N	22°37′W	0.46	0.57	0.65	0.63	0.47	0.50	0.60	0.70
O. S. V. "A"	62°00′N	33°00′W	0.70	0.55	0.62	0.60	0.43	0.46	0.60	0.70
O. S. V. "I"	59°00′N	19°00′W	–	0.52	0.65	0.66	–	0.43	0.59	0.71
O. S. V. "C"	52°45′N	35°30′W	0.60	0.47	0.61	0.66	0.36	0.44	0.57	0.71
West of Greenland and Greenland west coast:										
Alert, N.W.T.	82°30′N	62°20′W	−0.47	0.22	0.57	0.49	0.24	0.50	0.60	0.66
Eureka, N.W.T.	80°00′N	85°56′W	−0.74	0.24	0.56	0.47	0.55	0.50	0.60	0.65
Isachsen, N.W.T.	78°47′N	103°32′W	−0.67	0.28	0.56	0.47	0.28	0.45	0.60	0.64
Thule, Greenland	76°31′N	68°50′W	0.02	0.33	0.56	0.49	0.39	0.48	0.61	0.67
Resolute, N.W.T.	74°43′N	94°59′W	−0.44	0.26	0.55	0.45	0.34	0.46	0.60	0.65
Egedesminde, Greenland	68°42′N	52°52′W	0.16	0.44	0.58	0.56	0.26	0.53	0.62	0.71
Narssarssuaq, Greenland	61°11′N	45°25′W	0.37	0.53	0.59	0.58	0.44	0.52	0.58	0.70
O. S. V. "B"	56°30′N	51°00′W	0.64	0.44	0.57	0.60	0.23	0.47	0.59	0.71
Goose Bay, Newfoundland	53°19′N	60°25′W	0.02	0.26	0.55	0.59	0.50	0.57	0.58	0.71
Station Centrale (elevation 2,993 m)	70°55′N	40°38′W			0.04	0.59			(0.47	0.65)

* 1 = 850–1,000 mbar
2 = 700–850 mbar
3 = 500–700 mbar
4 = 300–500 mbar

computed for every ascent and then summarized for two seasons—winter (December–March) and summer (June–August). For the summer season, only nineteen observations were available for the lower layer and ten for the upper layer, compared with 104 and 97 observations in the two layers respectively during the winter season.

The mean temperature lapse rate in winter for the layer from the surface to 500 mbar was 0.04°C/100 m. This very stable condition above the ice cap is principally explained by the strong inversion prevailing in winter up to 1 km from the ice cap surface. The layer between 4,000 and 5,000 m in winter (November–March) shows a lapse rate of 0.57°C/100 m. The variability of the lapse rate in the layer below 500 mbar is great. The strongest negative lapse rate for this layer was recorded on February 22, 1950 (−0.90°C/100 m), and the strongest positive lapse rate on February 25, 1950 (0.84°C/100 m). Although the strongest lapse rates observed were negative in winter, positive lapse rates were observed more frequently: the ratio of the number of positive to negative lapse rates was 60:44. The frequency distribution of the lapse rates in winter (December through March) is shown in per cent in the following:

Lapse rate (°C/100 m)	−1.00 to −0.76	−0.75 to −0.51	−0.50 to −0.26	−0.25 to −0.01	0.01 to 0.24	0.25 to 0.49	0.50 to 0.74	0.75 to 0.99
Frequency (%)	2	5	15	20	23	25	7	3

68% of all cases occurred between the values −0.25°C/100 m and 0.49°C/100 m. With respect to the variation of frequency distribution from month to month, during the winters of the Victor expedition, it can only be noted that both Decembers showed the most stable conditions.

It is clear that the very stable layer over the ice cap contributes to the low mean temperature lapse rate up to the 500 mbar level. However, the 1,000 m thick layer immediately below the 500 mbar level shows a mean lapse rate which is apparently in accordance with the general distribution of lapse rates over the Greenland area in winter. The mean lapse rate for the layer between the 500 and 300 mbar levels (0.59°C/100 m) is in very good agreement with the mean values of the lapse rates over the west and east coasts of Greenland; e.g., Egedesminde shows 0.56°C/100 m and Kap Tobin, on the east coast, shows 0.63°C/100 m in winter (Table XXXIII). The extreme values for the 500–300 mbar layer were 0.88°C/100 m (February 1, 1950), and 0.03°C/100 m (February 21, 1950). The following shows frequency distribution of the lapse rates for this layer in winter (97 observations).

Lapse rates (°C/100 m)	0.00 to 0.24	0.25 to 0.49	0.50 to 0.74	0.75 to 0.99
Frequency (%)	4	21	64	11

We see that 75% of all cases are included between the values 0.50 and 0.99°C/100 m. In summer (August 1950, June and July 1951) there were only nineteen ascents which reached the 500 mbar level, and ten which reached the 300 mbar level. For this reason, only very tentative results are available.

The mean lapse rates for the layers, surface to 500 mbar and 500–300 mbar were 0.47 and 0.65°C/100 m respectively; the frequencies of the several lapse rates are given below.

Lapse rates	0.00 to 0.24	0.25 to 0.49	0.50 to 0.74	0.75 to 0.99	Number of observations
Surface–500 mbar	1	11	6	1	19
500–300 mbar			8	2	10

It seems that the influence of the ice cap on the vertical temperature distribution also in summer is noticeable up to the 500 mbar level though not as strongly as in winter. This influence, however, is apparently not mainly confined to the lower layer over the ice cap as it is in winter, but is evident up to the 500 mbar level: the lapse rate for the layer between 3,000 and 4,000 m was 0.43°C/100 m and between 4,000 and 5,000 m it was 0.46°C/100 m.

The measurement of lapse rates in the layer near the surface is especially difficult during the summer, when solar radiation is strong and diffusely reflected by the snow surface. In such cases it is difficult to provide both adequate screening and ventilation for conventional thermometers. HAMILTON (1953) used optical methods to determine the lapse rates in the lowest few meters above the snow surface at Northice (78°N 38°W) during June and July.

Generally, the morning and evening observations showed strong inversions; the maximum value was −0.27°C/m. Afternoon measurements, generally at 15h00 or 16h00Z, showed a decrease of temperature with height in 16 cases out of 29. The maximum lapse rate measured was 0.06°C/m.

LOEWE (1935) computed the mean lapse rates from temperature observations at Eismitte, Weststation, Umanak, and the two stations from Watkins' expedition (Table XXXIV).

TABLE XXXIV

TEMPERATURE LAPSE RATES (°C/100 m)
(After LOEWE, 1935)

	Jan.	Feb.	Mar.	Apr.	May	June	July	Aug.	Sept.	Oct.	Nov.	Dec.	Year
Eismitte–Weststation	1.20	1.12	0.82	0.72	0.76	0.79	0.67	0.84	0.90	1.08	*1.22*	0.92	0.92
Weststation–Umanak	1.00	1.03	0.49	0.56	0.45	0.60	0.79	0.77	0.85	1.10	*1.26*	1.15	0.84
Watkins' Ice Cap Station–Base Station	0.96	1.02	(0.90)	–	–	–	–	–	0.83	1.02	*1.10*	0.99	(0.98)

Maxima in italics.

The high winter time lapse rates are remarkable. A check was made by computing the lapse rates from continuous observations on the slope of the ice cap in the Thule area (Table XXXV). The stations are listed in the order of decreasing elevation. Generally, the lapse rates are high in every month, but not in all layers, and some are even superadiabatic. For example, from May through August 1960 the lower layer showed superadiabatic lapse rates, but the lapse rates decreased with elevation. From June through

TABLE XXXV

TEMPERATURE LAPSE RATE (°C/100 m) ALONG THE SLOPE OF THE ICE CAP IN THE THULE AREA[1]

	1960								1961				1962											
	May	June	July	Aug.	Sept.	Oct.	Nov.	Dec.	Jan.	Feb.	Mar.	Apr.	Jan.	Feb.	Mar.	Apr.	May	June	July	Aug.	Sept.	Oct.	Nov.	Dec.
Fistclench		–	–	–																				
		0.55	0.55	0.09																				
Century	–	–	–	–	–	–	–	–	–	–	–	–	–	–	–	–	–	–	–	–	–	–	–	–
	0.38	0.69	0.58	0.75									1.04	0.98	1.04	0.84	0.69	0.75	0.64	0.69	0.84	1.04	0.84	0.95
Tuto East													–	–	–	–	–	–	–	–	–	–	–	–
													0.71	0.71	0.71	0.89	0.71	1.24	1.24	1.08	1.08	0.71	0.89	0.71
Tuto III	–	–	–	–	0.93	0.75	0.73					0.98												
	1.02	1.15	1.22	1.15				1.06	0.69	0.71	0.80													
Tuto I	–	–	–	–	–	–	–					–	–	–	–	–	–	–	–	–	–	–	–	–
					−0.64	0.66	0.38					−0.30	0.0	0.24	0.0	−0.93	0.93	0.69	0.93	1.19	0.93	0.69	0.69	0.46
Tuto West					–	–	–	–	–	–	–	–	–	–	–	–	–	–	–	–	–	–	–	–

[1] Dashes indicate the layers taken: e.g., Aug. 1960, 0.75 between Century and Tuto III; June 1962, 1.24 between Tuto East and Tuto I.

September, 1962, the highest values, also super-adiabatic, were recorded in the layer between Tuto I and Tuto East. From January through April, and in December 1962, the air was very stable in the lower layer but unstable in the higher layer. During the transition months, September through November 1960, the air was more stable in the lower layer than it was in 1962. Instability increased with elevation in both years. As the area around Thule is quite turbulent, we can hardly expect the same consistency of the lapse rates with elevation and seasons elsewhere. These values do show, however, that the lapse rates are consistently high along the slope of the ice cap.

Inversions

One of the most persistent phenomena in both cold and temperate climates is the occurrence of surface inversions during the cold season, especially if the surface is covered with snow or ice. The surface inversion exists almost permanently over the Greenland ice cap, and it is possibly more pronounced than over other parts of the Arctic Basin. The following monthly means are shown in Table XXXVI: temperature, pressure, height of the base and of the top of the inversion, height of the top of the isothermal layer (when the inversion is surmounted by such a layer), and mean values of temperature and pressure at the surface during inversion conditions. The height at which surface observations were made was taken as 3,000 m above sea level (expedition data list the surface as 2,993 m). This table gives a rough picture of the average structure of the inversion layer, as well as of a mixing layer near the surface, and an isothermal layer above the principal inversion layer. It should be mentioned again that the values for the months May through September must be considered to be very uncertain, because of the small number of observations made in these months.

TABLE XXXVI

MONTHLY MEAN VALUES OF SURFACE INVERSION, STATION CENTRALE[1]
(After PUTNINS et al., 1959b)

Month	Surface		Base of inversion layer			Top of the inversion			Top of the isothermal layer		
	temp. (°C)	press. (mbar)	temp. (°C)	press. (mbar)	height (m)	temp. (°C)	press. (mbar)	height (m)	temp. (°C)	press. (mbar)	height (m)
Jan.	−36.4	678	−36.4	678	3,000	−26.8	648	3,329	−26.8	646	3,358
Feb.	−39.9	679	−39.9	679	3,000	−29.5	647	3,334	−29.5	642	3,388
Mar.	−35.8	679	−36.0	676	3,032	−26.8	640	3,431	−26.8	632	3,515
Apr.	−34.3	687	−34.3	687	3,000	−30.0	653	3,351			
May	−18.7	697	−18.7	697	3,000	−15.6	661	3,400			
June	−12.7	705	−13.6	694	3,125	−11.1	676	3,355	−11.1	668	3,425
July	−10.4	697	−16.1	635	3,705	−15.3	613	3,968			
Aug.	−11.2	704	−13.0	674	3,352	−9.8	650	3.622	−9.8	645	3,698
Sept.	−19.7	692	−21.4	677	3,160	−18.7	630	3,715			
Oct.	−31.9	686	−31.9	686	3,000	−23.9	635	3,555			
Nov.	−33.2	684	−33.5	677	3,086	−25.0	643	3,455	−25.0	638	3,513
Dec.	−37.5	684	−37.5	683	3,005	−25.9	656	3,303	−25.9	650	3,400

[1] P.E. Victor Expedition, 1949–1951.

TABLE XXXVII

SURFACE INVERSION, GREENLAND ICE CAP[1]
(After PUTNINS et al., 1959b)

Season	Height of inversion (m)				No. of obs. yes/no	Intensity (°C)				Remarks
	mean	max.	min.	variab.		mean	max.	min.	variab.	
Nov. 1949–Mar. 1950	384	1,307	67	147	87/12	9.6	24.2	0.5	4.8	
Nov. 1950–Mar. 1951	407	1,157	107	138	27/4	10.0	22.5	1.5	4.6	no report in March
Winter (9 months)	389	1,307	67	155	114/16	9.7	24.2	0.5	4.8	
Summer (3 months)	375	1,047	7	219	8/12	2.6	5.5	0.4	1.6	
Transition period (Apr., May, Sept. Oct., 6 months)	393	1,207	87	203	25/12	6.2	18.0	1.0	2.2	

[1] P.E. Victor Expedition, 1949–1951.

Table XXXVII shows mean values of the surface inversion by seasons: winter (November–March), summer (June–August) and transition season (April, May, September, October). The intensity of the inversion (temperature difference between the highest temperature in the inversion layer and the temperature of the base of the inversion) is, as expected, greatest in winter. Also, the maximum intensity is recorded in winter (24.2°C), as is the maximum variability of the intensity. The height of the inversion does not differ greatly from winter (114 observations) to the transition season (25 observations); however, the variability of the height is notably lower in winter (155 m against

TABLE XXXVIII

SURFACE INVERSION OVER THE GREENLAND ICE CAP, STATION CENTRALE[1]
(After PUTNINS et al., 1959b)

Month	Total number of soundings	Surface inversion present	Mixing layers present
Jan.	32	24	0
Feb.	29	28	0
Mar.	19	17	3
Apr.	13	11	0
May	9	3	0
June	8	4	1
July	7	1	1
Aug.	5	3	1
Sept.	6	2	0
Oct.	9	9	0
Nov.	25	22	4
Dec.	25	23	1
Year	187	147	11

[1] P. E. Victor Expedition, 1949–1951.

203 m). The two winters of the expedition, 1949/1950 and 1950/1951, were evaluated separately (see Table XXXVII). The mean values of the intensity of the inversion as well as the extreme and the variability of the intensity are very similar for the two winters. The same can be said about the height of the inversion.

The question of how frequently the surface inversion was encountered can be answered from Table XXXVIII. Here, given by months, are: the total number of soundings, the number of cases when a surface inversion was observed and the number of cases when a mixing layer was present. It is seen that 79% of all observations showed a surface inversion; in winter alone, 88% of all observations. A mixing layer was present only in 7% of the winter cases with a surface inversion.

The intensity of the surface inversion should increase with decreasing surface temperatures. Table XXXIX shows the relationship between the intensity of the inversion and the surface temperature.

TABLE XXXIX

SURFACE INVERSION, DEPENDENT ON SURFACE TEMPERATURE. GREENLAND ICE CAP, STATION CENTRALE[1]
(After PUTNINS et al., 1959b)

Intensity of inversion (°C)	Surface temperature (°C)									Total
	−15.0 to −19.9	−20.0 to −24.9	−25.0 to −29.9	−30.0 to −34.9	−35.0 to −39.9	−40.0 to −44.9	−45.0 to −49.9	−50.0 to −54.9	−55.0 to −59.9	
0.0– 0.9		1	1							2
1.0– 1.9		3	1							4
2.0– 2.9		4	2	1			1			8
3.0– 3.9		2	3	1			1			7
4.0– 4.9		2	4	3						9
5.0– 5.9		1	4	2	1		1	1		10
6.0– 6.9			1	3	1	2				7
7.0– 7.9				2	1					3
8.0– 8.9				2		2	1		1	6
9.0– 9.9				3	1	2	2			8
10.0–10.9				1	1	2	1	1		6
11.0–11.9			1	1	2	3	2			9
12.0–12.9									1	1
13.0–13.9			1	1	1		2			5
14.0–14.9				1	3	3	1			8
15.0–15.9					1	3	1			5
16.0–16.9					1	1				2
17.0–17.9						3	1			4
18.0–18.9							1			1
19.0–19.9									2	2
20.0–20.9							1	1		2
21.0–21.9							1			1
22.0–22.9								1	1	2
23.0–23.9										
24.0–24.9								1		1
25.0–25.9								1		1
Total:		13	18	21	13	21	17	6	5	114

[1] P. E. Victor Expedition, 1949–1951, winter (Nov.–March).

Cloudiness

The figures of Table XL showing daily variation of cloudiness, are based on observations reported by LOEWE (1935), who remarked that cloud observations were not very precise, particularly during the dark period.
Loewe assumes that the noontime maximum, observed during the dark period, may be wrong because there is hardly an explanation for such a phenomenon. The minimum at noon in summer, however, is apparently real.
These figures show that the cloudiness in the central part of the ice cap (Eismitte) is the same as at the edge of the ice cap (Weststation).
The values of Table XLI show the frequency of clear days (<2) and cloudy days ($\geqq 8$) (LOEWE, 1935).
In the averages for the year, there is no difference between the central part of the ice cap and the western edge of the inland ice. The similarity is even more pronounced when the frequency of extreme cloudiness is taken into account (0 and 10).
The mean cloudiness observed along the route of the British Trans-Greenland Expedition, during 80 days in June–September, 1934, was 5.6 (LINDSAY, 1935).
At Watkins' Ice Cap Station, the frequency of days on which the mean cloudiness for the day was within specified limits, from September through March, is shown in Table XLII (after MIRRLESS, 1934).
Apparently, the large cloud amount in the central part of the ice cap does not consist principally of high clouds. Basically there is no difference in the appearance of certain cloud forms over Eismitte and Weststation; Ast clouds over Weststation frequently appeared as St clouds over Eismitte.

TABLE XL

DAILY VARIATIONS OF CLOUDINESS IN TENTHS[1]
(After LOEWE, 1935)

Hour	Jan.	Feb.	Mar.	Apr.	May	June	July	Aug.	Sept.	Oct.	Nov.	Dec.	Year
Weststation:													
08h00	(6.5)	6.1	6.5	5.3	5.9	5.1	4.6	6.5	5.3	*6.8*	(4.3)	(5.6)	5.8 (Feb.–Oct.)
14h00	6.5	6.3	6.4	5.7	5.6	4.3	3.9	6.4	5.6	5.8	5.3	*6.9*	5.7 (Jan.–Dec.)
21h00	(5.2)	(4.9)	4.9	6.6	5.4	4.7	4.2	*6.7*	5.2	(5.2)	(4.6)	(6.6)	5.4 (Mar.–Nov.)
Eismitte:													
08h00	(5.1)	4.5	4.9	5.9	5.9	5.2	*8.1*	6.2	5.4	5.6	(4.1)	(4.2)	5.7 (Feb.–Oct.)
14h00	6.6	5.4	6.3	6.6	5.8	5.2	*7.4*	5.8	5.6	5.1	5.2	5.7	5.9 (Jan.–Dec.)
21h00	(4.1)	(3.0)	5.0	5.5	6.4	5.7	*7.7*	6.6	4.8	(4.5)	(3.4)	(4.6)	6.0 (Mar.–Nov.)
Watkins' Ice Cap Station:													
07h00	(3.7)	3.4	–	–	–	–	–	–	5.1	4.7	(4.3)	(3.1)	–
13h00	3.4	3.8	–	–	–	–	–	–	4.7	5.2	4.3	5.1	–
22h00	(4.4)	(3.3)	–	–	–	–	–	–	(4.8)	(4.2)	(3.7)	(2.9)	–
Mean:	3.9	3.6	–	–	–	–	–	–	4.9	4.6	4.4	4.9	–

[1] Numbers in italics show maximum values.

TABLE XLI

FREQUENCY (%) OF CLEAR AND CLOUDY DAYS

Station	Spring		Summer		Autumn		Winter		Year	
	clear	cloudy	clear	cloudy	clear	cloudy	clear	cloudy	clear	cloudy
Weststation	16	33	24	32	23	30	15	40	21	34
Eismitte	17	31	15	41	32	22	27	21	23	29

TABLE XLII

FREQUENCY OF MEAN CLOUDINESS AT WATKINS' ICE CAP STATION AND CLOUD TYPES AT EISMITTE AND WESTSTATION
(After MIRRLESS, 1934; LOEWE, 1935)

Cloud types	Frequency (%)		Mean cloud. (tenths)	Frequ. (%) at Watkins' Ice Cap Station
	Eismitte	Weststat.		
Cist	15	19	0	12
Ci	18	8	1–3	40
Cicu	2	2	4–6	17
Ast	6	18	7–9	18
Acu	4	3	10	11
St	36 }	28		
Ni	2 }			
Stcu	14	14		
Frcu	1	2		
Frst	1	–		
Cu	1	2		
Foehn	–	4		

Precipitation

The problem of measuring precipitation in the Arctic is difficult. Not only does precipitation occur mostly in the form of snow, measurement of which is troublesome, but it is usually associated with strong wind which causes the snow to drift.

In the central part of the ice cap, precipitation occurs almost entirely in the form of snow. If warm air masses do intrude into these regions rain can occur occasionally, but this would happen very seldom. Rain was never observed at Eismitte.

In spite of difficulties, precipitation has been determined with a certain degree of accuracy. Eismitte recorded 204 days (56%) with snow, while at Weststation snow was observed on 126 (34%) days. The zone with most frequent precipitation, approximately 100 km from the edge of the ice sheet, seems to have the same frequency of precipitation as markedly maritime areas.

Along with increasing frequency of precipitation southward along the coast, one might also expect an increase in this direction on the ice cap. Surprisingly, Watkins' Station showed a lower frequency than Eismitte (Table XLIII).

TABLE XLIII

FREQUENCY (%) OF PRECIPITATION
(After LOEWE, 1935)

Station	Spring	Summer	Autumn	Winter	Year
Weststation	35	21	33	50	34
Eismitte	63	56	53	50	56
Watkins' Station	–	–	40	23	–

Eismitte shows a very even distribution of precipitation frequency during the year. Weststation shows a pronounced maximum in winter and minimum in summer. Northward, the frequency of precipitation is smaller:

Eismitte: 56%
Centrale: 62%
Site 2: 36%
Northice: 24%

It is very likely that Northice, located a little east of the divide, is subjected to a different regime (HAYWOOD and HOLLEYMAN, 1961).
The annual increase of snow on the ice cap was determined from stakes between Weststation and Eismitte (Table XLIV).

TABLE XLIV

ANNUAL INCREASE OF SNOW BETWEEN WESTSTATION AND EISMITTE
(After LOEWE, 1935)

Distance from the edge of ice sheet (km)	25	60	120	180	240	300	Eismitte
Elevation (m)	1,450	1,800	2,200	2,500	2,700	2,850	3,000
Water content of annual increase of snow (cm)	40	52	45	45	35	31	31.5

The snow deposit from precipitation apparently increases between the west edge of the ice sheet and a certain zone inland in the region 65°–75°N. Snow deposit is smaller at the highest elevations and then increases again farther east. The eastern part of the ice sheet presumably receives less precipitation than the western part.
DIAMOND (1958) estimated the annual accumulation of snow on the ice cap. Fig.6 shows precipitation expressed in centimetres of water. According to his estimates, the area with maximum annual accumulation, north of the Arctic Circle, is located several hundred kilometres west of the central crest of ice with values in excess of 50 cm of water equivalent. Along the crest, the annual precipitation averages about 25 cm water equivalent.
Measurements made by the British Greenland expedition on their traverse of the ice cap indicate an almost constant snow accumulation of 12–14 cm water equivalent north of 78°N. Other studies, made in the summer of 1959, indicate that the annual accumulation north of 80°N may vary from 13 to 26 cm water equivalent. South of the Arctic Circle,

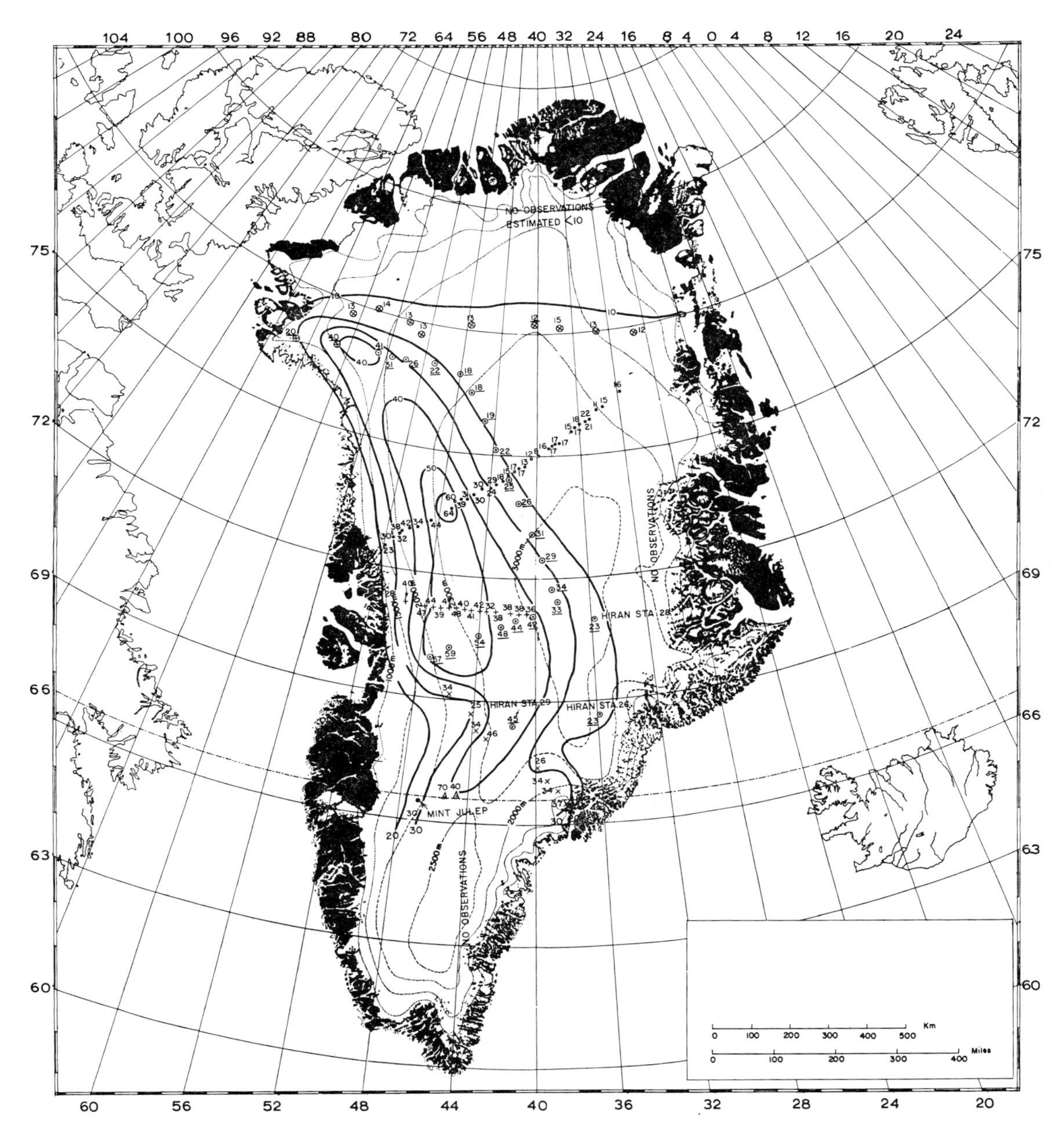

Fig.6. Annual mean accumulation on Greenland ice cap (cm of water equivalent). (After DIAMOND, 1958.)

the annual accumulation may be 5–7 times as large as in the area north of 80°N (GERDEL, 1961). BADER (1961) obtained a much smaller estimate.

A stratigraphic study of the snow in a deep pit permits the determination of annual layers, because there is usually a minor indication of melting and a density minimum. Investigations at Site 2, on the ice cap east of Thule, provided information concerning the accumulation of precipitation on the ice cap at this site (BADER et al., 1955). During 1908–1914 and 1919–1954, the annual mean precipitation amounted to 43.4 cm water equivalent. This amount counts the period from summer to summer as a precipitation year. The variability of the annual precipitation amount was 10.3 cm. During these periods, the maximum (74 cm) occurred in 1923/1924 and the minimum (22 cm) in 1933/1934.

In addition to snow, rime may also contribute to the total precipitation. The frequency (%) of occurrence of rime at Eismitte was as follows (LOEWE, 1935):

Spring	Summer	Autumn	Winter	Year
13%	13%	36%	34%	22%

Loewe estimated rime to contribute 3 cm of water-equivalent to the total precipitation- If correct, this would amount to about 10% of the total. In WEGENER's (1939b) estimation, the amount of water received in the form of rime is small. During nine months the air was, on the average, supersaturated with respect to ice, which means that conden. sation should occur. On the other hand, observations showed that rime seldom occurred. Apparently, very strong supersaturation is necessary for condensation to take place.

The water content of the air over the ice cap is very small when compared to the content in middle latitudes. At Eismitte, with an annual mean temperature below −30°C, saturated air contains only 0.2 g/m³, compared to 10 g/m³ at 15°C and 70% relative humidity. It is for this reason that Wegener completely neglects the contribution of condensation to precipitation.

Blowing snow (together with the attendant restrictions on visibility) is another phenomenon which is frequently observed on the ice cap. According to studies by DIAMOND and GERDEL (1957) and GERDEL (1961), blowing snow occurred at Site 1 (about 1,160 m and 190 km southeast of Thule) approximately 57% of the time during a 13 month period. During 36 months at Site 2 (about 2,075 m and 320 km east of Thule) there was blowing snow 38% of the time.

Blowing snow occurs most frequently in winter and is at a minimum from June through August. Blowing snow may occur with or without falling snow. Apparently there is little correlation between blowing snow and temperature—the temperature intervals given by Gerdel and Diamond for the most frequent occurrences of blowing snow at Sites 1 and 2 do not suggest any physical explanation. Blowing snow occurs most frequently at wind speeds between 8 and 12 m/sec. Most storms of blowing snow last less than one day. From July 7, 1953 to June 30, 1956, 88% of all storms with blowing snow at Site 2 lasted less than one day. One third showed a duration of 3 h or less (DIAMOND and GERDEL, 1957; GERDEL, 1961).

Most of the precipitation on the ice cap comes from ascending air of cyclonic origin. As a rule, this air destroys the surface cold layer, partly by mechanically removing the inversion and partly by heating due to precipitation falling into the cold surface layer. At Eismitte, heavy snowfall occurred in 40 of 48 cases associated with warming. On an average, the temperature was 6°C higher at the time of heavy snowfall than at the same time on the previous day. This effect is not as pronounced at the edge of the ice cap— Weststation only showed an average warming of 3.5°C, which occurred in four out of five cases. Apparently the strong katabatic flow continues during cyclonic precipitation (LOEWE, 1935).

Humidity

Humidity conditions on the ice cap, below the temperature inversion, are completely regulated by the temperature of the snow cover. The vapour pressure in the interior seldom falls below that of saturation over ice. At Eismitte the air is supersaturated with respect to ice in all months, except June, July and August. Even during these summer months, the air is supersaturated with respect to ice during the night. The mean relative humidity over ice at Eismitte (after LOEWE, 1935) was as follows:

Spring	Summer	Autumn	Winter	Year
110	97	114	119	110 (%)

According to Loewe this supersaturation is understandable, provided the air is clear (free of nuclei), because the higher layers are not saturated due to the temperature increase with height. For example, at Eismitte with a temperature of −30°C, mean pressure of 683 mbar, intensity of inversion of 10° and a pressure at the upper boundary of the inversion of 660 mbar, the relative humidity over the inversion layer will be 40% when there is 110% supersaturation over the ice.

Evaporation

Only sporadic observations have been made of evaporation. Table XLV shows some results obtained by the University of Michigan expedition (CHURCH, 1941).

TABLE XLV

MONTHLY RATES OF EVAPORATION AND CONDENSATION (mm)
(After CHURCH, 1941)

	Snow		Ice	
	evaporation	condensation	evaporation	condensation
Mount Evans, 370 m, 1927–28	24.6	9.6	20.1	7.4
Mount Evans, 1928–29	16.0	9.4	11.9	2.5
Inland Ice, 460 m, Feb.–Mar., 1928	89.4	4.83	45.7	6.4

Apparently the evaporation at the edge of the ice cap can be quite substantial in certain regions, especially under foehn conditions. For example, measurements made during a foehn at the ice cap during February 24–26, 1928, showed an evaporation rate of 17.8 mm/day.

Fog and visibility

Fog was reported at 15% of all observations at Eismitte, based on three observations per day. Two thirds of these fogs were due to presence of cloud at the ice surface, and the remaining one third were radiation fogs. At Weststation, 22% of all observations reported fog, but almost all were due to clouds at the surface. The sea fog of the outer coastal zone seldom extends to the edge of the inland ice.

Radiation fog does not always show a definite daily period in the daylight season. Generally, it forms during the night and dissipates at noon during the strongest insolation. Radiation fog is also frequently observed at the upper boundary of the inversion layer.

During the time when temperature decreases, August through January, fog was observed at Eismitte seven times as often as during the time when temperatures increase, February through July (LOEWE, 1935).
Visibility observations were made during the British Trans-Greenland Expedition, June 18–September 5, 1934. They were distributed as follows (204 observations): ca. 100 m: 24%; ca. 500 m: 16%; ca. 2.3 km: 60%.
A visibility of 100 m persisted for 24 hours, one day in twelve during June, twelve days in 31 in July, four days in 31 in August and one in five in September for a total of eighteen days in 79 (LINDSAY, 1935).

White-out

The term "white-out" is used to describe an atmospheric condition in which there is a lack of contrast between the sky and the snow surface. White-out is a phenomenon showing the absence of hydrometeors or blowing snow; thus it is limited to completely overcast sky conditions. The traveller on the ice cap learns that the classic white-out, produced by a solid overcast of stratus, frequently changes to fog white-out as the stratus descends and comes in contact with the snow. Since the cloud base commonly is less than 100 m above the snow surface, and often only 30 m high, a stratus cloud white-out at one point may be a fog type white-out a few miles away (GERDEL, 1961).
Sometimes the white-out appears as a low stratus or fog layer, up to 100 m thick, as a result of orographic lifting of air to a higher elevation and in spite of the constant flux of moisture from the fog to the snow surface along the air trajectory over the ice cap.
Observations of white-out were carried out for several summers, in the middle and late fifties, at Site 2 (Camp Fistclench), approximately 300 km from the edge of the ice cap east of Thule (GERDEL and DIAMOND, 1956; REIQUAM and DIAMOND, 1959).
GERDEL and DIAMOND (1956) distinguish between fog and stratus white-outs. Apparently there is a difference in the appearance of these two types. From a physical point of view, however, there is no difference because fog is only a stratus deck on the ground.
The white-outs observed on the ice cap were divided into the following types:
(*1*) *Overcast white-out* is associated with a complete cloud cover with light reflection between the snow surface and the cloud base. Judgment of distance is very limited and the actual horizontal visibility of *dark* objects is not materially reduced.
(*2*) *Water-fog white-out* is produced by those clouds containing supercooled, almost microscopic, water droplets with the cloud base usually in contact with the cold snow surface. Visibility, both horizontal and vertical, is affected by the size and distribution of water droplets suspended in the air.
(*3*) *Ice-fog white-out* is produced by clouds containing minute ice crystals, with the cloud base usually in contact with the snow surface. This white-out is frequently accompanied by brilliant spectral reflections from the small crystals. This type of white-out is also frequently integrated with the water-fog type.
(*4*) *Blowing snow white-out* is produced by fine blowing snow plucked from the snow surface and suspended in the lower few meters of air by winds of 10 m/sec or more.
(*5*) *Precipitation white-out.* Although all forms of falling snow reduce visibility, a storm characterized by very small wind driven snow crystals falling from low clouds, above which the sun is shining, produces a white-out condition. The confusion caused by mul-

tiple reflections of light between the snow surface and the cloud base is further complicated by spectral reflection from the snow flakes and obscuration of landmarks by the falling snow.

The relationship between weather with low visibility and the wind direction indicates that low visibility is generally associated with the advection of moist air—in the case of the ice cap east of Thule (Site 2) with southerly winds which transport moist air from Baffin Bay.

The fog-type white-out, which is quite dangerous for transportation, is mainly caused by the advection of warm maritime air which is lifted to the ice cap and cooled both adiabatically and through contact with the cold snow surface. Cooling of the snow surface by terrestrial radiation prior to the advection of maritime air furthers the formation of white-out (GERDEL and DIAMOND, 1956).

Observations show that the relative humidity during fog conditions is always less than 100% with respect to water. For water fog to be in equilibrium in such low relative humidities, the droplets must form on salt nuclei of the order of 5–10 μ in diameter (REIQUAM and DIAMOND, 1959).

Fog-type white-outs, which are the least predictable, were studied at Site 2 during the summers of 1956 and 1957. These white-outs were associated with large-scale synoptic disturbances which cross open water from the Northwest Territories. This is a necessary but not fully sufficient condition.

The following is a description of the appearance of a white-out at Site 2 (REIQUAM and DIAMOND, 1959): "The first indication of an approaching fog white-out was generally the appearance on the southwest horizon of a fog bank preceded by small patches of fracto-stratus... As the fog bank advanced, the southwest horizon completely disappeared and it became very difficult to judge the distance between the observer and the leading edge of the fog. As the fracto-stratus patches moved overhead and thickened to a fairly solid overcast, the fog enveloped the site and horizontal visibility was reduced to a few hundred feet... A fog of this type may persist for no more than one hour or two, or it may last for several days, depending upon how long the synoptic system remains able to supply moisture."

The temperatures during the fog white-outs observed at Site 2 were between −0.7 and −12.2°C.

Visibility during white-out, dependent on the property of the object, the brightness of the sky, the visual albedo, the turbidity of the air and the properties of the eye, is discussed by KASTEN (1960).

Sunshine and radiation

The amount of sunshine received on the ice cap is high. From February through November, Weststation and Eismitte received 2,300 and 2,430 hours of sunshine respectively (LOEWE, 1935). The percentage of possible sunshine showed a maximum in June with 63 and 65% at the two stations. Their means for the period were 50 and 55%. Loewe considers these values to be too low because a large part of the sunshine is not recorded with a relatively low sun. Considering only noontime observations (11h00–13h00) gave means of 64 and 74% at Weststation and Eismitte respectively, and maxima of 76% in June at Weststation and 86% in April at Eismitte (LOEWE, 1935).

TABLE XLVI

ANNUAL VARIATION OF INSOLATION AT EISMITTE AND WESTSTATION
(After LOEWE, 1935)

	Feb.	Mar.	Apr.	May	June	July	Aug.	Sept.	Oct.
Weststation	1	4	12	21	*23*	20	12	5	2
Eismitte	1	6	11	18	*20*	19	15	8	2

Maxima in italics.

Due to the relatively high number of hours of sunshine and the low water content of the air, insolation on the ice cap is high. Weststation received 170 kcal./cm^2 year and Eismitte recorded a little more. The high reflectivity of the snow may also possibly contribute to the large amount of incoming radiation. The annual variation of the insolation at Weststation and Eismitte is shown here (Table XLVI), expressed in per cent of the annual amount (LOEWE, 1935).

The high amount of radiation received on the ice cap, compared to other places, is apparently explained by the low water content of the air over the ice cap. GEORGI (1960) found that, because of the cloudiness, the loss of radiation over the ice cap amounts to 15% of its possible value but at Spitsbergen, 78% is lost due to cloudiness.

The maximum intensities of incoming radiation as measured on clear days at the three stations of Wegener's expedition, Weststation, Eismitte and Oststation, showed that the maximum intensity was reached before the sun reached its highest elevation. This was especially pronounced at Weststation (GEORGI, 1960).

Radiation was measured at Site 2, Camp Fistclench, east of Thule from July 6 to August 7, 1955 (DIAMOND and GERDEL, 1956). The temperature was always below 0°C and

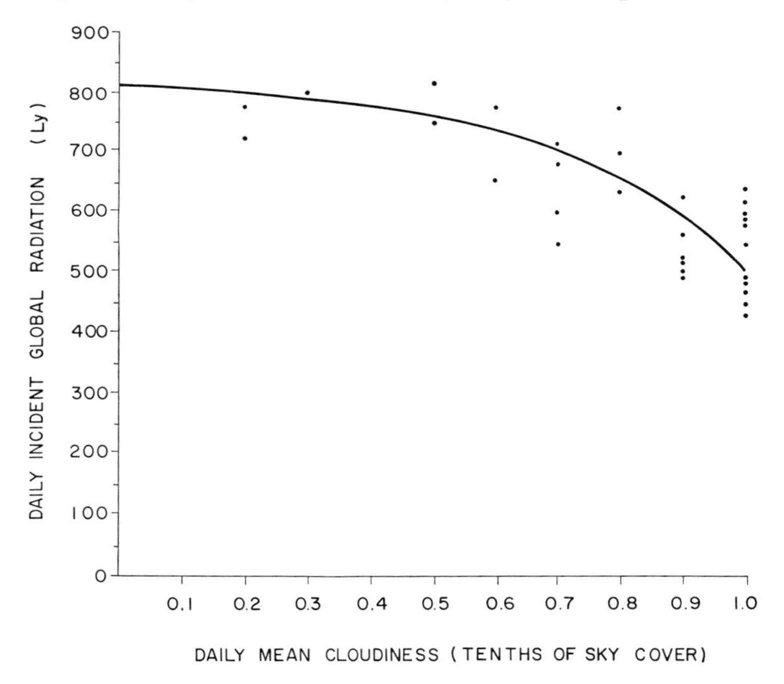

Fig.7. Daily incident global radiation vs. daily mean cloudiness, Greenland ice cap, east of Thule, 2,070 m. (After DIAMOND and GERDEL, 1956).

there was no visible evidence of melting at the snow surface. Drifting or blowing snow occurred almost every day. The relation between total direct and diffuse solar radiation and daily mean cloudiness is presented in Fig.7.

Incident global radiation was reduced only 6% when the cloudiness was 0.5 but was reduced 35% under fully overcast conditions. Such high values of global radiation with overcast skies over a large snow field have also been observed in the Antarctic. A relationship between global radiation and cloud cover, similar to that shown in Fig.7 was found by Wallén on the Kårsa Glacier in Sweden.

The albedo for the Greenland ice cap is estimated to be more than 80% for the entire year. The coastal region has the same albedo from November to April. Only in summer months, May through October, are there places in the coastal area where values of the albedo are even less than 20%. This is especially pronounced in July and August (LARSSON and ORVIG, 1961).

TABLE XLVII

MEAN VALUES OF ALBEDO IN THE MORNING AND AFTERNOON
(After DIAMOND and GERDEL, 1956)

Solar time:	05h00–11h00	13h00–19h00
Clear days (≤0.2)	0.77	0.87
Cloudy days (≥0.7)	0.80	0.86
All days	0.80	0.85

Table XLVII shows the dependence of the albedo on cloudiness.

The large amount of incoming solar radiation on the ice cap naturally represents strong melting power. According to LOEWE (1935), 1.5 m of ice is melted at 71°N and 1,000 m elevation during a year, and this is in addition to the winter precipitation. Without the great reflecting power of the snow, the Greenland ice cap could hardly exist (LOEWE, 1936).

AMBACH (1961) computed the heat balance for an ablation area on the ice cap for an eleven day period (June 27–July 7, 1959). He found that radiation makes a large contribution to the melting of the ice cap; the contribution from heat convection is insignificant. Evaporation is much more frequent than condensation. Nevertheless, the amount of evaporated ice is only 1.5% of the melted ice. About 10% of the energy available for melting is used to heat the ice at lower levels. Some measurements made of the radiation extinction in ice showed that radiation can penetrate to deeper levels (measurements were at depths as large as 50 cm).

The estimate of occurrences of melting on the ice cap during the summer season is also interesting. GERDEL (1961) computed the duration of 0°C isotherms at several elevations from maximum temperatures at Thule and at Site 2, as well as from maximum temperatures recorded at coastal stations (Table XLVIII), using a lapse rate of 0.6°C/100 m. These figures indicate that, in the northern part of the ice cap, melting can occur occasionally even at an elevation of 2,000 m, and considerable melting may be expected south of 67°N at elevations as high as 2,700 m.

TABLE XLVIII

NUMBER OF DAYS DURING JUNE, JULY AND AUGUST WHEN MAXIMUM AIR TEMPERATURE ON THE GREENLAND ICE SHEET EQUALS OR EXCEEDS 0°C, AS COMPUTED FROM SEA LEVEL TEMPERATURE
(After GERDEL, 1961)

Lat. (°N)	Elevation (m)								
	500	1,000	1,300	1,500	1,700	2,000	2,500	2,700	3,000
76	73	51	32*	22	14*	5	0	0	0
67	91	91	89	86	81	66	41	34	16
61	92	92	91	90	86	69	41	33	15

* Presence of ice masses indicates melting at this elevation and latitude.

General description of the ice cap climate

The climatic peculiarities of the inland ice depend basically on several factors. One of these is the surface of the inland ice. The large elevation furthers the loss of heat due to outgoing radiation. As the ice and snow surface reflects the solar radiation very strongly, a marked decrease in temperature takes place and a strong temperature gradient is established, directed from the coast toward the inland ice. Further, temperatures at the border of the ice cap are also diminished due to melting in summer. The low heat conductivity of the surface and the shallow layer of cold air are the causes of large temperature amplitudes. Due to strong cooling, super-saturation occurs and rime forms. Katabatic winds also cause snow to drift and thus remove a part of the inland snow to the edges of the ice cap. The crest, extending in a north–south direction, serves as a barrier, especially in the northern part, to prevent weather development from the surrounding areas (LOEWE, 1935).

Another factor influencing the climate of the ice cap is the permanent outflow of air which created the impression of a "glacial anticyclone." The role of the cold air over the ice cap is not yet clear. However, it is very likely that the outflow of cold air from the ice cap under favourable conditions could cause a strong vertical circulation, which can then contribute to the dynamics of cyclogenesis.

Migratory lows also exert a strong influence on the climate of the ice cap. These lows of the sub-polar zone carry moist and relatively warm air over the ice cap. As a result, cloudiness increases, the effective terrestrial radiation decreases, the temperature increases and the temperature amplitudes decrease.

The adiabatic cooling of ascending air gives precipitation in the form of snow over the ice cap, which contributes to the increase of ice on the cap. In summer, the melting process at the edges of the ice cap is frequently associated with rain caused by the updraft. During warm air advection, cold air is modified by mixing and sometimes completely removed by turbulence, so that sudden warming occurs. Strong winds in the northern part of the ice cap are almost always caused by such warm air advection. At boundary zones, the katabatic wind can be of gale force. However, air from the rear of lows only seems to reach the southern part of the ice cap.

The interior "anticyclonic" and exterior "cyclonic" forces determine the climate of the ice cap. The southern part and the boundary zone are more subject to outside influences than the northern part and the inland ice. It happens, however, that particular features of the inland climate are intensified by the outside forces. If the contrast between the coast and the inland ice is pronounced, the outflow will predominate entirely. It was found that the areas in the central part of Greenland are more subject to the influence of upper systems due to their smaller slope and high elevation. In some respects the particular inland climate is more disturbed than the climate of the boundary areas, and it is doubtful if any extended area of the ice cap is entirely subject to climatic factors determined by the orographic features (LOEWE, 1935).

According to MILLER (1956), the snow surface creates a peculiar climate in the interior of the ice cap, which occurs 40–50% of the time in spite of the convex terrain. In winter, this local climate is distinguished by long term storage of 100–150 cal./cm^2 in the three-layer environment (top layer of snow, the ground inversion, and the isothermal layer). This heat, conserved from periods of advection of maritime air into interstorm periods, supplies two-thirds of the heat loss. In summer, short term thermal storage consumes approximately 30 cal./cm^2 from the high-sun to the low-sun part of the day. During the daily cycle, the range in radiation balance is 17 cal./cm^2 h but, because the regional specific heat is small, the range of snow and air temperatures is relatively large. Quoting from MILLER (1956):

"During a storm, mixing of maritime air with the cold air of the inversion layer, and the consequent warming of the snow, represent an addition to heat the local environment, this is, a filling of the thermal reservoir. After the storms, air temperatures fall 15°–20°C, and snow temperatures as much or more, representing rapid withdrawal of heat from storage. The fall of temperature is accomplished within the first two days following a storm. Records of expeditions (Wegener and Victor) show that during the first day air temperature falls 9°C, during the second day 2°C, and thereafter the fluctuations are minor, probably in response to changes in flow aloft."

A general weather description, by seasons, is given by LOEWE (1935).

In summer the contrast between the boundary zones and the interior is the smallest. The katabatic wind is weak. Temperatures occasionally exceed the melting point at sites far in the interior, and rain can sometimes occur in some areas quite remote from the coast. At the edges of the ice cap, "snow marshes" and "ice lakes" are formed, until the thawing of winter precipitation causes outflow. Strong incoming radiation melts the surface of the névé and contributes to the change of snow to névé.

Autumn brings increasing cyclonic activity and snowfall on the inland ice. In the central part of the ice cap, the temperature decreases rapidly with the weaker incoming radiation; at the coast, the influence of open water delays the first cooling. Strong katabatic flow starts, raises the fresh snow into dense clouds which are drifted toward the coast, and cuts the surface in sharp edges. Cyclonic and anticyclonic influences change rapidly.

In winter the anticyclonic system develops with great force. Temperatures in the interior of the ice cap fall to the minimum. At the boundary zone, strong cooling is prevented by regular outflow. From time to time, strong cyclonic influences break through—more frequently at the edges of the inland ice than at the faces of glaciers in the valleys. Also, cold air over the ice cap is sometimes removed by strong winds; the temperature rises rapidly to that of the free atmosphere and then gradually the cold air is restored. Loewe

emphasizes that such warm air invasion, with the accompanying clouds and snow drift, is more dangerous than the radiation cold of the dark period.
In spring, noontime warming increases with the rising sun; the night temperatures increase more slowly. The temperature difference between the inland ice and the coast attains a minimum value due to the ice cover over the sea. High pressure prevails in the entire polar area and oceanic air masses play a smaller role. Incoming radiation and the temperature variation rise to very high values. In the boundary zone, the winter snow starts to sink. With increasing brightness and decreasing wind, the best time for travel in the interior begins in May.

Upper air conditions

In concluding this chapter on the climate of the Greenland ice cap, some summaries of upper air parameters will be given, taken from ascents made during the Victor expedition October 1949–July 1951 (Putnins et al., 1959b).
Table XLIX shows monthly mean values of heights and temperatures of standard isobaric surfaces. The minimum pressure was recorded in January or February (above 300 mbar in February) and the maximum pressure in August. A secondary maximum appeared in June. Due to the very limited number of observations in summer (June, July and August), it is difficult to decide in which month the maximum actually occurred. However, these values are in no way conclusive.
The maximum of temperature occurred, apparently, in July or August, and the minimum in February (in January at the 100 mbar level). A very pronounced secondary minimum was recorded in April up to the 300 mbar level, but perhaps this secondary minimum appeared only as a result of the warming that occurred in March, 1950.
The ascents were also summarized by seasons. In order to compare the results with coastal summaries, only soundings of December through March were included in the winter season, since November cannot be considered a winter month on the coast, due to the influence of open water.
It is interesting to compare upper air conditions over the ice cap with those over the coast of Greenland and the stations surrounding Greenland. For this purpose, all available upper air observations from coastal stations and from surrounding stations were summarized by months for the period of Victor's expedition, but only for those days when soundings were made at Station Centrale.
Table L shows the mean values of temperature at standard pressure levels (600, 500 and 400 mbar), as well as the number of observations. In the table the stations have been divided into two groups: those west and those east of Station Centrale. Within each group the stations are arranged in order of decreasing latitude. Unfortunately, there were no stations available north of Kap Tobin on the Greenland east coast.
In December the temperatures at all levels above the ice cap were lower than over Kap Tobin and the other stations on the east side. On the west side, these "ice cap" temperatures are higher than those over the northern part but lower than over Egedesminde and stations to the south. January shows even larger differences between the ice cap and the eastern part. However, the temperatures are higher than over the western part, except for the extreme southern station Narssarssuaq. The same holds for February and March. For the summer season (June–August), the temperatures over Kap Tobin at the 500

TABLE XLIX

MONTHLY MEANS OF THE HEIGHT AND TEMPERATURE OF STANDARD ISOBARIC SURFACES. STATION CENTRALE[1]
(After PUTNINS et al., 1959b)

	Means	Jan.	Feb.	Mar.	Apr.	May	June	July	Aug.	Sept.	Oct.	Nov.	Dec.
100	height	15,400	*15,335*	15,648	15,850	16,210	–	–	16,356	15,970	15,550	15,550	15,564
mbar	temp.	*−62.9*	−59.7	−53.7	−46.9	−40.8	–	–	−43.8	−49.1	−57.8	−59.6	−57.9
	no. of obs.	10	15	15	8	1	–	–	3	2	2	3	10
150	height	12,830	*11,510*	13,046	13,191	13,480	–	–	13,710	13,263	13,166	13,020	12,972
mbar	temp.	−56.4	*−59.8*	−53.3	−46.8	−44.4	–	–	−48.3	−50.0	−57.3	−57.1	−57.0
	no. of obs.	17	24	15	9	4	–	–	3	3	3	6	12
200	height	11,021	*10,982*	11,203	11,246	11,550	11,750	–	11,806	11,392	11,318	11,205	11,140
mbar	temp.	−57.4	*−61.6*	−54.5	−48.8	−47.5	−45.8	–	−53.8	−55.1	−57.7	−59.2	−58.1
	no. of obs.	23	27	16	10	4	1	–	3	4	5	7	15
250	height	10,018	*9,244*	9,782	9,810	10,346	10,270	–	10,356	9,987	9,903	9,775	9,736
mbar	temp.	−60.0	*−60.7*	−55.3	−50.4	−52.6	−43.7	–	−54.1	−58.5	−57.2	−58.2	−59.2
	no. of obs.	26	27	18–17	11–10	6	1	–	3	4	6	14–13	16
300	height	*8,144*	8,463	8,610	8,606	8,934	9,077	9,110	9,140	8,810	8,733	8,621	8,588
mbar	temp.	−57.3	*−59.2*	−52.5	−53.8	−49.8	−45.8	−46.0	−45.5	−54.4	−54.1	−56.3	−57.2
	no. of obs.	29	28	19	11–10	7	4	2	4	4	6	17–14	21
400	height	6,649	*6,624*	6,716	6,730	7,026	7,167	7,092	7,174	6,930	6,816	6,743	6,728
mbar	temp.	−47.2	*−49.5*	−42.9	−47.2	−38.5	−33.9	−32.4	−33.6	−41.6	−42.5	−46.2	−46.4
	no. of obs.	30	28	19	13–12	9	6	4	5	5	8	22–18	22
500	height	5,143	*5,131*	5,184	5,222	5,458	5,567	5,501	5,572	5,371	5,277	5,230	5,222
mbar	temp.	−37.0	*−39.9*	−34.1	−39.5	−27.9	−22.7	−21.9	−23.1	−30.6	−34.0	−34.9	−36.3
	no. of obs.	32	29	19	13	9	7	7	5	6	9	25	24
600	height	*3,861*	3,867	3,891	3,947	4,128	4,205	4,144	4,222	4,053	3,984	3,942	3,939
mbar	temp.	−29.7	−32.1	−27.8	*−32.9*	−20.2	−16.1	−15.7	−17.8	−22.4	−25.8	−28.0	−29.2
	no. of obs.	32	29	19	13	9	8	7	5–4	6	9	25	25
650	height	*3,290*	3,398	3,314	3,385	3,526	3,606	3,544	3,606	3,475	3,406	3,368	3,363
mbar	temp.	−28.3	*−30.3*	−29.9	*−30.4*	−17.8	−13.9	−13.3	−12.3	−20.6	−24.5	−26.6	−26.8
	no. of obs.	32–31	29	19	13	9	8	7	5	6	9	25–24	25
Sur-	pressure	*677*	*678*	678	686	698	703	697	704	692	686	683	685
face	temp.	−34.3	*−39.3*	−33.1	−33.6	−15.8	−11.8	− 9.2	−11.2	−19.4	−32.0	−33.2	−36.7
	no. of obs.	32	29	19	13	9	8	7	5	6	9	25	25

[1] P. E. Victor Expedition, 1949–1951.
N.B.: Maximum values shown by underlining; minimum values, in italics.

and 400 mbar levels are similar to or even a little lower than over the ice cap in June and July (data for August from Kap Tobin are missing). The temperatures over the northern part of the west side are lower throughout than over the ice cap.

In the transition months (April, May, September, October, November) the temperatures over the ice cap were lower at all three levels than corresponding temperatures at east side stations. The western part of the area was cooler only in its northern part.

Generally, one could conclude that the upper air over the ice cap is colder than the air at corresponding levels over the east coast and off the eastern coast of Greenland, but warmer than the air over the northern part of the west coast of Greenland and the area west of it. However, during the winter months, January through March, the upper air

TABLE L

MEAN TEMPERATURES ALOFT DURING THE DAYS WHEN ASCENTS WERE TAKEN AT STATION CENTRALE (VICTOR EXPEDITION)
(After PUTNINS et al., 1959b)

Month	Press. level (mbar)		Station west								Station east			
			A	B	C	D	E	F	G	H	I	J	K	L
Jan.	600	T	28.8	28.4	29.8	35.9	31.8	32.3	34.9	29.7	24.0	26.0	20.8	23.6
		n	25	30	32	23	31	32	10	32	30	27	29	29
	500	T	36.7	35.9	38.1	43.8	38.4	39.8	42.9	37.0	31.9	35.4	31.3	32.4
		n	23	27	32	23	30	30	10	32	30	27	28	26
	400	T	44.7	45.9	48.7	52.6	49.0	50.7	52.7	47.2	43.0	45.4	41.4	42.8
		n	22	24	32	22	28	28	10	30	30	26	28	26
Feb.	600	T	22.5	25.8	33.3	34.4	34.2	35.7	35.6	32.1	26.8	26.6	23.3	23.8
		n	7	25	27	29	29	28	6	29	28	27	29	25
	500	T	30.5	35.1	42.5	42.2	41.6	43.4	43.2	39.9	35.5	34.8	32.6	33.1
		n	6	24	27	28	28	26	6	29	28	27	28	25
	400	T	40.9	43.4	53.5	51.2	50.4	51.8	53.1	49.5	46.2	45.5	44.5	44.6
		n	5	24	27	27	27	26	6	28	28	26	28	24
Mar.	600	T	19.1	21.8	27.2	29.8	32.0	32.4		27.8	24.3	21.7	20.6	19.5
		n	14	17	19	19	19	19		19	19	19	19	19
	500	T	28.0	30.4	35.3	38.4	40.0	39.4		34.1	32.4	29.3	28.8	28.4
		n	14	17	19	19	19	19		19	19	19	19	19
	400	T	37.6	40.7	44.8	47.9	46.2	48.1		42.9	43.3	39.8	39.5	39.4
		n	14	16	19	17	19	19		19	19	19	19	19
Apr.	600	T	15.9	20.6	29.1	30.8	32.4	30.6	31.3	32.9	26.8	25.8	22.2	20.5
		n	12	9	13	12	13	12	9	13	13	13	12	7
	500	T	24.5	28.5	36.8	37.7	39.5	40.0	38.6	39.5	34.4	33.2	30.6	28.4
		n	12	9	13	12	13	12	9	13	13	12	12	7
	400	T	35.9	38.5	45.7	46.6	46.6	47.8	48.4	47.2	43.9	43.2	41.0	38.3
		n	12	9	13	12	13	12	9	12	13	12	12	7
May	600	T	15.7	17.7	18.6	22.3	21.8	22.0	20.7	20.2	15.4	15.9	12.5	15.5
		n	8	9	9	9	9	9	9	9	9	9	8	6
	500	T	23.6	26.3	27.9	30.7	30.1	30.2	29.4	27.9	23.6	24.6	21.4	25.1
		n	8	9	9	9	9	9	9	9	9	9	8	6
	400	T	35.5	37.2	40.0	39.3	40.2	41.0	39.7	38.5	35.0	36.0	32.5	35.6
		n	8	9	9	9	8	9	9	9	9	9	8	6
June	600	T	8.8	9.9	11.0	15.5	15.1	18.0	17.6	16.1	15.0		13.0	
		n	8	8	8	8	8	8	8	8	8		8	
	500	T	17.4	18.4	19.7	24.8	23.9	26.1	26.2	22.7	23.3		21.4	
		n	8	8	8	7	8	8	8	7	8		8	
	400	T	29.1	30.4	32.2	37.0	35.6	36.2	37.8	33.9	35.3		33.1	
		n	8	8	8	7	5	8	8	6	8		8	

TABLE L *(continued)*

Month	Press. level (mbar)		Station											
			west								east			
			A	B	C	D	E	F	G	H	I	J	K	L
July	600	T	9.0	12.5	10.7	15.2	13.8	15.5	16.3	15.7	12.8	10.2	9.7	8.2
		n	6	7	7	6	6	6	7	7	7	7	7	7
	500	T	18.1	21.6	19.5	24.7	22.9	23.8	24.7	21.9	21.5	18.3	18.2	16.8
		n	6	7	7	6	6	6	7	7	7	7	7	7
	400	T	30.1	32.4	31.4	36.0	34.3	35.1	35.7	32.4	32.9	30.0	29.1	28.7
		n	5	7	7	6	6	5	7	4	7	7	7	7
Aug.	600	T	8.6	11.7	12.5	19.0	17.5	18.0		17.8		12.7	13.1	15.0
		n	5	5	5	5	5	5		5		5	5	5
	500	T	17.6	19.8	21.4	27.0	24.4	25.8		23.1		21.7	22.4	23.8
		n	5	5	5	5	5	5		5		5	5	5
	400	T	28.5	30.9	33.2	37.7	34.9	34.7		33.6		33.4	34.2	34.8
		n	5	5	5	5	5	5		5		5	5	5
Sept.	600	T	13.4	18.1	20.1		22.4	25.6		22.4	15.6	19.4	15.5	18.4
		n	4	6	6		6	6		6	6	5	6	5
	500	T	21.5	26.1	28.7		29.9	33.7		30.6	24.3	27.5	24.9	27.2
		n	4	6	6		6	6		6	6	5	6	5
	400	T	34.9	37.0	38.2		39.3	41.4		41.6	36.3	39.5	36.3	38.2
		n	3	6	6		6	6		5	6	5	6	5
Oct.	600	T	16.4	20.8	25.7	28.9	26.2	26.0	26.1	25.8	20.5	22.8	19.4	24.1
		n	9	9	6	8	6	6	5	9	9	9	9	6
	500	T	23.8	29.4	33.2	37.5	33.8	34.7	34.1	34.0	29.6	32.0	28.0	32.9
		n	9	9	6	7	6	6	5	9	9	9	9	6
	400	T	34.0	40.1	43.6	49.1	44.0	45.2	42.9	42.5	39.6	42.6	38.3	43.0
		n	9	9	6	6	6	5	5	8	9	9	9	6
Nov.	600	T	19.4	23.3	31.5	30.2	29.9	30.7	32.5	28.0	22.3	22.2	19.4	20.1
		n	25	19	7	23	24	7	7	25	24	24	23	22
	500	T	27.2	31.6	38.3	38.0	37.7	38.7	40.4	34.9	31.1	30.7	28.2	28.5
		n	25	19	7	23	23	7	7	25	24	24	22	22
	400	T	37.5	40.8	46.3	47.7	47.2	47.6	49.3	46.2	46.1	41.5	38.4	39.7
		n	25	18	7	23	21	5	7	18	24	23	21	22
Dec.	600	T	24.4	25.2	27.7	32.4	31.3	30.6	33.5	29.2	27.2	24.7	22.6	24.5
		n	13	23	5	19	25	7	8	25	24	25	25	12
	500	T	33.1	33.2	35.6	40.3	39.6	40.8	42.0	36.3	34.6	33.4	32.1	33.7
		n	12	22	25	18	24	7	8	24	24	25	24	12
	400	T	41.9	43.2	45.9	50.0	49.2	50.6	51.6	45.4	45.7	43.9	42.4	43.8
		n	12	21	25	18	25	7	8	22	24	25	24	12

A = Ship "B"; B = Narssarssuaq; C = Egedesminde; D = Arctic Bay; E = Thule; F = Eureka; G = Alert; H = Station Centrale; I = Kap Tobin; J = Angmagssalik; K = Keflavik; L = Ship "A"; n = no. of observations; T = temperature —°C.

over the whole west coast, excepting only the southern coast of Greenland, is apparently colder than the air over the ice cap.

A qualitative measure of perturbations of weather occurrences, and also of air mass changes, is shown by the interdiurnal changes of pressure and the changes of thicknesses. Interdiurnal changes of surface pressure; 700, 500, and 300 mbar heights; and 500–700 mbar and 300–500 mbar thicknesses were computed from ascents made at Station Centrale during the Victor expedition. For comparison, the same quantities were obtained for Thule and Narssarssuaq for those days when soundings were made at Station Centrale. The expedition period was divided into three seasons: winter, summer and transition periods. Because of the influence of open water at the coastal stations in November, this month was included in the transition season. Thus the winter season includes December through March; summer includes June through August, and the transition period includes September, October, November, April and May. Since only three cases of interdiurnal changes of pressure and temperature were available during the summer months, this season was excluded from the final evaluation. The interdiurnal surface

TABLE LI

INTERDIURNAL CHANGES OF SURFACE PRESSURE, STANDARD ISOBARIC HEIGHTS AND THICKNESSES[1]
(After PUTNINS et al., 1959b)

	Surf. press. (mbar)	Height (m)			Thickness (m)		
		700 mbar	500 mbar	300 mbar	700–1000	500–700	300–500
Ice cap, Station Centrale:							
n_0	0		3	1		3	7
n_-	31		28	26		29	20
n_+	23		23	23		22	23
Mean	4.2		60.6	105.0		39.5	56.8
Max.	+13.4		+290	+450		+180	−210
Mean—	4.3		61.7	111.3		45.9	65.2
Mean+	4.1		66.7	103.5		38.6	67.0
Thule:							
n_0	1	0	1	0	0	2	1
n_-	28	26	30	26	34	31	24
n_+	25	28	22	18	20	20	19
Mean	7.6	46.3	58.9	64.8	35.6	30.9	32.0
Max.	+28	+190	+235	+279	+136	+118	+113
Mean+	8.4	44.6	65.3	65.9	46.8	35.4	36.6
Mean−	7.2	48.2	56.2	64.0	29.1	30.0	29.8
Narssarssuaq:							
n_0	4	0	1	1	2	4	2
n_-	18	20	18	13	25	18	13
n_+	24	27	21	11	20	19	10
Mean	6.7	50.8	62.1	78.6	38.9	30.4	30.5
Max.	−19	−158	+179	−188	−118	−84	−95
Mean+	7.5	47.3	60.4	73.6	41.0	30.6	32.7
Mean−	7.3	55.5	67.5	88.9	40.4	36.9	33.5

Period: October 1949–July 1951; winter: December through March.

pressure changes were evaluated only for those days when consecutive ascents were available (PUTNINS et al., 1959b).

Table LI shows the interdiurnal changes of pressure, height of isobaric surfaces, and thicknesses for winter (December–March). As expected, the mean interdiurnal change of surface pressure over the central part of the ice cap is smaller than over the coastal stations Thule and Narssarssuaq—4.2 mbar as against 6.7 mbar and 8.4 mbar at Narssarssuaq and Thule, respectively. Also, the maximum value of interdiurnal pressure change is smaller over the ice cap. At the 500 mbar level, however, the interdiurnal change of height is practically identical at all three stations; Station Centrale even recorded the highest maximum of interdiurnal change of 500 mbar height—290 m as against 177 m and 235 m at Narssarssuaq and Thule, respectively.

The variation at approximately 8,500 m (300 mbar) presents a quite different picture; Station Centrale shows a mean value 35–60% higher than at the coastal stations—105 m at Station Centrale compared with 78.6 m at Narssarssuaq and 64.8 m at Thule. The maximum values of interdiurnal height change of the 300 mbar level were also recorded

TABLE LII

INTERDIURNAL CHANGES OF SURFACE PRESSURE, STANDARD ISOBARIC HEIGHTS AND THICKNESSES[1]
(After PUTNINS et al., 1959b)

	Surf. press. (mbar)	Height (m)			Thickness (m)		
		700 mbar	500 mbar	300 mbar	700–1000	500–700	300–500
Ice cap, Station Centrale:							
n_0	0		0	0		3	0
n_-	7		7	3		7	4
n_+	5		5	1		2	0
Mean	3.1		42.6	87.5		35.8	62.5
Max.	+8		−110	−230		−110	−120
Mean+	2.8		30.2	20.0		70.0	–
Mean−	3.3		51.4	110.0		41.4	62.5
Thule:							
n_0	1	0	0	0	0	0	0
n_-	3	6	8	6	10	10	7
n_+	8	6	4	4	2	2	3
Mean	5.9	35.0	41.6	49.7	23.9	28.8	29.3
Max.	+18	−86	−94	−199	−79	−84	−105
Mean+	6.9	41.5	35.8	15.8	17.0	13.5	16.7
Mean−	5.3	28.5	44.5	72.3	25.3	31.8	34.6
Narssarssuaq:							
n_0	1	0	0	0	0	0	0
n_-	1	2	2	1	3	3	1
n_+	3	4	3	2	3	2	2
Mean	6.2	59.8	72.0	94.3	27.2	37.0	25.3
Max.	+10	+134	+150	+215	−51	−73	−148
Mean+	8.0	76.5	79.3	113.5	18.0	43.0	34.0
Mean−	7.0	26.5	66.0	56.0	36.3	33.0	8.0

[1] Period: October 1949–July 1951; transition period: April, March, September, October, November.

at Station Centrale (450 m). Even more pronounced is the interdiurnal change of thicknesses over the ice cap as compared with the coastal stations, with regard to mean values as well as to the extreme. This is especially pronounced for the 300/500 mbar layer. It seems that for the days with ascents at Station Centrale, the vertical circulation was more pronounced than over Thule and Narssarssuaq.
Table LII shows the values of interdiurnal changes for the transition period. The values in the table can be regarded as an illustration—they are not conclusive because of the very limited number of cases. Here also, the values of the interdiurnal changes are quite pronounced over the ice cap.

The tropopause over the ice cap

An investigation of tropopause heights can provide some information concerning the advection of air masses over a station.
According to ascents made during Victor's expedition, monthly mean tropopause heights decrease from November through February, when the tropopause is in its lowest position (8,230 m). The height increases again in March (8,670 m). The mean temperature of the tropopause is quite steady from November through February, with a minimum in February (−61.7°C). The highest tropopause temperature was observed in March (−39.8°C at 7,360 m), which is also the absolute maximum observed during the expedition (PUTNINS et al., 1959b).
Table LIII presents the seasonal values of tropopause height, temperature and pressure. The mean tropopause heights and temperatures are lower in winter, but the variations of height and temperature are then very large. This is confirmed by the high variability of tropopause height in winter—1,091 m as opposed to 937 m during the transition period

TABLE LIII

SEASONAL VALUES OF TROPOPAUSE HEIGHT, TEMPERATURE AND PRESSURE AT THE GREENLAND ICE CAP[1]
(After PUTNINS et al., 1959b)

	Tropopause (seasonal values)		
	winter (Nov.–March)	summer (June–Aug.)	transition season (April, May, Sept., Oct.)
Mean height (m)	8,652	9,506	8,876
Max. height (m)	11,750	11,080	11,870
Min. height (m)	5,100	7,630	6,870
Variability	1,091	1,155	937
Mean temp. (°C)	−60.1	−51.4	−56.8
Max. temp. (°C)	−39.8	−41.8	−46.0
Min. temp. (°C)	−73.8	−58.7	−68.4
Mean press. (mbar)	299	288	296
Max. press. (mbar)	486	374	392
Min. press. (mbar)	191	231	193
No. of observ.	108	5	27

[1] P.E. Victor Expedition, 1949–1951.

(there were only five ascents during the summer). One can draw conclusions about the strong interchange of air masses over the ice cap from the high variability of tropopause heights. The interchange of air masses should appear more evident in the interdiurnal change of tropopause heights. However, only 78 ascents could be used for the purpose: these provided 57 cases of interdiurnal changes of height (PUTNINS et al., 1959b).
The individual and mean values of interdiurnal height changes are shown in Table LIV. The mean was 1,090 m, and the number of cases (30) with negative changes (cooling) slightly exceeded the number of cases with positive changes (warming). The mean of the negative changes (1,175 m) was larger than the mean of the positive changes (990 m).
Sometimes very large changes of tropopause height can occur, as shown in Table LIV. The tropopause height over the ice cap increased from 5,100 m at 13h45Z on February 21, 1950, to 11,390 m at 18h00Z on the following day, i.e., a change of 6,290 m. An analysis of weather maps and time cross sections for Station Centrale, Egedesminde, Angmagssalik, Kap Tobin and Narssarssuaq suggests that this extreme change was apparently real and occurred due to the invasion of warm air from the south over the ice cap, especially in lower levels.

TABLE LIV

INTERDIURNAL CHANGES OF THE HEIGHT OF THE TROPOPAUSE AT STATION CENTRALE
(After PUTNINS et al., 1959b)

Date	Height (m)	Change (m)	Date	Height (m)	Change (m)	Date	Height (m)	Change (m)	Date	Height (m)	Change (m)
1949			1950			1950			1950		
Oct. 30	9,950		Jan. 12	9,910		Feb. 12	9,310		Mar.18	7,270	
Oct. 31	7,880	−2,070	Jan. 13	8,830	−1,080	Feb. 13	7,400	−1,910	Mar.19	7,840	+570
Nov. 7	7.620		Jan. 14	8,270	−560	Feb. 14	5,970	−1,430	Mar.20	8,870	+1,030
Nov. 8	7,990	+370	Jan. 15	9,310	+1,040	Feb. 15	8,150	+2,180	Mar.21	9,800	+930
Dec. 2	8,410		Jan. 17	9,460		Feb. 17	7,530		Mar.22	10,050	+250
Dec. 3	7,380	−1,030	Jan. 18	7,430	−2,030	Feb. 18	7,320	−210	Mar.23	8,220	−1,830
Dec. 10	8,420		Jan. 19	7,790	+360	Feb. 19	6,930	−390	Mar.25	7,000	
Dec. 11	7,790	−630	Jan. 20	9,220	+1,430	Feb. 20	7,350	+420	Mar.26	7,850	+850
Dec. 12	6,850	−940	Jan. 21	7,100	−2,120	Feb. 21	5,100	−2,250	Mar.27	9,570	+1,720
Dec. 25	8,920		Jan. 22	8,880	+1,780	Feb. 22	11,390	+6,290	Mar.28	10,470	+900
Dec. 26	7,740	−1,180	Jan. 24	7,000		Feb. 23	11,380	−10	Mar.29	8,790	−1,680
Dec. 27	8,900	+1,160	Jan. 25	5,520	−1,480	Feb. 24	9,210	−2,170	Mar.30	9,210	+420
Dec. 29	8,100		Jan. 27	8,570		Feb. 25	10,400	+1,190	Apr. 1	7,670	
Dec. 30	8,290	+190	Jan. 28	8,900	+330	Feb. 26	10,670	+270	Apr. 2	6,960	−710
Dec. 31	9,250	+960	Jan. 29	8,320	−580	Feb. 27	9,030	−1,640	Apr. 3	6,870	−90
1951											
Jan. 2	8,800		Feb. 1	8,930		Mar. 1	7,930				
Jan. 3	7,200	−1,600	Feb. 2	9,340	+410	Mar. 2	7,050	−880			
Jan. 7	10,400		Feb. 5	8,500		Mar. 3	7,970	+920			
Jan. 8	10,500	+100	Feb. 6	8,020	−480	Mar. 4	6,480	−1,490			
Jan. 9	9,140	−1,360	Feb. 7	8,400	+380	Mar. 14	10,450				
Jan. 10	9,590	+450	Feb. 8	7,880	−520	Mar. 15	9,540	−910			

Mean negative changes: 1,175.3 m.
Mean positive changes: 990.3 m.
Mean 1,090.5 m.
Maximum value is underlined.

The surface temperature at Station Centrale did not change much during these 28 hours, but rather on the next day when the cold surface air was removed. The temperature then increased more than 28°C in 24 hours. In the central part of the ice cap, the temperature increase was accompanied by an increase of surface pressure (15 mbar).
The increase of surface temperature was also well marked at the west coast (Jakobshavn recorded an increase of 18°C from February 21 to 22). The temperature increase was much less pronounced at the east coast (Angmagssalik). The temperature increase at the west coast was associated with a strong decrease of pressure—Jakobshavn and Umanak recorded a pressure decrease of 23 mbar in 24 hours (CREASI, 1961).
It might be mentioned that a similar large change of tropopause height was once recorded in the central Arctic (GAIGEROV, 1962).

The climate of the coastal region

Introduction

The analysis of observational data indicates some peculiarities of the climate of the Greenland coastal area: large variability of climatic elements, strong differences between neighbouring stations when they are orographically different, and the climatic similarity of stations quite remote from each other when the consequences of longitudinal differences are smoothed by orographic conditions.
The coastal zone of Greenland is exposed to the interaction of two air masses: the polar air mass of the ice cap and the maritime air mass from the ocean. This results in sharp changes of the meteorological parameters. The configuration of the terrain can delay or it can enhance the exchange of these two air masses and in some cases it can completely prevent an invasion by one of the air masses (PETERSEN, 1935).
A very marked difference does exist between the border of the coast and the inner parts of the fjords. This is especially striking in summer, when it becomes warmer and the weather more summer-like with increased distance from the coast and closer to the ice cap. In south Greenland there are even small "woods" of birches and also some pastures. But the beautiful summer weather is frequently interrupted by the advection of cool and moist air masses, whose movement is furthered by orographic conditions. Sometimes the orographic influence is so strong that the winds in the fjords flow only in or out of the fjord, e.g., at Pustervig, a secondary station during the Danmarkshavn Expedition; the frequencies of winds for the seven months' period were: 47% calm, 17% east-northeast and 27% west-southwest (PETERSEN, 1935).
The winds in the fjords frequently have a daily and an annual variation: during the day and in summer, the wind flows into the fjord; during the night and in winter, it blows out of the fjord. Sometimes the wind can flow with such force that all movement against it is prevented. On the other hand, there can also be a striking "lee action", e.g., the sea can be absolutely calm at Jakobshavn while there is a gale in the Gulf of Disko (PETERSEN, 1935).
Comparing Ivigtut with Nanortalik shows how climatic conditions can differ due to exposure at two stations having the same annual mean temperature. Both stations are located on the southwest coast, a little more than 100 km apart. Ivigtut is well protected,

while Nanortalik is exposed. The annual mean temperature at both stations is 1.8°C, but the annual range of temperature is 15.2°C at Ivigtut and 10.3°C at Nanortalik. Ivigtut shows frequent calms, 55% for the year, resulting in a warmer summer and a colder winter. Nanortalik, however, has only 11% of calms. The annual mean wind speed at Ivigtut is about half of that at Nanortalik. Ivigtut has 22 days with fog each year, while Nanortalik has 41.

The climate of the northern part of Greenland deserves special attention. Here the arctic desert displays its most characteristic features, which are especially pronounced in Peary Land.

The high latitude arctic deserts are characterized by violent wind erosion and strong evaporation that causes salt crustation and salt lakes. As a rule, the climate of these arctic deserts is extremely continental, with high temperatures, insignificant precipitation and strong wind erosion in summer, due to foehn activity. Freezing starts very suddenly in the autumn, and warm weather comes very quickly in springtime. There are few days during the year with alternate freezing and thawing. Precipitation amounts are small, seldom more than 100–125 mm and may be considerably less. Due to the small amount of snow, the land is not covered by a continuous snow cover and large snow-free areas are exposed to the wind (FRISTRUP, 1952b).

Air pressure

In general, pressure increases toward the north. Pressure on the east coast also appears to be slightly higher than that at the same latitude on the west coast. The pressure increase with increasing latitude is irregular, however. Thus, the annual mean pressure at Godhavn is 1,008 mbar, 1,006 mbar at Umanak, and 1,011 mbar at Upernavik. In general the pressure in the southern part of Greenland (65–68°N) is adjusted to the pressure in the adjacent areas. On the other hand, pressure over the northern part of Greenland is higher in all seasons than the pressure over the ocean (Jan Mayen and Svalbard–Green Harbour). Pressure differences between northern and southern parts of Greenland tend to disappear only in summer; at this time the increase of pressure with increasing latitude is the smallest.

Greenland coastal stations, especially stations in the northern part of the island, show a polar type of annual pressure variation with a maximum in May, a minimum in July, and a secondary minimum in autumn. However, the times of the secondary maxima and minima are not consistent for all stations and depend perhaps also on the period of record. At Brønlund Fjord there is a maximum in May, a secondary maximum in November, a minimum in August and a secondary minimum in January. Nord, the northernmost station on the east coast, has a maximum in April, a secondary maximum in November, a minimum in July, a secondary minimum in December, and a sharp decrease in pressure from April to July. These times at Nord are based on only four years of data, from 1953 to 1956.

The day to day pressure variability at Angmagssalik shows a simple annual variation, with a maximum in winter (10.0 mbar) and a minimum in summer (3.7–4.3 mbar). A marked pressure variability is characteristic of this region in general (PETERSEN, 1935). The three-hourly pressure variation is also large. SUNDE (1956) computed the three-hourly pressure falls, at six west coast stations, on those occasions from May through

October 1951 when a trough appeared over, or off, the west coast. The mean three-hourly pressure decreases were from 0.40 to 1.00 mbar, and maximum decreases ranged up to 4.5 mbar.

Wind

Wind directions on the coast of Greenland are highly dependent on orographic conditions. The frequency of calms is very high, 20–30% on an average, at exposed stations. At Godhavn and Godthaab, the frequency of calms decreases to 10–20% but at a well protected station such as Ivigtut, the frequency increases to more than 50%. Angmagssalik reports calms on 35% of all observations (see also PETERSEN, 1935).

Although wind directions are greatly influenced by orographic conditions (e.g., Godhavn, located on the steep coast of Disko Island, reports few north and northwest winds) some general features still appear in the frequency distributions of the wind directions.

In general, the wind directions in winter and summer are approximately in opposite directions, at some stations this is more pronounced than at others. For example, at Danmarkshavn, northwest winds are the most frequent in winter while in summer the most frequent direction is southeast. At Godthaab, the most frequent wind directions are east and northeast in winter, but in summer southwest winds occur most frequently. As a rule we can assume that the winds in south Greenland belong to the large wind system of the North Atlantic, in terms of both direction and the annual course. This is not valid, however, for winds in the northern part of Greenland.

Wind directions around the southern corner of Greenland and off the west coast show the highest frequency for a cyclonic circulation and the second highest frequency for an anticyclonic circulation. For example, as an average, a cyclonic circulation occurred in 36% of all cases and an anticyclonic circulation in 25% of all cases in April, August and November. The existing tendency for air masses to move around Greenland is associated, at the coast, with the outflow of air from the island in winter, and with the inflow of air toward the island in summer (RODEWALD, 1955).

The wind speeds are highest in winter. At northern stations the maximum appears towards the end of the year, while at the southern stations it appears in the beginning of the year. The lowest annual wind speeds are found at Ivigtut and Angmagssalik; these are stations with a high frequency of calms. The frequency of gales is generally at a maximum in winter.

The concept that the weather in the coastal area is very stormy is not at all justified by observations. Compared with northwest Europe, the coastal zone of Greenland can only be classed as "medium" with regard to storms, e.g., Iceland stations have 10–65 days per year with storm. However, the wind statistics for Greenland are not complete. They are based on observations made only three times daily, and many gales are only of short duration and occur between observation times. In addition, the fjord winds frequently have a daily variation with the maximum speeds also occurring between observation times.

Winds at the fjord stations are predominantly weak. Observations made during Wegener's expedition allow some comparison (Table LV), between wind conditions at the coast, at the edge of the ice cap, and in the central part of the ice cap.

TABLE LV

FREQUENCY OF ANNUAL WIND SPEED (%)
(After KOPP and HOLZAPFEL, 1939)

Station	0–3 m/sec	4–6 m/sec	>6 m/sec
Umanak	67	23	10
Weststation	26	16	58
Eismitte	29	53	18
Oststation	74	21	5
Scoresbysund	65	26	9

The frequencies of wind speeds are almost the same at Umanak (west coast) and at Oststation and Scoresbysund (east coast), with a strong preponderance of light winds and only a small percentage of strong winds (>6 m/sec). Weststation, at the edge of the ice cap, recorded three times as many strong winds as Eismitte. This was apparently due to the much greater influence of lows at Weststation. Only in summer were the frequencies of the several wind speeds at Weststation about the same as those at Eismitte. Local influences at Oststation are even greater than at Scoresbysund. In winter the percentage of light winds (0–3 m/sec) at Oststation is 70%, as opposed to 52% at Scoresbysund.

Generally, low wind speeds in the fjords are also seen in the results from the British East Greenland Expedition at Kangerdlugssuak (68°10′N 31°44′W) in 1935–1936. In spite of conditions favourable for air drainage from the ice cap in the large fjord at Kangerdlugssuak, the frequency of fjord gales is not great. "The importance of the check imposed by the development of an inversion layer is noteworthy, both in winter when the fjord has become firmly frozen, and in summer when the sea-breeze develops. . . The fjord gales which may be expected at some time on about one day in five, may be attributed generally in regard to perhaps four-fifths of all that occur, to katabatic causes in the presence of a favourable pressure gradient; a few occur when the centre of a depression crosses central Greenland and is accompanied by foehn, that is, high temperature and exceptional dryness. A particularly violent gale occurs if a depression develops in a polar current; this occasionally happens on the east Greenland coast. Some few occasions of local constriction, if a northeasterly or easterly current, may also give rise to a strong outflowing wind in the fjord" (MANLEY, 1938).

In the central part of the east coast, wind statistics from Myggbukta and Scoresbysund show that winds from the southeast and east are relatively frequent from June through August and relatively rare from October through February. The reverse is true with northerly directions. March through April are similar to winter, and May is more closely associated with summer. In September, an abrupt transition takes place from summer to winter conditions. The maximum wind speed occurs in winter at both stations. At Myggbukta the minimum occurs in spring, while at Scoresbysund it comes in summer. The frequency of calms is much greater at Scoresbysund than at Myggbukta. The frequency of gales is much larger over the sea than in the coastal zone, and much larger in winter than in summer. The greatest number of gales are from a northerly direction, i.e., from the rear of lows (HOVMØLLER, 1947).

Danmarkshavn, a more northerly station, also shows the highest wind speeds in winter

and autumn (4.7 and 4.6 m/sec) and the lowest in summer (3.0 m/sec). The maximum wind recorded during Wegener's expedition (August, 1906–July, 1908) was 30.0 m/sec.
Nord, located still further north, is apparently strongly influenced by local conditions. Both in winter and summer, southerly winds are the most frequent. The frequency of calms is very high all year. The maximum wind speed recorded from 1954 to 1958 was approximately 25 m/sec.
At Brønlunds Fjord, the northernmost station on the east coast with regular observations during two expedition years, the maximum monthly frequencies are confined to calms, east and west winds. The results of the two expedition years may be generalized as follows: at all observation times, east is the most frequently observed wind direction during May–August; during the remainder of the year the most frequently observed wind direction is west. Calms are most frequent in several individual months, but this is never the case from May to August.
The average speed, regardless of direction, was 5.1 m/sec. The lowest wind speed occurred in October and the highest in July and March. There is a great difference between the summer when easterlies (cyclonic type) prevail, and winter with its prevailing westerlies (anticyclonic type). The frequency diagrams for both directions are striking: in winter, calms prevail; wind speeds 6 m/sec or less have an almost constant frequency (about 6%) and the frequencies drop for higher speeds. In summer there are two maxima of wind frequency: calm and 1 m/sec and another around 5–7 m/sec. Both show a frequency of 15% (FRISTRUP, 1961).
At Brønlunds Fjord high winds occur throughout the year, especially from February to July. During the six-month period, gales (>14 m/sec) occur on an average of 50% of all days. The number of days with gale was as high as 105/year. The highest wind speeds were recorded from December to March. The absolute maximum sustained wind was 29 m/sec with gusts of at least 33 m/sec (STOERTZ, 1957).[1]
In the southeastern part of Greenland, winds of gale force can occur in any season of the year. Apparently, in all but the most sheltered stations in the southern part of Greenland, hurricane force winds can occur in winter or spring and gusts of 51 m/sec have been recorded at Atterbury Dome and Kap Dan (HASTINGS, 1960).
In case a depression approaches southwest Greenland, or a front or trough associated with a stationary depression is situated near the coast, then southeasterly winds appear along the southwest coast. The most exposed area in such cases is the sea off the coast at Nunarsuit–Frederikshaab. There the glacier extends almost to the coast and is as high as 2,400 m. Air crossing the glacier creates a pronounced pressure fall and, consequently, very strong supergeostrophic winds which can reach hurricane force (SUNDE, 1956).
In the northwestern part of Greenland, records from Thule were investigated for the occurrences of strong winds (>15.6 m/sec) for the period from September 1950 to August 1951 (U.S. AIR WEATHER SERVICE, 1952). During that year, seventeen periods with such strong winds were encountered.
All winds of this strength were from the quadrant between east-southeast and south-southeast. When the baric patterns accompanying these occurrences of strong winds were examined more closely, it was found that the pressure gradients were only sufficient

[1] It should be noted that the mean wind speeds and frequencies of gales can be misleading, because these speeds were obtained from observations made three times per day.

to produce a wind of about 5 m/sec. In no case was the pressure gradient strong enough to produce the observed surface winds. Apparently these strong winds were produced as a combination of the gradient wind and the induced gravity flow of cold air from the ice cap. LAMB (1957) calculated the mean number of gales per year for a number of stations. Britannia Sø showed the highest frequency (1952–1954) with 76 gales per year; then the base station of the British Air Route Expedition (1930–1931) and Brønlunds Fjord followed, with 42 and 20 gales per year, respectively. On the west coast, Godthaab showed fourteen and Upernavik only three gales per year (both for the period 1938–1950).

These numbers, however, are not conclusive. The periods of record are generally not uniform, and the pressure patterns are very changeable from year to year in the Arctic. Wind conditions will be different for different periods.

Foehn

DE QUERVAIN and MERCANTON (1920) defines as a "total foehn", a situation when the pressure distribution and the wind over the whole ice cap are influenced by a depression located on one side of the ice cap. A "double foehn" occurs simultaneously on the west coast, with southeast winds, and on the east coast, with northwest winds, and is caused by the outflow of air from the ice cap anticyclone.

Such situations could also occur when moist air is lifted from the coast to the ice sheet, releases part of its moisture due to adiabatic cooling, travels over the ice cap for a time and finally, perhaps due to a change in the pressure gradient, returns somewhere along the same coast as a foehn.

The direction of the foehn wind is generally towards the sea on both coasts. However, the direction is modified by orographic conditions. If these winds are not identified as foehn winds, they are associated with cold and dry weather in the northern part of Greenland. In the southern part of Greenland they can bring mild rainy weather in winter (PETERSEN, 1935).

The vertical extent of the foehn current is not great. Intermittent ascent and descent of the foehn current imparts the character of a "stream in the air". A. Wegener also observed so-called "foehn clouds" in the upper part of the boundary layer, which form when the foehn-stream is raised by the terrain.

Temperature increases due to the foehn can be very large. The occurrences of maxima $\geqq 10°$ from December through March is: Upernavik, 13 (45 years); Godthaab, 14 (50 years); Angmagssalik, 11 (30 years). PETERSEN (1935) assumes that these high temperatures cannot be caused by cold air pouring down from the ice cap. He is of the opinion that the foehn is of oceanic origin, i.e., relatively warm and moist air masses moving north are compelled to rise due to the configuration of the pressure field over the ice cap and are relieved of part of their moisture by condensation. The air masses later appear somewhere on the same coast, or after crossing the ice cap to the other coast, as a foehn. Strong foehns usually occur when the katabatic flow is strengthened by a pressure gradient in the same direction. The intensity of the foehn is highly dependent on orographic conditions, and the fjords extending in the northwest–southeast direction are especially subject to strong foehns.

"Most foehn gales starting at the heads of the fjords do not reach the coast, while other

foehns flow above the cold bottom layer and pass beyond the coastal mountains... In winter there may be perfect calm near the ground, while a hurricane may be heard thundering ahead a few hundred meters in the air... "(TRANS, 1955).
SCHNEIDER (1930) investigated some cases of foehn on the west coast, in the vicinity of Mount Evans (67°N 51°W; 395 m) in January, 1929. On one occasion the temperature rose approximately 15°C at Mount Evans. The analysis of the weather situation, and of the winds aloft, indicated that the air at the lowest levels at the ice margin was being impelled outward by inflowing air aloft. On another occasion, the maximum temperature at Mount Evans was an extraordinary high of 11°C, and remained above 5°C for 102 consecutive hours. Godthaab and Jakobshavn had maxima of 16°C and 15°C respectively—about 5°C higher than the next highest maximum for thirty years. In this case, a low at the west coast contributed substantially to the development of the foehn (see also ROSSMAN, 1950).
The following description of a foehn, observed by the Germans during their stay in northeast Greenland during the winter of 1943–1944, shows that a foehn can penetrate to quite a distance from the coast.
An unusually strong foehn was observed at Shannon Island, off the east coast of Greenland a little north of the 75th parallel, during February 20–23, 1944 (SCHATZ, 1951). On February 20 the temperature was −32°C. In the afternoon a strong south wind started. Drifting snow appeared in the form of patches on the ice field. The wind increased and changed its direction from south to west. Drifting snow from the heights of the island filled the accommodation sites. At this time the temperature was −3.0°C and some hours later it was +0.3°C—the temperature rose above the freezing point for the first time since September. At eleven o'clock the next morning the temperature was +6.0°C, and the relative humidity was 13%. The winter landscape of the island changed entirely: cloud patches moved rapidly in the dark blue sky, the mountains were red, and the air was unusually clear. There appeared 3–5 m wide open water channels in the ice field close to land, stretching far to the north and south. The snow became soft and water ran from the rocks. Far to the east, the icebergs started to float again—they had been fixed in position all winter—and the ship, which was about 8 km from the shore, turned slowly and drifted southward. The wind persisted but did not bring more drifting snow due to the fact that the surface had become spongy. The snow cornices became covered with ice and below them the snow had a temperature of −3° to −5°C. On the 22nd the wind speed decreased a little. The ship was stalled several miles south of the camp. On the 23rd the west wind increased and it was still warm. During the foehn period the temperature reached +8.4°C, and both temperature and relative humidity showed large variations. In the afternoon, the pressure rose 2.8 mbar in 15 min. Pilot balloon observations showed a wind of about 48 m/sec at 800 m. On the 24th the foehn ceased.

Temperature

Ice conditions off the coast of Greenland influence weather conditions on the coast, especially by stabilizing the weather after the building of a broad ice belt off the coast. Ice conditions also affect the temperature regime.
The annual mean temperatures on both coasts decrease with increasing latitude, which is natural. The decrease is not uniform, however. Some of the more northerly stations

have higher mean temperatures than do some of the more southerly stations: Ivigtut 1.8°C but Igaliko 2.2°C, Godthaab —0.7°C but Jakobshavn —3.9°C. Local orographic conditions could be the reason for this, e.g., orographic conditions could favour a development of local lows with more or less pronounced advection of warmer air from the south. Ice drift, which has a pronounced period, could be another cause. The boundary of the polar ice at the east coast of Greenland retreats in September to approximately 70°N (Scoresbysund), and then it moves southward again with the East Greenland polar current. On the average, the drift ice reaches Kap Farvel at the end of January. At Kap Farvel the ice first moves westward, but because of northerly and northeasterly winds it is kept quite a distance from the coast. Open water could be covered by newly formed ice, however. The northern boundary, approximately 65°N, is reached in June and July. After that, the ice disappears so rapidly that the coastal region becomes ice-free in August and September. The "storis" of the East Greenland polar current does not move beyond 65°N on the west coast. North of this latitude, the ice conditions are determined by the winter freezing of coastal waters and by the appearance of "west ice." North of Disko–Jakobshavn (70°N), freezing starts earlier, i.e., in November and December, and the newly formed ice, together with "west ice", forms an uninterrupted ice cover. This can frequently be destroyed by passing storms, however. Near Upernavik (73°N) there are still ice-free months (July–August to October–November), but the exact times depend greatly on conditions in particular years (Petersen, 1935). Apparently the ice cover could be destroyed as far north as Thule (75°N) under certain circumstances.

Summer temperatures at the coastal stations show a maximum in July (shown in italics) (Table LVI) except for stations around the southern corner of Greenland which have the maximum in August (Nanortalik, Sydpröven, Torgilsbu). At these stations, however, the mean temperatures in August do not exceed the July means by more than ½°C. The east coast shows a pronounced drop in summer temperatures (means for June–August) toward the north. On the west coast, however, the distribution of summer temperatures is very uniform; an expected decrease toward the north is not obvious. The highest mean temperatures in summer months are recorded at Igaliko, Ivigtut and Qornoq with 9.3, 9.0 and 8.7°C respectively. Godhavn and Umanak show mean summer temperatures (6.8° and 6.8°C) which are higher than those at the two southernmost stations, Nanortalik and Sydpröven (6.2° and 5.0°C). The average summer temperature at Upernavik is the same as that at Sydpröven. By comparing only the July temperatures at these stations, these differences are even more striking.

The reason for the relatively low summer temperatures on the southwest and west coasts of Greenland could also be the influence of the cold polar current, which appears at the southern part of the west coast as a continuation of the East Greenland current and in this region moves in a northerly direction from Kap Farvel.

Winter temperatures show a minimum (underlined values) in January at the more southerly stations, and in February at the more northerly stations. Nord has its minimum in March. Again, a decrease of winter temperatures toward the north is not consistent but rather depends on orographic conditions. Thus, Jakobshavn has a minimum monthly mean temperature of —14.4°C in February but at Godthaab, approximately 100 km further north, the minimum is —7.7°C in January. Agto has a minimum of —14.9°C in February but Godhavn, approximately 150 km to the north, has a minimum of —13.9°C, also in February.

TABLE LVI

MONTHLY MEAN TEMPERATURES, 1931–1956 (°C)*

Station	Jan.	Feb.	Mar.	Apr.	May	June	July	Aug.	Sept.	Oct.	Nov.	Dec.	Annual	Annual range
West Greenland:														
Nanortalik (9–15 yrs.)	−3.3	−2.4	−1.7	0.7	4.0	5.2	6.5	*7.0*	5.8	2.7	−1.0	−2.2	1.8	10.3
Sydpröven (3–13 yrs.)	−4.0	−3.6	−2.7	−0.3	2.7	3.8	5.4	*5.9*	4.3	2.0	−1.8	−3.5	0.7	9.9
Sletten (5–10 yrs.)	−5.1	−3.9	−3.9	0.1	4.4	6.3	*8.1*	7.9	5.8	1.7	−1.5	−3.6	1.4	13.2
Igaliko (9–14 yrs.)	−5.1	−4.4	−2.4	1.0	5.9	8.2	*10.3*	9.5	6.5	1.9	−1.6	−3.5	2.2	15.4
Ivigtut (19–22 yrs.)	−5.4	−4.5	−2.8	0.2	5.3	8.5	*9.8*	8.8	5.8	1.6	−1.8	−4.0	1.8	15.2
Jakobshavn (19–26 yrs.)	−13.5	−14.4	−12.5	−7.8	0.6	5.9	*8.2*	6.8	2.5	−3.7	−7.6	−10.8	−3.9	22.6
Godthaab (18–26 yrs.)	−7.7	−7.3	−5.8	−3.5	2.1	5.7	*7.6*	6.9	4.1	−0.3	−3.6	−6.2	−0.7	15.3
Kapisigdlit (3–19 yrs.)	−9.6	−9.9	−7.1	−3.4	4.3	9.4	*10.9*	8.7	3.8	−1.7	−5.7	−7.6	−0.7	20.8
Qornoq (14–24 yrs.)	−8.0	−7.2	−4.9	−1.8	4.1	8.2	*9.7*	8.1	4.2	−0.8	−4.2	−6.9	0.04	17.7
Agto (14–24 yrs.)	−13.3	−14.9	−13.3	−8.3	0.0	4.2	*6.7*	6.2	3.2	−1.6	−5.6	−9.4	−3.8	21.6
Godhavn (9–21 yrs.)	−11.8	−13.9	−12.7	−7.3	0.0	5.1	*8.0*	7.2	3.2	−2.2	−5.5	−8.9	−3.2	21.9
Umanak (19–25 yrs.)	−12.9	−15.3	−14.0	−10.0	−1.0	4.8	*7.8*	7.0	2.7	−2.0	−5.4	−9.2	−4.0	23.1
Upernavik (10–25 yrs.)	−17.0	−19.6	−18.4	−12.3	−2.4	3.0	*6.0*	5.7	1.5	−3.3	−7.3	−12.6	−6.4	25.6
East Greenland:														
Torgilsbu (7–8 yrs.)	−3.5	−3.8	−2.6	0.6	3.1	6.1	7.6	*7.9*	6.2	2.8	−0.3	−1.7	1.9	11.7
Angmagssalik (16–26 yrs.)	−6.8	−7.2	−5.7	−2.7	2.1	5.8	*7.7*	6.6	4.2	−0.05	−3.1	−5.1	−0.4	14.9
Scoresbysund (17–20 yrs.)	−15.3	−16.2	−16.1	−11.7	−2.9	2.4	*4.7*	3.7	0.8	−5.6	−10.6	−13.7	−6.7	20.9
Myggbukta (19–20 yrs.)	−20.2	−21.3	−20.3	−15.7	−5.6	1.4	*3.7*	3.1	−1.4	−9.6	−16.0	−18.5	−10.0	25.0
Nord (4–5 yrs.)	−29.6	−29.7	−32.5	−23.1	−10.9	−0.4	*4.2*	1.6	−7.8	−18.5	−24.3	−25.6	−16.4	36.7

* Figures in italics are maxima. Underlined figures are minima.

Temperature differences between east and west coast stations, Thule and Brønlunds Fjord in northern Greenland, are quite interesting. Thule has much milder winter temperatures (FRISTRUP, 1961). The climate at Thule is more maritime, apparently due to the greater influence of open water in winter. This is also clear from a comparison of winter temperatures at Thule with those at Alert and Eureka. The much higher winter temperatures at Thule cannot be explained by latitudinal differences alone.

In spite of some variations, the decrease of winter temperatures with increasing latitude is quite pronounced at both coasts, and the general drop in annual mean temperatures northward is explained principally by the decrease of winter temperatures. This is further illustrated on the east coast when winter temperatures decrease more sharply toward the north than do summer temperatures.

The east coast stations still show a sharp drop in temperature from October to November. This is especially pronounced along the northern part of the east coast. There is only a relatively small decrease from November to December and also from December to January. From November to January, there is a reduction in the rate of temperature decrease. This reduction is not observed at the west coast. There the temperature curves are much smoother, e.g., compare Myggbukta with Upernavik. This phenomenon appears not only on the east coast of Greenland, but also over the North Atlantic, at Bear Island and Green Harbour, and this may be sufficient reason to assume that the cause is the same for all these regions. PETERSEN (1935) assumes that the strong domination of the southwesterly flow over the sea between Greenland and northwest Europe may prevent the full development of the radiation process.

In the following are given some temperature values for the lower layers of the atmosphere. Ascents from 1949 to 1958 showed the following temperatures in winter, December through March, and in summer, June through August (PUTNINS et al., 1960):

(*1*) At 850 mbar in winter: mean temperatures along the west coast are a little lower than those along the east coast. Thule shows a temperature of −22.2°C, but Danmarkshavn −19.2°C; Egedesminde has a mean of −16.5°C, but −14.2°C at Kap Tobin. The west coast temperatures are lower by approximately 2° or 3°C. Also, the maximum relative frequency is nearer the lower mean value for Thule and Egedesminde. The mean isobaric temperature gradients along the northern part of the west and east coasts are approximately equal.

(*2*) At 850 mbar in summer: the mean temperatures at Thule and Danmarkshavn are almost the same, 1.1° and 0.9°C respectively. The temperature at Egedesminde, however, is a little higher, +1.6°C, than that at Kap Tobin, +0.9°C.

(*3*) At 700 mbar in winter: the mean seasonal temperature is 3°–4°C lower at the west coast. Thus, Thule shows −26.8°C but Danmarkshavn −22.7°C; Egedesminde has −22.8°C but Kap Tobin has −19.9°C.

(*4*) At 700 mbar in summer: Thule has a lower mean temperature, −8.5°C, than Danmarkshavn, −7.4°C. However, Egedesminde has a slightly higher temperature, −6.5°C, than Kap Tobin, −7.1°C.

The annual range increases toward the north, but this depends again strongly on local orographic conditions. For example, Jakobshavn has a larger range than Godthaab which is approximately 100 km further north. The northernmost west coast station with a long-year period, Upernavik, and east coast station, Nord, show the highest values, 25.6° and 36.7°C respectively (Table LVI).

The variability from year to year of the monthly mean temperatures is very large at stations in the coastal area, especially in the winter months. For example, during the period from 1884 to 1930, one February at Nanortalik was below −9°C and another was above +3°C. In west Greenland, the variation is even larger: from 1876 to 1930, the coldest February at Godthaab was −17°C and the warmest was +1°C. In summer, the variations are smaller: at Godthaab the warmest July was 9°C and the coldest was 4°C (RODEWALD, 1955). During 1876–1932, the following extreme monthly mean temperatures were found at Jakobshavn in winter: December, −2.6° and −22.3°C; January, −2.9° and −26.4°C; February, −3.6° and −29.5°C and March, −2.9° and −25.4°C. In the summer months, the differences between the extremes of monthly mean temperatures are much smaller, about 5°C for the same period. The annual distribution of extreme monthly mean temperatures at Upernavik and Jakobshavn from 1875 to 1958 present a similar picture (LINKE and BAUR, 1962).

On the east coast, Angmagssalik in winter recorded the following extreme monthly mean temperatures during 1895–1956: December, 0.5° and −13.7°C; January, −1.2° and −14.8°C; February, −0.2° and −16.9°C and March, −0.1° and −13.0°C.

Perhaps the daily mean temperature range (aperiodic variation: maximum minus minimum) is larger at the more northerly stations on the east coast. The east coast stations have the largest temperature range in winter (Scoresbysund, Nord) or in Spring (Angmagssalik), or in winter and spring (Myggbukta). On the west coast, the largest daily temperature range occurs in the spring or summer (Table LVII).

The periodic amplitudes are, of course, smaller than the aperiodic amplitudes. An evalu-

TABLE LVII

DAILY MEAN TEMPERATURE RANGE (°C)

Station	Spring (Mar.–May)	Summer (June–Aug.)	Autumn (Sept.–Nov.)	Winter (Dec.–Feb.)	Year
West Greenland:					
Nanortalik	6.1	7.4	5.1	5.0	5.9
Sydpröven	7.5	7.0	5.6	8.0	7.0
Sletten	8.7	7.4	6.6	7.4	7.5
Igaliko	7.6	7.2	6.2	8.2	7.3
Ivigtut	7.0	8.0	6.2	6.5	6.9
Jakobshavn	8.0	6.4	6.1	7.2	6.9
Godthaab	7.1	8.2	6.1	6.6	7.0
Kapisigdlit	8.7	9.8	7.5	8.1	8.5
Qornoq					
Agto					
Godhavn	7.4	5.8	4.7	6.5	6.1
Umanak	6.6	5.5	3.6	5.0	5.2
Upernavik	8.6	7.0	5.2	6.9	6.9
East Greenland:					
Angmagssalik	8.1	7.7	5.3	6.1	6.8
Scoresbysund	9.2	6.1	5.3	8.3	7.2
Myggbukta	11.8	5.8	8.1	10.3	9.0
Nord	7.6	5.0	6.6	8.3	6.9

TABLE LVIII

DAILY MEAN AMPLITUDES (°C)
(After BRAND, 1912a)

	Periodic	Aperiodic
Spring	4.1	6.7
Summer	2.5	5.3
Autumn	1.4	4.8
Winter	0.6	6.0

ation of thermograph records from Danmarkshavn for 1906–1908, showed the following mean daily amplitudes in °C; Table LVIII (after BRAND, 1912a).
The daily amplitudes are very dependent on orographic conditions. For example: Pustervig is only 60 km from Danmarkshavn in the same fjord. Parallel observations (Table LIX) were made at the two stations from November, 1907 to May, 1908 (BRAND and WEGENER, 1912).

TABLE LIX

DAILY MEAN AMPLITUDES (°C)
(After BRAND and WEGENER, 1912)

Station	Winter		Spring	
	period.	aperiod.	period.	aperiod.
Pustervig	1.3	7.5	5.8	10.3
Danmarkshavn	0.9	6.4	3.9	6.9

At Danmarkshavn, maximum temperatures generally occurred between 13h00 and 15h00 l.m.t., and the minima between 22h00 and 00h00. Mean, hour-to-hour, temperature changes for the year were greatest, 0.39°C, between 08h00 and 09h00, and least between 02h00 and 03h00 and between 13h00 and 14h00 (0.14°C).
Day-to-day temperature changes can be quite large on the coast. The following are mean interdiurnal temperature changes, by months, at Myggbukta. The changes were computed from morning observations made over a 6-year period (HOVMØLLER, 1947).

Jan.	Feb.	Mar.	Apr.	May	June	July	Aug.	Sept.	Oct.	Nov.	Dec.	
5.4	6.3	5.1	5.2	3.0	2.6	2.9	2.2	1.9	3.4	4.9	5.5	(°C)

Interdiurnal temperature changes $\geqq$5°C are most frequent in winter, about 50% of the time. Changes $\geqq$10°C are quite frequent at Myggbukta and especially in winter (17% in winter, 11.5% in spring, 2% in summer and 5.5% in the autumn). Even day-to-day changes of 15°–20°C are not unique there. The largest changes observed during the six year period were: the temperature rose 22°C in 5 h on February 2, 1932; 26°C in 13 h on January 10–11, 1938, and 31°C in 24 h on December 9–10, 1932.

TABLE LX

EXTREME TEMPERATURES

	Jan.		Feb.		Mar.		Apr.		May		June	
	max.	min.	max.	min.	max.	min.	max.	min.	max.	min.	max.	min.
West Greenland (south to north):												
Nanortalik (9–15 yrs.)	11.5	−17.2	11.6	−13.4	10.5	−14.3	15.0	−14.3	16.7	−6.5	20.1	−3.1
Sydpröven (3–13 yrs.)	12.2	−24.1	12.1	−18.0	10.1	−18.2	15.0	−16.0	15.2	−9.0	18.1	−5.0
Sletten (5–10 yrs.)	12.2	−22.0	12.5	−25.0	9.5	−26.5	15.5	−24.0	17.5	−10.5	18.5	−3.0
Igaliko (9–14 yrs.)	11.5	−23.2	15.0	−27.6	13.2	−21.1	17.0	−16.8	19.0	−13.5	20.4	−2.5
Ivigtut (19–22 yrs.)	12.0	−21.8	14.5	−22.2	14.0	−23.4	16.5	−19.1	19.5	−11.0	23.1	−1.0
Jakobshavn (19–26 yrs.)	8.3	−36.0	9.1	−34.8	10.2	−36.4	12.3	−28.2	17.1	−20.5	20.8	−5.8
Godthaab (18–26 yrs.)	12.0	−25.2	11.6	−26.0	14.2	−25.2	13.4	−22.0	18.5	−17.8	23.0	−4.3
Kapisigdlit (13–19 yrs.)	11.0	−30.5	13.5	−32.0	10.6	−30.5	15.6	−23.5	19.0	−13.5	21.5	−4.5
Qornoq (14–24 yrs.)	Missing											
Agto (14–24 yrs.)	Missing											
Godhavn (9–21 yrs.)	9.0	−33.2	8.8	−32.0	9.7	−33.8	10.8	−25.5	18.3	−17.5	17.9	−6.7
Umanak (19–25 yrs.)	9.2	−32.0	10.0	−35.0	10.5	−35.2	9.5	−30.0	14.3	−20.5	18.0	−7.0
Upernavik (10–25 yrs.)	8.0	−38.0	9.9	−40.0	11.5	−39.0	8.0	−35.0	18.0	−25.0	16.0	−13.0
East Greenland (south to north):												
Torgilsbu (7–8 yrs.)	5.2	−20.0	6.2	−26.2	5.9	−20.1	9.2	−15.9	11.8	−8.7	22.4	−2.6
Angmagssalik (16–26 yrs.)	6.6	−28.7	9.0	−26.0	8.5	−22.5	12.2	−20.3	17.9	−13.7	25.3	−8.6
Scoresbysund (17–25 yrs.)	3.0	−43.7	10.8	−41.6	4.5	−39.5	7.0	−32.5	11.8	−25.6	15.8	−7.5
Myggbukta (19–20 yrs.)	1.2	−50.9	5.0	−47.5	7.1	−47.5	7.0	−37.7	12.4	−30.2	22.1	−10.3
Nord (4–5 yrs.)	−8.8	−45.6	−0.1	−51.1	−17.8	−47.0	−3.6	−43.3	2.4	−27.9	13.6	−12.6

Extreme temperatures

From 1931 to 1956 all of the stations on the west coast, from Nanortalik in the south to Upernavik in the north, reported above freezing temperatures in all months of the year. There are quite a few missing data in the record, however. The three northern stations, Godhavn, Umanak and Upernavik, recorded absolute maxima between 7.4° and 10.0°C from December through February. During winter months, the more southern stations recorded temperatures between 8.2° and 15.5°C. Nanortalik even recorded 17.4°C in December (Table LX).

The same is true for Torgilsbu, Angmagssalik, Scoresbysund and Myggbukta on the east coast, though the range of absolute maximum temperatures is smaller, i.e., between 1.2° and 12.4°C during winter months. Only Nord (81°36′N) recorded subfreezing maximum temperatures in winter. However, the absolute maxima observed in December and February were very near the freezing point, −0.8° and −0.1°C respectively. As there is still only a short record from Nord, temperatures above freezing in winter can be expected to occur there also.

The absolute minima are also dependent on station exposure; thus the west coast stations at Igaliko, Ivigtut, Godthaab and Jakobshavn, reading from south to north, recorded absolute minima of −27.6°, −23.4°, −36.4° and −26.0°C respectively. On the northern part of the west coast, the absolute minima are around −35° to −40°C, and in the northern part of the east coast they are around −44° to −51°C.

July		Aug.		Sept.		Oct.		Nov.		Dec.		Annual	
max.	min.	max.	min.	max.	min.	max.	min.	max.	min.	max.	min.	max.	min.
18.5	−2.9	17.1	−0.7	16.7	−2.8	14.7	−6.0	13.7	−10.7	17.4	−13.2	20.1	−17.2
18.0	−4.1	18.0	−3.1	14.1	1.1	14.0	−7.0	13.0	−15.0	12.2	−16.2	18.1	−24.1
21.0	0.0	20.0	0.0	18.0	−2.5	16.0	−9.5	15.5	−19.5	13.5	−22.5	21.0	−26.5
20.2	−1.5	21.5	1.4	20.0	−3.3	16.3	−9.3	16.4	−16.1	15.3	−22.5	20.2	−27.6
22.4	1.4	19.2	−0.6	21.0	−7.2	17.0	−9.8	17.0	−16.4	15.5	−18.3	23.1	−23.4
19.4	0.0	18.9	−3.2	17.0	−11.1	12.5	−21.4	10.0	−26.5	8.2	−28.8	20.8	−36.4
20.2	−6.0	19.0	−5.2	20.4	−7.4	13.8	−13.9	12.0	−16.0	12.2	−23.0	23.0	−26.0
21.5	−0.5	21.0	−1.0	20.0	−14.0	16.5	−22.0	12.5	−22.5	10.5	−31.0	21.5	−32.0
17.3	−0.7	16.0	−0.2	14.6	−6.1	11.6	−12.1	7.6	−19.8	7.5	−26.7	18.3	−33.8
18.0	−1.5	16.3	−2.0	16.0	−6.3	17.0	−12.5	11.0	−18.0	9.2	−26.8	18.0	−35.2
19.0	−7.0	16.5	−9.0	14.7	−13.0	9.0	−17.0	8.0	−25.0	7.4	−34.0	19.0	−40.0
20.5	−0.6	21.6	−0.8	18.2	−4.0	14.0	−4.4	8.2	−10.1	12.4	−11.2	22.4	−26.2
22.9	−3.5	19.9	−3.5	20.2	−7.1	14.9	−18.3	10.2	−18.1	7.2	−21.9	25.3	−28.7
16.8	−5.0	13.0	−4.4	15.0	−11.5	12.6	−19.0	5.0	−33.0	4.5	−39.0	16.8	−43.7
21.2	−6.3	20.3	−6.7	14.5	−24.7	10.2	−30.3	7.0	−42.5	3.2	−42.5	22.1	−50.9
16.3	−5.3	14.2	7.2	−7.2	−25.0	−2.8	−34.5	3.5	−40.7	−0.8	−46.5	16.3	−51.1

On the southern part of the west coast, south of 65°N, the monthly mean minimum temperatures for December through March vary between −5 and −18°C and on the northern part of the west coast, at Godhavn, Umanak and Upernavik, they range from −19° to −24°C. On the southern part of the east coast, Torgilsbu and Angmagssalik, the range is from −7° to −11°C and on the northern part of the east coast, at Scoresbysund and Myggbukta, it is from −22° to −27°C. Nord, in March, shows a mean minimum of −37°C.

The ice disappears slowly in summer; its extent from shore is generally least in August and September, but even then, the sea is more or less filled with ice. As a result the water temperature is low all year round, and the summer heating is limited to a very narrow marginal zone. There, however, and particularly in somewhat enclosed places, relatively high temperatures can occur. These are very local and only comparatively small air masses take part in the heating process. Besides, the high temperatures are frequently only of short duration and the slightest breeze brings cooling. Hence, the temperature extremes at the coast are strongly influenced locally: a station exposed to the sea can reflect strong maritime influences while a sheltered location can result in a continental temperature regime, most surprising when considering the short distance from the sea (Petersen, 1928).

On the southern part of the west coast, approximately south of 65°N, summertime extreme temperatures lie in the lower twenties. This is true also at Torgilsbu and Angmagssalik, on the east coast. In June 1942, a temperature of 25.3°C was recorded at Angmagssalik. In the northern part of the west coast, i.e., at Godhavn, Umanak and Upernavik, the extreme summer maxima were 18–19°C. On the northern part of the east coast, Scoresbysund and Nord, extreme maxima of 16–17°C have been recorded. The maxima at Myggbukta, however, are in the lower twenties.

The monthly mean maxima on the southern part of the west coast are around 11°–14°C in summer, and around 9°–11°C on the northern part of the coast. At the east coast, however, maximum temperatures are around 6.5–7.5°C in the northern part. Angmagssalik's maximum is about 12°C.

Some synoptic-climatological investigations of weather conditions on the east coast of Greenland (HOVMØLLER, 1947) showed that the greatest temperature increases occurred when the ground inversion was destroyed and maritime air penetrated from the northeast, or when there was a northwest foehn. A strong temperature variation occurred at Myggbukta on December 27, 1929. At that time the station was under the influence of a high, centred only a little west of the station. Apparently, conditions suitable for intensive radiational heat loss developed, and the temperature decreased to −36°C in the course of the afternoon. The centre of the high moved northward during the night and Myggbukta came under the influence of northeasterly winds. The inversion was probably destroyed. On the morning of December 28, a temperature of −10°C was recorded at the station, along with a total overcast.

Very high temperatures may occur at both coasts in winter, e.g., the following temperatures were recorded at Upernavik: 13.7° and 10.3°C in February, 1892; 12.8°C in January, 1895; 15.6°, 13.4° and 15.3°C in February, 1895; and 13.7°C in February, 1901. On the east coast we find Angmagssalik reporting: 13.5°, 15.2°, 15.0° and 14.7°C in February, 1901 (PETERSEN, 1934).

Sometimes it is possible to explain the very high coastal temperatures as the result of relatively mild air from one coast crossing the ice cap and descending at another coast as a foehn. However, PETERSEN (1934) assumes that in the majority of cases the sharp increases in winter temperature are not due to air having crossed the ice cap, but are very likely caused by moist maritime air from a low over Davis Strait ascending over the west coast with heavy precipitation and then descending to the west coast with winds veering from east.

Temperatures depend on the properties of the air masses moving over the island, and therefore on the patterns of the general circulation. Thus, a strong low centred east of Greenland will reinforce the advection of cold arctic air southward and along the coasts of Greenland, while a strong high somewhere over eastern or central Europe will eventually supply Greenland with relatively warm maritime air moving northward over the Atlantic. One pronounced case occurred in February 1956, a month when very low temperatures were recorded in Europe. The monthly mean maps showed a very extensive high pressure system (1,040 mbar) centred over northwestern Siberia and covering all of Siberia. A ridge from this system reached into the Atlantic over the British Isles. A quite extensive low was centred south of Greenland. Pressure anomalies were positive over Europe and also over Greenland, but those over Greenland were only about one third as large as those over western Europe.

There were large negative temperature anomalies, around −11°C, over central Europe and below −4°C along the west coast of Norway, but positive anomalies (>4°C) in the vicinity of Iceland and the southeast coast of Greenland.

Generally, a positive correlation could be expected between pressure in Fennoscandia and temperatures on the east coast of Greenland.
Fennoscandia is usually a region affected by high pressure systems in winter, thus influencing weather conditions in eastern and central Europe. When eastern and central Europe is influenced by a high pressure system, that system is either a ridge of the Siberian high which has spread toward the west, or a high over northwestern Europe. In the first case, cold air from Siberia moves over east and central Europe. In the second case there is an outbreak of very cold air from the Arctic Basin which reaches eastern Europe via Fennoscandia. The cold February 1956 is an example of the first type, while the cold January and February of 1941 illustrates the second type.
Stations with long records were selected for a pilot study, i.e., temperature at Angmagssalik in Greenland and pressure at Stockholm, Bodø and Helsinki in Fennoscandia, and Stykkisholm in Iceland.
Since the correlations of simultaneous values are trivial, only correlations with time lags were computed. Several combinations were tried but the best results were obtained by correlating monthly mean pressure differences (Bodø minus Stykkisholm) with monthly mean temperatures at Angmagssalik during one of the succeeding months. For example, the correlation between the pressure differences in October and the mean temperature for the following December was 0.49, significant at the 1% level. Better correlations can be expected from pressure differences and temperatures than from the correlation of monthly mean pressures in Fennoscandia with monthly mean temperatures in east Greenland, because the intensity of a flow will depend on the pressure gradient (PUTNINS et al., 1959b). The correlations which were obtained are, of course, not conclusive. It could be expected, however, that correlations computed for shorter periods, such as ten days, could have some prognostic value.

Number of days with $t_{min} \leqslant 0°C$, $t_{min} > 0°C$, $t_{max} \leqslant 0°C$

The mean number of days with frost ($t_{min} \leqslant 0°C$) is large in the Greenland coastal area, about 200–300/year. The number of ice days ($t_{max} \leqslant 0°C$) is about 100–200/year. Days with $t_{max} \leqslant 0°C$ do not occur in summer, except on the northern part of the coast. On the west coast, at Upernavik, there are some days in June with $t_{max} \leqslant 0°C$ and on the east coast these temperatures appear at Myggbukta in June and August. At Scoresbysund and Nord, $t_{max} \leqslant 0°C$ is observed in all summer months.
The number of frost-free days ($t_{min} > 0°C$) decreases toward the north in the coastal area. For example, Scoresbysund averaged 100 frost-free days/year from 1925 to 1931. Mean dates of the first and last frost days were June 7 and September 14, respectively. During the same period, Angmagssalik averaged 175 frost-free days beginning April 18 and ending October 9. On the average, the frost-free period at Angmagssalik begins approximately 50 days earlier than at Scoresbysund (KOPP, 1939). From August 1906 to July 1907, there were 89 days with daily mean temperatures above freezing at Danmarkshavn. From August 1907 to July 1908, there were 76 days (WEGENER, 1911).
The northeast corner of Greenland is of great climatic interest. As noted before, the climate of Peary Land is decidedly continental and extremely arid. During two years of observations (August, 1948–July, 1950) at Brønlunds Fjord (82°N) the average July temperature was 6.1°C and the highest observed temperature was 18.0°C. Tempera-

tures above 10°C are very frequent in summer. Here the frost-free season is very long compared to that at other Greenland stations; there are 70 days during which temperatures remain positive. Since Peary Land is the northernmost land area in the world, this very long frost free season is very extraordinary. It is of great biological importance as it accounts for the large number of species of plants and animals found there (FRISTRUP, 1951). At Thule and Nord, however, there are no months which are frost-free.
FRISTRUP (1961) computed the continentality index, defined by Gorczynski, for Greenland coastal stations. The continentality is generally greater on the west coast and increases from south to north. There the continentality index ranges from 17–24 in the region between 63°N and 73°N. On the east coast, the continentality index increases from near 1 to 16 at 70°N. Brønlunds Fjord, at 82°N, has a continentality index of more than 48; this is the same as at Alert and Eureka in the Canadian Arctic Archipelago.

Humidity

Some coastal stations, such as Umanak, show a high saturation deficit. Apparently this is to be explained by the foehn influence. The same is observed on the east coast, e.g., at Myggbukta relative humidities <50% occur almost exclusively with northwesterly and northerly winds of moderate force. Temperatures are usually above normal at such times. Similar depression of the relative humidity is also observed at Scoresbysund.
At Brønlunds Fjord the humidity is low. Relative humidities less than 30% are frequent, and considerable evaporation was observed to take place from the snow and ice surface even at temperatures below zero (FRISTRUP, 1951).

Cloudiness

There will always be some doubt concerning the accuracy of cloud observations at Greenland coastal stations. PETERSEN (1935) says very little about cloudiness in his description of the coastal climate of Greenland. He assumes that the observations are not very reliable due to poor instruction of observers, infrequent inspections, etc.
Cloud structure in arctic regions is probably different from that in the temperate zone. According to VOSKRESENSKII and DERGACH (1961), the thickness of clouds in the Arctic is only about 200–300 m during the warm period. In some cases, when St clouds are associated with fog below the St layer, the vertical extent was 600–700 m.
With regard to cloud droplets, it was found that the greatest number of drops fell in the range with $r = 2$ to 25μ. Larger drops were extremely rare. Also, the droplet concentration is less in the Arctic than in the temperate zone: St clouds contain approximately 30, and Sc clouds about 23, drops per cubic centimetre. Therefore, the concentration of water in St and Sc clouds is much less: in the Arctic region, St contain 0.10 g/m^3 and Sc contain 0.14 g/m^3 but in the European part of the U.S.S.R., these types contain 0.23 and 0.29 g/m^3 respectively. Presumably these statements are more applicable to the northern part of Greenland. We can expect greater water concentrations in St and Sc clouds in the southern part of the island, and especially in the coastal zone.
Perhaps the comparison of cloudiness at stations in operation during A. Wegener's expedition with those at the coastal stations at almost the same latitude, i.e., at Umanak (west coast) and Scoresbysund (east coast) may be of interest. Table LXI shows the

TABLE LXI

PERCENTAGE OF CLEAR AND CLOUDY DAYS AT DIFFERENT STATIONS

Cloudiness (tenths)	Umanak	Weststation	Eismitte	Oststation	Scoresbysund
Winter (Oct.–Mar.):					
0–2	20	15	30	20	26
9–10	15	16	15	14	21
Summer (Aug.–Sept., Apr.–July):					
0–2	25	26	14	16	17
9–10	9	9	14	22	29
Year:					
0–2	23	21	22	18	22
9–10	12	13	14	18	25

percentage of clear (0–2) and cloudy (9–10) days (after KOPP and HOLZAPFEL, 1939). The frequency of clear days is the same at all stations throughout the year. The frequency of cloudy days increases from west to east, and Umanak, Weststation and Eismitte show the same frequency.

The difference between the west and east coasts is most striking in summer: the frequency of cloudy days at Scoresbysund is three times as large as at Umanak. The mean annual cloudiness at Umanak and Weststation is essentially the same; 5.4 and 5.7 respectively.

On the east coast, observations made during 1906–1908 showed cloud amounts of 4.4 at Pustervig and 4.9 at Danmarkshavn from November to May (BRAND and WEGENER, 1912). The annual mean cloudiness at Myggbukta from 1932 to 1937 was 5.7 (HOVMØLLER, 1947).

In the northern part of Greenland, Thule has an annual cloud amount of 5.5 while Brønlunds Fjord and Nord on the east coast both have 4.8. These northern stations show an increase in cloudiness in summer, to a maximum in August–September. The minimum occurs in December, January or even in February (FRISTRUP, 1961). At Myggbukta the maximum also occurs from May through August. The minimum, however, is shifted to some time from February through April (HOVMØLLER, 1947).

According to climatic tables for the period from 1931 to 1956 (many stations incomplete), maximum cloudiness occurs in autumn–winter on the west coast, from Umanak south to Godthaab. The east coast stations at Scoresbysund and Angmagssalik recorded their maxima at the same time. Ivigtut and Nanortalik on the southwest coast have maximum cloudiness in summer. In the northern part of Greenland, the coastal stations at Upernavik and Myggbukta show maxima primarily in summer, but the maximum at Nord comes in late summer or early autumn. It must be noted that the maxima are not very pronounced and the annual course of the cloudiness is quite uniform at most stations.

Fog

Fog is very frequent on the coast, about 50–60 days/year, and apparently it occurs about as frequently here as at other maritime stations in high latitudes such as Jan

Mayen, Icelandic coastal stations, and Bear Island. The annual variation of fog is very marked, with a maximum in summer and a minimum in winter. There is no pronounced variation in the frequency of fog with latitude. The frequency does decrease toward the outer ends of fjords. It is possible that the frequency is greater in places exposed to the sea. The figures also depend on the period of observation.
There were 40 days with fog at Scoresbysund and 17 days at Oststation from October through March, during Wegener's expedition. Angmagssalik shows 45 days/year with fog; summer brings the most with 7–9 days with fog reported each month from May to August.
The Base Station of Watkins' expedition, which was located in a fjord, rarely reported fog, i.e., only on 4 days from May to August. From November to February, fog was reported on 9 days. Coastal fog penetrated some distance up the fjord in summer and was visible from the Base Station as isolated patches (MIRRLESS, 1932).
In general, the appearance of fog is associated with specific wind directions. At Myggbukta, the visibility is either good or very poor. Visibilities in the range from 2 to 40 km are comparatively rare. Excellent visibility is normally associated with northerly or westerly winds and, at Myggbukta, is also common on calm days, whereas fog is very frequent on days with easterly and southeasterly winds (HOVMØLLER, 1947).

Precipitation

Precipitation amounts generally decrease with increasing latitude. There is a difference in the course of the annual variation on the east and west coasts: at northerly stations on the west coast, the maximum comes primarily in summer, e.g., at Thule. Upernavik also has its maximum during July to October. Umanak, about two degrees south of Upernavik, has its maximum from September through January. On the east coast the maximum occurs in winter. Nord shows a maximum in August and another during November through February.
The largest precipitation amounts are recorded on the southern coasts of Greenland: 1,300 mm/year at Ivigtut (26 years of record), 895 mm at Nanortalik (9 years), 1,940 mm at Torgilsbu (7 years).
Comparing east and west coast stations located at approximately the same latitudes, it is seen that larger annual precipitation amounts are generally recorded on the east coast, i.e., 515 mm at Godthaab (west coast) but 770 mm at Angmagssalik (east coast); 391 mm at Godhavn (west) but 428 mm at Scoresbysund (east); 1,308 mm at Ivigtut (west) but 1,940 mm at Torgilsbu (east).
The annual precipitation can vary widely from year to year. MANLEY (1938) stated when writing about the occurrence of precipitation during the British East Greenland Expedition of 1935–1936 at Kangerdlugssuak, located on the east coast at 68°N: "Precipitation at Kangerdlugssuak fall in large amounts on relatively few days. The year's total is likely to vary considerably in the latitude, depending on the frequency with which surface currents can reach the coast. It is likely to be heavier in ice free years, especially in autumn and spring."
It is difficult to estimate the amount of precipitation received in the coastal area of Greenland. This is due also to the fact that precipitation amounts are highly dependent on elevation. For example, during Wegener's expedition, Umanak recorded 103 mm

during the expedition year but Weststation, at 1,000 m on the edge of the ice cap, received 265 mm. These amounts were received during 90 days at Umanak and 140 days at Weststation (HOLZAPFEL, 1939). A rough estimate of precipitation on the east coast during the expedition year is about 200 mm at Oststation (on the fjord), but 350–400 mm at Scoresbysund.

The number of days with precipitation $\geqslant$0.1 mm is between 70 and 130/year, and decreases toward the north. The number is also generally larger on the east coast. At the northernmost west coast station, Upernavik, there are 120 days/year with precipitation.

Snow falls all year, though snow seldom occurs south of the Arctic Circle in summer. Snow cover is usually thick. The mean seasonal snow accumulation is generally greatest in April and least in August. Although the observations are inconclusive, some east coast stations probably average about 250 cm during spring. During a single year, Commanche Bay (about 65°N) had 300 cm of snow on the ground during April and May. Skjordungen (at 60°N), in three years of record, had an average depth of more than 250 cm from February to May (HASTINGS, 1960).

Finally, thunderstorms are few and the recorded occurrences are dependent on the period of observation, e.g., thunderstorms were never observed at Upernavik before 1931. Since then, three thunderstorms have been recorded in 25 years of record. There have been three thunderstorms at Jakobshavn during 62 years but none since 1931.

Climatic trends

The climate of Greenland was evidently less severe several centuries ago, compared to present conditions: there were some periods when agricultural enterprise was possible, although to a very limited extent. Very likely, these periods were followed by more severe ones, even more severe than the present.

There is some information concerning the Norse settlement in southwest Greenland about the period from the end of the 10th century to the middle of the 11th century (HOVGAARD, 1925).

Herjolfsnes, in southwest Greenland, was settled in 985. The Norsemen crossed the sea from Norway to Iceland, thence from Iceland to Greenland, and later directly from Norway to Greenland. The settlement existed about 400 years. But the vigorous northern race that originally colonized Greenland degenerated in the course of centuries under the influence of hard life conditions, isolation and race hygiene, and finally became a race of small people with defects and pathological conditions. It seems, however, that the most important factor in the decline and final extinction of the Norse colony was the deterioration in the climatic conditions. The Norsemen brought with them cattle and sheep which are known to have been raised successfully during the first centuries after settlement. In *The King's Mirror* we read: "It is reported that the pasture is good and there are large and fine farms in Greenland. The farmers raise cattle and sheep in large number and make butter and cheese in great quantities." They even tried to raise grain, which at present would be quite hopeless. It would also be impossible at present to obtain sufficient winter fodder for the need of cattle (HOVGAARD, 1925).

One fact which quite conclusively supports the increased severity of the climate is the finding of wooden objects and woven costumes in the cemetery of Herjolfsnes, which were preserved for some five hundred years. The soil cannot account for this—it is gravel and sand with small preservation power; but all the well preserved finds were exhumed from layers that at the present time are frozen throughout the year. Evidently, these objects must have been buried under conditions where the soil was thawed, at least at midsummer.

Many centuries elapsed, and only in the modern times have we some evidence of climatic conditions at Greenland's coast. Besides the direct temperature measurements at the coastal stations, there are also observations of ice conditions, glacier movements and the migration of fish which give us some clues to the climatic changes.

A general increase of temperature since the 1920's also influenced the ice conditions around Greenland. The drift ice moves along the east coast of Greenland, and at Kap Farvel it turns north along the west coast. It is known as "storis" (heavy, or polar, ice). Since 1920 the drift ice has been reduced, and its strongest decrease occurred in the 1930's. There have been some exceptions to the general decrease of ice, 1938 was a very heavy "ice year", not known before in Angmagssalik (FRISTRUP, 1952a).

Most of the glaciers in Greenland retreated in the decades since 1920. The glacier Egip Sermia reaches the west coast of Greenland at 69°47′N 50°15′W. This is a mighty ice stream which moves in the ice sheet up to the firn boundary. The general direction of movement is from east to west. Measurements in the years 1912, 1920, 1929, 1933, 1948–1953 showed that the glacier was advancing in 1912; this advance reached the highest rate in 1920. After that year the glacier front retreated, and in 1929 the position of the glacier was similar to that in 1912. The retreat continued until 1948. Between 1948 and 1953 the front was stationary, if not in a slight advance. The movement of the glacier at Jakobshavn shows a similar picture for the years 1912–1929 (BAUER, 1953). This glacier apparently continues to retreat. All glaciers are in continuous retreat also in the district of Umanak. This is the case, generally, also for the glaciers in south Greenland. Many of the north Greenland glaciers advanced up to 1920, and some of the glaciers were stationary up to that year. Since 1920 these glaciers have not advanced. In the 1930's for instance, a diminishing of ice in the glacier was recorded at Thule and some new nunataks appeared.

There are very few observations from the northern part of Greenland. However, the measurements in 1948–1950, during the Peary–Land expedition, indicated that the ablation of these glaciers exceeded the accumulation (FRISTRUP, 1952a).

The water of the fishing banks also showed a change. The period 1880–1895 was relatively warm. The water was quite cold in the present century, until 1926; since then the water temperatures have increased. Baffin Bay lost its extreme arctic character in the last 20 years or so and showed a similarity to south Greenland waters.

In connection with variation of the water temperature, changes also occurred in the migration of cod fish in Greenland waters. At Julianehaab, for instance, there was a period with rich catch up to about 1820. Since that time the amount of cod fish decreased and it even disappeared, only to reappear in the beginning of the century. Up to 1931 cod fish spread out northward beyond the 70° parallel (FRISTRUP, 1952a; RODEWALD, 1953).

An increase of temperatures in the Arctic might be associated with a stronger meridional air mass exchange, resulting from stronger cyclonic activity.

LAMB (1957) computed the long term mean of the number of days with gale for four stations on the west coast and one station on the east coast, for the periods 1890–1900, 1930–1935 and 1938–1950. There was a noticeable decrease of annual frequency of gales in the latter period. At Godthaab, for instance, there was a marked culmination of frequency of gales in the years 1925–1932. This could be explained by a more northward penetration of lows. According to the results of investigations done by the British Meteorological Office, the January depressions penetrated noticeably farther into the Spitsbergen area in the 1920's and 1930's than in the decades before and since. This holds apparently also for the whole winter (LAMB, 1957).
The direct temperature records show the following results:
LYSGAARD (1949) computed 30-year running means of temperatures. He found, for Jakobshavn, the mean for January changed from −19.0°C for the period ending 1911 to −14.6°C for the 30-year period ending in 1940. For July the temperature changed from 7.6°C (30-year period ending 1904) to 8.0°C for the period ending 1940. For the whole year the temperature changed from −7.0°C (period ending 1911) to −4.1°C (period ending 1940). Two Iceland stations also showed the same tendency, but less strongly pronounced.
More detailed temperature trends in Greenland were investigated by LANGE (1959). He computed overlapping 30-year and 10-year means. In July there is noticed a general increase of temperature. In 30-year overlapping means this increase appears only since 1926 (mean for the period ending 1926), the increase is approximately 1°C. In 10-year overlapping means the warming also occurred around 1926, but not uniform for all stations. Ivigtut and Angmagssalik also show a maximum around 1920, which does not appear in the overlapping mean temperature of northwest Greenland stations. However, Jakobshavn and Upernavik show a maximum around 1910.
In January the temperature variations are more pronounced. In the 30-year overlapping means the warming started around 1911, but a strong warming appears around 1925, especially pronounced at Jakobshavn and Upernavik. At more southerly stations (Godthaab and Ivigtut), the warming is noticed around 1920. In the overlapping 10-year means all stations show a start of a pronounced warming around 1920 but again, the time of the beginning of this trend is not the same for all stations.
For annual temperatures the warming trend appears in the 30-year overlapping means from 1911 with a small recess between 1917–1922. In the course of 10-year overlapping means the strongest warming is shown between the temperatures around 1925 and 1936.

TABLE LXII

ANNUAL MEAN TEMPERATURES (°C)
(After LANGE, 1959)

Station	I (1881–1910)	II (1914–1943)	II−I
Ivigtut	0.44	1.72	1.28
Godthaab	−2.24	−0.75	1.49
Jakobshavn	−5.97	−4.06	1.91
Upernavik	−8.85	−6.92	1.93

In Table LXII are given, according to Lange, some 30-year mean temperatures for some stations at the west coast. According to these data the warming increased toward the north.

Further, the temperature records at four west coast stations and one station on the east coast (Angmagssalik) with longer records were investigated for temperature fluctuations of shorter period (PUTNINS et al., 1959b).

In order to identify the trends, both long and short, in the period of record, January, July and annual mean temperatures were plotted year by year. An examination of the graphs presenting the annual temperatures year by year indicated some periods within which the temperatures showed some kind of trend. However, at times there were some differences from station to station; i.e., neither the type of fluctuation, nor the period of the temperature change is the same for every station. The period taken for Greenland stations was 1880–1955, during which most of the stations recorded temperatures. Within this period the following sub-periods could be encountered: 1880–1895, 1895–1909, 1909–1929, 1929–1943 and 1943–1955.

The monthly mean temperatures show a strong variability from year to year throughout, especially pronounced in the winter season and increasing toward the north. Although the changes of monthly and annual mean temperatures by years generally show a similar course, there are differences even in the sign of the trend. Thus, for instance, Upernavik shows a slight upward trend in July for the period 1917–1926; but Ivigtut, also located on the west coast of Greenland (but further south) shows a definite downward trend for the same period. The course of January temperatures at Upernavik and Jakobshavn is very similar for the period 1875–1955, although in some cases the drop or rise of temperature was delayed by one year for one of these stations. Comparing the annual temperatures, the trend is generally similar but not parallel throughout—in some cases a rise of temperature from year to year at one station corresponds to a drop in temperature at another station, and vice versa. This shows that even at the same coast of Greenland (west) the trends can be different for separate stations. For widely separated locations we can expect larger discrepancies.

To exclude the short term variations, 25-year overlapping means of annual, January and July temperatures were computed. The odd number of years (25) was chosen because of convenience, in order to place the computed mean values in the centre of the period. The gap in observations was offset by averaging the temperatures over a period less than 25 years, so that in some cases the mean value was derived from 24, 23 and even smaller numbers of monthly and annual temperature values.

Fig.8 shows annual temperatures. Ivigtut shows a slight upward trend with only small fluctuations. For two northerly stations, Godthaab and Jakobshavn, the means were computed for the period 1904–1955 and 1874–1925 respectively. For the first part of the period Jakobshavn, north of Godthaab, shows also a slight upward tendency of annual temperatures. Compared to Ivigtut, the fluctuations of temperatures are greater at Jakobshavn. Godthaab also shows a slight upward tendency for the second part of the period (1916–1943 in years of overlapping means), with a slight decreasing tendency at the end of the period. For Upernavik, the northernmost station (72°47′N), the 25-year means were computed only for the period 1875–1936. Here also the trend is upward. The difference in comparison with the more southerly stations is only that up to 1913 the increase was very slight, but marked from 1913.

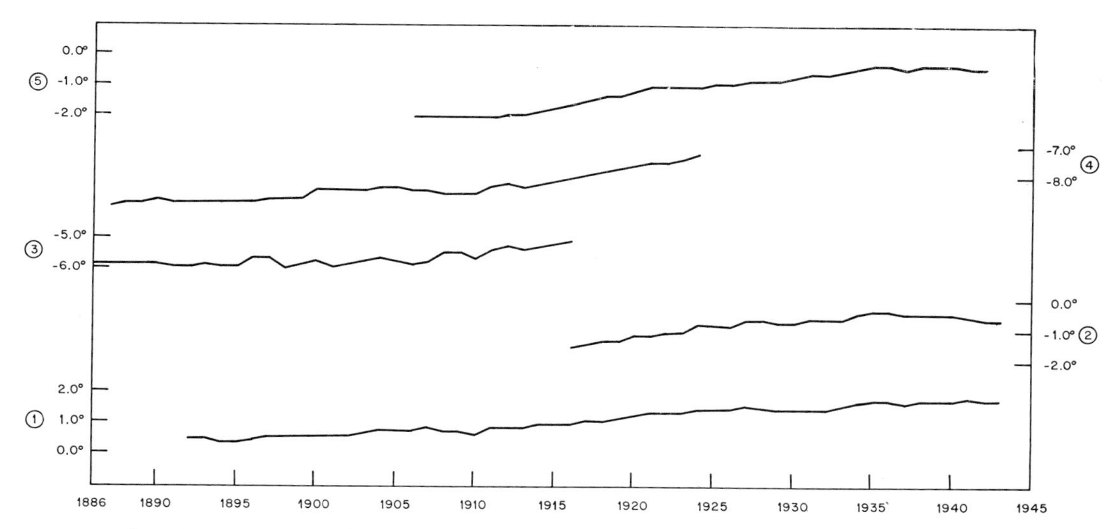

Fig.8. 25-year overlapping mean annual temperatures: (*1*) Ivigtut; (*2*) Godthaab; (*3*) Jakobshavn; (*4*) Upernavik; (*5*) Angmagssalik.

Fig.9 shows January temperatures. Ivigtut has had, beginning in 1910, an almost continuous upward trend with a slight retreat in the last years of the period. The same generally holds for the temperatures at Godthaab. A slightly different and more pronounced trend is seen at Jakobshavn, located about 560 km north of Godthaab. Up to 1908 there is a slightly decreasing trend, but beginning in 1908 a relatively strong upward trend occurs, showing the same decreasing tendency as Ivigtut and Godthaab in the last years of the period. Upernavik, located about 390 km north of Jakobshavn, shows a trend of January temperature similar to that of Jakobshavn.

Fig.10 shows July temperatures. Ivigtut shows a very even distribution of temperatures during the period under consideration, with a slight downward tendency during the last years of the period. Godthaab, about 320 km north of Ivigtut, shows an even distri-

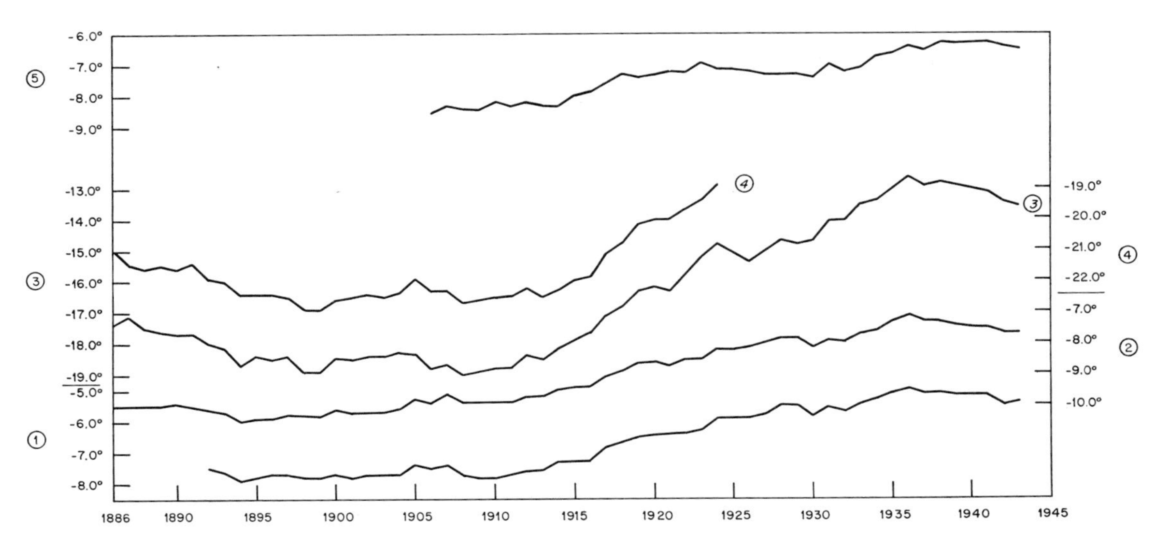

Fig.9. 25-year overlapping January mean temperatures (same stations as Fig.8).

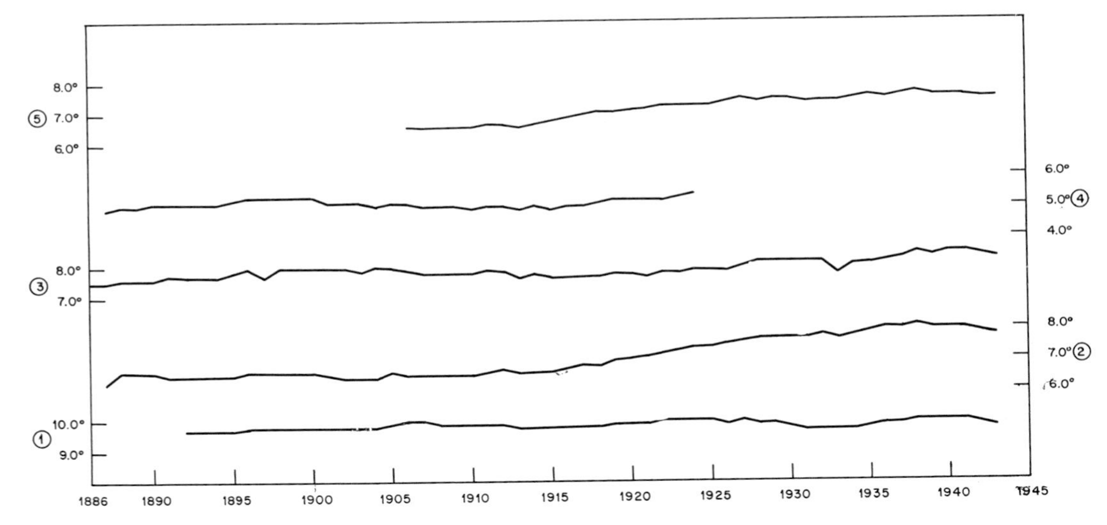

Fig.10. 25-year overlapping July mean temperatures (same stations as Fig.8).

bution of temperatures up to 1915 but a slight upward trend thereafter, with a decrease at the end of the period. Jakobshavn, north of Godthaab, again shows a very even distribution of July temperatures, similar to those recorded at Ivigtut. A slight decreasing tendency at the end of the period is also noted there. The same holds for Upernavik. Only one station, Angmagssalik, was available with a long period of record on the east coast of Greenland. The curves for Angmagssalik were compared with those for Godthaab, both stations being located in the vicinity of the 65° parallel. The annual temperatures for both stations show a similar slight upward trend, perhaps somewhat smoother at Angmagssalik.

For January, Angmagssalik shows a slight upward trend for values centred from 1906, but less pronounced than Godthaab. There also appear some differences in the trend. The downward tendency of the temperature at the end of the period is much more pronounced at Godthaab. The general trend of temperatures at Angmagssalik is more like that recorded at Ivigtut.

For July the upward trend of temperatures was more pronounced at Godthaab.

Similar investigations carried out for Icelandic stations show that temperature trends sometimes are quite different for sites situated close to each other (PUTNINS et al., 1959b).

To compare quantitatively the temperature changes in time for different stations, the temperatures were correlated with consecutive years and the regression coefficients were computed. The correlation coefficient indicates how close the temperatures are correlated with consecutive years. The statistical significance of the correlation coefficient is indicated by underlining with a solid line (at 1% level) and a dashed line (at 5%level). The regression coefficient (α) gives the increase (+) or decrease (—) of annual or monthly temperatures per year, in degrees centigrade. For most of the stations the period 1880–1955 was used. A deviation from this period is indicated in the tables.

Table LXIII gives the changes of annual, January and July temperatures for the period 1880–1955. The temperatures on the west coast show statistically significant correlation coefficients for annual and January changes. The increase of annual temperatures per year is strongest in the northern part of the west coast (Upernavik and Jakobshavn). In

TABLE LXIII

TEMPERATURE TRENDS, 1880–1955

Station	Annual		January		July	
	α*1	r*2	α	r	α	r
Ivigtut	0.02	0.493	0.04	0.326	−0.0002	−0.004
Godthaab	0.02	0.427	0.05	0.349	0.02	0.333
Jakobshavn	0.03	0.446	0.09	0.415	0.01	0.145
Upernavik	0.04	0.557	0.02	0.410	0.003	0.066
Angmagssalik*	0.04	0.260	0.006	0.037	0.002	0.037

* 1895–1955.
*1 α = regression coefficient (positive: increase; negative: decrease); r = correlation coefficient (see text).
*2 .... = significant at the 5% level; —— = significant at the 1% level.

January, Jakobshavn shows the strongest mean increase per year. In July, only Godthaab shows a statistically significant correlation coefficient. The increase of temperature per year is very small in July, and Ivigtut even shows a very slight decrease. Angmagssalik, on the east coast, shows a significant correlation coefficient only for the annual temperatures; the temperature increase in January and July is very slight there.

Table LXIV, LXV show the annual, January and July temperature fluctuations for the periods 1880–1895, 1895–1909, 1909–1929, 1929–1943 and 1943–1955, during which a more or less pronounced trend was encountered.

(1) 1880–1895 annual temperatures: increase of temperatures at the west coast, diminishing toward north. Upernavik shows a negative trend. January and July: a slight increasing trend more pronounced in the south.

(2) 1895–1909 annual temperatures show a slight decrease at Ivigtut and Angmagssalik, but an increase at Jakobshavn and Upernavik. In January there was a strong negative trend at Ivigtut, Godthaab and Jakobshavn, but a slight increase at Upernavik and Ang-

TABLE LXIV

TEMPERATURE TRENDS, ANNUAL[1]

Station	1880–1895		1895–1909		1909–1929		1929–1943		1943–1955	
	α	r	α	r	α	r	α	r	α	r
Ivigtut	0.07	0.284	−0.01	−0.059	0.02	0.134	−0.01	−0.026	0.02	0.092
Godthaab	0.07	0.344	–	–	0.10	0.532	−0.06	−0.283	0.02	0.102
Jakobshavn	0.002	0.344	0.11	0.353	0.12	0.508	−0.03*	−0.144*	0.03	0.115
Upernavik	−0.03	−0.071	0.02	0.114	0.13	0.535	−0.12	−0.445	−0.19	−0.488
Angmagssalik	–	–	−0.01	−0.073	0.09	0.599	−0.05*	−0.312*	0.02	0.118

[1] For explanations see Table LXIII.
* 1929–1936.

TABLE LXV

TEMPERATURE TRENDS[1]

Station	1880–1895		1895–1909		1909–1929		1929–1943		1943–1955	
	α	r	α	r	α	r	α	r	α	r
	January;									
Ivigtut	0.18	0.257	−0.28	−0.663	0.18	0.410	0.09	0.115	−0.23	−0.363
Godthaab	0.18	0.250	−0.21	−0.474	0.21	0.412	−0.24	−0.158	−0.23	−0.360
Jakobshavn	0.12	0.109	−0.23	−0.331	0.47	0.446	−0.05*	−0.045*	−0.32	−0.306
Upernavik	0.16	0.138	0.002	0.067	0.45	0.457	−0.83	−0.347	−0.18	−0.176
Angmagssalik	–	–	0.04	0.083	0.09	0.168	−0.03*	−0.052*	−0.05	−0.067
	July;									
Ivigtut	0.03	0.276	0.03	0.118	−0.04	−0.238	−0.05	−0.216	−0.10	−0.371
Godthaab	0.02	0.123	0.07	0.233	0.08	0.465	0.17	0.601	−0.07	−0.233
Jakobshavn	0.02	0.127	0.12	0.510	−0.002	−0.017	−0.10*	−0.429*	0.06	0.177
Upernavik	0.02	0.080	0.04	0.130	0.05	0.288	0.02	0.066	−0.006	−0.042
Angmagssalik	–	–	0.03	0.216	0.05	0.249	−0.05*	−0.251*	−0.02	−0.079

[1] For explanations see Table LXIII.
* 1929–1936.

magssalik. In July all stations show a slight increase of temperatures, more pronounced at Jakobshavn.

(3) 1909–1929. All stations show an increase of annual and January temperatures, especially pronounced in January at Jakobshavn and Upernavik. In July: a slight increase of temperatures at Godthaab, Upernavik and Angmagssalik, but a very slight decrease at Ivigtut and Jakobshavn.

(4) 1929–1943. All stations show a decrease of annual temperatures, perhaps more pronounced at Upernavik. In January, only Ivigtut recorded an increase of temperatures; all other stations showed a decrease, especially at Upernavik. In July, Ivigtut, Jakobshavn and Angmagssalik recorded a decrease, but Godthaab and Upernavik an increase of temperatures.

(5) 1943–1955. The annual temperatures show a slight increase, except Upernavik which recorded a relatively strong decrease. January shows for all stations a decrease of temperatures, more pronounced at the west coast. In July, temperatures decreased slightly, except at Jakobshavn (Table LXV).

Summarizing the results, we see that the general trend for different stations can not only be quite different in values for the same period, but even different in sign. Sometimes the trends are similar for different stations, but the drop or raise of temperatures begin later at some stations. The 25-year overlapping means show that the differences in trends are quite pronounced in some cases at the west coast of Greenland (PUTNINS, 1956; PUTNINS et al., 1959b).

The strongest increase of temperatures apparently occurred between the 1910's and 1930's. Table LXVI shows the computed temperature trends for the period 1914–1936 for four Greenland stations on the west coast and Angmagssalik on the east coast.

TABLE LXVI

TEMPERATURE TRENDS[1], 1914–1936

Station	Annual		January		July	
	α	r	α	r	α	r
Nanortalik*1	0.07	0.429	−0.004	−0.013	0.07	0.384
Ivigtut	0.03	0.183	0.06	0.133	0.01	0.061
Godthaab	0.10	0.566	0.08	0.168	0.14	0.748
Upernavik	0.16	0.700	0.35	0.396	0.15	0.702
Angmagssalik	0.10	0.661	0.0003	0.0006	0.07	0.411
Spitsbergen*3	0.20	0.661	0.37*2	0.672*2		

[1] For explanations see Table LXIII.
*1 1914–1934.
*2 Winter 1921/22–1938/39.
*3 1921–1938.

Jakobshavn was not taken because many years of data were missing. For comparison, the temperature trend for Spitsbergen was also computed.

A relatively large increase of temperatures was observed on the coast of Greenland. The increase was apparently more strongly pronounced in the northern part of Greenland. Upernavik showed the highest values of regression coefficients, especially for January temperatures. The smallest increase of temperatures was observed at Ivigtut. As the table shows, Spitsbergen recorded the strongest increase, both in annual and winter temperature. It was confirmed by PETROV (1959) in his investigations of climatic fluctuations in the Soviet Arctic that the warming and cooling periods are different for separate sites.

The warming trend observed at coastal stations of Greenland was possibly not so strongly pronounced over the ice cap. The snow temperature at 10 m depth increased from August 1931 (Eismitte) to August 1950 (Station Centrale) by 1.2°C when at coastal stations the temperature increase was larger. It is also possible that a lesser warming on the ice cap can occur due to the heat sink of the permanent snow cover, which has a maximum temperature of 0°C and therefore has an effect of reducing above-freezing changes in air temperatures (GERDEL, 1961).

An indication of climatic changes on the ice cap could be obtained from the ice measurements. Detailed studies were made at "Northice". Accumulation of ice was measured by erecting stakes, by sounding and excavation pits. The results of measurements showed a more pronounced accumulation noticed since the beginning of this century up to 1910; from the middle 1920's until the end of the 1930's a prevailing ablation is noticed, but from the 1940's again an increase of accumulation (HAMILTON, 1956).

The ice profile studies could perhaps give better conclusions with regard to annual precipitation. The investigation at Site 2 (Fistclench) indicated a gradual decrease of precipitation between 1920 and 1954; also Upernavik showed a downward trend in annual precipitation beginning about 1921. In the central part of the ice cap (Eismitte) the profile indicated that annual precipitation decreased there between 1920 and 1931 (DIAMOND, 1956).

References

AMBACH, W., 1961. Investigations of the heat balance in the area of ablation on the Greenland ice cap. In: N. K. JACOBSEN (Editor), *Physical Geography of Greenland, Intern. Geograph. Congr., 19th, Norden,—Folia Geograph. Danica*, 9: 9–12.

ARNASON, G. and VUORELA, L., 1955. Two-parameter representation of the normal temperature distribution of the 1,000/500 mbar layer. *Tellus*, 7: 189–203.

BADER, H., 1961. The Greenland ice sheet. *U.S. Army, Corps Engrs., Cold Regions Res. Eng. Lab., Res. Rept.*, I-B2: 17 pp.

BADER, H., WATERHOUSE, R. W., LANDAUER, J. K., HANSEN, B. L., BENDER, J. A. and BUTKEVICH, T. R., 1955. Excavations and installations at SIPRE test site, Site 2, Greenland. *U.S. Army, Corps Engrs., Snow, Ice, Permafrost Estab., Res. Rept.*, 20.

BAUER, A., 1953. Frontverschiebungen des Gletschers Egip Sermia, West-Grönland, 1912–1953. *Polarforschung*, 3: 234–235.

BAUER, A., 1955. The balance of the Greenland ice sheet. *J. Glaciol.*, 2: 456–462.

BAUMANN, G. H., 1933. Grönland-Flug von Gronau 1931. In: R. BECKER und G. H. BAUMANN (Herausgeber), *Beiträge zur Meteorologie des Luftweges über Grönland—Arch. Deut. Seewarte*, 52: 31–48.

BELMONT, A. D., 1954. Apparent diurnal and seasonal variations of upper air temperatures at Narssarsuak, Greenland. *Univ. Calif., Dept. Meteorol. Sci. Rept.*, 1: 66 pp. (Contract AF 19(122)–228).

BRAND, W., 1912a. Stündliche Werte des Luftdrucks und des Temperatur an Danmarks-Havn. Danmarks-Ekspeditionen til Grønlands Nordöstkyst 1906–1908. *Medd. Grønland*, 42: 361–445.

BRAND, W., 1912b. Einige Ergebnisse der Registrierung von Luftdruck und Temperatur auf der Dänemarkexpedition nach Nordgrönland. *Wiss. Beilage Jahresber. Oberrealschule Marburg Lahn*, 39 pp.

BRAND, W. and WEGENER, A., 1912. Meteorologische Beobachtungen der Station Pustervig. Danmark-Ekspeditionen til Grønlands Nordöstkyst 1906–1908. *Medd. Grønland*, 42(6): 451–462.

CHURCH, J. E., 1941. Climate and evaporation in alpine and arctic zones. *Rept. Greenland Expeditions, Univ. Mich.*, 2: 7–59.

CREASI, V. J., 1961. An analysis of an extreme case of air mass change over the Greenland ice cap. In: *Some Further Studies on the Meteorology of the Greenland Area—Final Rep., Nov. 1, 1960–Jan. 31, 1961*. Sponsored by U.S. Army Signal Research and Development Laboratory, Fort Monmouth, N.J., U.S. Weather Bureau, Washington, D.C., pp.35–67.

DE QUERVAIN, A. und MERCANTON, P. L., 1920. *Ergebnisse der Schweizerischen Grönlandexpedition 1912–1913—Denkschr. Schweiz. Naturforsch. Ges.*, Bd. 53, Teil 1–4, 402 pp.

DIAMOND, M., 1956. Precipitation trends in Greenland during the past 30 years. *U.S. Army, Corps Engrs., Snow, Ice, Permafrost Res. Estab., Res. Rept.*, 22: 4 pp.

DIAMOND, M., 1958. Air temperature and precipitation on the Greenland ice cap. *U.S. Army, Corps Engrs., Snow, Ice, Permafrost Res. Estab., Res. Rept.*, 43: 9 pp.

DIAMOND, M. and GERDEL, R. W., 1956. Radiation measurements on the Greenland ice cap. *U.S. Army, Corps Engrs., Snow, Ice, Permafrost Res. Estab., Res. Rept.*, 19: 22 pp.

DIAMOND, M. and GERDEL, R. W., 1957. Occurrence of blowing snow on the Greenland ice cap. *U.S. Army, Corps Engrs., Snow, Ice, Permafrost Res. Estab., Res. Rept.*, 25: 5 pp.

DORSEY, H. G., 1945. Some meteorological aspects of the Greenland ice cap. *J. Meteorol.*, 2: 135–142.

FRISTRUP, B., 1951. Climate and glaciology of Peary Land, north Greenland. *U.G.G.I., Assoc. Intern. Hydrol. Sci., Gen. Assembly, Brussels*, 1: 185–193.

FRISTRUP, B., 1952a. Die Klimaänderungen in der Arktis und ihre Bedeutung besonders für Grönland. *Erdkunde*, 6: 201–212.

FRISTRUP, B., 1952b. Wind erosion within the Arctic deserts. *Geograf. Tidsskr.*, 52: 51–65.

FRISTRUP, B., 1961. Climatological studies of some high Arctic stations in north Greenland. In: N. K. JACOBSEN (Editor), *Physical Geography of Greenland. Intern. Geograph. Congr., 19th, Norden—Folia Geograph. Danica*, 9: 67–78.

GAIGEROV, S. S., 1962. *Problems of Aerological Structure, Circulation and Climate of the Free Atmosphere in the Central Arctic and Antarctic. Results of Researches of the International Geophysical Year.* Section II of the I.G.Y. Program, No. 4, Moscow, 320 pp.

GEORGI, J., 1929. Starke Temperaturumkehr an der grönländischen Küste (vorläufige Mitteilung). *Ann. Hydrograph. Maritim. Meteorol.*, 57: 86–88.

GEORGI, J., 1933. Greenland as a switch for cyclones. *Geograph. J.*, 81: 344–345.

GEORGI, J., 1939. Das Klima des grönländischen Inlandeises und seine Einwirkung auf die Umgebung. *Abhandl. Naturw. Ver. Bremen*, 21(1): 408–467.

GEORGI, J., 1953. Bemerkungen zur Klima von "Eismitte" (Grönland). *Ann. Meteorol.*, 6: 283–295.

GEORGI, J., 1959. Das Klima des Nordgrönlandischen Inlandeises. *Ann. Meteorol.*, 8: 259–264.

GEORGI, J., 1960. Die Absolutwerte der 1929/1931 in Grönland ausgeführten Strahlungsmessungen. *Polarforschung*, 30:58–63.

GERDEL, R. W., 1961. A climatological study of the Greenland ice sheet. In: N. K. JACOBSEN (Editor), *Physical Geography of Greenland. Intern. Geopgraph. Congr., 19th, Norden 1960—Folia Geograph. Danica*, 9: 89–106.

GERDEL, R. W. and DIAMOND, M., 1956. Whiteout in Greenland. *U.S. Army, Corps Engrs., Snow, Ice, Permafrost Res. Estab., Res. Rept.*, 21: 11 pp.

HAMILTON, R. A., 1956. British North Greenland Expedition 1952–1954: Scientific results. *Geograph. J.*, 122: 203–240.

HAMILTON, R. A., 1958a. The determination of temperature gradient over the Greenland ice sheet by optical methods. *Proc. Roy. Soc. Edinburgh, A, 1953–1957*, 64: 381–397.

HAMILTON, R. A., 1958b. The meteorology of north Greenland during the midsummer period. *Quart. J. Roy. Meteorol. Soc.*, 84: 142–158.

HAMILTON, R. A. and ROLLIT, G., 1957a. Climatological tables for the site of the expedition's base at Brittania SØ and the station on the inland-ice, "Northice", Greenland—British North Greenland Expedition, 1952–1954. *Medd. Grønland*, 158(2): 83 pp.

HAMILTON, R. A. and ROLLIT, G., 1957b. Meteorological observations at "Northice" Greenland. British North Greenland Expedition, 1952–1954. *Medd. Grønland*, 158(3): 45 pp.

HASTINGS, A. D., 1960. Environment of southeast Greenland. *U.S. Army. Quartermaster Res. Eng. Command, Tech. Rept.*, EP-140: 62 pp.

HAYWOOD, L. J. and HOLLEYMAN, J. B., 1961. Climatological means and extremes on the Greenland ice sheet. *U.S. Army, Corps Engrs., Cold Regions Res. Eng. Lab., Res. Rept.*, 78: 22 pp.

HOLZAPFEL, R., 1939. Diskussion der Ergebnisse der meteorologischen Beobachtungen im Bereich der Weststation. *Wiss. Ergeb. Deut. Grönland-Expedition Alfred Wegener 1929 und 1930/1931*, 44: 137–173.

HOLTZSCHERER, J. J. and DE ROBIN, G., 1954. Depth of polar ice caps, Greenland. *Geograph. J.*, 120: 193–202.

HOVGAARD, W., 1925. The Norsemen in Greenland. Recent discoveries at Herjolfsnes. *Geograph. Rev.*, 15: 605–616.

HOVMØLLER, E., 1947. Climate and weather over the coastland of northeast Greenland and the adjacent sea. *Medd. Grønland*, 144: 1–208.

JACOBS, I., 1958. 5-respective 40 year monthly means of the absolute topographies of the 1,000 mbar, 850 mbar, 500 mbar and 300 mbar levels, and of relative topographies 500/1,000 mbar and 300/500 mbar over the Northern Hemisphere and their month to month changes. *Inst. Meteorol. Geophys. Freien Univ., Berlin*, 2: 121 pp.

JENSEN, K. M., 1961. Methods of forecasting of stream lines in the Greenland region. In: N. K. JACOBSEN (Editor), *Physical Geography of Greenland. Intern. Geograph. Congr., 19th, Norden 1960—Folia Geograph. Danica*, 9: 154–155.

KASTEN, F., 1960. Über die Sichtweite im Polar Whiteout. *Polarforschung*, 30: 41–44.

KEEGAN, T. J., 1958. Arctic synoptic activity in winter. *J. Meteorol.*, 15: 513–521.

KOPP, W., 1939. Diskussion der Ergebnisse der Oststation im Scoresbysund. *Wiss. Ergeb. Deut. Grönland-Expedition Alfred Wegener 1929* und *1930/1931*, 4: 1–86.

KOPP, W. und HOLZAPFEL, R., 1939. Beiträge zum Mechanismus des Witterungsverlaufs über Grönland. *Wiss. Ergeb. Deut. Grönland-Expedition Alfred Wegener, 1929 und 1930/1931*, 4: 274–325.

KUHLMAN, H., 1959. Weather and ablation observations at Sermikavsak in Umanak district. *Medd. Grønland*, 158:20–50.

LAMB, H. H., 1957. On the frequency of gales in the Arctic and Antarctic. *Geograph. J.*, 123: 287–297.

LANGE, R., 1959. Zur Erwärmung Grönlands und der atlantischen Arktis. *Ann. Meteorol.*, 8: 265–303.

LARSSON, P. and ORVIG, S., 1961. Atlas of mean monthly albedo of Arctic surfaces. *Meteorology, McGill Univ.*, 45: 34 pp.

LINDSAY, M., 1935. The British Trans-Greenland Expedition. *Geograph. J.*, 86: 242–246.

LINKE, F. und BAUR, F., 1962. *Meteorologisches Taschenbuch*. I. Band, 2. Aufl., Geest und Portig, Leipzig, 806 pp.

LOEWE, F., 1933. Die Bedeutung des Schneefegens für den Massenhaushalt von Inlandeisen. *Meteorol. Z.*, 50: 434.

LOEWE, F., 1935. Das Klima des Grönlandischen Inlandeises. *Handbuch der Klimatologie*, 2(K): 67–101.

LOEWE, F., 1936. The Greenland ice cap as seen by a meteorologist. *Quart. J. Roy. Meteorol. Soc.*, 62: 359–377.

Loewe, F., 1938. The amount of rime and snow drift as factors in the mass balance of glaciers. *Trans. Meetings Intern. Comm. Snow Glaciers, Assoc. Intern. Hydrol., Edinburgh*, 23: 415–421.

Loewe, F., 1954. The lowest temperature recorded in Antarctic and Greenland. *Polar Record*, 7: 231.

Lysgaard, L., 1949. Recent climatic fluctuations. *Folia Geographica Danica*, 5, 215 pp.

Maede, H., 1956. Über einige Beziehungen zwischen der Lage des Kältepols und der Zirkulation über Mitteleuropa. *Z. Meteorol.*, 10: 193–206.

Manley, G., 1938. Meteorological observations of the British East Greenland Expedition, 1935–1936 at Kangerdlugssuak 68°10′N 31°44′W. *Quart. J. Roy. Meteorol. Soc.*, 64: 253–276.

Matthes, F. E., 1946. The glacial anticyclone theory examined in the light of recent meteorological data from Greenland. Part I. *Trans. Am. Geophys. Union*, 27: 329–341.

Matthes, F. E. and Belmont, A. D., 1956. The glacial anticyclone theory examined in the light of recent meteorological data from Greenland. Part 2. *Trans. Am. Geophys. Union*, 31: 174–182.

Miller, D. H., 1956. The influence of snow cover on local climate in Greenland. *J. Meteorol.*, 13: 112–120.

Mirrless, S. T. A., 1932. The weather on a Greenland air route. *Geograph. J.*, 80: 15–30.

Mirrless, S. T. A., 1934. Meteorological results of the British Arctic Air Route Expedition, 1930–1931. *Meteorol. Office, London, Geophys. Mem.*, 7 : 61 pp.

Namias, J., 1958. Synoptic and climatological problems associated with the general circulation of the Arctic. *Trans. Am. Geophys. Union*, 39: 40–51.

Obukhov, A. M., 1949. On the question of geostrophic wind. *Bull. Acad. Sci., U.S.S.R.*, 13: 281–306.

Öpik, E. J., 1964. Ice ages. In: D. R. Bates (Editor), *The Planet Earth*. Pergamon, London, pp. 164–194.

Petersen, H., 1928. The climate of Greenland. In: *Greenland*. Comm. Direct. Geol. Geograph. Investigations Greenland, Copenhagen and London, I: 257–276.

Petersen, H., 1934. Extrem hohe Temperaturen und Föhn in Grönland. *Meteorol. Z.*, 51: 289–296.

Petersen, H., 1935. Das Klima der Küsten von Grönland. *Handbuch der Klimatologie*, 2(K): 33–65.

Petrov, V. L., 1959. The pattern of climatic fluctuation structure in the Arctic during the last decades. *Vestn. Leningr. Univ.*, 6: 132–136.

Petterssen, S., 1956. *Weather Analysis and Forecasting*. McGraw-Hill, New York, N.Y., 1: 428 pp.

Putnins, P., 1942. Über das Vorauseilen der Kaltluftmassen in der Höhe. *Meteorol. Z.*, 59: 218–224.

Putnins, P., 1956. Climatic changes in the Eurasian Northland. In: P. Putnins and N. A. Stepanova, *Climate of the Eurasian Northlands*. OPNAV P03-30: 95–104.

Putnins, P., 1961. A case of cyclogenesis at the west coast of Greenland, March 15–16, 1950. In: *Some Further Studies on the Meteorology of the Greenland Area—Final Rept., Nov. 1, 1960–Jan. 31, 1961*. Sponsored by U.S. Army Signal Research and Development Laboratory, Fort Monmouth, N.J., U.S. Weather Bureau, Washington, D.C., pp.77–110.

Putnins, P., 1963. An outlook on some problems of the atmospheric circulation around the southern part of Greenland. In: *Studies on the Meteorology of Greenland—Second Interim Rept., June 15, 1962–March 15, 1963*. Sponsored by U.S. Army Electronics Material Agency, Fort Monmouth, N.J., U.S. Weather Bureau, Washington, D.C., pp.75–113.

Putnins, P. and Choate, M., 1960. The distribution of temperature lapse rates over the arctic regions. *Quart. Rept., Febr. 1–Apr. 30, 1960*. Sponsored by U.S. Army Signal Research and Development Laboratory, Fort Monmouth, N.J., U.S. Weather Bureau, Washington, D.C., pp.1–53.

Putnins, P., Schallert, W. and Choate, M., 1960. Seasonal means of heights, temperatures and relative humidities for standard levels at Greenland and surrounding stations. *Quart. Rept., Nov. 1, 1959–Jan. 31, 1960*. Sponsored by U.S. Army Signal Research and Development Laboratory, Fort Monmouth, N.J., U.S. Weather Bureau, Washington, D.C., 54 pp.

Putnins, P., Schallert, W. et al., 1959a. Some aspects of the meteorological dynamics in the Greenland area. *Quart. Rept., Jan. 1–March 31, 1959*. Sponsored by U.S. Army Signal Research and Development Laboratory, Fort Monmouth, N.J., U.S. Weather Bureau, Washington, D.C., 84 pp.

Putnins, P., Schallert, W. et al., 1959b. Some meteorological and climatological problems of the Greenland area. *Final Rept., June 20, 1958–July 31, 1959*. Sponsored by U.S. Army Signal Research and Development Laboratory, Fort Monmouth, N.J., U.S. Weather Bureau, Washington, D.C., 262 pp.

Rastorguev, V. I. and Alvarez, J. A., 1958. Description of the Antarctic circulation observed from April to November 1957 at the IGY Antarctic Weather Central Little America Station. *I.G.Y. Gen. Rept. Ser.*, 1: 10 pp.

Reiquam, H. and Diamond, M., 1959. Investigations of fog whiteout. *U.S. Army, Corps Engrs., Snow, Ice, Permafrost Res. Estab., Res. Rept.*, 52: 18 pp.

Rodewald, M., 1953. Der Grönlandkabeljau und die Schwankung der Eisdrift. *Fisherwirtschaft*, 10: 241–242. (Abstract in: *Polarforschung*, 3: 309.)

RODEWALD, M., 1955. *Klima und Wetter der Fischereigebiete West- und Südgrönlands.* Deutscher Wetterdienst, Hamburg, 98 pp.

ROSENDAL, H. E., 1961. Correlation between some upper air elements at the ice cap and Greenland coastal stations. In: *Some Further Studies on the Meteorology of the Greenland Area—Final Rept., Nov. 1, 1960–Jan. 31, 1961.* Sponsored by U.S. Army Signal Research and Development Laboratory, Fort Monmouth, N.J., U.S. Weather Bureau, Washington, D.C., pp.69–76.

ROSSMAN, F., 1950. Über den Föhn auf Spitzbergen und Grönland. *Polarforschung*, 2: 347–353.

SCHALLERT, W. L., 1960. Further circulation studies in the Greenland area based on observed winds. *Suppl. Third Quart. Rept., May 1, 1960–July 31, 1960.* Sponsored by U.S. Army Signal Research and Development Laboratory, Fort Monmouth, N.J., U.S. Weather Bureau, Washington, D.C., 74 pp.

SCHALLERT, W. L., 1962. Mean monthly diurnal variations of pressure, temperature and wind at stations on the slope of the Greenland ice cap east of Thule. In: *Studies on the Meteorology of Greenland—First Interim Rept., Dec. 15, 1961–June 15, 1962.* Sponsored by U.S. Army Signal Research and Development Laboratory, Fort Monmouth, N.J., U.S. Weather Bureau, Washington, D.C., 43 pp.

SCHALLERT, W. L., 1963a. Harmonic components of the monthly mean diurnal variation of the surface wind speed. In: *Studies on the Meteorology of Greenland—Second Interim Rept., June 15, 1962–March 15, 1963.* Sponsored by U.S. Army Electronics Material Agency, Fort Monmouth, N.J., U.S. Weather Bureau, Washington, D.C., pp.52–74.

SCHALLERT, W. L., 1963b. The diurnal variation of the monthly resultant wind at Camp Century. In: *Studies on the Meteorology of Greenland—Third Interim Rept., March 15, 1963–September 15, 1963.* Sponsored by U.S. Army Electronics Material Agency, Fort Monmouth, N.J., U.S. Weather Bureau, Washington, D.C., pp.1–40.

SCHALLERT, W. L., 1964a. Frequency distributions of upper winds in the Greenland area. In: *Studies on the Meteorology of Greenland—Fourth Interim Rept., Sept. 15, 1963–March 15, 1964.* Sponsored by U.S. Army Electronics Material Agency, Fort Monmouth, N.J., Environmental Science Services Administration, Rockville, Md., 31 pp.

SCHALLERT, W. L., 1964b. The diurnal variation of the vector surface wind at several ice cap stations. In: *Studies on the Meteorology of Greenland—Fifth Interim Rept., March 15, 1964–Sept. 15, 1964.* Sponsored by U.S. Army Electronics Material Agency, Fort Monmouth, N.J., Environmental Science Services Administration, Rockville, Md., 82 pp.

SCHATZ, H., 1951. Ein Föhnsturm in Nordostgrönland. *Polarforschung*, 3: 13–14.

SCHNEIDER, L. R., 1930. Greenland west-coast foehns: a discussion based on the foehn of January, 1929. *Monthly Weather Rev.*, 58: 135–138.

SCHUMACHER, N. S., 1962. *The tropopause (Maudheim 71°03′S; 10°56′W). Norwegian–British–Swedish Antarctic Expedition, 1949–1952. Scientific Results, 1(1) (Aerology).* Oslo, 123 pp.

STEPANOVA, N. A., 1960. Surface temperature regime in Greenland. *Quart. Rept., Febr. 1–Apr. 30, 1960.* Sponsored by U.S. Army Signal Research and Development Laboratory, Fort Monmouth, N.J., U.S. Weather Bureau, Washington, D.C., pp.54–95.

STOERTZ, G. E., 1957. *Investigation of Ice-free Sites for Aircraft Landings in Northern and Eastern Greenland and Results of Test Landing of C-124 at Brønlunds Fjord, North Greenland.* Geophysics Research Directorate, U.S. Air Force, Cambridge Research Center, Mass., 40 pp.

SUNDE, A., 1956. Studies of winds off western Greenland. *Meteorol. Ann. (Oslo)*, 4: 37–54.

THOMSON, H. A., 1962. Temperature normals, averages, and extremes in the Yukon and northwest territories. *Arctic*, 15: 308–312.

TOYLI, M. J., 1954. Forecasting problems in the Arctic with special reference to northern Greenland. *Proc. Toronto Meteorol. Conf., 1953*, pp.74–77.

TRANS, P., 1955. Report from the weather service. *Medd. Grønland*, 127: 1–40.

TRAVNICEK, F., 1942. Verteilung und Änderung der mittleren jährlichen Häufigkeit von Windstillen. *Meteorol. Z.*, 59: 335–342.

TVERSKOI, P. N., 1962. *Kurs meteorologii (Fizika atmosfery).* Gidrometeorologicheskoje Izdatel'stvo, Leningrad, 700 pp.

VAN EVERDINGEN, E., 1926. Gibt es stationäre Antizyclonen? *Ann. Hydrograph. Maritim. Meteorol.*, 54: 18–19.

VOROBJEVA, E. V., 1962. *Association of Atmospheric Processes in the Northern Hemisphere.* Gidrometeorologicheskoje Izdatel'stvo, Leningrad, 106 pp. (in Russian).

VOSKRESENSKII, A. I. and DERGACH, A. L., 1961. Microphysical characteristics of St and Sc type clouds in the Arctic in the warm period of the year. In: *Issledovanie oblakov, osadkov i grozovogo elektrichestva—Dokl. Mizhduvedomstv. Konf.* Akad. Nauk S.S.S.R., Moskva, 6 pp.

VOWINCKEL, E. and ORVIG, S., 1962. Water balance and heat flow of the Arctic Ocean. *Arctic*, 15: 205–223.

WALDEN, H., 1958. Ein Tief mit Warmesektor über dem Grönlandmassiv. *Z. Meteorol.*, 12: 147–154.

WALDEN, H. 1959. *Statistisch-synoptische Untersuchung über das Verhalten von Tiefdruckgebieten im Bereich von Grönland.* Deutscher Wetterdienst, Seewetteramt, 20: 69 pp.

WATKINS, H. G., 1932. The British Arctic Air Route Expedition. *Geograph. J.*, 79: 353–367, 466–501.

WEGENER, A., 1911. Meteorologische Terminbeobachtungen am Danmarks-Havn. *Medd. Grønland*, 42: 129–355. (*Danmark-Ekspeditionen til Grønlands Nordøstkyst 1906–1908*, B. II, Nr. 4.)

WEGENER, K., 1936. Die stationären Hoch- und Tiefdruckgebiete. *Gerlands Beitr. Geophys.*, 48: 225–228.

WEGENER, K., 1939. Vorläufige Zusammenfassung der meteorologischen Ergebnisse. *Wiss. Ergeb. Deut. Grönland Expedition Alfred Wegener 1929, 1930–31*, 4: 363–380.

WEGENER, K., 1939. Ergänzungen für Eismitte. *Wiss. Ergeb. Deut. Grönland Expedition Alfred Wegener 1929, 1930–31*, 4: 87–136.

WEISS, G., 1952. Beobachtungen und Erfahrungen auf den deutschen Grönland Expeditionen 1942–43 und 1944. *Polarforschung*, 3: 162–168.

Other sources

AIR MINISTRY GREAT BRITAIN, 1939. Greenland from the standpoint of synoptic meteorology. *Atlantic Meteorol. Rept.*, 5, 61 pp.

DANSKE METEOROLOGISK INSTITUT. *Summaries of Weather Observations at Weather Stations of Greenland, 1954–1958.* Danske Meteorologisk Institut, Charlottenlund, 248 pp.

DANSKE METEOROLOGISK INSTITUT. *Meteorologisk Årbog, 2. Grønland.* Danske Meteorologisk Institut, Charlottenlund (various years).

Expéditions Polaires Françaises. Missions Paul-Emile Victor. Expéditions Arctiques, Résultats Scientifiques. Les Observations Météorologiques de la Station Française du Groenland, 70°55′03″N 40°38′22″W, alt. 2,993 m.

(a) *Conditions Atmosphériques en Surface du 5 septembre 1949 au 20 juin 1950* (Fasc. 1). Paris, 1954, 289 pp.

(b) *Conditions Atmosphériques en Surface du 21 juin 1950 au 15 août 1951* (Fasc. 1). Paris, 1956, 432 pp.

(c) *Conditions Atmosphériques en Altitude du 17 septembre 1949 au 10 août 1951.* Paris, 1954, 119 pp.

METEOROLOGICAL DEPARTMENT. *Greenland Station Data.* U.S. Army Meteorol. Team Data. U.S. Army Electronics Research and Development Activity, Fort Huachua, Arizona.

Norsk Meteorologisk Årbok. Norsk Meteorologisk Institutt, Oslo.

U.S. AIR WEATHER SERVICE, 1952. *The Occurrence of Strong Winds at Thule—Tech. Rept. 105-91.* Headquarters Air Weather Service, Washington D.C., 8 pp.

TABLE LXVII

CLIMATIC TABLE FOR UPERNAVIK (1931–1956; 10–25 years)
Latitude 72°47′N, longitude 56°10′W, elevation 35 m

Month	Mean sta. press. (mbar)	Temperature (°C)				Mean vap. press. (mm)	Precipitation (mm)	
		daily mean	mean daily range	extremes			mean	max. in 24 h.
				max.	min.			
Jan.	1,003.12	−17.0	6.9	8.0	−38.0	1.1	9	6.0
Feb.	1,007.78	−19.6	7.9	9.9	−40.0	1.2	11	10.8
Mar.	1,011.52	−18.4	9.2	11.5	−39.0	1.2	9	14.0
Apr.	1,013.65	−12.3	9.3	8.0	−35.0	1.7	11	14.0
May	1,014.05	−2.4	7.3	18.0	−25.0	3.7	11	18.0
June	1,010.58	3.0	6.9	16.0	−13.0	5.2	9	13.0
July	1,008.05	6.0	7.5	19.0	−7.0	6.5	21	33.5
Aug.	1,007.38	5.7	6.6	16.5	−9.0	6.3	24	26.8
Sept.	1,006.72	1.5	5.3	14.7	−13.0	4.7	30	35.0
Oct.	1,004.18	−3.3	5.1	9.0	−17.0	3.2	23	28.5
Nov.	1,004.45	−7.3	5.2	8.0	−25.0	2.3	17	37.0
Dec.	1,007.52	−12.6	5.8	7.4	−34.0	1.5	11	10.3
Annual	1,010.70	−6.4	6.9	19.0	−40.0	3.2	186	37.0

Month	Number of days with				Mean cloudiness (oktas)	Wind	
	precip.	thunder-storm	fog	wind ≧ 15 m/sec		preval. dir.	speed (m/sec)
Jan.	13.5	0	1.3	0.2	6.2	E, NE	1.9
Feb.	10.5	0.09	1.1	0.2	5.7	E, NE, C	1.8
Mar.	12.2	0	2.2	0.3	4.9	E, C, NE	1.6
Apr.	9.2	0.05	1.9	0.2	6.1	NE, E, C	1.7
May	10.5	0	3.3	0.2	7.1	C, N, NE	1.7
June	3.9	0	8.4	0.2	7.1	C, N	1.6
July	5.1	0	9.6	0.4	6.8	C, N	1.7
Aug.	5.1	0	4.9	0.1	7.3	C, E	1.8
Sept.	9.1	0	1.2	0.4	7.4	E, C, NE	1.8
Oct.	12.7	0	0.6	0.7	7.4	E	2.0
Nov.	15.2	0	0.2	0.3	7.4	E, NE	2.1
Dec.	13.2	0	0.5	0.2	6.8	E, NE	2.0
Annual	120.2	0.14	35.2	3.4	6.7	E, C, NE	1.8

TABLE LXVIII

CLIMATIC TABLE FOR UMANAK (1931–1956; 19–25 years)
Latitude 70°41′N, longitude 52°07′W, elevation 8 m

Month	Mean sta. press. (mbar)	Temperature (°C)				Mean vap. press. (mm)	Precipitation (mm)	
		daily mean	mean daily range	extremes			mean	max. in 24 h.
				max.	min.			
Jan.	1,003.12	−12.9	5.0	9.2	−32.0	1.7	25	16.6
Feb.	1,004.98	−15.3	6.2	10.0	−35.0	1.3	15	15.8
Mar.	1,010.32	−14.0	6.6	10.5	−35.2	1.4	12	11.0
Apr.	1,011.78	−10.0	7.0	9.5	−30.0	1.9	13	9.0
May	1,012.72	−1.0	6.2	14.3	−20.5	3.7	12	11.0
June	1,009.38	4.8	5.9	18.0	−7.0	4.7	12	16.8
July	1,007.38	7.8	5.7	18.0	−1.5	5.7	12	13.2
Aug.	1,007.25	7.0	4.8	16.3	−2.0	5.4	12	18.6
Sept.	1,005.25	2.7	3.7	16.0	−6.3	4.3	21	19.5
Oct.	1,003.65	−2.0	3.5	17.0	−12.5	3.3	18	15.0
Nov.	1,003.12	−5.4	3.6	11.0	−18.0	2.6	25	13.0
Dec.	1,003.38	−9.2	3.9	9.2	−26.8	2.0	24	12.0
Annual	1,006.86	−4.0	5.2	18.0	−35.2	3.2	201	19.5

Month	Number of days with				Mean cloudiness (oktas)	Wind	
	precip.	thunderstorm	fog	wind ≧ 15 m/sec		preval. dir.	speed (m/sec)
Jan.	8.0	0	3.0	1.0	6.1	SE,C,NW	2.1
Feb.	6.2	0	4.4	1.5	6.0	SE,C,NW	1.9
Mar.	5.9	0	6.3	0.9	5.2	C,NW,SE	1.4
Apr.	6.1	0	6.7	0.5	5.5	C,NW,SE	1.4
May	5.7	0	8.6	0.3	6.0	C,NW,SE	1.5
June	3.3	0	9.5	0.2	6.0	NW,C,SE	1.7
July	4.5	0	8.3	0.3	6.2	C,NW,SE	1.3
Aug.	4.0	0	5.9	0.2	6.3	SE,C,NW	1.7
Sept.	6.6	0	2.6	1.3	6.9	SE,C,NW	2.2
Oct.	5.8	0	1.0	1.7	6.6	SE	2.5
Nov.	8.3	0	1.3	1.3	6.9	SE	2.5
Dec.	8.1	0	2.2	0.8	6.6	SE	2.2
Annual	72.5	0	59.8	10.0	6.2	SE,C,NW	1.9

TABLE LXIX

CLIMATIC TABLE FOR GODHAVN (1931–1956; 14–21 years)
Latitude 69°14′N, longitude 53°31′W, elevation 11 m

Month	Mean sta. press. (mbar)	Temperature (°C)				Precipitation (mm)	
		daily mean	mean daily range	extremes		mean	max. in 24 h.
				max.	min.		
Jan.	1,004.45	−11.8	5.8	9.0	−33.2	12	17.4
Feb.	1,008.05	−13.9	7.8	8.8	−32.0	22	32.0
Mar.	1,011.65	−12.7	7.6	9.7	−33.8	15	22.3
Apr.	1,014.45	−7.3	7.5	10.8	−25.5	15	16.5
May	1,014.18	0.0	7.0	18.3	−17.5	32	40.4
June	1,009.92	5.1	6.1	17.9	−6.7	30	22.2
July	1,008.05	8.0	6.2	17.3	−0.7	47	36.6
Aug.	1,008.18	7.2	5.0	16.0	−0.2	45	42.1
Sept.	1,006.05	3.2	4.6	14.6	−6.1	60	102.3
Oct.	1,003.65	−2.2	4.4	11.6	−12.1	53	54.1
Nov.	1,005.25	−5.5	5.0	7.6	−19.8	38	31.4
Dec.	1,002.72	−8.9	5.8	7.5	−26.7	22	21.3
Annual	1,008.05	−3.2	6.1	18.3	−33.8	391	102.3

Month	Number of days with				Mean cloudiness (oktas)	preval. Wind dir.
	precip.	thunder-storm	fog	wind ≧ 15 m/sec		
Jan.	6.8	0.1	1.5	0.3	6.4	E,C,NE
Feb.	5.5	0.1	1.1	0.6	6.2	C,E,NE
Mar.	4.7	0	0.9	0.1	5.2	C,E,NE
Apr.	4.0	0	1.8	0.2	5.9	C,E,NE
May	6.7	0	2.7	0.3	6.1	E,C,W
June	5.8	0	5.2	0.1	5.8	W,E,C
July	7.5	0	5.8	0.2	6.0	W,E,C
Aug.	7.3	0	5.1	0.0	6.4	E,W,C
Sept.	8.5	0	1.1	0.4	6.6	E,C
Oct.	6.8	0	0.2	0.9	6.5	E,NE,C
Nov.	7.7	0	0.1	0.5	7.0	E,NE,C
Dec.	6.5	0.1	0.3	0.5	6.8	E,NE,C
Annual	77.8	0.3	25.8	4.1	6.2	E,C,W,NE

TABLE LXX

CLIMATIC TABLE FOR JAKOBSHAVN (1931–1956; 19–26 years)
Latitude 69°13′N, longitude 51°03′W, elevation 31 m

Month	Mean sta. press. (mbar)	Temperature (°C)				Precipitation (mm)	
		daily mean	mean daily range	extremes		mean	max. in 24 h.
				max.	min.		
Jan.	998.32	−13.5	7.1	8.3	−36.0	10	10.7
Feb.	1,002.58	−14.4	7.9	9.1	−34.8	13	13.0
Mar.	1,005.92	−12.5	8.3	10.2	−36.4	14	22.2
Apr.	1,007.78	−7.8	8.8	12.3	−28.2	15	15.6
May	1,009.25	0.6	7.0	17.1	−20.5	20	15.6
June	1,006.18	5.9	6.5	20.8	−5.8	19	43.2
July	1,004.45	8.2	6.6	19.4	0.0	35	39.0
Aug.	1,004.18	6.8	6.0	18.9	−3.2	34	33.6
Sept.	1,002.32	2.5	6.0	17.0	−11.1	41	23.8
Oct.	994.85	−3.7	6.0	12.5	−21.4	29	27.2
Nov.	1,000.32	−7.6	6.3	10.0	−26.5	21	29.0
Dec.	998.45	−10.8	6.6	8.2	−28.8	18	32.0
Annual	1,002.88	−3.9	6.9	20.8	−36.4	269	43.2

Month	Number of days with				Mean cloudiness (oktas)	Wind	
	precip.	thunderstorm	fog	wind $\geqq$ 15 m/sec		preval. dir.	speed (m/sec)
Jan.	9.7	0	0.4	0.9	6.0	E,SE	3.2
Feb.	9.6	0	1.9	0.7	6.0	E,SE	2.9
Mar.	8.8	0	1.8	0.3	5.3	E,SE	2.7
Apr.	10.1	0	1.9	0.0	6.0	E,SE	2.7
May	9.3	0	2.5	0.0	6.0	N,C,E,SW	2.3
June	6.6	0	4.7	0.1	5.9	W,N	2.4
July	8.9	0	5.8	0.3	6.2	W,C,N	1.9
Aug.	8.6	0	3.5	0.1	6.2	C,W,E	2.1
Sept.	12.0	0	1.5	0.4	6.6	E,SE,C	2.6
Oct.	10.0	0	0.5	0.2	6.4	E,SE	3.1
Nov.	11.2	0	0.5	0.5	6.6	E,SE	3.3
Dec.	9.4	0	0.3	1.1	6.3	E,SE	3.4
Annual	114.2	0	25.3	4.6	6.1	E,SE,C	2.7

TABLE LXXI

CLIMATIC TABLE FOR AGTO (1931–1956; 14–24 years)
Latitude 67°56′N, longitude 53°38′W, elevation 4 m

Month	Mean daily temp. (°C)	Number of days with			Mean cloudiness (oktas)	Preval. wind dir.
		thunder-storm	fog	wind ≧ 15 m/sec		
Jan.	−13.3	0	7.5	0.4	7.2	E,NE,SW
Feb.	−14.9	0	7.1	0.5	6.8	NE,E
Mar.	−13.3	0	6.1	1.3	6.3	NE,E
Apr.	−8.3	0	3.4	0.6	6.8	NE,E,N
May	0.0	0	4.5	0.9	7.1	SW,NE,NW
June	4.2	0	8.0	0.3	6.8	SW,NW,C,N
July	6.7	0	9.5	0.3	6.8	SW,C,NW
Aug.	6.2	0	7.3	0.2	7.0	SW,C,NW,N
Sept.	3.2	0	3.6	1.1	7.2	SW,NE,C
Oct.	−1.6	0	0.3	1.4	7.2	E,NE,C,SW
Nov.	−5.6	0.1	1.2	1.5	7.8	E,NE
Dec.	−9.4	0	2.3	0.5	7.5	E,NE
Annual	−3.8	0.1	60.8	9.0	7.0	NE,SW,E

TABLE LXXII

CLIMATIC TABLE FOR QORNOQ (1931–1956; 14–24 years)
Latitude 64°26′N, longitude 50°58′W, elevation 3 m

Month	Mean daily temp. (°C)	Precipitation (mm)		Number of days with				Mean cloudiness (oktas)	Preval. wind dir.
		mean	max. in 24 h.	precip.	thunder-storm	fog	wind ≧ 15 m/sec		
Jan.	−8.0	20	33.0	6.4	0	1.4	0.1	6.5	E
Feb.	−7.2	14	23.2	6.3	0	1.7	0.2	7.0	E
Mar.	−4.9	15	24.5	4.6	0	2.9	0	5.9	E
Apr.	−1.9	18	46.5	3.8	0	4.7	0	6.4	E,C,W
May	4.1	19	31.9	4.1	0	7.4	0	6.4	C,W,E
June	8.2	24	26.8	4.1	0	7.8	0	6.1	W,C
July	9.7	33	45.8	7.1	0	7.0	0.2	6.1	W,C
Aug.	8.1	48	47.7	5.2	0	7.4	0	6.6	C,W,E
Sept.	4.2	50	50.0	7.8	0	4.5	0	6.9	C,E,W
Oct.	−0.8	50	49.1	7.9	0	1.6	0.1	6.5	E,C
Nov.	−4.2	28	36.1	6.5	0	2.4	0	6.6	E
Dec.	−6.9	16	19.0	6.1	0	2.0	0.1	6.2	E
Annual	0.03	335	50.0	69.9	0	50.8	0.7	6.4	E,C,W

TABLE LXXIII

CLIMATIC TABLE FOR KAPISIGDLIT (1939–1956; 13–19 years)
Latitude 64°20′N, longitude 50°15′W

Month	Temperature (°C)				Precipitation (mm)		Number of days with		
	daily mean	mean daily range	extremes		mean	max. in 24 h	precip.	thunder-storm	fog
			max.	min.					
Jan.	−9.6	8.1	11.0	−30.5	10	12.0	5.5	0	1.5
Feb.	−9.9	8.7	13.5	−32.0	10	12.5	4.1	0	1.4
Mar	−7.1	8.8	10.6	−30.5	8	9.2	3.6	0	1.1
Apr.	−3.4	8.2	15.6	−23.5	12	34.2	3.4	0	2.4
May	4.3	9.0	19.0	−13.5	14	15.1	2.5	0	5.9
June	9.4	10.3	21.5	−4.5	20	21.3	3.2	0	5.4
July	10.9	9.7	21.5	−0.5	37	32.1	4.7	0	5.7
Aug.	8.7	9.4	21.0	−1.0	41	37.0	5.7	0	6.5
Sept.	3.8	7.9	20.0	−14.0	39	42.0	6.9	0	4.7
Oct.	−1.7	7.4	16.5	−22.0	30	29.7	6.4	0	1.5
Nov.	−5.7	7.3	12.5	−22.5	16	23.7	4.7	0	0.9
Dec.	−7.6	7.5	10.5	−31.0	18	19.9	4.0	0	1.5
Annual	−0.7	8.5	21.5	−32.0	255	42.0	54.7	0	38.5

TABLE LXXIV

CLIMATIC TABLE FOR GODTHAAB (1931–1956; 19–26 years)
Latitude 64°10′N, longitude 51°45′W, elevation 20 m

Month	Temperature (°C)				Precipitation (mm)	
	daily mean	mean daily range	extremes max.	extremes min.	mean	max. in 24 h
Jan.	−7.7	6.7	12.0	−25.2	26	49.5
Feb.	−7.3	6.8	11.6	−26.0	24	31.5
Mar.	−5.8	7.0	14.2	−25.2	18	48.8
Apr.	−3.5	7.0	13.4	−22.0	25	26.5
May	2.1	7.3	18.5	−17.8	29	30.6
June	5.7	8.3	23.0	−4.3	46	50.5
July	7.6	8.5	20.2	−6.0	59	35.0
Aug.	6.9	7.7	19.0	−5.2	69	80.9
Sept.	4.1	6.5	20.4	−7.4	84	66.7
Oct.	−0.3	5.9	13.8	−13.9	71	62.5
Nov.	−3.6	5.8	12.0	−16.0	44	74.0
Dec.	−6.2	6.3	12.2	−23.0	20	44.0
Annual	−0.7	7.0	23.0	−26.0	515	80.9

Month	Number of days with				Mean cloudiness (oktas)	Wind	
	precip.	thunderstorm	fog	wind ≧ 15 m/sec		preval. dir.	speed (m/sec)
Jan.	7.7	0.04	1.1	1.1	7.3	E,NE	3.1
Feb.	6.9	0	1.3	0.9	7.6	NE,E	3.3
Mar.	5.6	0.04	1.8	0.7	6.6	NE,E	3.0
Apr.	5.6	0	2.6	0.8	7.0	NE,SW,N	2.8
May	5.9	0.1	7.6	0.8	7.1	N,SW	2.2
June	6.1	0	12.2	0.9	7.0	SW,N,C	2.2
July	9.9	0.2	13.4	1.3	7.2	SW,C,N	2.1
Aug.	10.3	0.2	13.9	0.9	7.5	SW,C,N	2.2
Sept.	12.2	0	6.4	1.8	7.4	SW,C,E	2.4
Oct.	10.2	0.1	1.6	1.3	7.0	E,SW,NE	2.8
Nov.	8.5	0.1	1.2	0.9	7.0	E,NE,SW	3.0
Dec.	7.2	0	2.0	0.5	6.8	E,NE	3.0
Annual	96.1	0.8	65.1	11.9	7.1	SW,E,NE	2.7

TABLE LXXV

CLIMATIC TABLE FOR IVIGTUT (1931–1956; 19–22 years)
Latitude 61°12′N, longitude 48°10′W, elevation 30 m

Month	Mean sta. press. (mbar)	Temperature (°C)				Precipitation (mm)	
		daily mean	mean daily range	extremes		mean	max. in 24 h
				max.	min.		
Jan.	994.45	−5.4	6.3	12.0	−21.8	92	72.0
Feb.	998.72	−4.5	7.1	14.5	−22.2	129	169.4
Mar.	1,000.58	−2.8	7.4	14.0	−23.4	87	59.1
Apr.	1,005.52	0.2	7.0	16.5	−19.1	79	80.2
May	1,007.78	5.3	7.6	19.5	−11.0	89	60.2
June	1,006.18	8.5	8.4	23.1	−2.6	96	168.2
July	1,005.25	9.8	8.2	22.4	1.4	82	73.2
Aug.	1,004.32	8.8	7.3	19.2	−0.6	97	75.2
Sept.	1,001.92	5.8	6.5	21.0	−7.2	162	88.7
Oct.	999.65	1.6	5.9	17.0	−9.8	172	128.2
Nov.	998.85	−1.8	6.1	17.0	−16.4	146	99.0
Dec.	994.59	−4.0	6.2	15.5	−18.3	77	72.6
Annual	1,001.48	1.8	7.0	23.1	−23.4	1308	169.4

Month	Number of days with				Mean cloudiness (oktas)	Wind	
	precip.	thunderstorm	fog	wind ≧ 15 m/sec		preval. dir.	speed (m/sec)
Jan.	10.7	0	0.2	1.5	6.3	C,(SE)	1.5
Feb.	10.8	0	0.1	0.8	6.5	C,(SE)	1.6
Mar.	10.5	0	0.2	0.8	5.8	C,(SE)	1.2
Apr.	9.4	0	0.8	0.7	6.3	C,(NW)	1.2
May	8.8	0	2.3	0.1	6.7	C,(NW,SE)	1.1
June	7.0	0	3.9	0.3	6.3	C,(NW)	1.1
July	10.3	0.1	6.2	0.1	6.5	C,(NW)	0.8
Aug.	9.8	0.1	5.1	0.0	6.6	C,(NW)	0.8
Sept.	12.2	0	2.6	0.6	6.6	C,(NW)	0.9
Oct.	11.1	0	0.6	0.7	6.4	C,(NW,SE)	0.9
Nov.	10.8	0	0.3	0.5	6.3	C,(SE)	1.1
Dec.	9.7	0	0	1.3	6.0	C,(SE)	1.2
Annual	121.1	0.2	22.3	7.4	6.4	C,(NW,SE)	1.1

TABLE LXXVI

CLIMATIC TABLE FOR IGALIKO (1933–1946; 9–14 years)
Latitude 60°59′N, longitude 45°30′W, elevation 7 m

Month	Temperature (°C)				Precipitation (mm)		Days with fog
	daily mean	mean daily range	extremes		mean	max. in 24 h	
			max.	min.			
Jan.	−5.1	8.6	11.5	−23.2	37	46.0	0.1
Feb.	−4.4	8.5	15.0	−27.6	46	58.2	0.3
Mar.	−2.4	8.3	13.2	−21.1	32	30.6	0.4
Apr.	1.0	6.6	17.0	−16.8	54	54.0	1.0
May	5.9	7.8	19.0	−13.5	35	38.9	4.4
June	8.2	7.7	20.4	−2.5	76	140.0	4.1
July	10.3	7.1	20.2	−1.5	79	80.6	4.9
Aug.	9.5	6.7	21.5	1.4	91	63.2	4.9
Sept.	6.5	6.1	20.0	−3.3	109	64.6	3.9
Oct.	1.9	5.9	16.3	−9.3	116	95.6	2.9
Nov.	−1.6	6.6	16.4	−16.1	78	107.6	1.1
Dec.	−3.5	7.6	15.3	−22.5	47	52.5	0.7
Annual	2.2	7.3	21.5	−27.6	800	140.0	28.7

TABLE LXXVII

CLIMATIC TABLE FOR SLETTEN (1933–1942; 5–10 years)
Latitude 60°35′N, longitude 45°26′W, elevation 5 m

Month	Temperature (°C)				Precipitation (mm)		Number of days with	
	daily mean	mean daily range	extremes		mean	max. in 24 h	thunder-storm	fog
			max.	min.				
Jan.	−5.1	7.0	12.2	−22.0	35	27.0	0	1.3
Feb.	−3.9	7.4	12.5	−25.0	43	40.3	0	2.0
Mar.	−3.9	9.5	9.5	−26.5	27	21.0	0	2.3
Apr.	0.1	8.8	15.5	−24.0	43	52.2	0	3.2
May	4.4	7.9	17.5	−10.5	38	32.2	0	10.0
June	6.3	7.7	18.5	−3.0	116	259.2	0	11.2
July	8.1	7.4	21.0	0.0	55	45.3	0	13.2
Aug.	7.9	7.1	20.0	0.0	85	53.0	0	11.5
Sept.	5.8	6.6	18.0	−2.5	99	49.6	0	10.0
Oct.	1.7	6.5	16.0	−9.5	128	93.1	0	4.0
Nov.	−1.5	6.7	15.5	−19.5	66	49.0	0	2.4
Dec.	−3.6	7.4	13.5	−22.5	38	33.1	0.2	2.6
Annual	1.4	7.5	21.0	−26.5	773	259.2	0.2	73.7

TABLE LXXVIII

CLIMATIC TABLE FOR SYDPRÖVEN (1934–1946; 3–13 years)
Latitude 60°28′N, longitude 45°38′W, elevation 15 m

Month	Temperature (°C)				Precipitation (mm)		Number of days with	
	daily mean	mean daily range	extremes		mean	max. in 24 h	thunder-storm	fog
			max.	min.				
Jan.	−4.0	9.5	12.2	−24.1	26	26.1	0	1.3
Feb.	−3.6	7.9	12.1	−18.0	60	60.5	0	2.7
Mar.	−2.7	7.7	10.1	−18.2	32	22.0	0	3.2
Apr.	−0.3	8.0	15.0	−16.0	44	51.4	0	6.4
May	2.7	6.9	15.2	−9.0	34	38.0	0	9.9
June	3.8	6.9	18.1	−5.0	53	30.0	0.7	15.1
July	5.4	7.1	18.0	−4.1	76	54.5	0.3	14.7
Aug.	5.9	7.0	18.0	−3.1	85	36.0	0	12.8
Sept.	4.3	5.3	14.1	0.0	96	55.5	0	9.5
Oct.	2.0	5.2	14.0	−7.0	121	62.8	0	4.0
Nov.	−1.8	6.3	13.0	−15.0	73	27.0	0	2.3
Dec.	−3.5	6.7	12.2	−16.2	54	90.0	0.7	2.1
Annual	0.7	7.0	18.1	−24.1	754	90.0	1.7	84.0

TABLE LXXIX

CLIMATIC TABLE FOR NANORTALIK (1931–1946; 9–15 years)
Latitude 60°08′N, longitude 45°11′W, elevation 7 m

Month	Mean sta. press. (mbar)	Temperature (°C)				Precipitation (mm)	
		daily mean	mean daily range	extremes		mean	max. in 24 h
				max.	min.		
Jan.	996.85	−3.3	4.8	11.5	−17.2	64	42.0
Feb.	1,002.18	−2.4	5.2	11.6	−13.4	71	60.8
Mar.	1,004.72	−1.7	5.7	10.5	−14.3	41	25.2
Apr.	1,010.18	0.7	6.2	15.0	−14.3	59	40.3
May	1,009.92	4.0	6.5	16.7	−6.5	45	27.5
June	1,009.78	5.5	7.6	20.1	−3.1	80	104.1
July	1,008.58	6.5	7.9	18.5	−2.9	53	30.0
Aug.	1,008.05	7.0	6.7	17.1	−0.7	92	50.6
Sept.	1,005.92	5.8	6.0	16.7	−2.8	119	143.1
Oct.	1,003.92	2.7	4.7	14.7	−6.0	125	72.7
Nov.	999.78	−1.0	4.7	13.7	−10.7	94	75.4
Dec.	997.38	−2.2	5.1	17.4	−13.2	52	37.8
Annual	1,004.77	1.8	5.9	20.1	−17.2	895	143.1

Month	Number of days with				Mean cloudiness (oktas)	Wind	
	precip.	thunder-storm	fog	wind ≧ 15 m/sec		preval. dir.	speed (m/sec)
Jan.	9.4	0.1	0	1.8	6.3	N,W	3.8
Feb.	11.2	0.3	0.6	1.8	6.9	N,NE,W	3.6
Mar.	10.7	0.1	0.1	0.3	5.9	N,NE,W	2.7
Apr.	9.5	0	1.5	0.2	6.3	N,W	2.9
May	9.7	0	5.8	0	7.2	W,N,NE	2.5
June	6.8	0.1	7.2	0.2	6.3	W,N	2.5
July	11.2	0.2	9.3	0.2	6.4	N,W	2.1
Aug.	9.9	0.2	9.6	0.1	6.6	W,N	2.4
Sept.	13.6	0	4.1	0.1	6.8	N,W	2.6
Oct.	12.4	0.1	1.9	1.4	6.5	N,W	3.0
Nov.	9.5	0.1	0.5	1.0	6.5	N,W	3.2
Dec.	10.6	0.1	0.6	2.0	6.3	N	3.3
Annual	124.5	1.3	41.2	9.1	6.5	N,W	2.9

TABLE LXXX

CLIMATIC TABLE FOR NORD (1952–1956; 4–5 years)
Latitude 81°36′N, longitude 16°40′W, elevation 35 m

Month	Mean sta. press. (mbar)	Temperature (°C)				Precipitation (mm)	
		daily mean	mean daily range	extremes		mean	max. in 24 h
				max.	min.		
Jan.	1,013.92	−29.6	7.4	−8.8	−45.6	23	28.7
Feb.	1,013.52	−29.7	9.3	−0.1	−51.1	20	23.4
Mar.	1,017.65	−32.5	7.9	−17.8	−47.0	8	3.6
Apr.	1,020.32	−23.1	8.1	−3.6	−43.3	5	4.3
May	1,017.52	−10.9	6.8	2.4	−27.9	3	3.2
June	1,012.05	−0.4	5.1	13.6	−12.6	5	5.6
July	1,007.12	4.2	6.0	16.3	−5.3	12	14.0
Aug.	1,009.25	1.6	4.0	14.2	−7.2	19	17.9
Sept.	1,012.05	−7.8	5.5	7.2	−25.0	21	18.3
Oct.	1,010.05	−18.5	6.4	−2.8	−34.5	16	9.3
Nov.	1,012.05	−24.3	7.9	3.5	−40.7	35	33.4
Dec.	1,008.32	−25.6	8.1	−0.8	−46.5	37	17.9
Annual	1,012.82	−16.4	6.9	16.3	−51.1	204	33.4

Month	Number of days with				Mean cloudiness (oktas)	Preval. wind direct
	precip.	thunderstorm	fog	wind ≧ 15 m/sec		
Jan.	11.7	0	0.7	1.0	4.4	S,C,SW
Feb.	8.2	0	1.2	1.0	3.8	SW,C,S
Mar.	8.2	0	0.5	0.2	3.7	SW,C,S
Apr.	4.7	0	0	0	3.7	SW,C,S
May	4.7	0	3.0	0	4.3	C,SW
June	3.0	0	7.4	0	4.7	C,SW,S
July	3.4	0	5.6	0	4.7	SW,S,E
Aug.	6.6	0	9.6	0.2	5.6	SW,C,E
Sept.	10.2	0	6.0	0	5.9	SW,C,S
Oct.	12.4	0	6.2	0.8	5.8	SW,C
Nov.	11.6	0	4.2	0.4	4.5	C,SW
Dec.	10.8	0	1.4	0.8	4.8	SW,C,S
Annual	95.5	0	45.8	4.4	4.6	SW,C,S

TABLE LXXXI

CLIMATIC TABLE FOR MYGGBUKTA (1931–1939, 1947–1958; 19–20 years)
Latitude 73°29′N, longitude 21°34′W, elevation 2 m

Month	Mean sta. press. (mbar)	Temperature (°C)				Precipitation (mm)	
		daily mean	mean daily range	extremes		mean	max. in 24 h
				max.	min.		
Jan.	1,007.8	−20.2	9.8	1.2	−50.9	44	60.7
Feb.	1,013.8	−21.3	10.9	5.0	−47.5	30	42.5
Mar.	1,017.4	−20.3	11.8	7.1	−47.5	24	45.6
Apr.	1,017.8	−15.7	13.9	7.0	−37.7	15	34.5
May	1,019.8	−5.6	9.6	12.4	−30.2	9	9.3
June	1,015.4	1.4	5.5	22.1	−10.3	13	38.0
July	1,012.4	3.7	5.2	21.2	−6.3	20	26.1
Aug.	1,011.1	3.1	6.6	20.3	−6.7	29	29.2
Sept.	1,009.1	−1.4	6.2	14.5	−24.7	21	30.0
Oct.	1,011.4	−9.6	8.0	10.2	−30.3	23	29.2
Nov.	1,011.1	−16.0	10.2	7.0	−42.5	31	28.0
Dec.	1,010.0	−18.5	10.3	3.2	−42.5	39	43.2
Annual	1,013.1	−10.0	9.0	22.1	−50.9	298	60.7

Month	Number of days with				Mean cloudiness (oktas)	Wind	
	precip.	thunderstorm	fog	wind ≧ 15 m/sec		preval. dir.	speed (m/sec)
Jan.	9.9	0	1.6	5.5	4.8	N,C,NW	2.7
Feb.	7.8	0	1.3	3.6	4.5	N,C,NW	2.4
Mar.	8.6	0	2.3	2.0	4.5	N,C,NW	2.2
Apr.	7.6	0	1.7	1.3	4.5	C,NW,N,SE	1.8
May	6.8	0	8.8	0.3	5.5	SE,C,E	1.6
June	5.6	0	10.4	0.4	5.7	SE,E,C	1.9
July	7.2	0	11.3	0.1	5.7	SE,(E)	2.1
Aug.	7.1	0.1	10.7	0.1	5.5	SE,E	2.1
Sept.	6.1	0	2.9	1.2	5.1	N,SE	2.3
Oct.	9.0	0	1.2	2.5	5.3	N,NW,C	2.5
Nov.	8.5	0	1.7	4.1	5.0	N,C,NW	2.5
Dec.	10.0	0	1.8	3.6	4.7	C,N,NW	2.5
Annual	94.2	0.1	55.7	24.7	5.1	N,SE,C	2.2

TABLE LXXXII

CLIMATIC TABLE FOR SCORESBYSUND (KAP TOBIN) (1931–1956; 17–26 years)
Latitude 70°25′N, longitude 21°58′W, elevation 17 m

Month	Mean sta. press. (mbar)	Temperature (°C)				Precipitation (mm)	
		daily mean	mean daily range	extremes		mean	max. in 24 h
				max.	min.		
Jan.	1,000.98	−15.3	8.7	3.0	−43.7	29	17.2
Feb.	1,006.45	−16.8	9.1	10.8	−41.6	29	22.8
Mar.	1,010.98	−16.1	9.8	4.5	−39.5	23	17.0
Apr.	1,011.25	−11.7	10.6	7.0	−32.5	21	15.5
May	1,015.12	−2.9	7.3	11.8	−25.6	12	13.1
June	1,010.72	2.4	6.4	15.8	−7.5	26	34.1
July	1,007.65	4.7	6.3	16.8	−5.0	38	30.0
Aug.	1,005.78	3.7	5.7	13.0	−4.4	33	22.8
Sept.	1,006.98	0.8	4.6	15.0	−11.5	53	38.0
Oct.	1,006.75	−5.6	5.0	12.6	−19.2	56	114.0
Nov.	1,007.78	−10.7	6.2	5.0	−33.0	44	42.5
Dec.	1,006.32	−13.7	7.0	4.5	−39.0	64	70.0
Annual	1,008.07	−6.7	7.2	16.8	−43.7	428	114.0

Month	Number of days with				Mean cloudiness (oktas)	Wind	
	precip.	thunder-storm	fog	wind ≧ 15 m/sec		preval. dir.	speed (m/sec)
Jan.	9.7	0	3.3	2.2	5.8	C,NE,N	1.7
Feb.	8.0	0	4.6	1.8	4.8	C,NE,N	1.8
Mar.	7.1	0	3.6	1.1	5.0	C,NE	1.8
Apr.	7.1	0	6.0	0.5	4.5	C,NE	1.4
May	4.3	0	11.1	0.2	5.7	C,(E)	1.1
June	5.0	0	14.1	0.3	5.7	C,E,SE	1.2
July	6.3	0	12.4	0.3	5.7	C,E	1.2
Aug.	5.9	0	13.8	0.4	5.5	C,E	1.5
Sept.	6.6	0	5.4	1.1	5.3	C,(NE,N)	1.8
Oct.	11.2	0	3.6	1.1	5.8	C,NE,N	1.7
Nov.	11.4	0	2.5	2.2	5.7	C,NE,N	1.7
Dec.	12.5	0	3.6	1.9	5.6	C,NE,N	1.6
Annual	95.1	0	84.0	13.1	5.4	C,(E,N)	1.5

TABLE LXXXIII

CLIMATIC TABLE FOR ANGMAGSSALIK (1931–1956; 16–26 years)
Latitude 65°37′N, longitude 37°39′W, elevation 29 m

Month	Mean sta. press. (mbar)	Temperature (°C)				Mean vap. press.	Precipitation (mm)	
		daily mean	mean daily range	extremes			mean	max. in 24 h
				max.	min.			
Jan.	995.52	−6.8	6.1	6.6	−28.7	2.4	58	44.0
Feb.	1,002.05	−7.2	6.6	9.0	−26.0	2.3	82	66.6
Mar.	1,004.72	−5.7	7.9	8.5	−22.5	2.5	62	32.0
Apr.	1,008.72	−2.7	8.8	12.2	−20.3	3.2	53	83.1
May	1,012.18	2.1	7.5	17.9	−13.7	4.4	54	38.0
June	1,009.38	5.8	7.7	25.3	−8.6	5.3	44	37.5
July	1,007.25	7.4	8.1	22.9	−3.5	6.0	35	40.0
Aug.	1,005.38	6.6	7.2	19.9	−3.5	5.9	62	45.3
Sept.	1,004.32	4.2	6.0	20.2	−7.1	4.8	76	79.0
Oct.	1,000.58	−0.05	4.9	14.9	−18.3	3.7	90	71.3
Nov.	999.78	−3.1	4.9	10.2	−18.1	3.0	86	70.0
Dec.	997.38	−5.1	5.6	7.2	−21.9	2.6	68	84.0
Annual	1,003.94	−0.4	6.8	25.3	−28.7	3.8	770	84.0

Month	Number of days with				Mean cloudiness (oktas)	Wind	
	precip.	thunderstorm	fog	wind ⩾15 m/sec		preval. dir.	speed (m/sec)
Jan.	11.6	0	1.4	1.0	7.0	C,(W,E)	2.0
Feb.	12.6	0.12	1.7	0.8	6.9	C,(W,NE)	1.7
Mar.	12.7	0	1.7	0.5	6.5	C,(W,N)	1.6
Apr.	10.8	0	2.1	0.6	6.4	C,(W)	1.5
May	9.5	0	7.4	0.3	6.5	C,(S)	1.3
June	9.1	0	7.8	0.0	6.1	C,S,(E,SW)	1.4
July	7.5	0.16	11.6	0.3	5.8	C,S,E	1.3
Aug.	8.5	0	8.9	0.2	6.0	C,S	1.1
Sept.	10.7	0	4.5	0.5	6.6	C,(S)	1.4
Oct.	12.5	0	3.3	0.6	7.1	C,(NE,N)	1.6
Nov.	12.8	0	2.9	0.8	7.1	C,NE(W)	1.7
Dec.	11.7	0.04	1.6	0.7	7.2	C,(NE,W)	1.8
Annual	130.0	0.32	54.9	6.3	6.6	C,(E,W)	1.5

TABLE LXXXIV

CLIMATIC TABLE FOR TORGILSBU (1933–1940; 7–8 years)
Latitude 60°32′N, longitude 43°11′W, elevation 9 m

Month	Temperature (°C)			Precipitation (mm)		Number of days with		
	daily mean	extremes		mean	max. in 24 h	precip.	thunder-storm	fog
		max.	min.					
Jan.	−3.5	5.2	−20.0	151	95.6	14.6	0	1.0
Feb.	−3.8	6.2	−26.2	224	123.3	15.1	0.4	0.2
Mar.	−2.6	5.9	−20.1	123	36.0	13.9	0	0.7
Apr.	0.3	9.2	−15.9	119	64.4	12.9	0.7	1.5
May	3.1	11.8	−8.7	235	61.1	18.2	0	2.5
June	6.1	22.4	−2.6	159	68.9	12.7	0	2.7
July	7.6	20.5	−0.6	71	35.7	11.1	0	3.2
Aug.	7.9	21.6	−0.8	96	42.8	12.2	0	3.6
Sept.	6.2	18.2	−4.0	239	103.9	14.4	0.1	1.9
Oct.	2.8	14.0	−4.4	192	128.0	13.3	0	2.0
Nov.	−0.3	8.2	−10.1	166	67.3	13.5	0	0.5
Dec.	−1.7	12.4	−11.2	165	41.7	14.6	0	0.8
Annual	1.9	22.4	−26.2	1940	128.0	166.5	1.2	20.6

Month	Mean cloudiness (oktas)	Wind	
		preval. dir.	speed (m/sec)
Jan.	5.7	NE,SW	3.3
Feb.	6.4	NE,SW	3.4
Mar.	6.2	NE,C,NW	2.9
Apr.	6.5	NE,C,SW	2.9
May	7.6	NE,SE	2.9
June	6.1	SE,NE,C	2.3
July	6.2	SE,C,NE	1.9
Aug.	6.1	SE,C,NE	1.9
Sept.	6.6	NE,C,SE	2.5
Oct.	6.1	SW,NE,C	2.6
Nov.	6.1	NE,SW	2.9
Dec.	6.3	NE,SW	3.5
Annual	6.3	NE	2.7

Chapter 3

The Climate of the North Polar Basin

E. VOWINCKEL AND S. ORVIG

Introduction

The North Polar Basin and the various subdivisions of the Arctic waters, shown in Fig.1, were delimited by consideration of the bottom contours. Limits of the seas are here given according to the INTERNATIONAL HYDROGRAPHIC BUREAU (1953). Climatologically, no such series of lines are particularly important in the Arctic, except where open water meets an ice covered surface. There the meteorological environment changes drastically. Generally, the choice of boundaries remains arbitrary for all climatic regions, since only few sharply defined boundaries exist.

In the oceanographic literature two boundaries are used for the Arctic Ocean:

(*1*) As used by SVERDRUP et al. (1942): Eurasian coast–Bering Strait–coast of North American continent (excluding Hudson Bay)–coast of Greenland–Denmark Strait–Shetland Islands–Norway. The area delimited by this boundary is 14,090,000 km^2.

(*2*) As frequently used by Soviet authors: mainly as under (*1*), but from Greenland east to Spitsbergen–north end of Novaya Zemlya–east coast of Novaya Zemlya to mainland. The area thus defined amounts to 9,906,000 km^2. To avoid confusion, this area will in the following be called "Polar Ocean".

The Polar Ocean therefore comprises the Arctic Ocean less the Norwegian Sea (2,705,000 km^2) and Barents Sea (1,479,000 km^2), but including the peripheral seas.

In the present text the boundary of the Arctic Ocean is taken as 65°N from Greenland to Norway. This includes the area within the Arctic Circle, where the annual variation in solar radiation reaches its maximum—as it becomes zero for a time in winter and of very high values in summer (at least on top of the atmosphere, although much energy is lost before the radiation reaches the surface).

In addition, geographical subdivisions are introduced, according to the distribution of ice and characteristic water masses:

(*1*) *Central Polar Ocean:* the area of the Polar Basin where less than 10% open water is found in summer.

(*2*) *Peripheral seas:* the areas with winter ice cover and more than 10% open water in summer.

(*3*) *Norwegian–Barents Sea:* generally open water.

The various divisions used in the following are shown in Fig.2.

Around the middle of the 1930's it was not possible to give a reasonably detailed description of the climate of the North Polar regions, as only few series of observations were available. In order to present the most important features of the Polar Ocean climate,

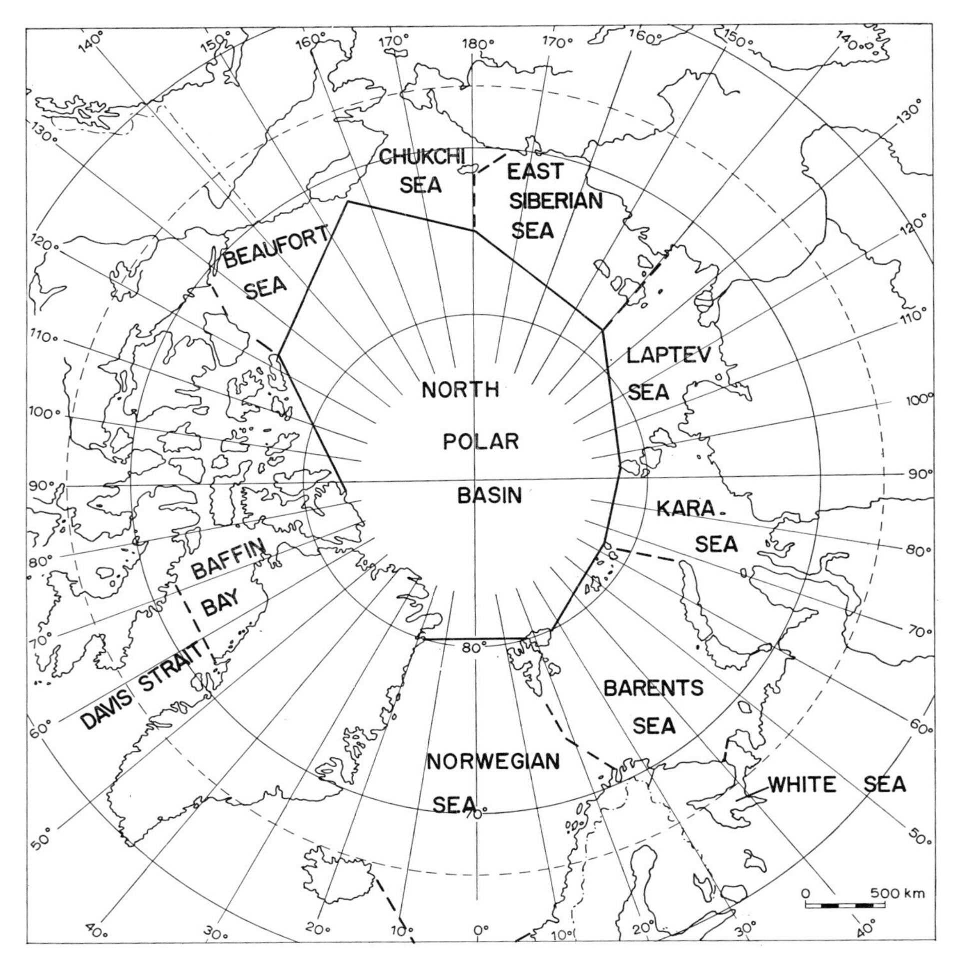

Fig.1. The Arctic seas.

SVERDRUP (1935) was forced to rely on his own observations and experience from six years' travel in the Arctic.

A new era opened with the first Soviet "North Pole" station, established on the pack ice by O. Schmidt and I. D. Papanin in May, 1937. Since that time, new knowledge has accumulated at an astonishing rate, both from the measurements at a series of floating ice stations and from new land bases along the shores of the Arctic Ocean. Soviet scientists have occupied, until 1968, eighteen such floating stations, and Americans have been engaged on six of their own.

These stations located on floes of sea ice usually last a year or two, as they are troubled by leads opening across them. The stations placed on the thicker, more stable ice islands last for many years. Fletcher's Island ("T–3"), whose thickness at one time was measured to be 43 m, has been occupied almost continuously since 1952. The floating stations are excellent platforms for many kinds of geophysical observations. The fact that they have been occupied at different times, however, is a disadvantage in attempts at obtaining statistical information about the Polar Ocean climate. They also move with the currents

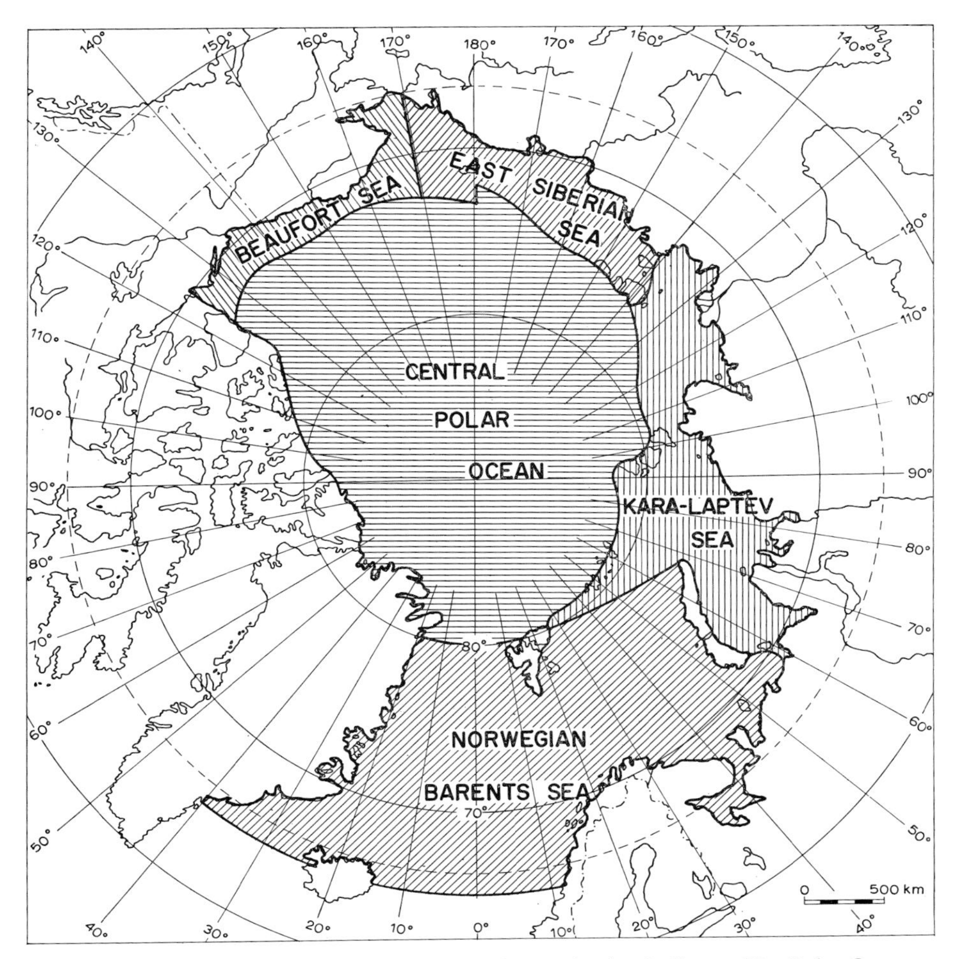

Fig.2. Regions investigated. All shaded areas make up the Arctic Ocean. The Polar Ocean comprises the Arctic Ocean less the Norwegian–Barents Sea, i.e., it consists of the central Polar Ocean plus the marginal seas.

and wind, although this is less important as the various areas sampled have rather uniform conditions, at least in the central Polar Ocean. A small number of observing stations may give useful data for a large area where the surface characteristics are uniform for great distances. However, if small-scale processes are involved, such as for example, local reflection of solar radiation and heat conduction in the ice, then caution must be exercised in the use of station observations. The site characteristics may not be representative for a larger area. The ice floes are selected for occupation because they are large and thick and they represent, therefore, a systematic selection of certain extreme conditions. This should be taken into account when evaluating the observations from the drifting stations in a regional climatology.

The floating stations, in any case, move rather slowly with the pack ice. The drift data obtained on Station Alpha (one of the American stations) give a mean speed for the period 1957–1958 of 6.2 km/day (REED and CAMPBELL, 1960). Information on the drift of some of the Soviet "North Pole" stations (GORDIENKO and LAKTIONOV, 1960; GORDIENKO, 1962) may be summarized as in Table I.

TABLE I

DRIFT SPEED OF SOME "NORTH POLE" STATIONS

Station	Period	Mean drift speed (true distance) (km/day)	
N.P. 1	June 1937–February 1938	7.5	
N.P. 2	April 1950–April 1951	6.9	
N.P. 3	April 1954–April 1955	4.9	
	April 15–June 20, 1954		2.4
	June 20–September 20, 1954		3.8
N.P. 4	April 1954–April 1955	6.8	
	April 8–June 21, 1954		4.5
	June 21–September 20, 1954		4.3
N.P. 4	April 1955–April 1956	6.8	
	April 10–August 8, 1955		3.3
	August 8–September 30, 1955		4.2
N.P. 4	April 1956–April 1957	5.3	
N.P. 5	April 1955–April 1956	6.9	
	April 20–September 21, 1955		3.8
	April–October 1956		6.4
N.P. 6	April 1956–April 1957	7.1	
	April 1957–April 1958	7.2	
	April 1958–April 1959	6.8	
	April–September 1959		6.2
N.P. 7	April 1957–April 1958	5.4	
	April 1958–April 1959	4.4	
N.P. 8	April 1959–April 1960	7.3	
	April 1960–April 1961	5.6	
	April 1961–April 1962	3.0	
N.P. 9	April 1960–March 1961	8.0	
Average speed (not including N.P. 1):		6.2	4.3

These figures indicate that over the year the average drift speed is about 6.2 km/day (true distance) and that the drift varies both from month to month and also from year to year. There is no indication that the summer drift is faster than the winter drift. By keeping careful track of the positions of the floating stations, their data have been very useful in many studies of the Polar Ocean climate. Data from many stations, including various Soviet "North Pole" stations, have been used in preparing the following climatology.

Surface conditions

It is apparent that the surface conditions, mainly the presence of either ice, snow or water, is of great climatic importance. Three physical characteristics are mostly responsible: surface reflectivity; thermal diffusivity; heat content of the ice.

Surface reflectivity (albedo)

This gives the fraction of the solar radiation which is reflected from the surface. Its value depends both on the wave length of the radiation and on the type of surface. For

TABLE II

FREQUENCY DISTRIBUTION (%) OF ALBEDO (TWO "NORTH POLE" STATIONS)

Albedo (%):	95–91	90–86	85–81	80–76	75–71	70–66	65–61	60–65
March	10	17	17	27	25	4		
April	8	29	29	9	11	7	5	2
May	2	15	45	32	4	2		

long wave radiation from the atmosphere its value is small for all surfaces—only a few per cent. For the heat radiated from the atmosphere to the ground the surface type is unimportant. For the short wave lengths of solar radiation, however, the albedo varies drastically with the surface conditions. Although a clear dependency on wave length does exist, the average albedo over all wave lengths of solar radiation will be considered in the following.

Over water surfaces the albedo is uniformly low, as long as the solar elevation remains reasonably high. With the formation of ice and snow the albedo increases sharply. This, however, will not introduce a new, constant, albedo of a higher value, but the albedo

TABLE III

THE ALBEDO OF DIFFERENT SURFACES FROM DATA OF DRIFTING STATIONS
(After BRIAZGIN, 1959)

Structure	Water content and colour	Albedo		
		average	max.	min.
Freshly fallen snow	dry bright-white clean	88	98	72
Freshly fallen snow	wet bright-white	80	85	80
Freshly drifted snow	dry clean loosely packed	85	96	70
Freshly drifted snow	moist grey-white	77	81	59
Snow, fallen or drifted 2–5 days ago	dry clean	80	86	75
Snow, fallen or drifted 2–5 days ago	moist grey-white	75	80	56
Dense snow	dry clean	77	80	66
Dense snow	wet grey-white	70	75	61
Snow and ice	dry grey-white	65	70	58
Melting ice	wet grey	60	70	40
Melting ice	moist dirty grey	55	65	36
Snow, saturated with water (snow during intense thawing)	light green	35	—	28
Melt puddles in first period of thawing	light blue water	27	36	24
Melt puddles, 30–100 cm deep	green water	20	26	13
Melt puddles, 30–100 cm deep	blue water	22	28	18
Melt puddles covered with ice	smooth grey-green ice	25	30	18
Melt puddles covered with ice	smooth ice, covered with icy white hoar frost	33	37	21

becomes very variable, depending on the crystalline structure of the ice and snow. This, in turn, is a function of temperature and, in the case of snow, its age. Soviet observations from floating stations give as limits for snow albedo, 55–100%, and for sea ice, 20–60%. UNTERSTEINER (1961) reported an albedo for clean melting sea ice of 62–70%. An example of the variability is given by Table II, giving the frequency distribution of the albedo during early summer for two Soviet stations. The variations are both short and long-term and are determined by the weather sequence and the seasonal temperature variation. Also, a marked diurnal variation is generally present.

During the melting period in mid summer, extended meltwater puddles form over the pack ice. Such surfaces will have lower albedo. On ice island T–3 it was measured to be about 38% (HANSON, 1961), and at "North Pole" 6 (BRIAZGIN, 1959) it was 20–22%. The extent of these puddles is highly variable, depending on the topography of the pack ice. They can certainly cover over half the area. The main surface types of the pack ice, and their albedo, are given in Table III.

Observations from aircraft are by far the best for average albedo values over an area, which are required in climatology. An example of such observations (HANSON, 1961) is given by the following figures for the line Cape Bathurst, N.W.T.–North Pole, for July and August 1958:

								(water)
Latitude (°N)	90	87.5	85	82.5	80	77.5	75	72.5
Albedo (%)	40	48	50	47	42	51	40	5

Fig.3 consists of six stereograms showing major seasonal and latitudinal changes of albedo. Areas without daylight remain blank in the diagrams (LARSSON, 1963).

From January through May the winter pattern of albedo distribution persists. It is characterized by extremely high and low albedos—a reflection of the contrast between snow or ice and forest or open water that is so striking to the observer flying over arctic regions.

The stereograms for June and August show, in lower latitudes, the gradual disappearance of snow cover from the land, producing a concentration of low albedos. Toward the end of the period, open water reaches its maximum areal extent, adding to the low albedo maximum. In the highest latitudes the greater homogeneity of natural surfaces is apparent from the progressive march of albedo frequencies from higher to middle values. This is brought about by the melting of the snow surface and by puddling on sea ice. These middle values of albedo are to a certain extent false, since they represent areal means of low reflection from water bodies and higher values of reflection from ice.

September and October are associated with seasonal transition. Ground vegetation dies off and, except in close-crown coniferous forest, produces a small rise in albedo. The first powderings of snow appear both on land and on thin new sea ice. The distribution quickly changes to the "black and white" winter pattern already stabilized in November.

Thermal diffusivity

This coefficient determines the amount of heat which, with a given temperature gradient, can be transported to or from the surface. It is highly variable in water, depending on the thermal stability and turbulence in the water. Observations from "North Pole" 4 gave

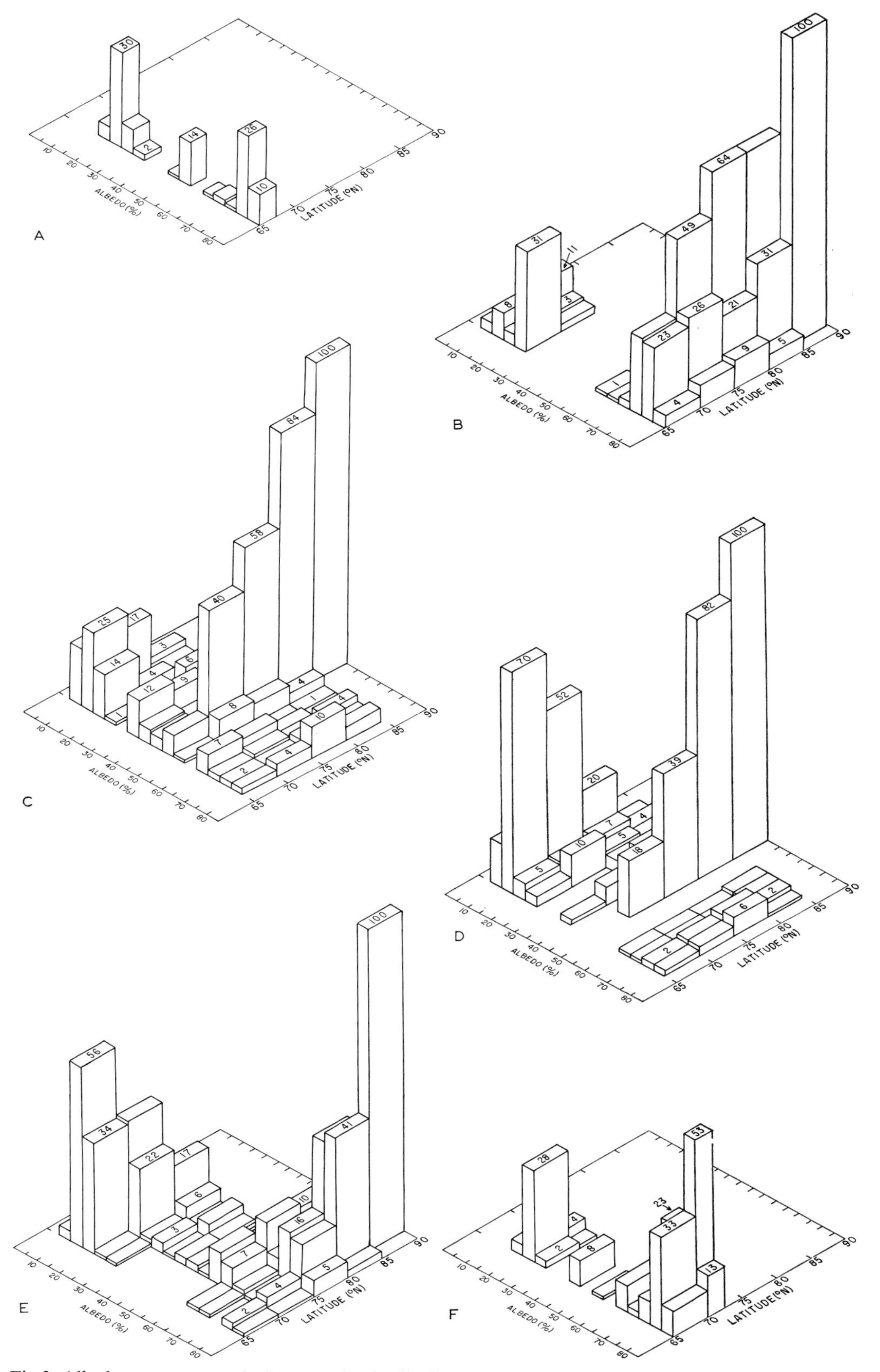

Fig.3. Albedo stereograms. A. January; B. April; C. June; D. August; E. September; F. November. (After LARSSON, 1963.)

values of about 20 cm^2/sec. For the northern Chuckchi Sea summer values of 0.1–1 cm^2/sec have been reported. Such low summer values must be explained by slow ocean current speed and the great stability due to production of fresh water by the melting ice. From oceanographic considerations, values of 0.2–1.1 cm^2/sec have been obtained in the Beaufort Sea (COACHMAN and BARNES, 1961).

It is apparent that this quantity will be highly variable with locality and season for open water, being lowest in late summer with much fresh water and a stable thermal stratification. It will be larger on the European–Asiatic side, where the ocean currents are stronger, than in the central Polar Ocean and on the American side.

The diffusivity changes sharply with the formation of ice. Since the water is now solid no turbulence takes place, but the conductivity varies with the temperature and salinity of the ice. Both the temperature and the salt content in the pack ice are variable, since the brine pockets in new ice gradually drain to the bottom. The result is that ice becomes less saline with age, and its age therefore becomes climatologically important. If a snow layer is found on the ice, the diffusivity is still further reduced, its value lying between 0.006 and 0.003 cm^2/sec. Sea ice, without snow cover, with a salinity of 2‰ and temperature of $-8°$, will have a thermal diffusivity of about 0.009 cm^2/sec (SCHWERDTFEGER, 1963).

Heat content of the ice

Since freezing of water releases about 80 $cal./cm^3$, the variations in amount of ice give the energy released or absorbed at the surface; this is an element of great significance for the climate. The mass of ice is given by its areal extent, which is fairly well known, and its thickness, which is only approximately known. Furthermore, the temperature distribution and the salinity are required, since the heat content of the ice is given by the following equation (UNTERSTEINER, 1961):

$$Q = \varrho \Delta T \left(0.5 + \frac{4.1S}{T_1 T_2}\right)$$

where: Q = the heat required to raise the temperature of the ice from T_1 to T_2; S = salinity in ‰; ϱ = density of ice (0.913 g/cm^3), and ΔT = temperature difference $(T_2 - T_1)$.

It is impossible, at present, to represent all physically significant surface parameters in the form of climatic maps. For the albedo areal evaluations of the relevant parameters have been attempted, but for thermal diffusivity and heat content no similar attempts have been made. Little more can be said about the conductivity of the open water, apart from the following statements:

(*1*) The values will be significantly higher in winter than in summer. The variation is larger than in normal ocean areas since, in addition to the stabilising effect of surface temperature increase in summer, the accumulation of fresh water on the surface, from melting and run off, will work in the same direction.

(*2*) It will be highest over the Gulf Stream in the Norwegian Sea and in the strong currents between Greenland and Spitsbergen. The lowest values will probably be reached in the slow moving water masses of the Beaufort Sea.

(*3*) A band of sharply increased water turbulence is found along the Eurasian shelf line, created by the submarine topography. Evidence of this is the relatively thin ice cover in this area and the frequent formation of leads and polynya (SYCHEV, 1960).

For the ice and the ocean as a whole, only indirect evidence is available: ice extent, thickness and age. These can be qualitatively used as indications for conductivity and heat content.

The mean areal extent of the pack ice is reasonably well known. Maps, giving extent, and concentration, have been published by various agencies (DEUTSCHES HYDROGRAPHISCHES INSTITUT, 1950; U.S. NAVY HYDROGRAPHIC OFFICE, 1958; SWITHINBANK, 1960). For individual years and months, data are published by the Danish Meteorological Service (DANSK METEOROLOGISK INSTITUT, 1900–1939, 1946–1956, 1960–1968), mainly for the Atlantic sector, and more recently by Canada (DEPARTMENT OF TRANSPORT, 1961–1968) for the American sector.

The total ocean area, north of 65°N, is about $14 \cdot 10^6$ km^2. The area covered by ice, of any concentration, amounts at its maximum to 83% and at its minimum to 54% of the total. To obtain the real amount of ice, the ice concentration must be considered. The seasonal variation of total ice area of concentration 1.0 is given in Table IV. The ice area remains rather constant from January to the end of May, while the rest of the year is characterized by rapid changes leading to a minimum in August–September, with a subsequent rapid rise.

Table IV also gives the monthly ice areas for certain sections of the Polar Ocean. As will become apparent in subsequent discussion, the area of open water within the pack ice shows pronounced features of its own in its energy budget. Table IV therefore also presents the open water area found within the pack ice belt.

The second important ice parameter is thickness. For the pack ice of the Polar Ocean, the main information of the variations of thickness with time are observations from the "Maud" expedition and the drifting stations. These data have been considered by various authors (PETROV, 1954–1955; SCHWARZACHER, 1959), and the results can be summarized as follows:

(*1*) The pack ice does not reach its final equilibrium thickness within one year, but only after 5–6 years.

(*2*) The rate of accretion during the freezing period is dependent on the ice thickness in September and is fastest in the first year.

(*3*) Summer melting takes place mainly at the top of the pack ice, and freezing at the bottom. The result is that ice moves gradually through the pack ice from the bottom to the top, an individual ice crystal having a lifetime of about 5–6 years.

(*4*) The period of ice accretion is considerably longer than the melting period, lasting until May or even June, so that bottom accretion still takes place at a time when the pack ice area is already diminishing.

Using data from the drifting stations on terminal thickness and time of maximum and minimum thickness, and a linear interpolation, the monthly ice thicknesses for ice older than 1 year are given in Table V. (After VOWINCKEL, 1964a.)

For one-year old ice, i.e., ice which does not last longer than about 12 months, the age-thickness relation is quite different. To calculate heat conduction and changes in ice volume, the areal extent of the different age groups of the ice has to be estimated. The melting in the central Polar Ocean, in summer, amounts to about 50 cm and in the

TABLE IV

ICE AREA (I) OF CONCENTRATION 1.0, AND OPEN WATER (W) WITHIN THE PACK ICE ($km^2 \cdot 10^3$)

Region	Jan.		Feb.		Mar.		Apr.		May		June		July		Aug.		Sept.		Oct.		Nov.		Dec.	
	I	W	I	W	I	W	I	W	I	W	I	W	I	W	I	W	I	W	I	W	I	W	I	W
Norwegian–Barents Sea	1,600	150	1,756	85	1,901	77	1,736	140	1,600	147	1,290	305	856	510	256	362	214	217	613	322	1,223	205	1,475	105
West Siberian coast	472	15	472	15	472	15	472	15	472	15	472	15	383	59	182	140	152	100	370	92	472	15	472	15
East Siberian coast	743	23	743	23	743	23	743	23	743	23	741	25	635	105	318	148	481	150	670	65	743	23	743	23
Beaufort Sea	407	13	407	13	407	13	407	13	407	13	382	38	357	21	151	89	134	78	349	41	407	13	407	13
Canadian Archipelago	941	29	941	29	941	29	941	29	932	38	915	47	871	68	665	175	607	115	886	69	937	33	941	29
Davis Strait–Baffin Bay	693	20	722	22	726	20	722	34	631	47	479	72	242	116	63	120	30	159	149	50	355	51	663	39
Central Polar Ocean	5,320	61	5,320	61	5,320	61	5,320	61	5,320	61	5,320	61	5,209	172	5,193	188	5,190	191	5,208	173	5,320	61	5,320	61
East Greenland Sea	757	31	807	21	908	31	907	41	733	48	718	44	499	117	230	194	242	103	334	45	391	44	527	33
Total Arctic Ocean	10,933	342	11,168	269	11,418	269	11,248	356	10,838	392	10,317	607	9,052	1,168	7,058	1,416	7,050	1,113	8,579	857	9,848	445	10,548	318

TABLE V

THICKNESS OF ICE (cm) OF DIFFERENT AGE GROUPS, CENTRAL POLAR OCEAN

	Sept.	Oct.	Nov.	Dec.	Jan.	Feb.	Mar.	Apr.	May	June	July	Aug.
1 year old	0	34	67	102	135	168	203	236	270	258	245	233
2 year old	220	230	240	250	260	270	280	290	300	288	275	263
3 year old	250	259	268	276	287	296	304	312	320	308	295	283
4 year old	270	277	284	291	297	304	311	318	325	313	300	288
5 year old	275	282	289	296	302	309	316	323	330	318	305	293

TABLE VI

AREA COVERED BY ICE OF VARIOUS AGES (PER CENT), CENTRAL POLAR OCEAN

Ice	Area (%)
1 year old	11.6
2 year old	10.3
3 year old	9.1
4 year old	8.1
5 and more years	60.9

The oldest ice is about 19 years old—2% of the area is occupied by this age group.

peripheral areas towards the American and Eurasian continents probably not more than 1.5–2.0 m. The areas of the Polar Ocean, ice free in summer, can therefore on the average contain only 1-year old ice, as 2-year old ice would be too thick to melt within the summer. This is substantiated by reports from the Canadian side (SWITHINBANK, 1960) that the bulk of the ice in the temporarily ice free areas is only 1-year old ice. However, the old ice, which is left at the end of summer, is not all older than 5 years since ice export takes place from the Polar Ocean. If it is assumed that this export is non-selective between new and old ice, it can be calculated how much of the area is covered by ice of the various age groups. The resulting age distribution for the area of permanent ice cover is given in Table VI. (After VOWINCKEL, 1964a.)

Using these data, the volume of ice can be calculated. The results are given in Table VII. It should be noted that these volume data refer to flat ice surfaces. In reality, however, large areas of the Polar Ocean are covered by hummock ice, the formation of which was described by ZUBOV (1943). Over the open ocean the ice can be piled up to 6–7 m above the water surface, and in the areas near coasts and shallows the pile-up can reach up to 13 m. Unfortunately, no areal estimates of these hummock regions are available. It seems, however, that more than 10% of the area might be covered by such ice. Since only 1/10 of the ice is above the water, it will be appreciated that the total volume of ice in the Polar Ocean will be significantly influenced by the assumptions made about extent and thickness of these areas.

Some information about the great irregularities of the ice thickness has been obtained from submarine cruises. Cross sections of ice draft in the Polar Ocean indicate that uniform ice thicknesses, of the magnitude expected from normal freezing, are the ex-

TABLE VII

ARCTIC OCEAN ICE VOLUME ($km^3 \cdot 10$); FLAT ICE SURFACE

	Jan.	Feb.	Mar.	Apr.	May	June	July	Aug.	Sept.	Oct.	Nov.	Dec.
Volume	2,406	2,627	2,720	2,740	2,699	2,456	2,043	1,755	1,678	1,708	1,967	2,178
Monthly change	+221	+93	+20	−41	−243	−413	−288	−77	+30	+259	+211	+228

ception rather than the rule. The estimate of volume given in Table VII is therefore certainly a lower limit. The average thickness of ice, obtained from a mean submarine cross section, is 3.7 m in August—at least 1 m more than indicated by the observations from the drifting stations. For August this would amount to approximately 7,000 km^3 more ice, i.e., between 30 and 40% more than the volume given in Table VII.

The amount of this additional hummock ice will most certainly vary with the season, being at its maximum in winter-spring and at its minimum in fall. Without more data no further assessments can be made. It is most likely that these hummock ice areas have a significantly lower thermal conductivity than normal ice, due to air spaces between the ice blocks. However, no measurements are available for these areas.

In summary: information about conductivity and ice volume, on an areal basis, is tentative and refers only to simplified conditions. Conclusions about the role of the ice in the energy budget, based on spot observations, can therefore be no more than approximate statements.

Clouds and radiation

The radiational climate of the Arctic (including solar, terrestrial, and atmospheric radiation) generally is much less well known than is the case for lower latitudes. As recently as the 1930's in the U.S.S.R., and in the 1950's in other countries, the measurements of radiation were performed on various expeditions but not as regular meteorological elements at permanent stations.

In the Soviet Arctic there are now more than five actinometric stations with an observational period of up to 20 years and many more with a period of less than 10 years. For the Polar Ocean proper, however, observational records are sparse and confined to particularly solid ice floes. In this region it is therefore necessary to use also indirect indications of the radiation climate. The most important of these are cloud amount and type.

Cloud amount and type

Climatic investigations of cloud conditions are relatively rare in meteorological literature. This is unsatisfactory as the clouds give, better than most meteorological elements, an indication of climatic conditions and processes in the atmosphere and are important in the radiation balance. In polar areas the observation of cloud amount and type is even more difficult than in other latitudes, for two reasons:

(*1*) The long polar night, when distinction between cloud types is very difficult.

(*2*) The low temperatures, which produce clouds of rather low density. This renders difficult the distinction of cloud types and amounts. From evaluations of observations it seems that it is especially difficult to separate stratus and altostratus (Vowinckel, 1962; Vowinckel and Orvig, 1962a).

Two basically different causes for clouds and their distribution can be distinguished:

(*1*) *Dynamic causes.* The vertical motion associated with pressure systems favours cloudiness under cyclonic and clear sky under anticyclonic conditions.

(*2*) *Geographic causes.* These usually work via the water—or radiation balance. Moist

surfaces with high evaporation rates favour cloud formation due to the high moisture content of the air, even under anticyclonic conditions, while the opposite holds for dry surfaces. A strong positive radiation balance favours cumulus development, to a large degree independent of the dynamic conditions.
A variety of cloud patterns will be caused by the interaction of these two influences. The following main types can be distinguished in the Arctic.

Norwegian-Sea type

In the west wind zone over the ocean all influences favour cloud formation. The cloudiness is high all through the year, with a slight maximum in summer (Fig.4). It is noteworthy that, even in this area which is so clearly governed by west wind cyclones, the maximum cloudiness is experienced in summer, the season with the lowest cyclone

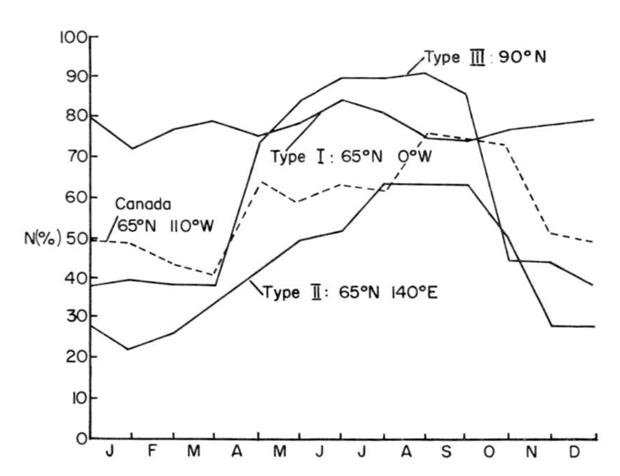

Fig.4. Regional cloudiness types.

frequency (65°N 0°W: 84% in July; 72% in February). The cloud types of this area also show peculiar behaviour. In winter the cumulus group reaches frequencies of 30–40%, more than that observed in any season anywhere else in the Arctic. The observations from the Norwegian coast indicate that the most important cloud in this group is the very intense cumulonimbus. In summer this dominance disappears, mainly in favour of altocumulus plus altostratus.
The reason for this development lies in the warm water of the North Atlantic Drift and the frequent outbreaks of arctic air masses over this area during the winter, in the rear of cyclones. Extreme instability is created, with a consequent development of convective cloud. As convectional clouds produce less cloudiness than do stratiform clouds, the cloudiness tends to diminish. Warm water and cold air masses combine to reduce the winter cloudiness below the conditions typical for the west wind zone.
In summer, with the water relatively much colder, and with no extremely cold air available, the cumulus frequency and cloudiness approach the more normal patterns for the west wind zone: cloudiness around 80%. Due to less cyclonic activity, however, a great proportion of the clouds is found in the medium cloud layers with altocumulus.

East-Siberian type

The whole of Siberia is dominated in winter by anticyclonic influences. The region of especially clear skies is not found in the zone of highest anticyclonic frequency at 100°E, but rather towards the east, around 140°E. This is contrary to the conditions in the maritime sub-tropical cells, where the eastern region of the high cell shows the highest cloud amounts. The distance from sources of moisture must be responsible. The Polar Ocean is no source of moisture in winter, no penetration takes place from the Pacific, and the moisture from the Atlantic source has been exhausted. Therefore, the characteristic stratus or stratocumulus formations under an inversion are conspicuously absent in this area. The clouds which exist are overwhelmingly cirrus (nearly 60%); even at the height of middle clouds the moisture content drops to low values, and only the odd cyclone is able to cause condensation.

The maximum cloudiness is reached in summer. Anticyclonic conditions are then much less frequent, and the Polar Ocean has stretches of open water and serves as a moisture source. The frequency distribution of cloud types is rather uniform with a predominance of the convective types in the low and medium layers, a development characteristic of all continental arctic areas in summer.

The change from winter to summer conditions is gradual, while the decrease in autumn is abrupt. This difference between spring and autumn has little to do with the pressure distribution. The pressure curve (and also the temperature curve) is symmetrical in spring and autumn. The reason must be the availability of moisture. As the snow cover is very thin in this area, the main moisture source during spring and summer will be the Polar Ocean. The melting of the ice at the coast, and the formation of water puddles on the remaining ice, is a gradual process and accordingly the cloud amount increases gradually in eastern Siberia. In autumn the freeze-over is quick and a sharp decrease in moisture supply occurs rapidly.

It is noteworthy that this type is not as clearly developed in Canada. Fig.4 shows that, in spite of the large land area and the sheltering effect of the Rocky Mountains, the winter minimum is not as pronounced as in eastern Siberia. Contrary to the Siberian conditions, however, the increase during spring is quite rapid. This is not caused by extensive stratus layers as over the Polar Ocean (see below), but by clouds of the medium types which would indicate a dynamic reason for this development.

Polar-Ocean type

Over the Polar Ocean the cloud amount is lowest in winter and spring and at its maximum in summer. Dynamic factors have their least influence in this area. The polar frontal zone is, in fact, best developed in winter when the cloud amount in this area is low. During winter the water content of the very cold air is too low for cloud formation, irrespective of the dynamic conditions.

In summer, the surface characteristics determine the cloud conditions, causing the most outstanding feature of the Polar Ocean area: the high frequency of stratus and stratocumulus. It rises to 80%, and stratus predominates. The reason is the continuous cooling of air to 0°C over the pack ice fields. These summer clouds are extremely uniform, extending as large sheets over much wider areas than other clouds. The mean horizontal extent during June–July is (DOLGIN, 1960): stratus, 460 km; stratocumulus, 430 km; altostratus, 370 km; altocumulus, 290 km.

In extreme cases 2,000 km are reached. The mean thickness of stratus and stratocumulus is 350 and 540 m respectively (DOLGIN, 1960), rather high values when the manner of their formation is considered. For radiation climatology it is important to note that the water content of these clouds shows a pronounced decrease from the coast towards the Pole.

A further characteristic of this area is that the seasonal change in cloudiness is restricted to a very short transitional period, while rather constant cloudiness prevails during the rest of the year. Similar abrupt changes are found only in monsoon areas with a complete changeover in circulation pattern.

Seasonal cloud conditions

Fig.5–10 show cloud amounts and types. Winter is the season with the greatest mean deviation from latitudinal average.

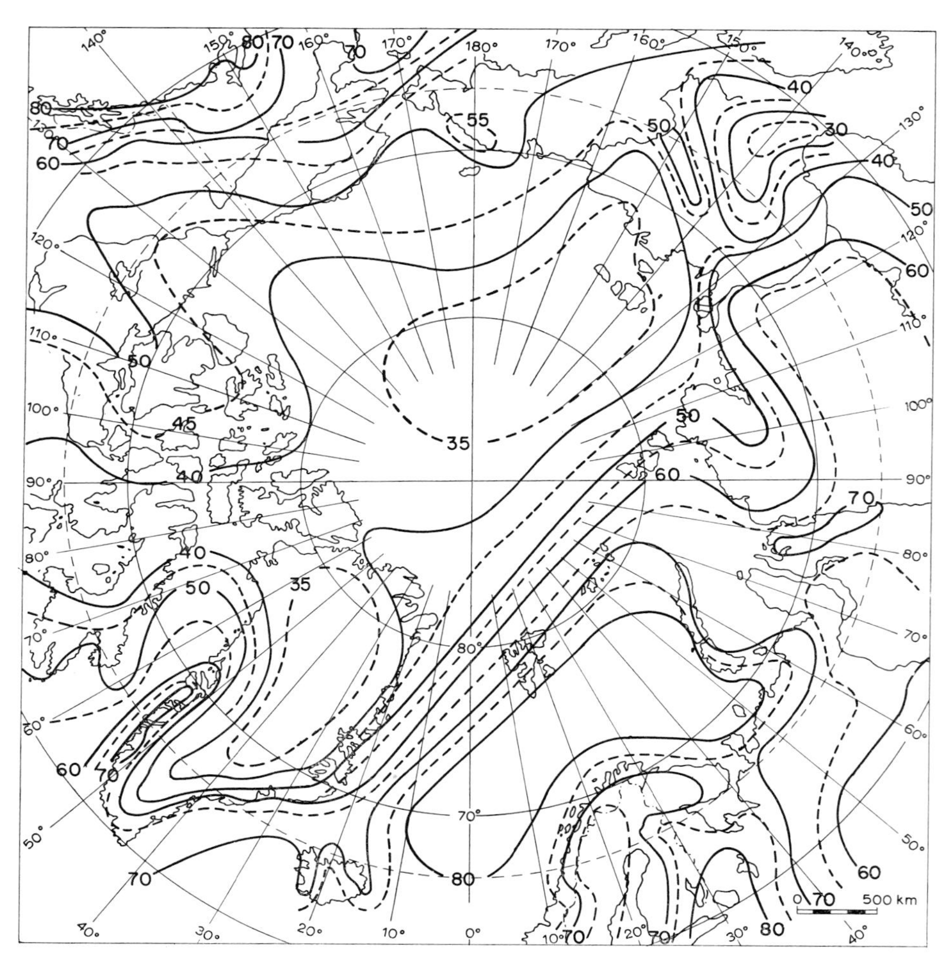

Fig.5. Mean cloud amount (%) in January.

An outstanding phenomenon of winter is the zone of high cloudiness in the Norwegian Sea. It extends along the Eurasian coast far into the Polar Ocean, until it gradually loses its identity. From its centre line, Thorshavn–south of Bear Island, where the described phenomena of the zone are most conspicuous, it changes its characteristics towards the east in the Polar Ocean. Cumulus clouds disappear in the Barents Sea, and stratus and stratocumulus become less frequent. From the Laptev Sea onwards, middle cloud layers are most important, with altostratus being especially frequent. With practically no moisture supplied from the frozen ground, cyclones penetrating to the east are only able to cause condensation in the higher levels. From the Norwegian Sea northwestwards to the ice rim, cumulus is replaced by stratus and stratocumulus as the water becomes colder and the time available to create unstable air becomes shorter.

Disregarding the discontinuity created by the Scandinavian mountains, Europe and western U.S.S.R. show very high cloud amounts, with a domination of low clouds. Medium clouds become prevailing in western Siberia, much farther west than in the Polar Ocean. They are replaced by cirrus from central Siberia onwards. The reason for

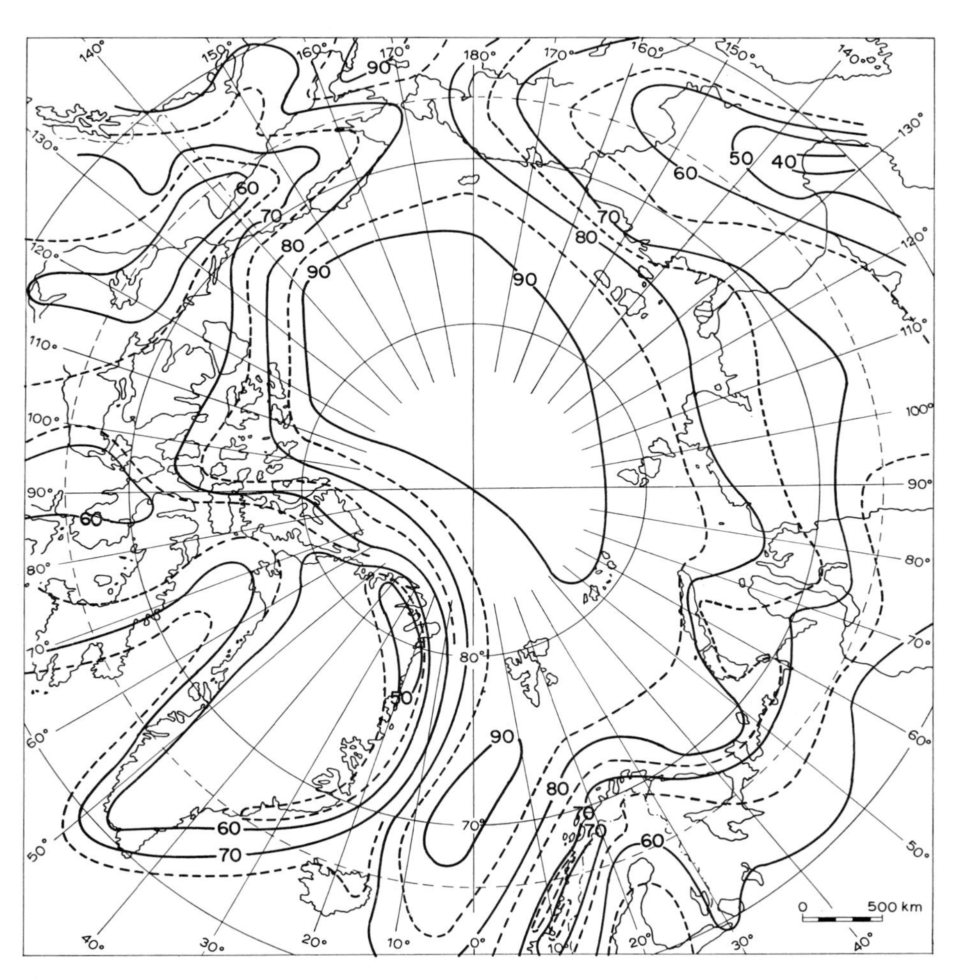

Fig.6. Mean cloud amount (%) in July.

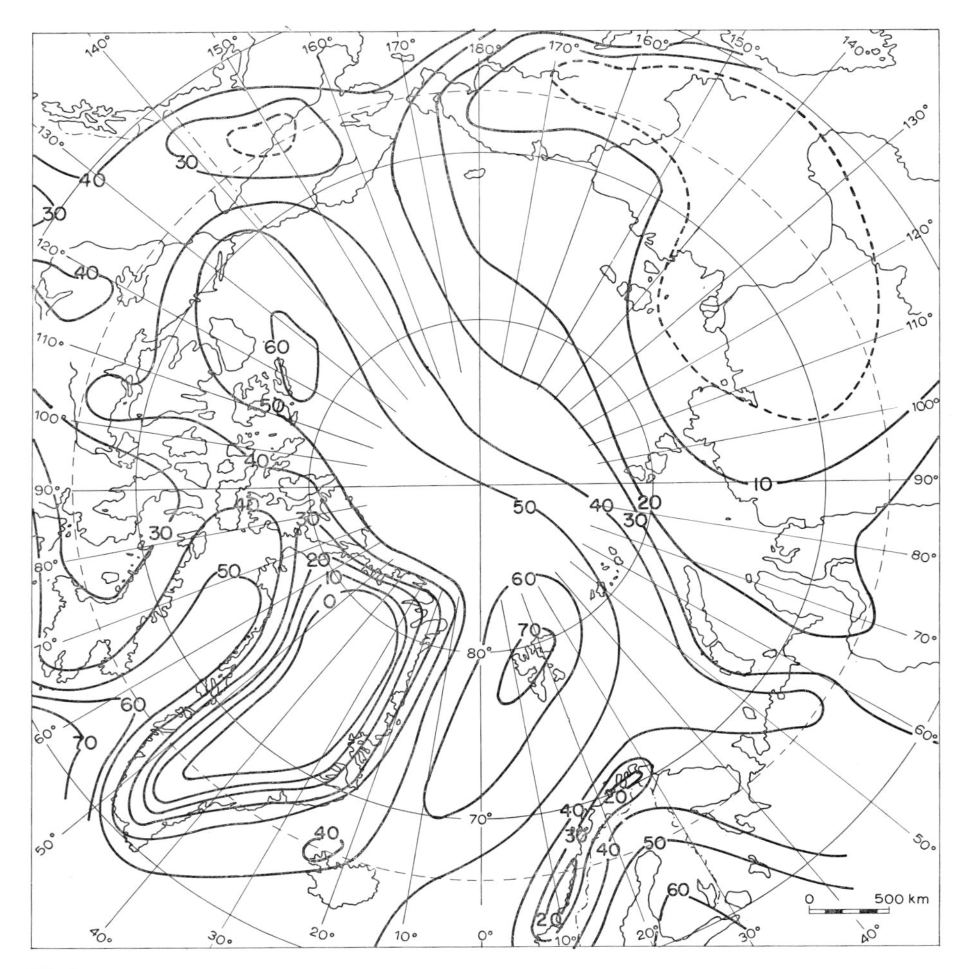

Fig.7. Frequency of St and Sc (%) in winter.

this rapid thinning of clouds over the continent is the position of the Siberian anticyclones, which divert the disturbances towards the Polar Ocean.

When considering this distribution of cloud amount and type, it will be seen that it is practically the idealised picture of a cross section through a middle latitude cyclone on an immensely extended scale: cumulus in the rear, a large sector of broken stratiform clouds merging gradually into medium and high layers farther east, until hardly any clouds are left in eastern Siberia.

During winter, when outgoing radiation predominates, this distribution of clouds gives an even more negative radiation balance in the east than the cloud amount alone would indicate. The Atlantic sector has not only a high cloud amount but thick clouds as well, while the thickness decreases to the east with diminishing cloud amount.

The Stanovoi Mountains form the greatest climatic barrier in the Arctic. Moving from eastern Siberia with clear skies, the Bering and Okhotsk seas again show very high cloud amounts with a dominance of low cloud types. No such clear pattern develops over the American continent. Apart from minor irregularities, the main feature is the gradual decrease in cloudiness to the north. There is a zone with few low clouds along

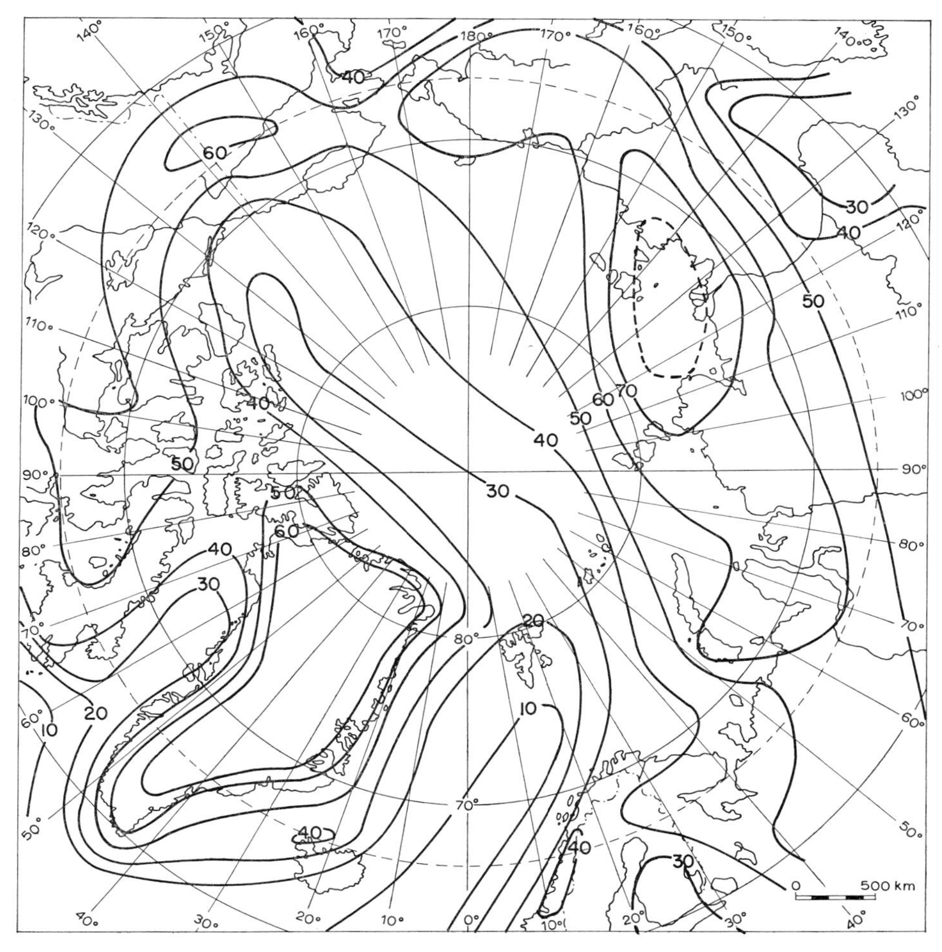

Fig.8. Frequency of As and Ac (%) in winter.

the 65°N parallel, with a minimum in the interior Yukon Valley of Alaska. Mountain ranges there hinder free flow of air in the meridional direction. Another minimum occurs over the Canadian Archipelago, where the lowest cloud amounts are observed. This same zone is characterized by a maximum in the medium clouds, with a much higher frequency of altocumulus than in the Eastern Hemisphere.

The influence of the Aleutian low and the Pacific frontal zone is hardly noticed in the Arctic. The mountain ranges form an effective barrier, and over the frozen surfaces of northern Canada the air does not pick up sufficient moisture, after the descent to the plains, to show another increase in cloud amount.

With the approach of summer the Atlantic zone loses much of its individuality. It appears merely as an extension of the large area of cloud now covering the whole Polar Ocean. It can be distinguished from the latter region only by the higher frequency of cumulus-type clouds. Although the reasons for the formation of these zones are different, in appearance there is not much to choose between the two, both having extremely dull and monotonous skies.

At the continental shores both influences come to a rapid end. The farther south, the

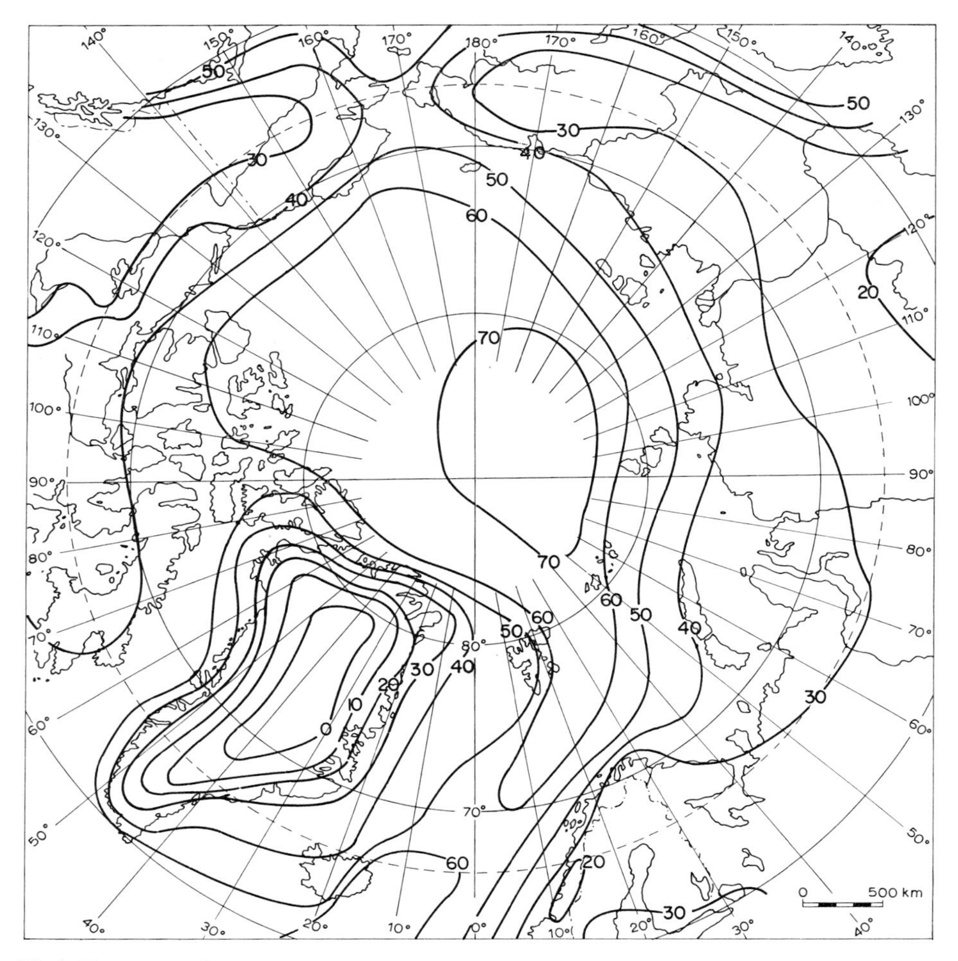

Fig.9. Frequency of St and Sc (%) in summer.

less cloudy the skies and the more is stratus and stratocumulus replaced by cumulus. As over the Polar Ocean, the maximum cloudiness over the continent is reached in summer. The difference from the Polar Ocean is firstly that the mean cloud amount remains lower, and secondly that a much greater diversity of cloud types exists.

The cloud maximum in summer is a peculiarity of the Arctic and contrary to the conditions in temperate latitudes. The reason does not lie so much in the different dynamic conditions as in the ground influences. In the Arctic the ground is frozen everywhere in winter and, due to the absence of sufficient incoming radiation for evaporation, the supply of moisture to the atmosphere is limited. In summer, the incoming energy and the largely moist surfaces permit high evaporation, and sufficient moisture is available for cloud formation. In temperate latitudes solar energy is received also in winter, and evaporation is not suppressed to the same extent. In summer, the incoming energy is high in more southerly latitudes and the evaporation will dry the soil, with the result that the relative humidity of the air is lower than in winter and the cloud amount is decreased.

Fig.11 shows the latitudinal means of cloudiness for three latitudes. While the northern

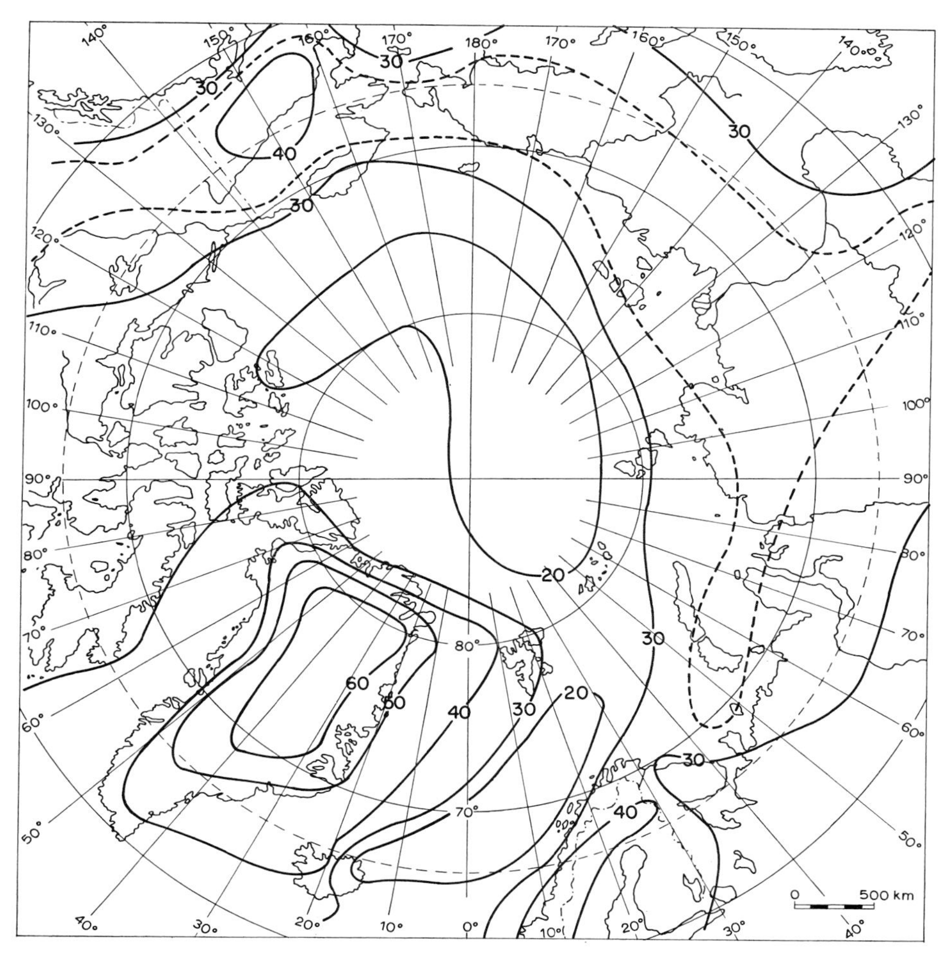

Fig.10. Frequency of As and Ac (%) in summer.

values are entirely dominated by the Polar Ocean type, the southern values are more complicated, reflecting mainly the continental behaviour of the East Siberian type. The Norwegian Sea type is too small in areal extent to have much influence. The seasonal variability is seen to increase from south to north. This is the result of the increasing influence of the peculiar cloud regime of the Polar Ocean. Latitudinal means of cloud type frequency show that the north has also the highest seasonable variability of cloud type. The farther north, the higher the cloudiness in summer and the more frequent the low and dense cloud types, while in winter the cloudiness diminishes northwards and the frequency distribution of types is more uniform.

A complete changeover of north–south gradient in cloudiness is produced by this high variability in the north and the rather stable conditions in the south. During the winter there is a continuous decrease to the north, while an increase is found in summer. The changeover takes place very rapidly, with the result that only during the short transition period in spring (May) and in autumn (October) is there little difference in the average cloud amount over the whole arctic sector.

These results seem to be unfavourable for the radiation balance, as at the time of highest solar radiation the north is definitely at a disadvantage. In reality this is not the case, as the higher transparency of the clouds in the north counteracts the cloudiness, and the ratio of clear sky radiation to actual radiation received at the ground varies only to about 75°N and remains fairly constant further polewards.

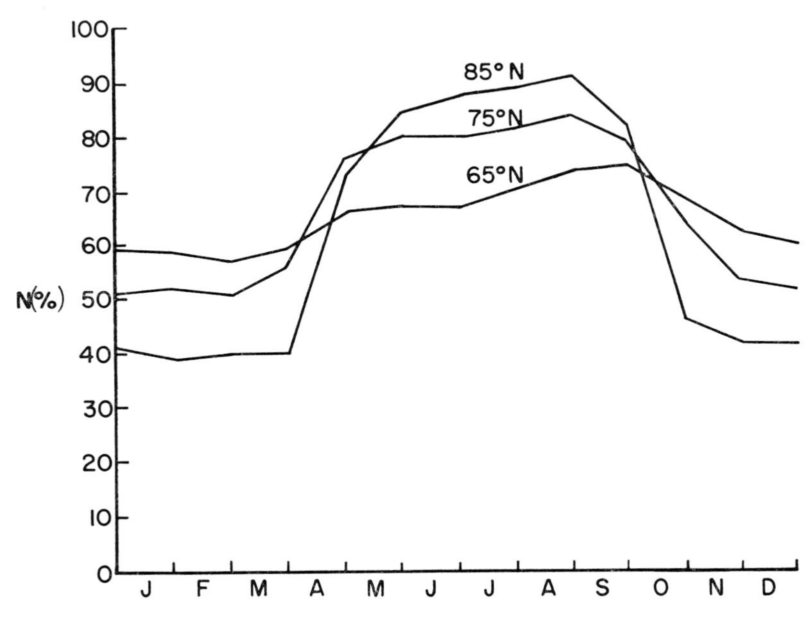

Fig.11. Latitudinal means of cloudiness.

Cloud effects on solar radiation

Various influences determine the transmissivity of clouds. Very important are albedo of the cloud top and the solar altitude. For cloud albedo, values show wide variations with significant changes depending on cloud thickness. Arctic data show, for all stations and cloud types examined, a marked increase in radiation income at the ground with increasing solar elevation.

In general, this decrease in the depletion of radiation with increasing solar elevation must be caused by a decrease in the albedo of the clouds with increasing altitude (smaller angle of incidence). The effect is much too large to be accounted for by a decrease in the absorption in the cloud. As the effect is great in the Arctic and increases towards the Polar Ocean, and is greatest in winter, it is possible that the cause may be associated with lower cloud temperatures. In this case, it should also be apparent in temperate latitudes when comparing summer and winter conditions. All available comparisons show that a slight seasonal change does take place in temperate latitudes. However, the difference remains far smaller there than at the Arctic coast or over the Polar Ocean. Therefore, although temperature seems to play a role, it cannot be the only cause of such change in depletion.

In general, the cloud top albedo varies with solar altitude, as does the albedo of a water surface, with the result that the transmitted radiation increases with solar altitude. The cloud top albedo is dependent upon the cloud temperature, being lower with low temperature. This is seen clearly in high latitudes. In temperate latitudes, especially in continental areas, convectional processes during the day tend to decrease the transmitted energy due to greater optical thickness. Conditions over the ocean are similar to those in the Arctic, and the small variations over the continent in temperate latitudes are caused by diurnal variations in cloud thickness.

The optical thickness of clouds must increase sharply from the Polar Ocean to the south, and from winter to summer. The extensive single-layer stratus fields are common in the Arctic, while farther south a greater proportion of the clouds are caused by dynamic processes which favour cloud layers of greater depth. The variation in transmissivity is lowest in cirrus, and the range in transmissivity is greatest where the thickest clouds occur. Therefore, for calculation of cloud transmissivity, a knowledge of the distribution of different cloud types is particularly important in areas of thick clouds. The following points are important:

(*1*) The transmissivity of a particular cloud type varies markedly with the seasons, being considerably higher in winter.

(*2*) The transmissivity of the clouds shows a high variability from place to place. For a particular cloud type, this geographical variation is greater than differences between various cloud types (low and medium) at one particular station.

(*3*) The transmissivity increases northwards.

(*4*) The transmissivity increases with solar elevation. This increase is small in middle latitudes and greater in high latitudes.

TABLE VIII

COMPARISON OF RADIATION AT THE ARCTIC COAST AND OVER THE POLAR OCEAN (AVERAGES FOR SOLAR ELEVATIONS FROM 7.5° TO 22.5°). ALSO, THE COMPARISON OF RADIATION IN SUMMER (JUNE–SEPT.) AND NON-SUMMER, ARCTIC COAST

Cloud type:	Ci.	Ac.	As.	Sc.	St.	
Polar Ocean	26	21	20	18	15	(1/100 Ly/min)
Arctic coast, in summer	20	16	13	12	10	(1/100 Ly/min)
Arctic coast in per cent of Polar Ocean	77	76	65	67	67	(%)
Arctic coast, non-summer	26	23	20	17	15	(1/100 Ly/min)
Arctic coast, summer in per cent of non-summer	77	69	65	71	67	(%)

Table VIII shows mean radiation income for various cloud types for two regions with different ground albedo. Due to multiple reflection between ground and cloud, this element will also determine the actual radiation income.

The effect of various cloud types on the clear sky radiation can be judged from Table IX for different latitudes and months.

TABLE IX

APPROXIMATE MEAN RADIATION INCOME, IN PER CENT OF CLEAR SKY RADIATION, WITH 10/10 CLOUD COVER

Lat. (°N)	Feb.	Mar.	Apr.	May	June	July	Aug.	Sept.	Oct.	Feb.	Mar.	Apr.	May	June	July	Aug.	Sept.	Oct.
	Stratus:									*Stratocumulus:*								
90			39	56	58	57	50	19				45	65	67	66	54	33	
85			41	55	59	58	50	26				47	63	66	65	56	37	
80			48	56	62	60	51	40				56	61	65	64	59	45	
75		42	53	56	57	54	49	42	27		50	60	61	60	58	55	50	39
70	38	50	55	56	53	49	45	40	36	52	56	63	63	57	53	50	46	45
	Altostratus:									*Altocumulus:*								
90			55	72	72	72	63	43				54	78	79	78	68	35	
85			57	70	72	72	65	47				57	76	82	80	69	41	
80			64	70	72	71	66	55				68	77	82	81	70	55	
75		61	70	70	65	63	61	57	51		64	76	79	79	77	74	66	51
70	62	68	75	72	59	55	53	52	55	73	75	82	82	77	74	72	67	63
	Cirrus:																	
90			72	91	91	91	83	60										
85			75	89	91	91	84	63										
80			83	88	90	89	85	74										
75		76	88	90	89	87	85	81	69									
70	79	81	92	94	90	87	84	80	78									

Insolation and absorbed solar radiation

The total incoming energy in the form of solar radiation depends also on daylength and solar altitude (Table X). The various latitudes therefore receive different amounts of insolation throughout the year.
The regional differences in solar radiation are quite marked, as the incoming radiation is affected by cloud type and amount which vary considerably from place to place. It is convenient to obtain first the solar radiation for cloudless sky and thereafter the actual conditions dependent on cloud conditions. Due to the paucity of observing stations, clear sky radiation has been calculated (VOWINCKEL and ORVIG, 1964a) for the various sections of the Arctic Ocean. A comparison of observed and calculated values is given in Table XI.

TABLE X

APPROXIMATE VALUES OF DAYLENGTH (a) AND SOLAR ALTITUDE AT NOON (b) FOR THE FIRST DAY OF EACH MONTH

		Latitude (°N): 60	65	70	75	80	85	90
Jan.	(a)	6 h 00 min	3 h 45 min	—	—	—	—	—
	(b)	6° 50′	1° 50′	—	—	—	—	—
Feb.	(a)	8 h 00 min	6 h 15 min	4 h 15 min	—	—	—	—
	(b)	12° 30′	7° 30′	2° 30′	—	—	—	—
Mar.	(a)	10 h 30 min	9 h 45 min	9 h 15 min	8 h 00 min	5 h 30 min	—	—
	(b)	22° 20′	17° 20′	12° 20′	7° 20′	2° 20′	—	—
Apr.	(a)	13 h 15 min	13 h 15 min	13 h 45 min	13 h 45 min	15 h 30 min	19 h 30 min	24 h
	(b)	34° 30′	29° 30′	24° 30′	19° 30′	14° 30′	9° 30′	4° 30′
May	(a)	16 h 00 min	17 h 00 min	18 h 15 min	22 h 00 min	24 h	24 h	24 h
	(b)	45° 00′	40° 00′	35° 00′	30° 00′	25° 00′	20° 00′	15° 00′
June	(a)	18 h 15 min	20 h 15 min	24 h	24 h	24 h	24 h	24 h
	(b)	52° 00′	47° 00′	42° 00′	37° 00′	32° 00′	27° 00′	22° 00′
July	(a)	18 h 45 min	22 h 15 min	24 h	24 h	24 h	24 h	24 h
	(b)	53° 10′	48° 10′	43° 10′	38° 10′	33° 10′	28° 10′	23° 10′
Aug.	(a)	16 h 45 min	18 h 30 min	22 h 00 min	24 h	24 h	24 h	24 h
	(b)	48° 00′	43° 00′	38° 00′	33° 00′	28° 00′	23° 00′	18° 00′
Sept.	(a)	14 h 15 min	14 h 45 min	15 h 30 min	16 h 45 min	20 h 00 min	24 h	24 h
	(b)	38° 20′	33° 20′	28° 20′	23° 20′	18° 20′	13° 20′	8° 20′
Oct.	(a)	11 h 30 min	11 h 15 min	11 h 15 min	10 h 45 min	10 h 00 min	7 h 15 min	—
	(b)	26° 50′	21° 50′	16° 50′	11° 50′	6° 50′	1° 50′	—
Nov.	(a)	8 h 45 min	7 h 45 min	6 h 45 min	2 h 30 min	—	—	—
	(b)	15° 40′	10° 40′	5° 40′	0° 40′	—	—	—
Dec.	(a)	6 h 30 min	4 h 30 min	—	—	—	—	—
	(b)	8° 10′	3° 10′	—	—	—	—	—

TABLE XI

COMPARISON OF OBSERVED AND CALCULATED SOLAR RADIATION FOR CLOUDLESS SKY[1]

Station	Mar.	Apr.	May	June	July	Aug.	Sept.	Oct.
Resolute	103	107	104	101	103	105	113	139
Aklavik	108	104	105	101	98	99	100	102
"North Pole" 85°N		105	109	105	103	96		
"North Pole" 80°N		106	112	106	104	97		
Isachsen Plateau (West Spitsbergen)					101			
Sveanor (Spitsbergen)					104			
"Maud"		108	108	109	118	101		

[1] Observed values expressed in per cent of calculated values.

Generally, the calculated values tend to be a little too low when compared with the observations. When considering the many uncertainties, however, one must regard the two sets of figures as corresponding reasonably well.

The cloudless sky radiation figures must next be corrected for effects of clouds. The corrected insolation values represent the direct solar and diffuse sky radiation as observed by instrument and received at the ground. Fig.12 and 13 show monthly mean maps of global radiation for March and June.

The minima of insolation are found, during all months, over the Norwegian Sea with its high cloud amount and cloud density and its high atmospheric water vapour content. The maxima are mostly over eastern Siberia between 140° and 160°E. It is remarkable that this eastern Siberian maximum exists through the whole year, although the anticyclone disappears during the summer; the extreme remoteness from open oceans is responsible. A secondary radiation maximum is experienced over Canada but without the same degree of stability. At 65°N, the maximum near 150°W is the last northward extension of the clear skies of the prairies. The main Arctic maximum, however, is found farther to the northeast, over the Canadian Archipelago. The main feature of the Polar Ocean is the markedly higher radiation in the western part, especially in the Greenland section. The global radiation values have rather limited significance, especially in polar regions, as the sum of direct solar and diffuse sky radiation indicates only the potentially available energy. Of greater interest is the amount of energy which is actually absorbed at the surface. It is obtained by multiplying the global radiation value by (1 — albedo). For most areas of the world this absorbed radiation will differ from insolation by a rather constant, relatively small amount as the albedo values are often almost constant for large areas. Compared to the albedo of snow and ice surfaces, all other albedos are small. However, in all regions with permanent or temporal snow cover, insolation and absorbed radiation will be significantly different, and this difference will vary sharply from season to season. Only absorbed radiation is of climatic significance. Smaller irregularities of the albedo over the continents and over island-water areas were smoothed out by overlapping averages for the grid points in such areas (LARSSON, 1963). Fig.14–18 give the distribution of short wave radiation actually absorbed at the surface.

The smallest difference between incoming and absorbed radiation is found in July

August and September along 65°N where the absorbed radiation is only 5–6% smaller than the insolation. Towards winter, and northwards, the differences become very great and the albedo becomes a factor of equal or greater importance than all other depletion factors.

The outstanding feature of all insolation maps is a pronounced minimum over the Norwegian Sea. This minimum disappears completely in the maps of absorbed radiation. In the latter maps, the main phenomenon is a band of extremely steep gradients which is apparent in all months but especially well developed in spring and early summer. Over the Atlantic sector this zone of high gradient follows the ice limit as is to be expected. More complicated conditions exist over the continental areas. The large zones of coniferous forests with low albedo cause rather higher absorbed radiation amounts to be found far northwards, beyond the snow line. Over the American continent, for example, where the area examined is essentially north of the tree line, the snow cover becomes significant. A pronounced maximum of absorbed radiation is found in the western part of the continent, where the coniferous forest reaches the coast. As surface conditions over the Polar Ocean are rather uniform, with only slight meridional variations, the wide region with slight radiation gradient in the insolation maps is sustained.

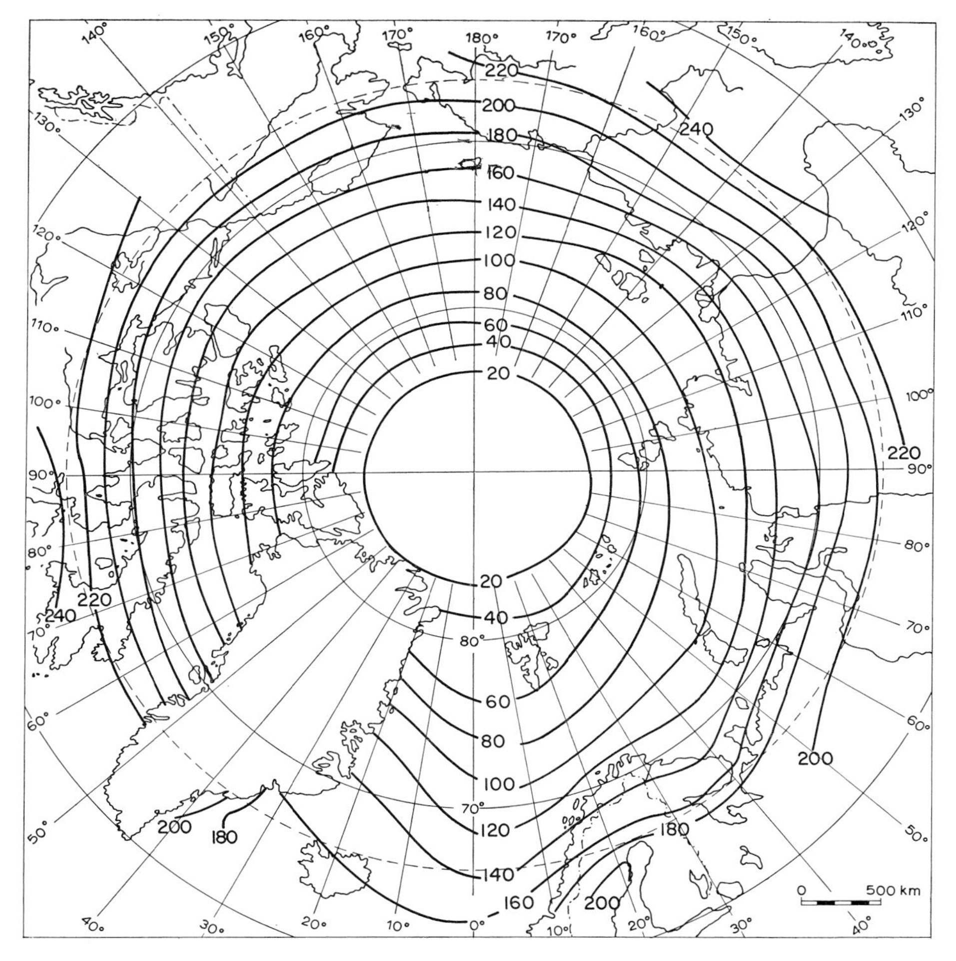

Fig.12. Global radiation of sun and sky (Ly/day) in March.

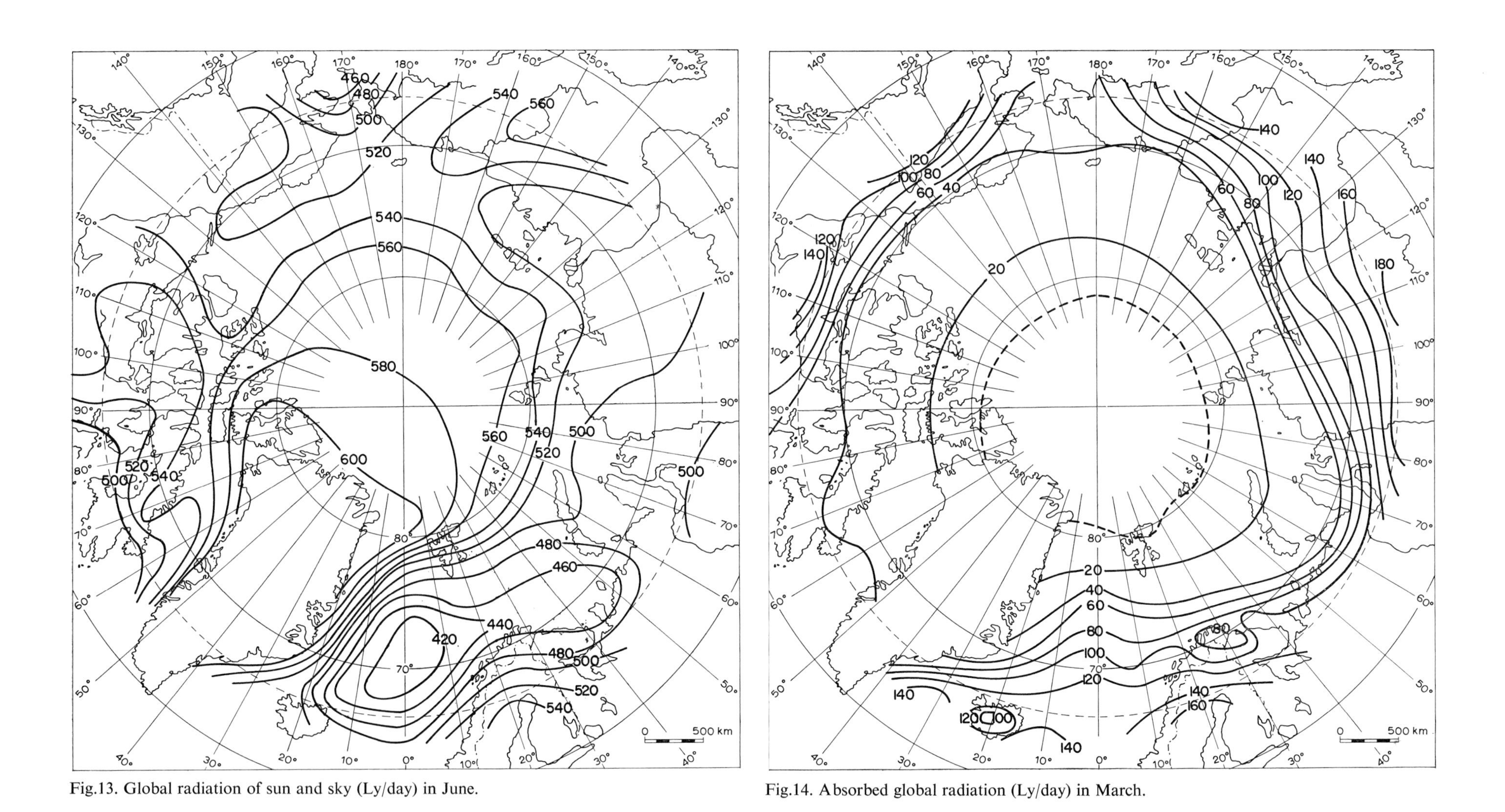

Fig.13. Global radiation of sun and sky (Ly/day) in June.

Fig.14. Absorbed global radiation (Ly/day) in March.

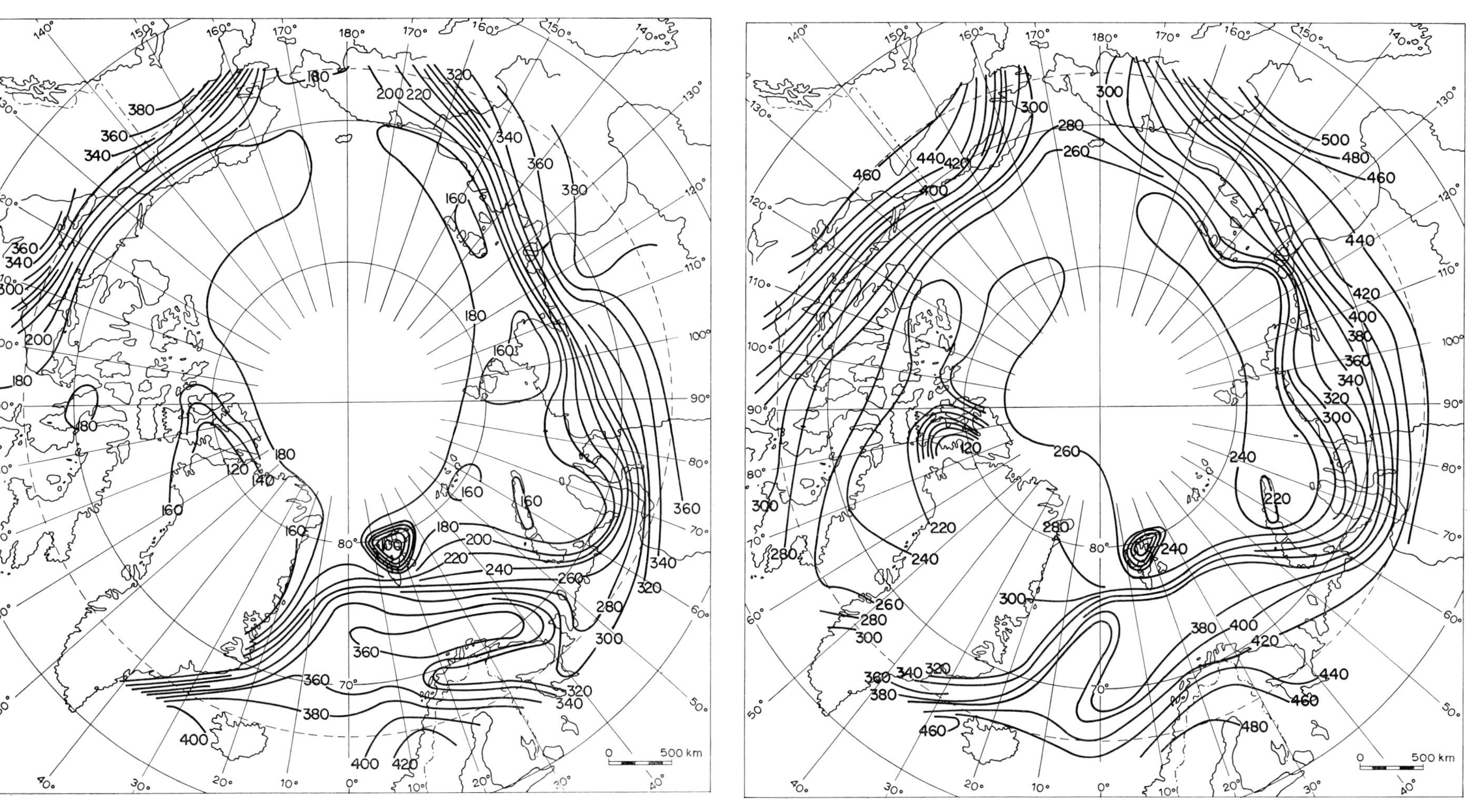

Fig.15. Absorbed global radiation (Ly/day) in May.

Fig.16. Absorbed global radiation (Ly/day) in June.

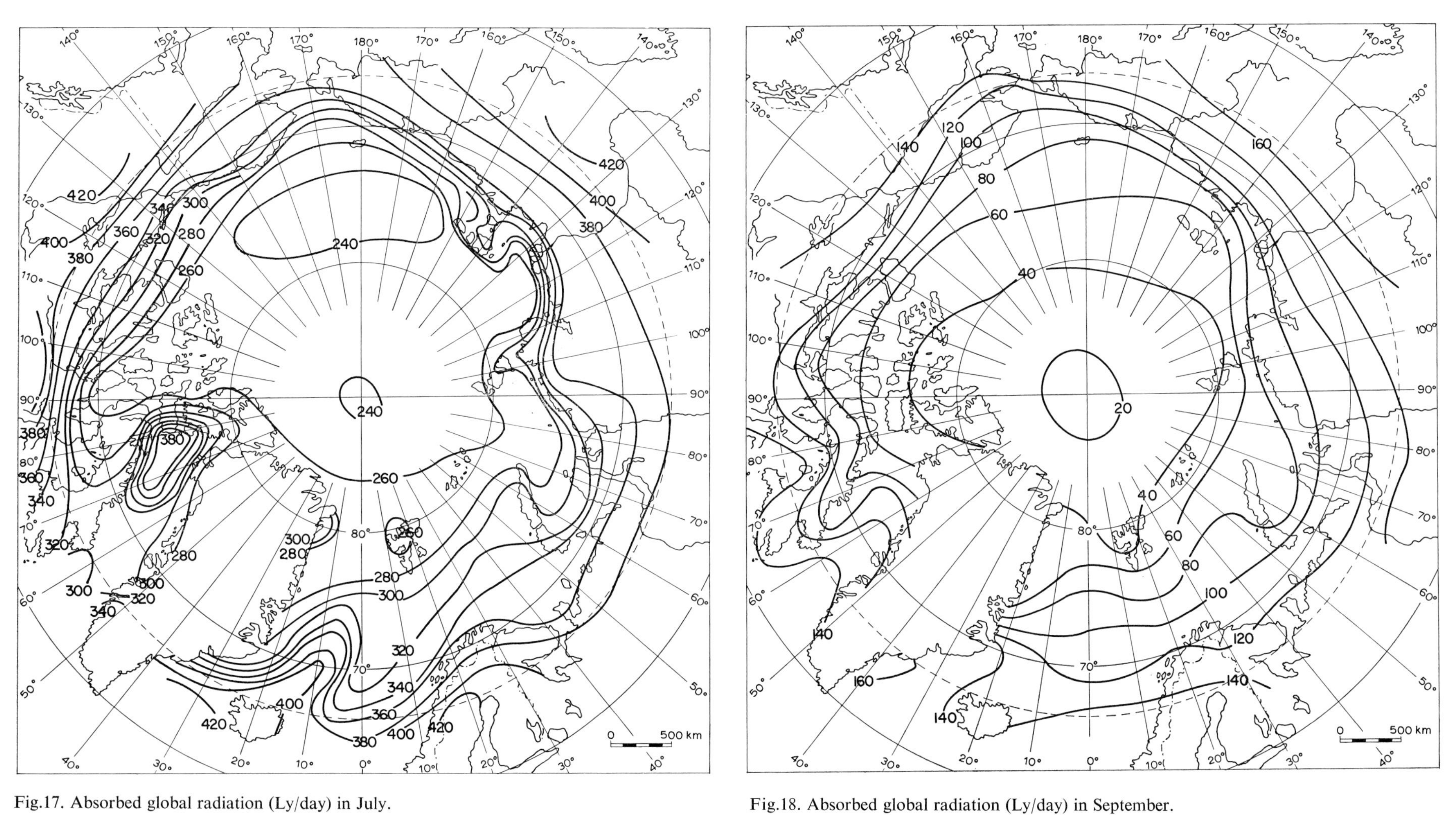

Fig.17. Absorbed global radiation (Ly/day) in July.

Fig.18. Absorbed global radiation (Ly/day) in September.

Regional differences are much greater in absorbed radiation than in incoming solar radiation. These differences along the latitudes are, in fact, higher than the meridional ones. As an example, the difference along 65°N for March is 136 Ly/day and for June 213 Ly/day, while the differences in the latitudinal mean for the two months for 65° and 70°N are only 55 and 74 Ly/day, respectively. In addition to these large-scale longitudinal variations, the variations from point to point are also significantly higher for absorbed radiation than for insolation. In this connection it must be kept in mind that the calculated values are based on smoothed albedo values, so that in reality the local differences are greater than indicated on the maps. Reasonably correct latitudinal means of radiation elements cannot be obtained from one or a few cross sections, but must be based on a rather dense network of stations or grid points.

Long wave radiation

The radiation energy budget comprises both short wave radiation and long wave radiation fluxes, the latter being infrared emission from the ground and similar emission from the atmosphere. The atmospheric radiation is also absorbed at the surface, representing another heat source. This long wave downward radiation ("counter-radiation") is particularly important at high latitudes, where the insolation becomes negligible during the winter season. However, this component of the radiative gain at the surface is numerically dominant even in summer, as shown by Table XII.

Even in midsummer the long wave contribution is higher than that of the solar radiation, and it is obvious that the atmospheric radiation must become the only radiative heat source for the surface during the polar night.

The long wave radiation received at the ground is determined by the temperature and moisture conditions of the atmosphere and the amount and height of clouds. The accuracy of the calculations of this radiation component will depend on the reliability of the observations. The moisture content is less important in this connection than the possible temperature errors, which are particularly critical in the lowest layers of the atmosphere. Investigation has shown (VOWINCKEL and ORVIG, 1964b) that the most critical factor is the emissivity of the clouds, of which little is known at high latitudes. Soviet research (MARSHUNOVA, 1959) indicates that arctic clouds do not radiate as a black body at the same temperature. The calculated values contained in the following should be considered to be estimates.

The long wave radiation emitted from the ground depends on the surface temperature.

TABLE XII

PER CENT CONTRIBUTION BY INSOLATION AND COUNTER-RADIATION TO TOTAL SURFACE RADIATION INCOME IN JUNE

	Latitude (°N): 65	70	75	80	85	90
Long wave (%)	60	64	68	69	69	69
Short wave (%)	40	36	32	31	31	31

Unfortunately, this is not equal to the temperature observed in the instrument shelter, and the actual temperature of the radiating surface is rather difficult to determine. Few reliable observations are available. Hence, for a calculation of terrestrial radiation, the areal surface temperature distribution can only be obtained by considering both surface conditions and observed air and surface temperatures, where they exist. Considering further that there are regions with open water or very thin ice cover which will radiate significantly more than the surrounding ice, the climatic values of terrestrial radiation used in the present discussion have been calculated rather than obtained from the existing observations, although due regard has been paid to the measurements.

The greenhouse effect

The atmosphere is nearly transparent for solar radiation, but the terrestrial radiation is partially trapped by water vapour and carbon dioxide, and completely by clouds. Some of this absorbed heat is radiated back to the earth's surface. This process is generally called the "greenhouse effect" of the atmosphere.

The numerical value of the greenhouse effect is not equal to the total atmospheric radiation. The atmosphere receives energy from several sources: absorbed long wave radiation from the ground, sensible heat transport from the ground, latent heat from condensation, and advection of sensible heat.

It is possible to calculate the amount of radiation at the top of the troposphere, caused directly by surface emission. Such values are given in Table XIII, which also gives the per cent of surface emission escaping through the troposphere for two areas. The figures vary very little from month to month and from one area to another. The explanation lies in the increase in water content over the Polar Ocean in summer, which suppresses the summer values so that they only become 140% of the January value over the central Polar Ocean. At the same time the radiation from the ground increases to nearly 170% of the January value. The amount of precipitable water increases 500–600% from winter to summer, so no large increase in radiation loss could be expected. A slight increase is found, however, because the absorption efficiency of water vapour does not go on increasing with higher water vapour content: it rises from winter values over the Polar

TABLE XIII

TERRESTRIAL RADIATION TRANSMITTED THROUGH THE ATMOSPHERE AT 300 MBAR FOR CLOUDLESS SKY (cal./cm^2 month · 10)

	Jan.	Feb.	Mar.	Apr.	May	June	July	Aug.	Sept.	Oct.	Nov.	Dec.
Central Polar Ocean	347	325	381	390	353	420	484	481	447	440	402	363
Norw.–Barents Sea	468	409	456	444	465	423	450	437	432	450	441	456
Kara–Laptev Sea	391	367	388	405	477	417	450	428	393	450	438	381
East Siberian Sea	353	322	375	372	409	441	471	471	426	437	408	375
Beaufort Sea	357	336	384	381	477	423	453	434	423	437	396	406
Per cent of surface emission												
Central Polar Ocean	28	29	30	25	26	22	23	23	25	27	29	28
Norwegian–Barents Sea	25	24	24	23	22	20	21	20	21	22	23	23

TABLE XIV

THE GREENHOUSE EFFECT UNDER ACTUAL CLOUD CONDITIONS, CENTRAL POLAR OCEAN

	Jan.	Feb.	Mar.	Apr.	May	June	July	Aug.	Sept.	Oct.	Nov.	Dec.	Year
G_a*	0.60	0.60	0.62	0.71	0.65	0.67	0.66	0.66	0.66	0.65	0.62	0.61	0.64
G_s**	0.56	0.48	0.52	0.50	0.56	0.59	0.60	0.60	0.60	0.57	0.50	0.49	0.55

* G_a: $\frac{L_a\downarrow}{L\downarrow}$ $L_a\downarrow$: that part of the back radiation which is caused by absorption in the atmosphere of terrestrial radiation.

$L\downarrow$: total back radiation.

** G_s: $\frac{L_a\downarrow}{L\uparrow}$ $L\uparrow$: terrestrial radiation.

Ocean to summer values, 150–160% of the former, in the tropics, while the water content increases about 50–60 times. Summer conditions are the most effective in reducing the portion of terrestrial radiation which is allowed to escape through a cloudless atmosphere. Water vapour alone can trap 3/4 of the terrestrial radiation and is the dominating influence on the direct loss of heat by radiation from ground to space. The effect of clouds in trapping heat radiated from the surface is a further factor of importance over the Polar Ocean, which is almost completely cloud covered in summer. As little as 3% of the terrestrial radiation will escape through the troposphere from July to September over the central Polar Ocean. Such values are likely to be the lowest figures anywhere on earth.

Table XIV presents values pertinent to the greenhouse effect under actual cloud conditions.

The G_a values show a slight seasonal variation with a tendency for maximum in summer. LONDON's (1957) calculations can be used to evaluate G_a for a meridional cross section. The following figures give the results for winter:

Lat. (°N):	0–10	10–20	20–30	30–40	40–50	50–60	60–70	70–80	80–90
G_a:	0.69	0.68	0.67	0.67	0.67	0.66	0.64	0.65	0.66

It is apparent from these figures that conditions are quite similar in all latitudes. It seems that there exists a balancing mechanism in the energy exchange: the polar latitudes are characterized by large advectional heat import, while in the tropics the main process is, directly or indirectly, one of turbulent transfer. These different processes are responsible for the additional energy radiated from the atmosphere back towards the ground.

Any positive heat advection will subdue the turbulent heat exchange: generally, increasing one process will decrease the other. Also, changes in any of the atmospheric heat-gain processes will be followed by adjustments in the atmospheric circulation, maintaining the value of G_a at its "normal" level. This value seems to be determined by the physical peculiarities of our atmosphere and radiation budget.

The values of G_s are just as stable as those of G_a. It seems most significant that the very sharp difference in cloud cover over the central Polar Ocean, between February and August, changes G_s by only 0.12. The conservative nature of the ratio shows the sensitivity of the thermal conditions near the surface to small changes in the terms of the energy balance.

It is of interest to compare the G_S values with evaluations from LONDON's (1957) calculations. If it is assumed that London's area 80–90°N is representative for the central Polar Ocean, one obtains an annual mean value of 0.55 from London, the same as that in Table XIV (VOWINCKEL and ORVIG, 1967a). The two evaluations were based on different observational material, and different formulae were used to evaluate the various radiation terms.

The meridional cross section from London's values indicates a somewhat lesser winter greenhouse effect in polar than in tropical latitudes. In summer there is no difference (Table XV).

TABLE XV

THE GREENHOUSE EFFECT
(After LONDON, 1957)

Latitude (N) :	0–10°	10–20°	20–30°	30–40°	40–50°	50–60°	60–70°	70–80°	80–90°
G_S (winter)	0.59	0.56	0.54	0.55	0.56	0.56	0.53	0.51	0.50
G_S (summer)	0.61	0.61	0.58	0.57	0.58	0.60	0.61	0.61	0.62

The difference in cloud amount (less in the tropics) is generally balanced by the difference in water vapour content (less in the Arctic).

It is possible also to visualize conditions with quite different G_S values. It can be done by merely lowering the water vapour content. With the present high temperatures of the atmosphere this would not be effective. However, if the average temperature decreased by 30°C (from +15°C to −15°C), the precipitable water content of the atmosphere would sink to 10–20% of the present value. Under such conditions G_S would decrease markedly in the polar regions. The effective shielding of the atmosphere lessens, progressively more rapidly, with lower temperature and moisture content. It seems that the atmospheric greenhouse effect is remarkably stable in all latitudes and climatic zones. This is so because the general moisture content of the atmosphere is sufficiently high, so that fluctuations in the amount have relatively small effects on heat absorption. Also, the greater cloud amount in high latitudes compensates for the smaller moisture content.

The radiation balance

At the surface

For radiation balance considerations the main difficulty lies in the fact that the net value is a very small residual of large components of incoming and outgoing radiation. If, in analogy to economics, the total of incoming and outgoing radiation irrespective of sign is called "turn-over", the net gain or loss expressed in per cent of this figure is as in Table XVI.

It is evident that the balance figures will be highly sensitive to slight inaccuracies in the incoming or outgoing radiation; these inaccuracies may be quite small with respect to the totals but must gain high significance for balance considerations.
The radiation balance is the result of so many opposing and interacting influences that it is quite impossible to come to a physical understanding from the areal representation

TABLE XVI

NET VALUES OF INCOMING OR OUTGOING RADIATION AS PER CENT OF RADIATION TURN-OVER

	Latitude (N): 65°	70°	75°	80°	85°	90°
Jan.	8.5	9.5	10.5	9.5	10.5	11.0
July	14.0	12.5	12.5	14.0	13.0	12.5

of the radiation balance for the individual months. An analysis of individual grid poin ts however, shows that there are a few quite distinct types of radiation regime. The predominance or interaction of these types result in the complicated pattern of the actual areal distribution.

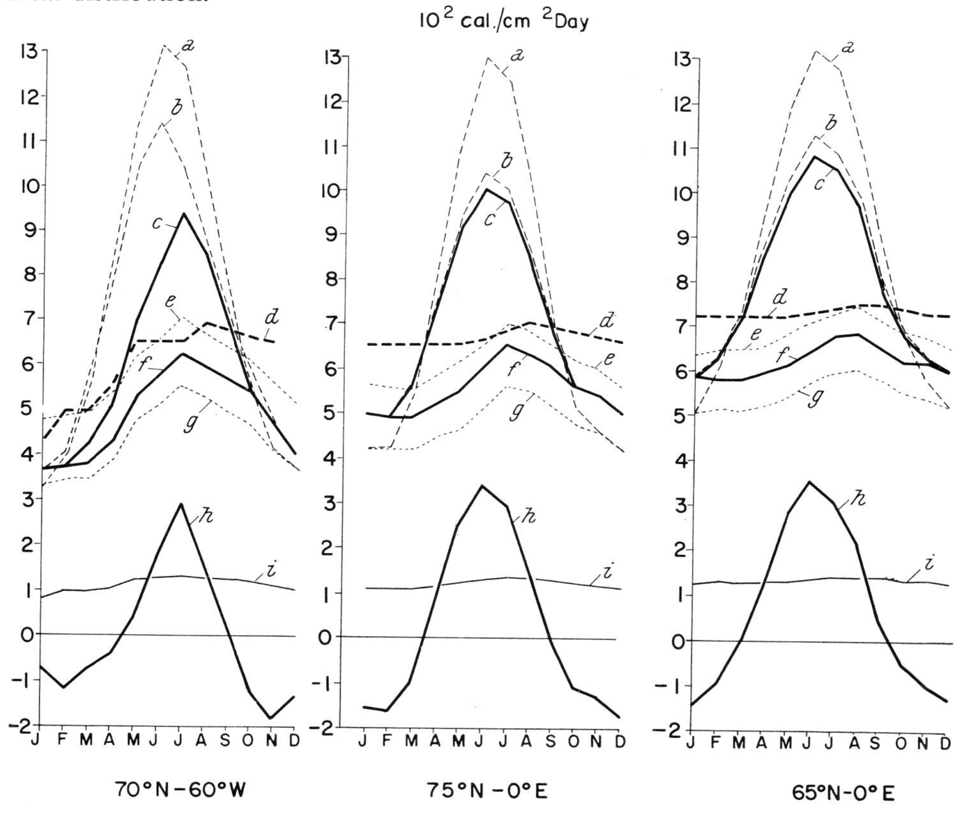

Fig.19. Radiation regimes—Norwegian Sea type: a = total incoming radiation, cloudless sky; b = actual total incoming radiation; c = actual total radiation absorbed on the ground; d = long wave radiation from the ground; e = long wave incoming radiation, overcast sky; f = actual long wave incoming radiation; g = long wave incoming radiation, cloudless sky; h = actual radiation balance; i = long wave radiation by CO_2.

Norwegian-Sea type

This type is characteristic for the year over open ocean areas north of the Arctic Circle (Fig.19).

It is significant in this type that a high positive balance during summer is compensated by large negative values of winter, which are rather high for the latitude. The different components of the balance show the reason clearly: the relatively warm ocean surface results in very high outgoing radiation in winter. This radiation reaches its maximum in late summer, just at the time when the global radiation begins to drop sharply. It must be emphasized that it is rather unimportant in polar areas whether a cold or a warm current is present in the ocean, as long as the water remains open. The difference in temperature between cold and warm currents is insignificant compared to the difference between water and solid ice. Open water surfaces in polar regions are under all circumstances very strong heat sources for the atmosphere in winter. Even changes in circulation and weather pattern cannot alter this. The greatest change would be caused by changes in the cloud amount, but even with complete overcast the negative surface balance could not be eliminated.

On the other hand, it is easy to increase sharply the negative balance of winter, for instance by decreasing the cloud amount. Therefore, anticyclonic cold spells will especially increase the negative radiation balance. The predominantly cyclonic conditions over these areas must therefore be regarded as a significant retarding factor for the radiation loss.

From the middle of March, to October, the positive radiation balance is solely determined by the global radiation term: the clearer the sky, the more favourable the balance. Before and after that time, the balance is not improved by cloudless skies, because the global radiation is too weak to counterbalance the increased long wave loss. The large difference between the curve for cloudless sky and actual radiation conditions shows the importance of clouds in summer. Contrary to winter conditions, therefore, anticyclonic conditions will be most favourable for the radiation balance. It should further be noted that the stable, cool summer surface temperatures of the ocean are favourable for the radiation balance by keeping the long wave loss low, compared both to global radiation and long wave back radiation.

Pack-ice type

This is characteristic for the ocean areas which are only temporarily free of ice (Fig.20). The best example of this type is found over Baffin Bay, but the type is widely represented all around the Polar Ocean in the temporarily ice free areas.

Although the ice cover for all practical purposes is complete in winter, the outgoing radiation is not reduced to values as low as those found over interior continental surfaces. A value of 350 Ly/day can be regarded as representative for continental conditions, 700 for the open oceans and about 420 for the areas now under consideration. Although the outgoing radiation approaches the continental rather than the oceanic value, the negative balance lies somewhere in the middle between ocean and continent. The reason is, firstly, that the cloud cover is small and ineffective in winter (the actual back radiation is very near to the clear sky curve). Secondly, the inversion is less intense, so that even overcast conditions would give a positive long wave balance only for a short period in the year.

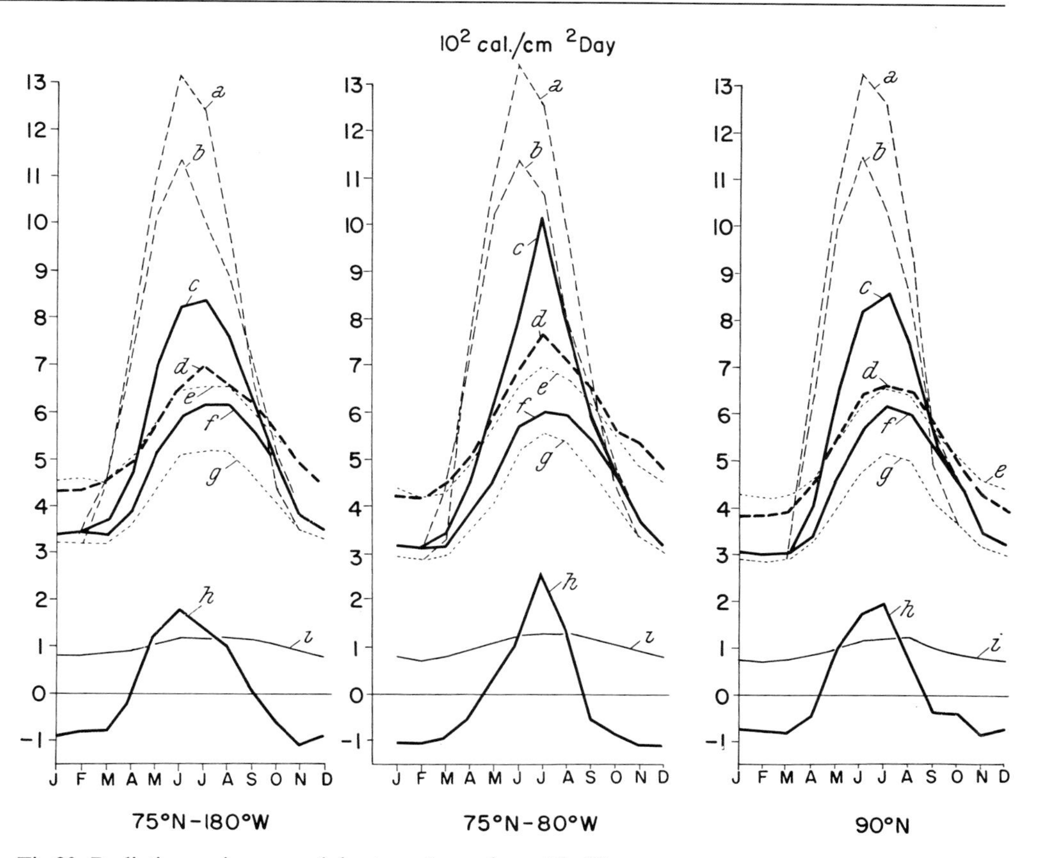

Fig.20. Radiation regimes—pack ice type. Legend: see Fig.19.

It is noteworthy that the influence on the long wave radiation of complete cloud cover does not vary in a significant amount in time or space over the whole Polar Ocean: it is always around 150 Ly/day. Its effect on the radiation balance will therefore be relatively greater, the lower the radiation balance value. Its influence in winter is more pronounced in continental areas, with a more balanced heat budget.

Spring brings only a very gradual change in the radiation balance in the pack ice areas. In spite of the almost clear sky, the gain in radiation is reduced to insignificance due to the very high albedo. Over the open water, the strong gain is dissipated downwards into the water without causing any increase in surface temperature, but at the ice surface, the gain is used to raise the low surface temperature. This is especially disadvantageous, as neither air temperature nor moisture content increases to the same extent, with the result that the radiation balance hardly improves from February to March.

The higher the global radiation becomes in spring, the higher will be the absolute amount of energy lost due to albedo. Towards summer the albedo decreases gradually, but when the ice finally disappears in July–August the short wave radiation has already passed its maximum, and the resulting energy gain remains relatively small. This condition is intensified by the gradual increase in cloudiness, which starts to deplete short wave radiation at the time when albedo conditions improve. The increase in long wave radiation due to higher cloudiness cannot change this picture.

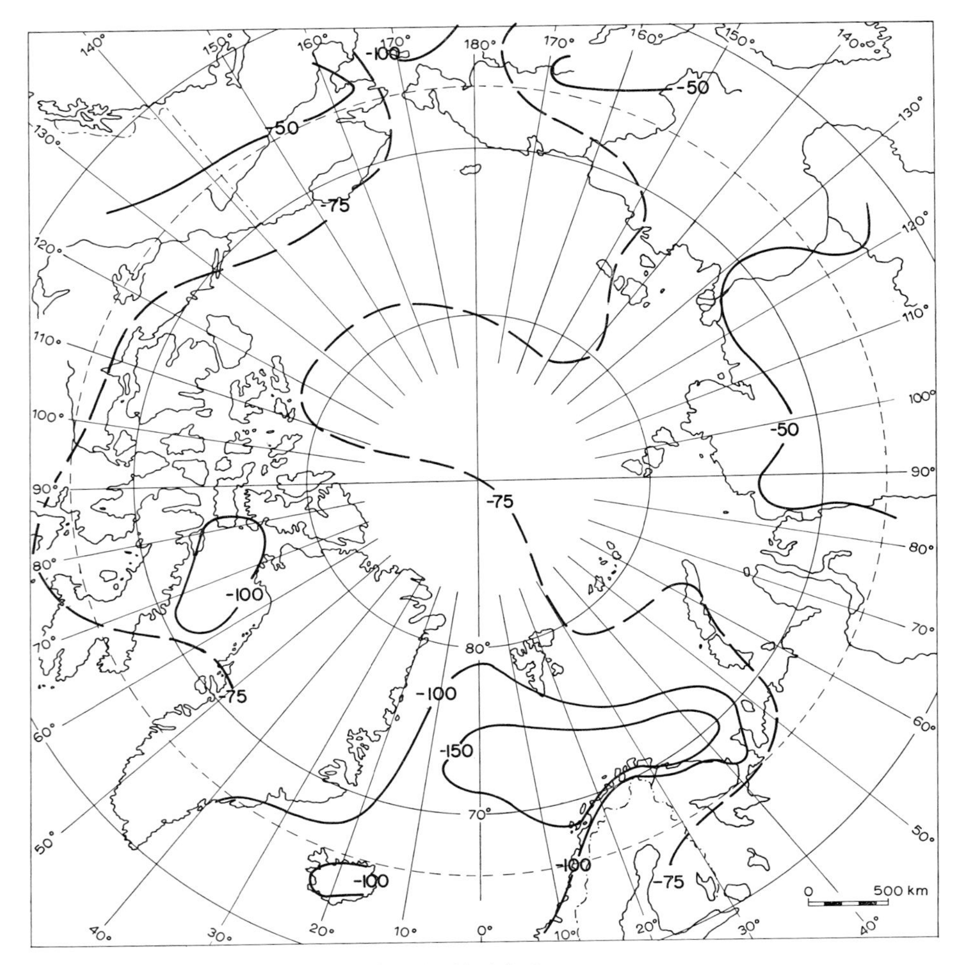

Fig.21. Total radiation balance, surface (Ly/day) in January.

The result of these summer conditions is that the maximum positive radiation balance is reached after the culmination of the sun. Considering all facts it is apparent that the pack ice areas are the least favourable in the whole Arctic.

It is perhaps justified to single out the central Polar Ocean area. Although the albedo decreases towards summer it remains relatively high during the polar day, with the result that the summer peak in the radiation balance becomes less clearly pronounced. It is interesting that the positive radiation balance in summer is not significantly less at the Pole. This is so because the clouds become thinner towards the north and hence the short wave radiation is progressively less depleted by the increasing cloudiness of summer, and also because of the low surface temperatures, which do not rise above 0°C. Thus, the terrestrial radiation is kept somewhat lower than over Baffin Bay for instance, where a temperature of 3–4°C is reached in August. In autumn the radiation balance does not become as strongly negative as farther south, since the rapid freezing-over reduces the surface temperature significantly.

Fig.21 and 22 show monthly maps for January and July of the total radiation balance at the surface in the Arctic. In winter, the maximum radiation loss is not in the polar basin proper but just to the south of it where the ice free ocean begins. It appears that the

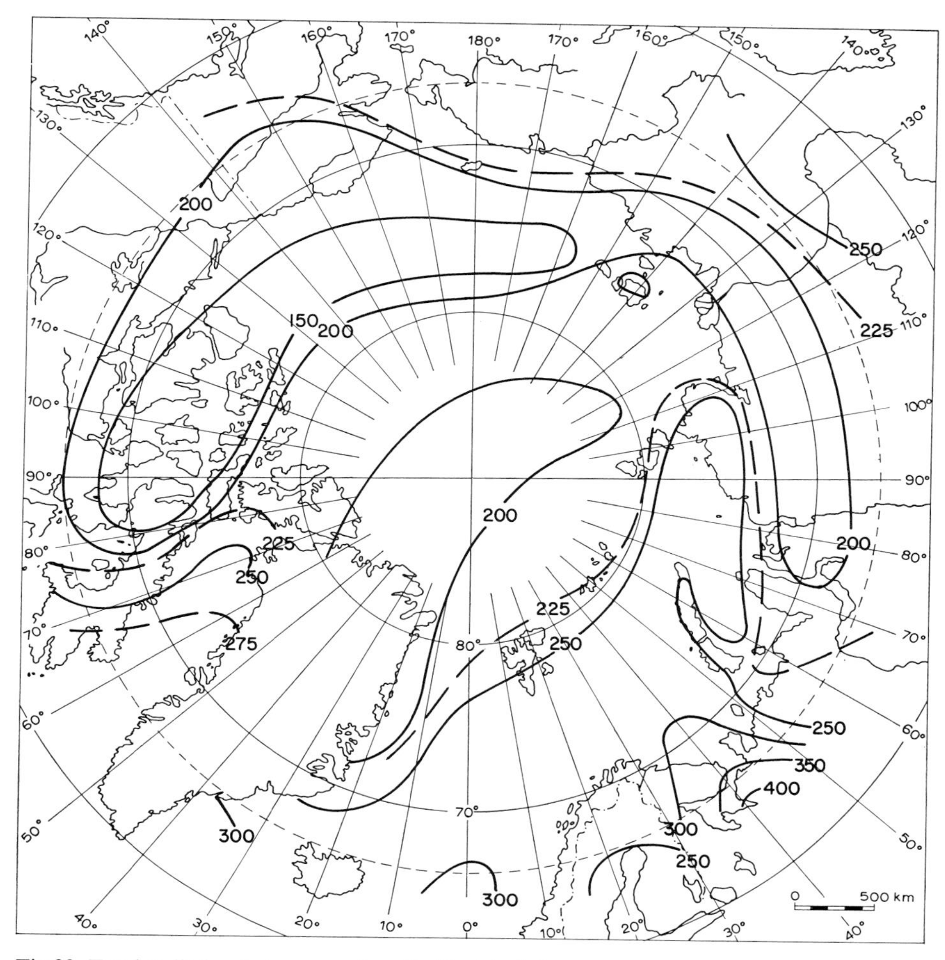

Fig.22. Total radiation balance, surface (Ly/day) in July.

Norwegian Sea and the adjoining oceans are the major radiational heat sinks of the earth's surface. Compared to these values the energy deficits over the Polar Ocean and the continental areas seem minor, being one half or less of the Norwegian Sea value. Towards spring, a zone of steep gradient develops in the radiation balance, caused by the rapid increase of the balance to high positive values from March to May over the ocean: the global radiation is utilised nearly to its full extent, while farther north the high albedo keeps the values down. Towards mid-summer the zone of sharp contrast disappears with the establishment of more uniform albedo values.

The radiation balance of the troposphere

A comparison between the radiation balance at the surface and at the top of the troposphere permits statements to be made about the energy budget of the atmosphere, and conclusions can be drawn about the magnitude of non-radiative processes compared to the radiative ones, and their variation in space and time. The radiation loss to space is the most difficult to determine and the uncertainties are greatest, mainly arising due to the paucity of observations of cloud top heights and of numbers of cloud layers present simultaneously. Attempts have been made to calculate the tropospheric radiation balance for grid points (VOWINCKEL and ORVIG, 1964c), and it is evident that the most pronounced difference from the surface radiation balance is the much smaller annual variation. This results from the fact that the short wave component is minor compared to the stable long wave components. The balance is everywhere negative. In essence it is characteristic for all areas that the smallest balance values are reached in spring–early summer, and the largest negative values in late summer–autumn.

The individual components of the radiation balance all show their maximum value in summer. The most outstanding feature is the extreme constancy throughout the year of the long wave radiation to space. The long wave radiation loss from atmosphere to earth is subject to considerably larger seasonal fluctuations. The variations in the energy loss of the atmosphere are therefore for the most part caused by changes in the radiation loss to the surface.

The existence of clouds improves the long wave radiation balance for the earth's surface. This is not the case for the atmosphere: although the long wave radiation to space is, on the average, slightly diminished due to the lower temperatures of the clouds compared to those of the ground, the sharp increase of radiation loss from clouds to ground over-compensates for this with the result that the overall energy loss of the atmosphere is increased. This is not compensated for on the positive side by increased short wave absorption. Thus, the energy balance is more negative under actual cloud conditions than with cloudless skies.

The long wave radiation to space is generally lower with clouds than without, if averages are considered. The Arctic, where overcast conditions over large areas are quite frequent, is probably one of the very few areas on the globe where the reverse is true for significant spells in winter. Then, the large anticyclonic areas show a substantial increase in radiation loss with clouds. This is caused by the semi-permanent inversion which shows higher temperatures of the cloud than of the surface. It is further noteworthy that this effect reaches even to the middle cloud layer. The downward radiation must also be increased. Under these conditions the radiation balance of the troposphere becomes even more negative than otherwise. During winter, the time of the strongest development of the inversion, global radiation is either non existent or plays only a minor role. It thus becomes apparent that overcast conditions in the arctic winter anticyclones produce a more negative radiation balance for the earth and atmosphere as a whole.

TABLE XVII

ENERGY REQUIREMENT OF THE TROPOSPHERE BY NON-RADIATIVE PROCESSES, IN PER CENT OF TOTAL ENERGY LOSS

Latitude (°N)	Jan.	Feb.	Mar.	Apr.	May	June	July	Aug.	Sept.	Oct.	Nov.	Dec.
65	35	31	26	21	17	13	13	18	24	27	32	34
70	34	31	28	22	19	14	15	20	25	30	29	32
75	33	32	29	23	20	16	17	21	27	30	32	32
80	36	35	33	26	19	16	20	23	29	33	35	35
85	37	36	34	23	19	17	20	23	31	36	35	36
90	37	36	34	24	19	17	21	24	31	37	36	37

The difference between the total outgoing and incoming radiation is the amount of energy which is required by the troposphere via non-radiative processes. In Table XVII this quantity is expressed as percentage of the total energy loss of the atmosphere. It is evident that by far the larger portion of the energy exchange is effected by radiative processes. Even in winter over the Arctic, with no solar radiation, the other processes contribute only one third of the energy requirements, and in summer the value sinks to less than 20%.

It must be kept in mind, however, that these statements refer only to the mode of energy transfer and not to the source of energy. In winter, for instance, the bulk of the radiation from the surface is in fact derived from advected energy, although this energy was transferred to the ground by radiative processes.

Total radiation balance of surface and troposphere

It is, finally, useful to construct the radiation budget for the whole system, including the surface and the troposphere, as the results will indicate the degree to which the North Polar Basin depends on heat import from southerly latitudes.

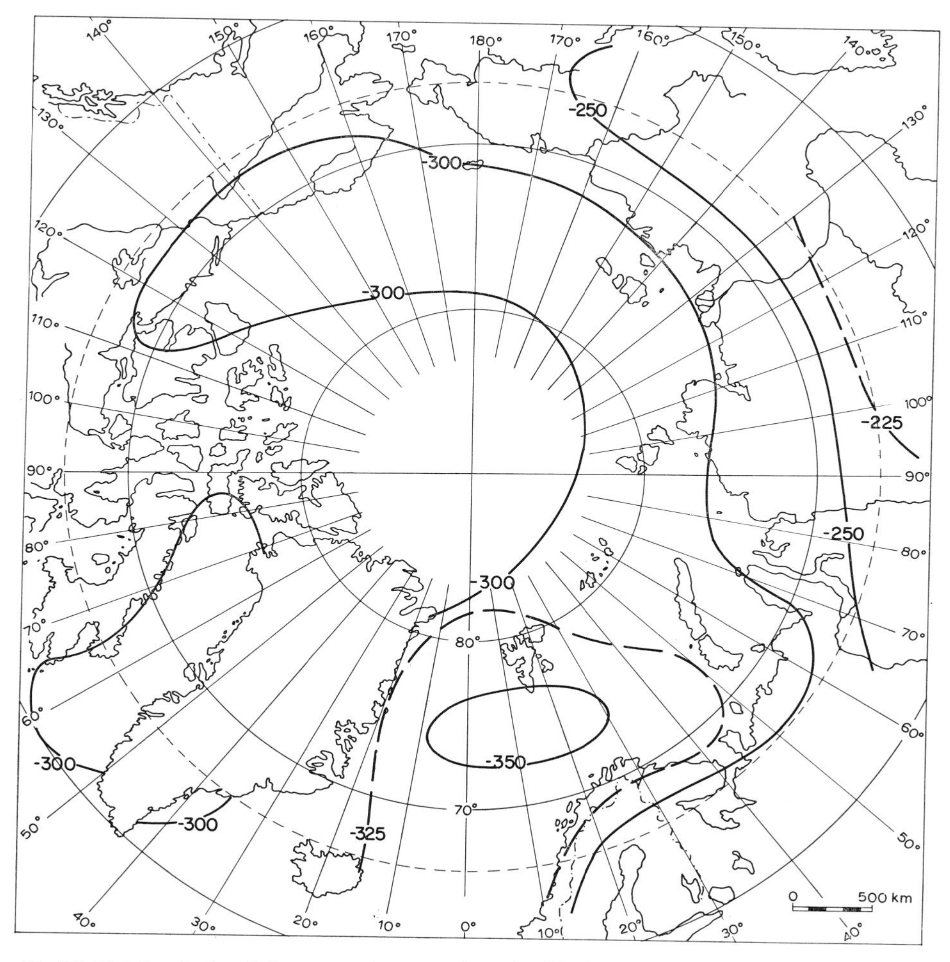

Fig.23. Total radiation balance, earth–atmosphere (Ly/day) in February.

The monthly maps of total radiation balance (Fig.23 and 24) show rather slight latitudinal variation in winter. The main difference is found between the oceanic and continental areas, the difference amounting to over 100 Ly/day in February along 65°N. Towards late winter, the maximum of radiation loss over the Atlantic shifts gradually northwards with a decrease in intensity: in December–January it is on 65°N or farther south, in February on 74°N, in March on 78°N and in April on 82°N. This is caused by the gradual increase in global radiation in the south. While the maximum loss is found in mid-winter over the open waters of the Norwegian Sea, towards spring it moves in the direction of—and over—the pack ice. The reason is that in mid-winter the terrestrial radiation from the warm surface is decisive for the high balance values, while towards spring the high surface albedo becomes critical by diminishing the energy income. The same reason is responsible for the fact that the difference between the highest negative value and the polar value—being representative for the central Polar Ocean—becomes less: in January, 91, February, 55, March, 23 and April, 14 Ly/day. Especially in March–April there is hardly any difference in surface albedo over the Polar Ocean, and the differences in surface temperature are also slight. The small differences which still exist are caused by the higher cloud cover in the vicinity of the Norwegian Sea, which reduces the incoming radiation more than the outgoing.

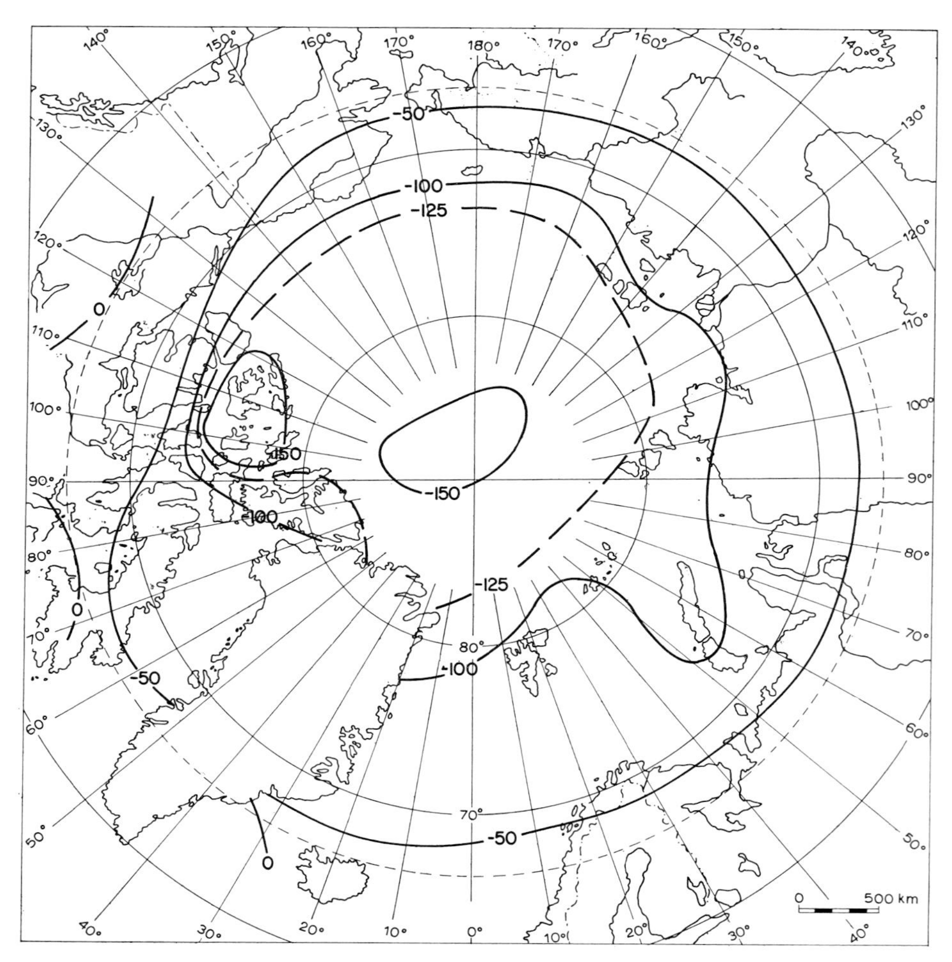

Fig.24. Total radiation balance, earth–atmosphere (Ly/day) in August.

It is of interest that the maximum radiation loss does not reach the area around the Pole until late summer (August). In May–June the maximum is found over Baffin Bay and the Canadian Archipelago. Relatively high cloud amounts are experienced in the eastern Canadian Arctic and, most important, the clouds in these southerly latitudes are less transparent than those farther north. The result is that, even with the same cloud amounts, incoming radiation is reduced more over Baffin Bay–Canadian Archipelago than over the Polar Ocean.

Steep gradients in radiation balance develop in the southern part of the Arctic in spring, resulting in increased meridional gradients. Although the gradients are much weaker than at the surface they are influenced significantly by the surface albedo and from mid-summer onwards by the extended cloud decks over the Polar Ocean.

The differences between land and ocean, a dominant feature in winter, disappear in summer and around mid-summer–early autumn the regional and latitudinal differences begin to diminish. August and September are the nearest representation of ideal distribution of the energy balance, with the greatest loss at the Pole and a gradual decrease southwards. In the dark season the non-radiative processes must obviously represent 100% of the energy expenditure (disregarding the small storage of heat in the ground, from summer surplus) and 100% of the radiative energy turnover. In this connection the slight latitudinal differences in radiation balance in winter are of interest. The difference between the highest and lowest value in the whole Arctic is as follows:

Jan.	Feb.	Mar.	Apr.	May	June	July	Aug.	Sept.	Oct.	Nov.	Dec.	
115	145	200	255	280	260	170	160	155	120	110	110	(Ly/day)

As the only source of energy in winter lies outside the Arctic, it might be expected that the amount of energy made available by advection should become less at a greater distance from the southerly source. This is not substantiated, however, and this suggests that the energy transport by advection of the air is extremely strong, so that the heat content of the advected air is not fully exhausted during the stay over the Polar Ocean. The role played by non-radiative processes becomes progressively less towards summer. This is shown by the following figures which give, for two points, the percentage contribution of non-radiative processes to the turn-over of energy:

	Jan.	Feb.	Mar.	Apr.	May	June	July	Aug.	Sept.	Oct.	Nov.	Dec.	
65°N 0°E	100	73	36	4	8	15	10	20	32	62	92	100	(%)
North Pole	100	100	100	48	10	9	22	25	79	100	100	100	(%)

Evaporation and sensible heat exchange

Apart from terrestrial radiation, the only means of transporting energy from the surface into the atmosphere are evaporation and sensible heat flux. They are therefore vital components in all climatic considerations. However, they are also the most elusive elements in the heat budget. While radiation calculations can be compared to actual observations, no accurate method exists of measuring and observing evaporation and sensible heat flux directly. Attempts at calculating these fluxes have been made (UNTERSTEINER, 1964; VOWINCKEL and TAYLOR, 1965).

The average evaporation and sensible flux values for the various areas of the Polar Ocean will depend on the extent of ice cover and open water in particular months. There is a lack of detailed information on this, and many values have been suggested. Soviet literature has indicated the following percentages of open water in the Polar Ocean:

Jan.	Feb.	Mar.	Apr.	May	June	July	Aug.	Sept.	Oct.	Nov.	Dec.	
0.1	0.0	0.2	2.8	6.7	9.3	10.0	9.6	7.9	5.2	2.4	0.8	(%)

These values are probably on the high side, due to observations from the Siberian side of the ocean. More recent observations from airplane flights have indicated the possibility that there may be about 6% open water in June, increasing to around 10% in late summer, and that areas of thin ice or open water may account for about 10% in winter. Accurate information would be of very great importance to our understanding of the climate of the Polar Basin, because the heat loss from an open water surface is two orders of magnitude greater than that from old ice. The following discussion is based on the assumption of about 4% open water in the central Polar Ocean in September. LYON (1961) mentions total area of open water as being perhaps 5%, based on submarine cruise observations. The results will therefore only be approximate estimates of the rate of heat exchange between the surface and the atmosphere. The averages for each area and for the whole Polar Ocean were obtained by weighting the ice and water extents according to the ice distribution in the particular month.

Evaporation

Table XVIII gives the evaporation values for the different areas. For convenience in later comparisons with the sensible heat flux, evaporation is expressed in its heat equivalent: cal./cm^2 per unit time (month or year). The evaporation in mm can be obtained by division by 60.

Fig.25 shows some characteristic curves of the annual march of evaporation. The complete reversal of the annual trend for ice and water surface is striking. The former has a pronounced double maximum, in spring and autumn, and the main minimum in winter.

TABLE XVIII

EVAPORATION (HEAT) OVER THE POLAR OCEAN (cal./cm^2 per month and per year)

	Jan.	Feb.	Mar.	Apr.	May	June	July	Aug.	Sept.	Oct.	Nov.	Dec.	Year
Central Polar Ocean average	−40	0	0	70	500	440	160	660	690	90	−90	0	2,480
Kara–Laptev Sea average	0	100	110	340	670	1,310	140	160	1,010	900	80	0	4,820
East Siberian Sea average	0	0	0	340	520	480	80	630	580	970	0	0	3,600
Beaufort Sea average	0	0	70	230	630	110	590	150	1,100	510	0	70	3,460
Total Polar Ocean average	−70	20	30	150	540	620	170	540	770	320	50	0	3,140

The low winter values are caused by the predominance of the surface temperature inversion at that time. The fact that no higher negative values are experienced is the result of the low temperatures and hence low water content of the air. Towards spring, with the lessening of the inversion and the increase of surface temperatures, evaporation increases rapidly. The summer minimum is the result of the persistent surface temperature of 0°C over the melting ice, while the air temperature still increases. In autumn, with dropping air temperatures, a renewed increase in evaporation takes place when the air temperature falls below 0°C. Later, the decrease of the surface temperature brings the drop to winter values, due to the lower saturation water vapour pressure.

The increase in air temperature above 0°C, in spite of the surface temperature remaining near zero, shows a predominance of advection terms in the climate even in the period of maximum insolation. Thus, it can be expected that in the peripheral areas, nearer the source regions for warm air, the appearance of a summer minimum in evaporation will be even more pronounced. This is evident in Fig.25.

The evaporation from the open water areas shows a pronounced summer minimum with condensation and a broad maximum in winter. This is explained by the fact that the sea surface temperatures are conservative: lower than the air temperatures in summer and very much higher in winter, resulting in extremely high evaporation values in the latter season. The values are smaller over the central Polar Ocean than in the peripheral areas, partly because of somewhat lower relative humidity at the periphery, but mainly because of significantly higher wind speed at the Eurasian coast.

Actually, the open areas in summer have a lessening influence on the evaporation. However, even small parts of open ocean during autumn, or during the whole winter, are very significant. Table XIX shows this. The energy loss increases very rapidly, and it is apparent that even slight inaccuracies in estimates of the extent of ice cover, or slight variations from year to year, will have significant influences on the whole energy budget.

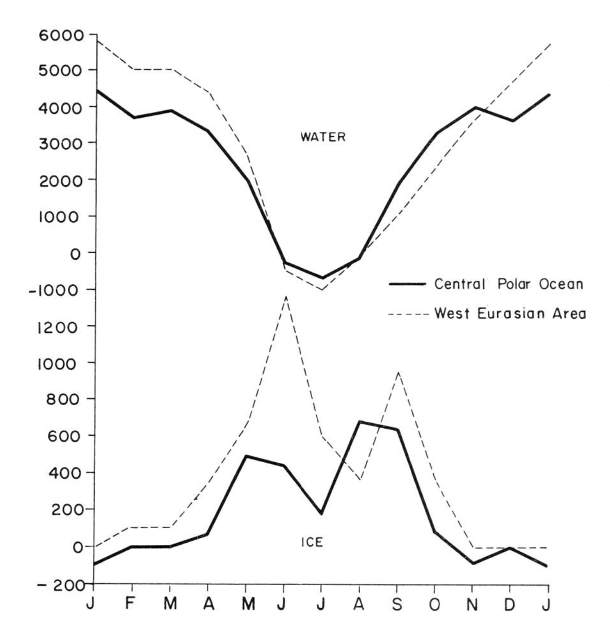

Fig.25. Evaporation, Polar Ocean (cal./cm² month).

TABLE XIX

HEAT LOSS FROM PARTLY OPEN OCEAN

	Open area from Sept. to May			
	0%	1%	2%	5%
Heat loss (kcal./cm²)	2.4	2.7	3.0	3.8
Heat loss (in per cent of loss from complete ice cover)	100	112	125	158

Some of the characteristic annual evaporation curves for the Norwegian–Barents Sea are shown in Fig.26. The areal averages show a gradual transition from pack ice to oceanic conditions: at 75°–80°N the typical pack ice shape still exists, except that the maxima occur earlier in spring and later in autumn and the summer minimum becomes the main minimum. The curve for 65°–70°N shows the practically undisturbed oceanic conditions, as the ice cover even in late winter becomes too small to have any important influence on the average.

The question arises as to what constitutes a purely oceanic condition. It is generally agreed that the oceanic type is characterized by an evaporation maximum in winter. The predominance of the summer minimum becomes most pronounced in the far north of the Norwegian–Barents Sea, which accordingly shows the best development of the oceanic type.

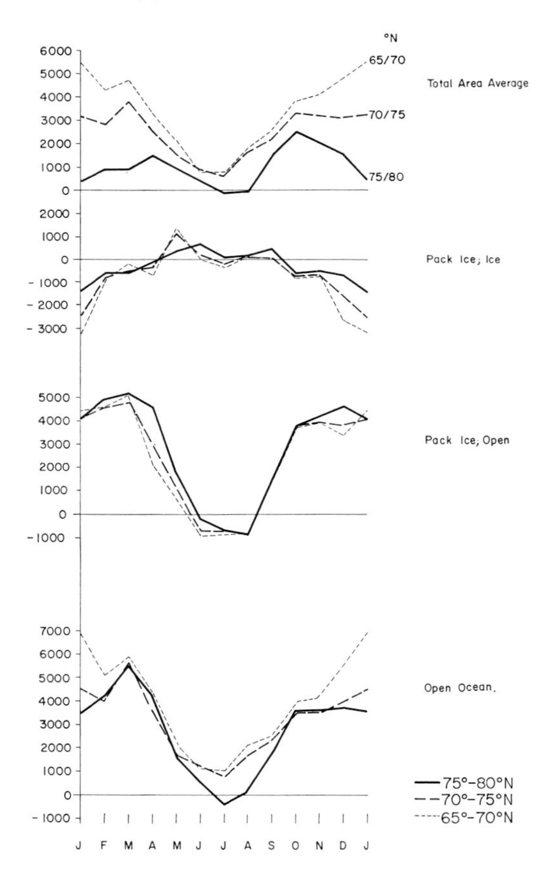

Fig.26. Evaporation, Norwegian–Barents Sea (cal./cm² month).

The opposite to this type could be termed the continental type. For its best development it would require shallow waters which could be heated substantially during summer. This type would probably be best realised in swamps with an ice cover in winter. The heating in summer would produce an evaporation maximum at that time, while the winter evaporation would be suppressed by the insulating properties of ice. This type will be best developed in subpolar latitudes. The pack ice type of the central Polar Ocean will appear as a transitional type, characterized by incomplete melting in summer and complete freeze-over in winter. The winter and transitory seasons represent a continental pattern, which is interrupted by the appearance of oceanic conditions in summer.

The effective evaporation is the net result of positive and negative values. This is of considerable importance over the Arctic Ocean. The values in Table XVIII give only an incomplete picture of this positive and negative transport. This is shown by the following values, giving mean turn-over and mean net transport at 75°N in the Norwegian Sea, in July:

Mean turn-over	Mean net evaporation
0.24 units	+0.05 units

In critical months, when the net evaporation is near zero, this might be important even over areas with a uniform surface. It must be of great significance where ice and water surfaces are present together in the same area. In winter there would then be strong evaporation from the open water, and sublimation on the ice. Such turn-over is irrelevant for the water budget, since only net evaporation represents withdrawal of water from the surface. For the energy transfer, however, it is important as it is unlikely that all the heat released by sublimation of hoar frost will be returned to the surface. As part of the heat may be diffused into the air, the gross evaporation would be the interesting value for the actual heat transfer. The following figures give a rather extreme example of the flux over the pack ice in the 80°–75°N region of the Norwegian–Barents Sea:

Net evaporation:	1.68 units
Turn-over of water:	6.69 units (400% of the net)

This extreme is caused by the presence of water and ice in the pack ice belt. The more one type of surface predominates, the less important is the difference. In the Polar Ocean one surface type becomes so predominant that the difference becomes insignificant. However, this does not indicate anything about the influence of day-to-day fluctuations, about which little can be stated quantitatively so far.

The conclusion is that the calculated energy fluxes may be rather lower than the real fluxes, but the absolute error is uncertain. The error caused by our insufficient knowledge of wind speeds will most likely also work in the same direction.

Sensible heat flux

Much less is known about the sensible heat flux than about evaporation. The reason is probably that evaporation is much more important numerically over temperate and tropical oceans. Such is not the case in polar latitudes.

Table XX gives the monthly flux values for the different parts of the Polar Ocean.

TABLE XX

SENSIBLE HEAT FLUX OVER THE POLAR OCEAN, (cal./cm² per month and per year)
(After SHULEIKIN's formulae, 1953)

	Jan.	Feb.	Mar.	Apr.	May	June	July	Aug.	Sept.	Oct.	Nov.	Dec.	Year
Central Polar Ocean average	−120	−60	−60	900	1,000	650	−40	560	990	0	−60	−60	3,700
Kara–Laptev Sea average	−150	−80	−90	900	930	870	−120	−110	810	1,200	220	−90	4,290
East Siberian Sea average	−150	−60	−60	900	930	890	−170	0	160	1,820	−60	−60	4,140
Beaufort Sea average	−120	−60	−60	900	930	−110	−280	700	1,080	470	−60	−60	3,330
Total Polar Ocean average	−130	−60	−70	900	980	680	−70	400	930	340	0	−70	3,830

The characteristic sensible heat flux curves are quite similar to the evaporation curves: over the ice is again found the double maximum, for which the same processes are responsible as in evaporation.

The temperature gradients in spring and fall are the most important in the energy transfer. A very exact determination of the difference between surface and air temperature is required for these critical months.

Of particular interest is the very high sensible heat flux from the open water areas in the Polar Ocean. An annual value of 160 kcal./cm² is calculated for an open central Polar Ocean: a value of similar magnitude to the long wave radiation components in the energy balance. It is apparent that small areas of open water in winter will have a very great effect on the energy budget. One per cent open water, from October to May, would increase the heat flux over the central Polar Ocean to 5.2 kcal. from the 3.7 kcal./cm² year obtained with complete winter ice cover. Open areas must remain small, as there would be a need for enormous amounts of energy to maintain such fluxes. The tempera-

TABLE XXI

TOTAL HEAT LOSS BY EVAPORATION AND SENSIBLE HEAT FLUX (cal./cm² per month and per year)

	Jan.	Feb.	Mar.	Apr.	May	June	July	Aug.	Sept.	Oct.	Nov.	Dec.	Year
Norwegian–Barents Sea	5,340	5,500	6,730	4,690	2,050	180	−130	90	2,750	4,690	5,630	5,800	43,320
Central Polar Ocean	−220	−60	−60	970	1,500	1,090	130	1,220	1,690	90	−150	−60	6,140
Peripheral Polar Ocean	−150	0	10	1,230	1,580	1,810	40	220	1,720	2,060	210	60	8,790

ture difference between air and water, causing this flux, is so great that the results are quite insensitive to slight inaccuracies in the air temperature used.
As a final summary, Table XXI gives the monthly and annual values of the total heat loss for the different areas.
It should be emphasized that, due to the paucity of observational material, the results cannot be expected to be more than a reasonable approximation. However, if the aim is a better understanding of the water and energy budgets of the earth, these uncertainties cannot be avoided. The infinite variety of natural conditions can never be accounted for completely, and generalizations and simplifications are the price paid.

Ocean circulation and heat transport

The central Polar Ocean is covered by a layer of water about 50 m thick which, in winter, is cooled to its freezing point of about −1.6°C, depending on its salinity which is around 30‰. In summer the open water leads show temperatures about 0°C, and temperatures slightly above the freezing point of sea water are found to near 50 m. Below a depth of 100—150 m there is only a small annual temperature variation but a temperature increase with depth.
The Atlantic water below the Arctic water, from about 200 m down to 900 m, has temperatures up to +1°C. Finally, the bottom water has almost uniform salinity (about 34.95‰) and temperature, on the Eurasian side around −0.75°C and on the American side about −0.35°C. The difference is due to the Lomonosov Ridge, about 1,500 m below the surface.
The circulation of the Arctic water is caused by both wind and density differences. The general flow is a slow surface motion, a few centimetres per second, from the eastern side of the ocean and across the Pole. Gradually the speed increases until the water passes into the Greenland Sea and down the coast as the cold East Greenland current. In the western part of the Polar Ocean the surface water tends to be driven in an anticyclonic gyre, moving southwestward along the Canadian Archipelago and northward toward the Pole.
The underlying warmer Atlantic water enters the Polar Ocean along the west coast of Spitsbergen. Having cooled in the Norwegian Sea, this water is dense enough to sink below the Arctic water. In the Arctic Basin it cools from about +3° to +0.5° by the time it has spread across the basin, the motion being mainly along the Eurasian continental slope.
The heat transported into the Polar Ocean is only slowly made available at the surface. For the Arctic Ocean as a whole the ocean heat transport is particularly important in the Norwegian–Barents Sea, which is mainly ice free throughout the year. To some degree it also influences the peripheral seas. Compared to the radiative means of heat exchange, the ocean advection is small in the central Polar Ocean.
The general method for the calculation of energy obtained or released by the ocean is to observe the energy influx and efflux for a given area. For this purpose the temperature and velocity profiles to the sea floor must be known. If these elements were given for a sufficient number of ocean stations, maps could be constructed and the energy balance could be obtained for each desired area. However, observations are far from sufficient

for such an undertaking. Especially are velocity observations or calculations lacking. The available data permit the best statements to be made about influx and efflux through the relatively narrow straits. The Arctic Ocean has, in this respect, rather favourable conditions as it is a mediterranean ocean bordered by continents and has relatively narrow connections with other oceans. Furthermore, a subdivision between the Arctic and Polar oceans is possible.

The influx and efflux values and the mean temperatures are discussed individually in the following.

The Atlantic Inflow

The warm current entering from the Atlantic Ocean is by far the most important source as far as water volume and heat is concerned. The main part of this water flows through the Faeroes–Shetland Channel, and only minor parts between Faeroes and Iceland and, as the Irminger current, on the west side of Iceland.

SVERDRUP et al. (1942), when giving a water balance for the Arctic Ocean, consider only the water entering the Arctic Ocean to the northwest of Shetland, group together all net inflow under this heading, and obtain $3.0 \cdot 10^6$ m^3/sec or 94,600 km^3/year. TIMOFEYEV (1956), using Soviet sources, presents the following figures:

Atlantic inflow	400,000 km^3/year
Outflow	248,000 km^3/year
Net gain	152,000 km^3/year

These results should be long term averages. Other Soviet researchers have estimated the net flux to be 128,500 km^3/year and 140,000 km^3/year.

A large number of observations are available of the temperature of this current, generally lying between 6.7°C and 8.1°C.

As the Atlantic inflow is by far the largest in volume, and hence the most important in heat transport, even small differences in temperature make a significant difference to the total. The best substantiated values seem to be 7.5° and 8.1°C. In the present discussion the mean between the two has been adopted, i.e., 7.8°C (VOWINCKEL and ORVIG, 1962b).

Bering Strait

Several independent investigations are available. SVERDRUP et al. (1942) quote an inflow of $0.3 \cdot 10^6$ m^3/sec. A further estimate originates with N. N. Zubov and D. B. Karelin (GORDIENKO, 1958b), giving 20,000 km^3/year. TIMOFEYEV (1960) gives the following annual values for water and heat flux:

Water flux:	36,125 km^3
Heat flux:	$+33{,}490 \cdot 10^{15}$ cal.

A further investigation, taking into account many stations with flux observations, originated with MAKSIMOV (1944). He obtained for the summer months, a flux of 4,100 km^3/month and for the winter 369,800 m^3/sec. He calculated the annual flux on the assumption of summer flow for the ice free months, an intermediate flow for the months

with some ice, and a winter flow with complete ice cover. The annual inflow derived in this way is 30,354 km³. His assumption of the dependence of the flux on the ice cover seems realistic, as the bulk of the flux takes place near the surface layers.
COACHMAN and BARNES (1961), after discussing the Soviet and more recent American observations, obtained a yearly inflow of $1 \cdot 10^6$ m³/sec, or 31,536 km³/year; a figure very near that of Maksimov. A more recent investigation (COACHMAN and AAGAARD, 1966) gives a typical August transport as $1.4 \cdot 10^6$ m³/sec, also observing both irregular and regular current speed fluctuations.
SVERDRUP et al. (1942) estimated the mean heat content of the current as zero. (A negative value would indicate flow of water at a mean temperature less than 0°C.) When converting Soviet heat flux figures into temperature, a mean temperature of 4.45°C is obtained for September, the month with the highest temperature, and the annual mean would be 0.93°C. These figures seem to be in conformity with the actual temperature observations published.

Runoff and precipitation

A less important source of water and heat for the Arctic Ocean is the discharge from the rivers.
For the whole continental drainage into the Arctic Ocean, ANTONOV (1958) gives the results in Table XXII.

TABLE XXII

ARCTIC RIVER DRAINAGE

Region	Drainage (km³/year)
Northwest Scandinavia, extreme northwest of U.S.S.R.	153
European U.S.S.R., excluding Pechora	359
Siberia, including Pechora	2,442
North America, including Yukon	1,053
Greenland (tentative estimate)	373

From this figure has to be deducted the Yukon River with 240 km³, flowing into the Bering Sea, and the Greenland figure, as this contribution will mainly consist of ice of low temperature; this would leave an inflow of 3,767 km³/year.
For the Ob, ANTONOV (1936) gives the amount of heat carried as $4{,}249 \cdot 10^{15}$ cal. and for the Yenisey $2{,}849 \cdot 10^{15}$ cal., which gives mean temperatures of 9.6° and 7.0°C respectively. Both these rivers have a very large drainage area, reaching far southward into warmer regions. Lena and Mackenzie have a rather cooler drainage area, and all coastal rivers are probably colder. It seems best to accept Antonov's result for Ob and Yenisey; for Lena and Mackenzie to use an estimated 5°C, and for the rest of the Arctic rivers 3°C. This would then lead to a heat transport of $18{,}413 \cdot 10^{15}$ cal./year.
The net excess precipitation must be considered as incoming water. An estimate by Sverdrup is $0.09 \cdot 10^6$ m³/sec, or 2,800 km³/year. Precipitation figures over oceans are

very uncertain, and this holds particularly for arctic areas, where most of the precipitation falls as snow. As precipitation is small in comparison to the water transport of the large ocean currents, it is possible to disregard this element completely. Furthermore, from energy considerations the contribution of precipitation will be very small indeed as the temperature difference between precipitation and ocean water is small.

Denmark Strait export

The most important avenue for outflow from the Arctic Ocean is Denmark Strait. SVERDRUP et al. (1942) give the outflow as $3.55 \cdot 10^6$ m³/sec, or 111,953 km³/year. TIMOFEYEV (1956) gives details of four profiles on which his outflow values are based: Aug. 13–17, 1929:—11.68 km³/h; Mar. 25–28, 1933:—57.92 km³/h; July 30–31, 1933:—2.10 km³/h; Aug. 21–22, 1933:—1.64 km³/h.
The difference between one observation and another is very great. Timofeyev uses the arithmetic mean and obtains around 161,000 km³/year. Other Soviet authors give similar values, about 162,000 km³/year.
A mean temperature of between 0°C and -1°C seems the best supported assumption for this current.

Davis Strait

The second channel carrying outflow from the Arctic Ocean is Davis Strait. Timofeyev obtains this outflow as a residual in his general budget calculation. His figure is 31,400 km³/year. Actual observations are published by SMITH et al. (1937). The observations used extend over several years. Using his out- and inflow figures from the Labrador Sea to Baffin Bay, the net result is 31,536 km³/year, almost exactly the same as Timofeyev's assumption.
The efflux of heat can also be calculated from Smith's results. He gives for the West Greenland current to Baffin Bay $1,576 \cdot 10^{15}$ cal./year and for the Baffin Island current $-3,784 \cdot 10^{15}$ cal./year. The net result is therefore $-2,208 \cdot 10^{15}$ cal./year.

The water balance and heat flux

The Polar Ocean is separated from the rest of the Arctic Ocean by a long stretch of sea between Novaya Zemlya and Greenland. Most authors are of the opinion that no significant flux takes place across this border between Novaya Zemlya and Spitsbergen.
Many estimates exist of the inflow of Atlantic water west of Spitsbergen. Soviet sources give values from as low as 50,000 km³/year to about three times that amount. As with the Atlantic inflow farther to the south, marked seasonal and annual variations seem to exist.
Results from TIMOFEYEV (1957) are available for the temperature of the West Spitsbergen current. He found the mean temperature to be 1.62°C. Timofeyev made a detailed analysis of this current, employing temperatures and velocities for each 50 m interval, and in this way obtained a heat content of $214,357 \cdot 10^{15}$ cal./year. In a later study it was pointed out that 34% of this heat gain is returned southward directly by mixing of the water masses of this current with the East Greenland current.

The outflow of polar water in the East Greenland current has only recently been observed. Summer ice breaker cruises in 1964 and 1965, as well as observations from the American ice island Arlis II, during its two-month drift in the East Greenland current in the winter 1964–1965 have recently been reported (AAGAARD and COACHMAN, 1968). Although the marked fluctuations in the current make it difficult to estimate an average flux, it seems from these recent observations as if the East Greenland current may well carry at times, up to ten times as much water as previously thought. The April–May measurements indicated a flow of about $8 \cdot 10^6$ m³/sec of polar water, 21 of Atlantic intermediate, and $2–3 \cdot 10^6$ m³/sec of deep water. This would give an East Greenland current flow of about 960,000 km³/year. At the surface, and down to about 200 m, the temperature is less than 0°C. This is the major outflow from the Polar Ocean. However, it is probable that only about one third of the East Greenland current consists of polar water. The greater bulk is deeper water from a cyclonic circulation in the Greenland–Norwegian Sea. Nevertheless, the export of Polar Ocean water would still be four times as large as previously estimated. The real outflow figures for the East Greenland current are not known. Soviet authors have given values from 80,000 to 118,000 km³/year. It has been supposed in various investigations that the inflow through Bering Strait plus river discharge should about balance the outflow through Davis Strait, and the Atlantic inflow should equal the outflow in the East Greenland current.

The following estimates of water and heat balances are based on estimates of flow commonly accepted before the two-month late winter observations done from Arlis II. The flux and balance calculations for the Polar Ocean should be regarded as only approximate (VOWINCKEL and ORVIG, 1962b).

Heat gain by ice export

A further heat gain for the Arctic Ocean results from the formation of ice. All ice that is not melted again but is exported from the area represents an actual heat gain for the ocean. Large quantities of ice are exported from the Arctic Ocean, mostly between Greenland and Spitsbergen, and an energy budget for the Arctic Ocean cannot disregard this energy source.

Available estimates of the ice export mostly go back, directly or indirectly, to Soviet investigations, and range from 1,300 km³ in Denmark Strait to 10,000 km³/year across the Greenland–Spitsbergen border.

Investigations have shown large fluctuations from year to year in the ice export between Greenland and Spitsbergen. As an annual average for the period 1933–1944 an export of 1,036,000 km² has been given (GORDIENKO and KARELIN, 1945). With an average thickness of 3 m this would total 3,108 km³; with 2.5 m ice thickness, 2,590 km³. A rather similar estimate was obtained independently (VOWINCKEL, 1964b). A more recent Soviet estimate (1966) for the ten years 1954–1964 gives a mean transport of ice into the Greenland Sea of about 950,000 km²/year.

The heat gained in the Arctic by the export of ice depends on the temperature of the ice. This has a pronounced seasonal variation as shown by all temperature profiles measured in the pack ice. No ice temperature profiles are available from Denmark Strait or the Greenland–Spitsbergen border. The latter is near the closed pack ice of the Polar Ocean, an area in which the local temperature variations are relatively small. As a first approximation, the temperature observations from the Polar Ocean can be taken to be representative also for the Greenland–Spitsbergen border.

The ice temperature profiles in the Polar Ocean depend both on the year of observation and the thickness of ice. For the present discussion two sets of observations have been used, the former taken in relatively thin pack ice and the latter in a thick ice floe. Thereby an average is obtained that includes at least some of the diverse ice conditions (MALMGREN, 1933; YAKOVLEV, 1955). The amounts of heat liberated by the ice layer existing in particular months would give as a total for the year:

Greenland–Spitsbergen border:	21,000 cal./cm^2 year
Denmark Strait:	16,000 cal./cm^2 year

The heat gain by ice export is obtained by multiplying the total yearly ice export by the mean annual figure for heat release. The results are, if the same heat content is used for Davis Strait as for Denmark Strait:

Greenland–Spitsbergen border:	218,000 · 10^{15} cal./year
Denmark Strait:	95,000 · 10^{15} cal./year
Davis Strait:	39,500 · 10^{15} cal./year

The heat gain for the Polar Ocean should therefore be 257,500 · 10^{15} cal., and for the whole Arctic Ocean 134,500 · 10^{15} cal., this figure being smaller as considerable melting takes place in the Norwegian Sea. The heat loss for this sea, due to melting of imported ice, would be 123,000 · 10^{15} cal./year.

TABLE XXIII

PROBABLE ANNUAL WATER AND HEAT BALANCE—ARCTIC OCEAN (EXCLUDING ICE)*

	km^3	°C	cal. · 10^{15}
Bering Strait	32,500	0.93	+30,225
Atlantic (Soviet est.)	140,000	7.8	+1,092,000
Runoff	3,800	9.6, 7, 5, 3	+18,413
Denmark Strait	140,000	0, −1	0 (a)
			−140,000 (b)
Davis Strait	36,300		−2,541
	Σ heat gain (a) +1,143,000 or		
	(b) +1,283,000		
Bering Strait	32,500	0.93	+30,225
Atlantic (European est.)	112,500	7.8	+877,500
Runoff	3,800	9.6, 7, 5, 3	+18,413
Denmark Strait	112,500	0, −1	0 (c)
			−112,500 (d)
Davis Strait	36,300		−2,541
	Σ heat gain (c) +928,500 or		
	(d) +1,041,000		

* Negative values in column 3 indicate flow of water at a temperature less than 0°C.
(a) Based on Soviet estimates of Atlantic flow, and a water temperature of 0°C in Denmark Strait.
(b) As above, with a Denmark Strait water temperature of −1°C.
(c) Based on European estimates of Atlantic flow, and a water temperature of 0°C in Denmark Strait.
(d) As above, with a Denmark Strait water temperature of −1°C.

TABLE XXIV

PROBABLE ANNUAL WATER AND HEAT BALANCE—POLAR OCEAN (EXCLUDING ICE)

	km^3	°C	cal. · 10^{15}
Bering Strait	32,500	0.93	+30,225
West Spitsbergen	93,500	1.62	+151,470 (e)
	117,500		+190,350 (f)
Runoff	3,300		+16,100
Davis Strait	36,300		−2,541*
East Greenland	93,500	−0.5	−46,750 (e)*
	(117,500)		(−58,750) (f)*
	Σ heat gain (e) +247,000 or		
	(f) +298,000		

(e) Based on West Spitsbergen current transport of 93,500 km^3/year.
(f) Based on West Spitsbergen current transport of 117,500 km^3/year.
* Negative values in column 3 indicate flow of water at a temperature less than 0°C.

The total heat gain

The various possibilities of water and heat transport, summarized from the literature, are given in Tables XXIII and XXIV. A first glance at these tables shows remarkable differences. The main reason for these discrepancies is that the oceanic observations extend over short periods and are not synoptic. The great differences in the seasonal and yearly fluctuations do in fact not permit strict comparisons to be made between the different sets of observations. A difference from the long-term mean, for a particular year, of 15–20% in the heat gain from ice and currents is quite likely for the Arctic

TABLE XXV

TOTAL HEAT GAIN AND HEAT GAIN PER CM^2 (ANNUAL)

	Gain by currents (cal. · 10^{15})	Gain by ice (cal. · 10^{15})	Total heat gain (cal. · 10^{15})	Gain per cm^2 (cal.)
Arctic Ocean				
(a)	1,143,000	134,500	1,277,500	9,000
(b)	1,283,000	134,500	1,417,500	10,000
(c)	928,500	134,500	1,063,000	7,500
(d)	1,041,000	134,500	1,175,500	8,500
Polar Ocean				
(e)	247,000	257,500	504,500	5,000
(f)	298,000	257,500	555,500	5,500

(a) Based on Soviet estimates of Atlantic flow, and a water temperature of 0°C in Denmark Strait.
(b) As above, with a Denmark Strait water temperature of −1°C.
(c) Based on European estimates of Atlantic flow, and a water temperature of 0°C in Denmark Strait.
(d) As above, with a Denmark Strait temperature of −1°C.
(e) Based on West Spitsbergen current transport of 93,500 km^3/year.
(f) Based on West Spitsbergen current transport of 117,500 km^3/year.

In light of the most recent observations of greater outflow in the East Greenland current, it may be that a greater than previously estimated inflow of Atlantic water takes place in the West Spitsbergen current. If so, the heat gain figures will be minimum values.

Ocean. However, until many more simultaneous observations become available, one must try to use these figures.
The mass flux figures can be transformed into heat fluxes. These are summarized in Table XXV for the Arctic and Polar oceans, and the heat gain by ice formation is also considered. From this table, it will be seen that the different assumptions about the flux of Atlantic water amount to a difference of 17–20% in the heat gain per cm^2 for the Arctic Ocean. Less serious is the difference in assumed water temperature which produces a difference of about 10% in the heat gain per cm^2.
These differences are more important for the Arctic Ocean as a whole than for the Polar Ocean. There, about one half of the total heat gain arises from the formation and export of ice. A wrong estimate of the ice export will be felt seriously.

Heat flux through the pack ice

It is of interest to ascertain if the estimated heat made available in the Polar Ocean water corresponds to possible conduction through the ice. The heat flux both in ice covered and open areas must be considered, and this requires knowledge of ice density, temperature, thickness and salinity, as well as the areal extent of ice cover. Further, the contributions of the different ice thicknesses to the total ice cover must be determined for each month.
An investigation of these problems (VOWINCKEL, 1964a) showed that, assuming about 4% open water in the central Polar Ocean at the end of the summer, the age groups of ice in the central Polar Ocean would be represented in amounts as shown in Table VI. The large amount of 5-year old, and older, ice (60.9%) means that an equilibrium thickness is reached after about five years. An ice floe can therefore exist for a long time, although its matter is not retained for long. A mean value for surface ablation in summer is around 0.5 m. As was seen in Table V, showing ice age and thicknesses, the pack ice cannot under normal conditions melt in the Polar Ocean once it has survived the first summer.
Table XXVI shows that the annual heat flux through one-year old ice is significantly higher than that in old ice. It is apparent that the percentage of one-year old ice is decisive for the annual heat flux in the Polar Ocean.
It is evident that it would lead to completely erroneous results to extrapolate flux values for the whole ocean from the findings at one station—or even a few. The only way to obtain areal averages is the consideration of regional distributions of various ice types. The flux values obtained can be checked against the import via ocean currents. However,

TABLE XXVI

HEAT FLUX THROUGH ICE IN THE CENTRAL POLAR OCEAN

	Heat flux (cal./cm^2 year)
Through 1 year old ice	18,500
Through 2 year old ice	6,800
Through 3 year old ice	5,900
Through 4 year old ice	5,400
Through 5 year old, or older, ice	5,300

allowance must first be made for the flux from the open water areas. The energy flux per average cm^2 for the whole Polar Ocean would be (VOWINCKEL, 1964a):

Flux through the ice	7,700 cal./cm^2 year
Flux through open water	820 cal./cm^2 year
Total flux	8,520 cal./cm^2 year

The energy available from ocean currents, including export of ice, is about 5,000–5,500 cal./cm^2 year. To this must be added the amount of energy returned to the ocean in the form of run-off of meltwater from the melting pack ice. UNTERSTEINER (1961) estimated this to be 2,300 cal. from the results at Station Alpha. Another additive term is the net radiation absorbed in the open areas in the peripheral regions of the ocean. This can be estimated as 1,200 cal./cm^2 (VOWINCKEL and ORVIG, 1964b). The total flux value must therefore balance against:

Gain by ocean currents	5,000 cal./cm^2 year
Gain by run-off	2,300 cal./cm^2 year
Gain by radiation	1,200 cal./cm^2 year
Total gain	8,500 cal./cm^2 year

or, if 5,500 is used for the ocean currents: 9,000 cal./cm^2 year.
The correspondence between input and output is very close indeed. This does not imply that the assumptions are correct and all processes sufficiently known. It may be accidental and only shows that the orders of magnitude obtained for the different processes are correct.

Atmospheric circulation and advection

The Polar Ocean region must import heat from southerly latitudes due to the net radiation loss to space from the top of the atmosphere. It therefore influences the circulation over adjacent continents and oceans as well as being the prime energy sink in the large scale circulation of the atmosphere. The possibility of studying the circulation over the Arctic only became a reality with the establishment of upper air observing stations in the 1930's and the subsequent planting of weather stations on the northernmost shores and on the polar ice itself.
The expected anticyclonic conditions proved to be much less firmly established than one had speculated, and it soon became clear that the central Polar Ocean was a region of frequent and sometimes quite intense cyclonic activity (DZERDZEEVSKII, 1945). Long period charts of storm track observations show that the winter cyclones move mainly from the North Atlantic, both via Baffin Bay and via the Norwegian and Barents Seas. This results in well developed troughs of low pressure over these regions in winter. During summer there forms a depression near the Pole.
Cyclones and anticyclones over the Arctic Ocean behave much like those found in middle latitudes. The upper tropospheric characteristics also resemble similar disturbances in temperate latitudes. There are few barriers to the interchange of air between the Arctic and regions to the south. Greenland and the mountains of northern Alaska do prevent free flow in the north–south direction, but there is otherwise no real reason

for considering the meteorology of the Arctic to be separate from that of the temperate latitudes. The differences are to be found in frequency and motion rather than in basic characteristics.

In general, the arctic atmosphere can be defined as the hemispheric cap of fairly low kinetic energy circulation, lying north of the main course of the planetary westerlies. Obviously this definition leads to no simple line on the map separating arctic from mid-latitude circulations. The limit varies from day to day, from season to season and from level to level. At times—during low-index phases of the westerlies—it may be said that the "arctic atmosphere", as so defined, is displaced entirely from arctic latitudes, lying instead over eastern Canada or northeast Siberia. Hence no firm limit can or should be applied. Since the arctic circulation is intimately connected with the planetary westerlies, it is plainly undesirable to draw any rigid limits. Climatologically, however, there is much to be said for regarding the 70°N parallel as the average limit of the westerlies. The strong zonal circulation usually terminates at about 70°N, at least on the American side of the Pole. In general, the sub-Arctic lies within the normal belt of the tropospheric westerlies, but the Arctic lies near the floor of the circumpolar vortex and displays in its tropospheric dynamic climate the somewhat stagnant behaviour that might be expected from such a situation.

The idea that the mid-latitude westerlies circulate about a vast low pressure centre in the north polar region goes back to FERREL (1889), who put the centre at the Pole itself. Subsequent changes of view have been summarized by DORSEY (1951), who placed in perspective the apparently conflicting views of the Ferrel vortex and the "polar high" of later periods.

Below the tropopause, the polar atmosphere *is* dominated by the Ferrel vortex, whose main thermal energy is concentrated in the baroclinicity of the temperate westerlies. Over the polar area itself, circulation is somewhat sluggish, since there is little internal drive. Above the tropopause, however, the cooling of the winter polar stratosphere itself creates a new drive for the westerlies. The latter extend throughout the lower stratosphere, increasing with height and becoming progressively more truly zonal.

The synoptic picture

Winter maps show several cyclones over the Arctic Ocean, on the average four to six per month over the Norwegian–Barents–Kara Seas. Further east, over the East Siberian Sea, about four cyclones are observed in a winter month. In the central parts of the basin, the most frequent invasion of winter cyclones centres on latitude 87°N, longitude 0°, where the January maps show an average of four cyclones during the month. On the Bering Strait side the central Polar Ocean is practically free from January cyclones. However, the storms which enter the Arctic from temperate latitudes are usually old storms in their dying phase, and they gradually fill and disappear at the surface. They are often seen at higher elevations, above 700 mbar (3 km), where they may persist for a long time as thermally symmetric "cold lows". They are usually small in extent and not very intense. Little bad weather is associated with them, the winter precipitation taking the form of ice crystals. The area of the Polar Ocean affected by an individual storm is not large, and the active life of the surface cyclones over the high latitudes is usually short—a few days as a rule.

During the summer months an area of relatively low pressure lies over the central polar regions, and the cyclones move into this from various directions, but most often from the North Atlantic and Bering Strait. In July, the cyclones converge in the central Polar Ocean, with a mean number of four in the eastern part (Alaskan side) and as many as six over the Chukchi Sea (RAGOZIN and CHUKANIN, 1959). As in winter, their life span in the Arctic is short.

There is a difference in the climatic importance of synoptic weather systems when the middle latitudes are compared with the Polar Ocean. In the former regions, synoptic systems control the weather and climate by an ever-changing sequence of different air masses. Over the Polar Ocean, cyclonic patterns generally do not influence the surface climate except so far as they control the turbulence in the inversion layer. Warm and cold spells in temperate latitudes are most often air mass phenomena (advection); over the Polar Ocean cold spells are caused by radiation, not by advection, while milder surface conditions require vertical transport downward of warmer air from above the inversion. Sporadic invasions of warm maritime air do occur, however, sometimes all the way to the Pole. An example of such advection is given in Fig.27, which shows a time-cross section of atmospheric temperature, based on observations at station "North Pole" 6 during the week of December 5–12, 1958. The position of the station was ap-

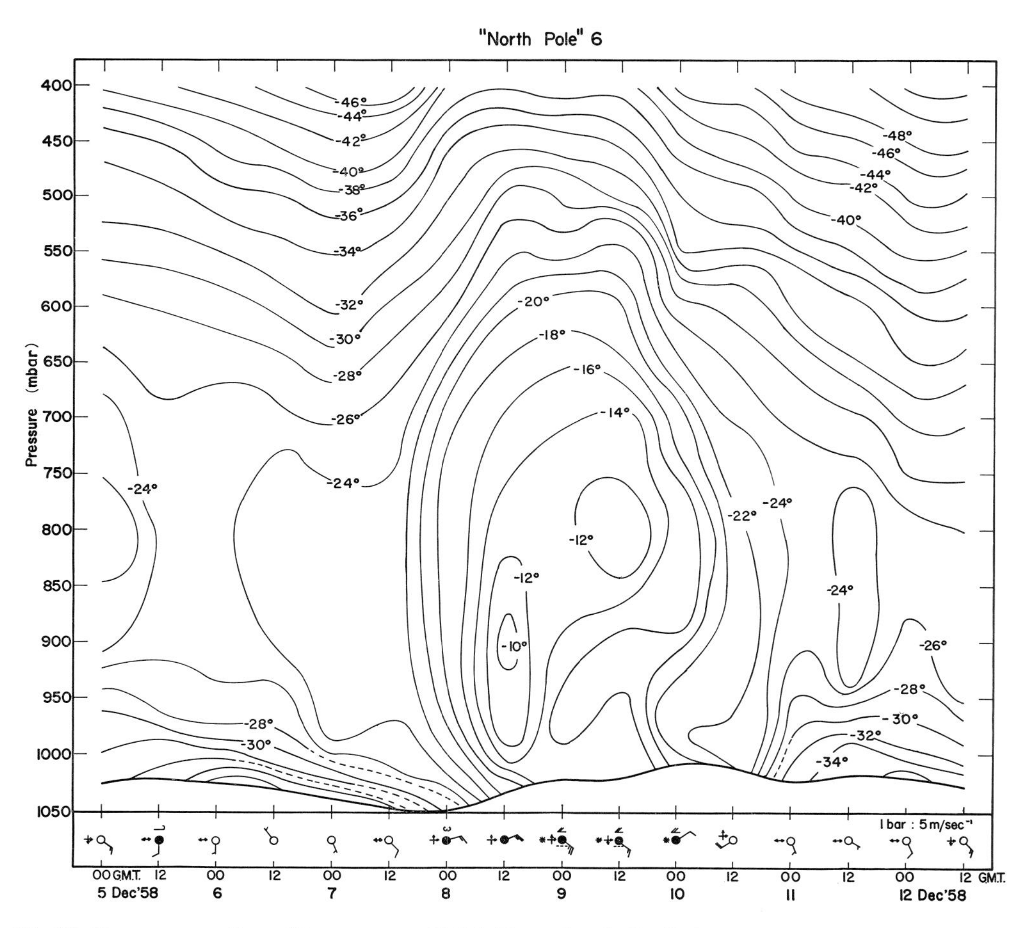

Fig.27. Time cross-section of temperature, N.P.6, December 5–12, 1958.

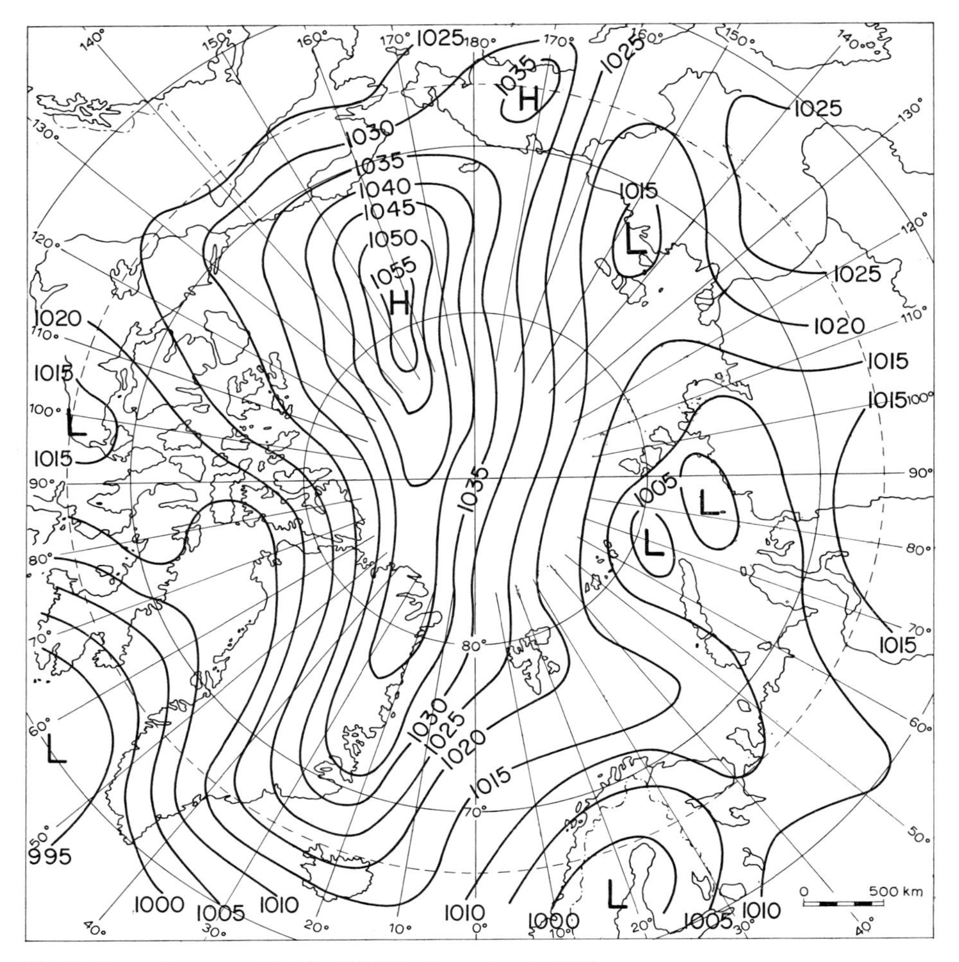

Fig.28. Synoptic map, midnight G.M.T., December 9, 1958.

proximately 86°N 100°E. Fig.28 shows the synoptic map for 00h00 G.M.T., Dec. 9, 1958.

The middle of the period experienced a warming at Station N.P. 6, caused by the remnants of cyclone systems moving eastward along the Eurasian coast. In the free atmosphere this resulted in a temperature increase of about 10°C. Associated with this warming was a cloud layer, appearing as medium cloud on the 8th and as Nimbostratus on the 9th and 10th. The cyclone also brought an increase in surface wind speed from 3–5 to 12 m/sec.

The result of these developments was, as shown in Fig.27, a gradual destruction of the surface temperature inversion, most pronounced on the 8th when the wind increase was strongest. The turbulence became sufficiently effective to break down the surface inversion completely on the 9th and 10th. The temperature decreased slightly with height in the first 500–1,000 m.

Clouds and turbulence act together to destroy the surface inversion in cases such as that shown; the former prohibit extreme radiational heat loss from the surface, the latter stirs the surface air and establishes a vertical lapse rate due to adiabatic cooling.

As the cloud deck disappeared and the wind dropped, the temperature started to fall

sharply at the surface, and the normal surface inversion was quickly re-established.
It is clear from Fig.27 that the lower tropospheric temperatures were warmer than those at the surface during the whole eight-day period. The surface temperature field is to a large extent separated from that higher in the troposphere. The effects of the free air disturbances are transmitted to the ground by turbulent mixing and by the associated clouds, which change the radiation balance.
It can also be seen, from the left-hand part of Fig.27, that the degree of coldness at the ground is actually independent of the type of air mass present aloft. The relation to upper air conditions is different for cold and warm spells in the arctic winter.
It is not usual to observe appearances of warm air advection as strong and well pronounced as that examined here. A more representative picture of Polar Ocean conditions would show a persistent surface inversion, varying in intensity with wind and clouds caused by weak disturbances.
In addition to moving lows without associated fronts, baroclinic development also occurs, mainly along the continental coasts during summer. Strong temperature contrasts develop due to the different absorption of solar radiation by land and ice. The land surface warms appreciably, but the ice absorbs only a small part of the insolation and cannot warm above the melting point. The summer storms generally develop in the continental coastal boundary but are not confined to any narrow strip. A broad zone of thermal contrasts may sharpen into fronts, the locations of which are determined both by the circulation and by the presence of heat sources and sinks (U.S. NAVY WEATHER RESEARCH FACILITY, 1962). Such summer frontal cyclones bring the heaviest precipitation to the Polar Ocean. The fronts are generally more distinct some distance above the surface. The cold air in contact with the ice often masks the existing temperature differences, especially in summer when a uniform surface temperature is found in all sectors. Above the surface layers, arctic disturbances show similar features of cloud distribution to the usual models of middle latitude frontal-type cyclones.
Traditionally it was thought that anticyclones play a major role in the lower arctic troposphere. It was widely believed that such systems were always cold anticyclones, deriving their excess mass from low temperatures in the bottom 2–4 km of the troposphere. It has become clear, however, that the strongest and most persistent arctic and subarctic anticyclones are neutral or have warm cores and are often preceded by advection of very warm air into the region at higher levels. Arctic anticyclones are as inhomogeneous in the vertical structure as those of middle latitudes, and frontal surfaces divide their arctic air from warm air, originating in lower latitudes.
In January, the erratic tracks of slow moving anticyclones originate in the Greenland high and also in the area of semi-permanent anticyclonic conditions over the Polar Ocean, centred at 78°N 165°W. The latter location is favoured by the import of migratory anticyclones emerging from the wintertime Siberian high. On the average only one to three anticyclones move across the basin during January. In summer, the Siberian high disappears and there is only a remnant of the semi-stationary anticyclone over the eastern portion of the Polar Ocean. One or two anticyclones move across the basin in July.
The accurate description of the geographical distribution of anticyclonic activity in the Arctic is difficult. Many of the most persistent systems are sprawling, formless belts of high pressure, lacking any distinct centre. Moreover, individual systems are often very slow moving. Hence the tracks of centres are often of limited interest.

In addition to the adiabatic processes in connection with vertical motion in the weather systems, large scale diabatic processes continually modify arctic air streams and contribute to development. Diabatic processes are especially important in extreme climates. They are of two main kinds—the radiative flux divergence of the free air itself, and the energy exchange at the surface boundary layer. The latter leads to modification of the surface air masses. Influences of heat sources like the warm surface waters of the North Atlantic and Pacific and heat sinks, like the Greenland ice cap, are quasi-conservative; they show up on the mean maps of the circulation and profoundly influence the behaviour of perturbations. Diabatic influences tend to contribute to cyclonic development in air moving over warmer surfaces, and to anticyclonic development in air moving over colder surfaces. In high northern latitudes, diabatic processes are particularly effective in winter, when very cold air often travels over regions with open, warm water.

The troposphere

Analyses of observations from the Soviet drifting stations in the central Polar Ocean (GAIGEROV, 1962) indicate that the monthly mean wind speed at the surface is not high, about 4–5 m/sec. However, in well developed cyclones the individual wind observations show speeds of 25–27 m/sec. The mean wind speed increases rapidly with height in the surface inversion layer. This is particularly true in the strong winter inversions. On the average, the tropospheric wind speed increases to a maximum just below the tropopause. At heights of 7–8 km the highest monthly mean wind speed may reach 20–25 m/sec. The tropospheric temperature gradients above the surface inversion may be large in individual cases but are small on the average. According to the thermal wind relationship, the tropospheric wind speed therefore changes only slowly with height. In winter, weak easterlies prevail up to about 5 km and relatively feeble westerlies above. Mean temperature and wind gradients are small in the summer troposphere, and the easterlies near the surface shrink in depth to less than 2 km (U.S. NAVY WEATHER RESEARCH FACILITY, 1962).

The thermal structure of the troposphere is rather variable. Considerable horizontal temperature differences occur in polar air masses, and the monthly mean heights of certain isotherms also show prominent annual variations. Over the central Polar Ocean in summer the 0°C isotherm generally appears to be associated with inversions, at heights close to 1 km. Along the coasts of the Arctic Ocean the 0°C isotherm may, in summer, reach heights of over 4 km. Advection of warm and cold air masses causes sharp changes in the heights of isotherms. The Soviet drifting stations have observed maximum and minimum monthly mean temperatures at an altitude of 5 km, north of 75°N, as follows (GAIGEROV, 1962): highest: −16.1°C (July, 1954); lowest: −41.7°C (January, 1955). The average of observations at five stations during the years 1954–1958 shows the 5 km temperatures to be highest in July (−18.3°) and lowest usually in January, although the lowest monthly mean may be experienced from November to March (−40.2°).

The strong negative radiation budget of the surface and troposphere in the central parts of the Arctic necessitates heat import from southerly latitudes. Part of this requirement is fulfilled by import via ocean currents, the main amount is advected in the atmosphere. As is the case with most energy transformations in nature, the net heat import into the Arctic is a small residual of a large turn-over. An example of this is given by the following measures of sensible heat transport in the layer between 700 mbar and 500 mbar, across the 70°N parallel of latitude, in February, 1961 (the values are non-dimensional and only intended to give relative magnitudes):

monthly import:	3,300 units
monthly export:	2,633 units
for a net import of:	667 units

The advected heat will be in the form of sensible as well as latent heat, the former being far more important in the Arctic, however.

The latent heat advection for the various parts of the Arctic Ocean are given in Table XXVII. Fig.29 shows the values for the central Polar Ocean. Apart from a very short spell in the autumn, the advection of latent heat is always positive into the Polar Ocean region, contributing as much as 65% of the annual precipitation. The major requirements for water vapour is locally supplied only for a time in fall and spring. The period of highest precipitation, in summer, is almost totally dependent on the import of water vapour. It is of further interest to note that, due to the small amounts of water vapour involved in the water budget, the storage of water vapour and its release is of importance for the annual march of precipitation. The period January–June accounts for 36% of the annual precipitation, and July–December for 64%. It is apparent from Fig.29 that, without changes in storage, the precipitation curve would be equal to the sum of evaporation (E) and moisture advection (A_E) and the difference in precipitation between the first and second half of the year would be sharply reduced. This phenomenon is solely dependent on the temperature variations and independent of the influence of vertical motion. Stronger vertical motion, with a corresponding increase in precipitation, would be associated with a higher supply from A_E as is in fact observed in mid-summer.

The same deduction can be made from the interesting fact that the small amount of water vapour export in September takes place at a time when $E + A_E$ is smaller than the precipitation and is therefore supplied solely from storage. This shows that the combined

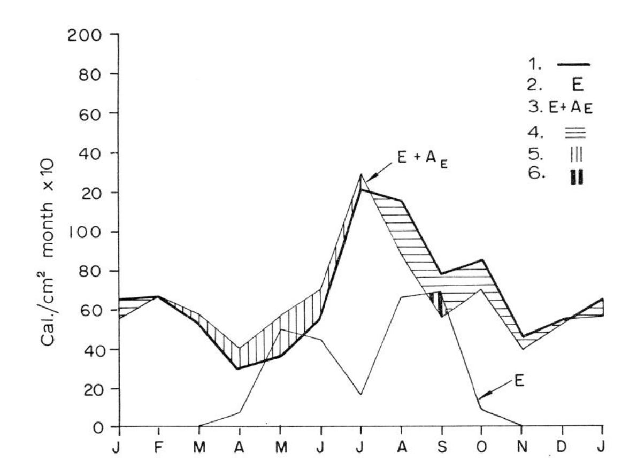

Fig.29. Precipitation, evaporation and moisture advection, central Polar Ocean. *1* = precipitation; *2* = evaporation; *3* = total evaporation and advection; *4* = water drawn from storage; *5* = water added to storage; *6* = negative advection (export).

TABLE XXVII

PRECIPITATION (P), EVAPORATION (E) AND LATENT HEAT ADVECTION (A_E) (cal./cm² month × 10)

	Jan.	Feb.	Mar.	Apr.	May	June	July	Aug.	Sept.	Oct.	Nov.	Dec.	Year	mm
Central Polar Ocean:														
P	66	66	54	30	36	48	120	114	78	84	45	54	795	133
E	-9	0	0	7	50	44	17	66	69	9	-9	0	244	41
A_E	65	66	57	33	6	26	109	22	-13	61	40	52	524	87
Norw.–Barents Sea:														
P	318	246	210	204	174	156	186	222	326	330	378	354	3,104	518
E	319	277	323	247	155	72	46	120	209	325	317	329	2,739	456
A_E	-3	-29	-113	-35	37	106	149	90	109	-9	50	18	370	62
Kara–Laptev Sea:														
P	115	108	96	102	96	102	162	198	150	114	114	108	1,465	244
E	0	10	11	34	67	131	14	16	101	90	9	0	483	81
A_E	111	98	89	79	54	-2	153	166	25	7	98	104	982	164
East Siberian Sea:														
P	60	54	54	48	60	120	168	162	156	102	108	72	1,164	194
E	0	0	0	34	52	48	8	63	58	97	0	0	360	60
A_E	59	56	59	27	34	94	165	83	71	-14	100	70	804	134
Beaufort Sea:														
P	66	66	54	48	72	84	174	180	156	150	72	72	1,194	199
E	0	0	7	23	63	11	59	15	110	51	0	7	346	58
A_E	66	66	51	40	33	95	125	150	16	78	64	64	848	141
Total Polar Ocean:														
P	77	72	66	48	54	66	132	138	102	96	48	66	965	161
E	-7	2	3	15	54	62	17	54	77	32	5	0	314	52
A_E	76	73	64	42	19	26	111	60	-1	47	55	65	637	106

water vapour supply from evaporation and from storage is too large to be condensed by the available vertical motion in the troposphere.

The maxima of effective water vapour advection are found in summer in all areas. This is the time when the circulation is at its minimum, which shows quite clearly that the advective term does in fact say very little about the actual amount of transport taking place. The magnitude of A_E is only partially determined by the rate of advective turnover; more important is the releasing mechanism, i.e., the vertical motion of the air. Hence the amount of heat gain by latent heat advection is a function of the magnitude of the wind component into the area, the vertical motion in the area, as well as of the initial saturation deficit of the air. With no vertical motion, no release of latent heat takes place in any case, regardless of the amount of advection or evaporation. Even with vertical motion, the release will obviously only take place if the ascent is strong enough to lift the air above the condensation level. Export of latent heat, on the other hand, can only take place if the rate of lifting does not cancel the moisture gain by evaporation.

The trade wind areas with cold air advection, which stimulates evaporation, and with no significant ascending motion, are the ideal latent heat export areas. Over the Norwegian–Barents Sea export does take place, in spite of strong lifting. This can only happen if the sensible heat flux from the surface simultaneously warms the air, overt compensating the cooling by vertical uplift. It should therefore be expected that the ratio of sensible heat flux to evaporation ($H : E$) is higher over the arctic latent heat export areas than over those of the sub-tropics. It should further be expected that during the export months the ratio is higher than during the import months. This is substantiat) ed by the following figures:

	Jan.–Apr.	May–Sept.
$H : E$	0.90	0.45

Table XXVII shows, in the last column, the annual amounts of water equivalent (mm corresponding to the heat figures in the table, for precipitation, evaporation and laten heat advection in the various areas.

The annual latent heat transport into the region of the central Polar Ocean is around 5.2 kcal./cm^2. This is close to the annual ocean heat transport, 5.9 kcal./cm^2. The atmospheric net advection of sensible heat is considerably greater, about 66 kcal./cm^2 year. The monthly and annual values for various areas are given in Table XXVIII. The advection is very effective, due to the steep horizontal temperature gradient between middle and high latitudes.

It is evident that sensible heat transport is far more important than latent heat advection in Arctic areas. The ratio $A_E : A_H$ is 1 : 13.6 in the central Polar Ocean, and 1 : 8.3 in the Norwegian–Barents Sea. This is in sharp contrast to the results of research in the Caribbean (COLON, 1960), where the ratio, regardless of sign, is 1 : 0.6 for December and January. It is apparent that the role played by latent heat transport diminishes sharply towards polar latitudes. The main reason is probably the smaller water holding capacity of the air at lower temperatures. The large vaporization heat of water, compared to the heat capacity of dry air, results in a great influence of relatively small temperature changes on the ratio $A_E : A_H$.

There is only a small seasonal variation in the sensible heat advection over the Polar Ocean. Over the Norwegian–Barents Sea there is a pronounced summer maximum. The reason lies in the mechanism of sensible heat release. Apart from an insignificant fraction, which is transferred in turbulent exchange to the surface, the bulk of the release takes place via atmospheric long wave radiation. The heat loss by radiation can only proceed until equilibrium is reached between the heat gain from the surface and the loss at the top of the atmosphere. The possible loss of sensible heat is, in the first instance, determined by the temperature excess of the advected air over the equilibrium temperature of the area into which the air is advected, rather than by its absolute temperature. The temperature gradient between 70° and 85°N is about 4.3°C in January and 4.5°C in July. The difference is very small indeed, and it can thus be expected that no great difference should be found between the summer and winter values of sensible heat advection.

The radiative release of sensible heat can only take place in an amount which is not already supplied by other mechanisms. The release of sensible heat from the surface and of latent heat of evaporation or advected latent heat are not governed by the radiative

TABLE XXVIII

ADVECTION OF SENSIBLE HEAT (A_H) AND TOTAL ADVECTIVE TERM (cal./cm² month × 10)

	Jan.	Feb.	Mar.	Apr.	May	June	July	Aug.	Sept.	Oct.	Nov.	Dec.	Year
Central Polar Ocean:													
A_H	667	557	617	382	516	429	500	484	510	684	639	628	6,613
$A_H + A_E$	732	623	674	415	522	455	609	506	497	745	679	680	7,137
Norw.–Barents Sea:													
A_H	186	100	80	181	339	439	488	469	303	308	87	71	3,051
$A_H + A_E$	183	71	−33	146	376	545	637	559	412	299	137	89	3,421
Kara–Laptev Sea:													
A_H	623	550	533	490	419	347	543	419	430	452	575	583	5,964
$A_H + A_E$	734	648	622	569	473	345	696	585	455	459	673	687	6,946
East Siberian Sea:													
A_H	689	597	643	473	578	360	366	522	460	339	580	654	6,261
$A_H + A_E$	748	653	702	500	612	454	531	605	531	325	680	724	7,065
Beaufort Sea:													
A_H	617	589	673	480	429	413	254	290	391	481	661	677	5,955
$A_H + A_E$	683	655	724	520	462	508	379	440	407	559	725	741	6,803
Total Polar Ocean*:													
A_H	659	561	600	483	491	405	496	461	485	612	626	622	6,501
$A_H + A_E$	735	634	664	525	510	431	607	521	484	659	681	687	7,138

* Not including Norwegian–Barents Sea.

requirements of the atmosphere. The former is dependent on wind speed and temperature differences in the boundary layer, and the latter on vertical motion. The sensible heat release is thus dependent on processes which are only indirect functions of the energy budget. It thus seems that the advected sensible heat generally is a passive rather than an active element in the energy budget. It can be regarded as a reserve fund, from which deficits can be covered and to which reserves can be moved. To a certain degree this is also true for advected latent heat. The difference is, however, that the energy from sensible heat is always directly available while latent heat only becomes available after its transformation into sensible heat.

Since the amount of withdrawal of energy from advected sensible heat determines the horizontal temperature gradient, and since this in turn determines the wind speed and rate of advection, it is clear that an increased requirement for advected heat automatically causes a greater supply—the atmosphere is a heat engine where differential heating causes differential pressure and subsequent flow to balance it.

The tropopause and lower stratosphere

Wherever frontal zones occur, high level strong winds, "jet streams", appear above them. Such bands of high winds are a regular feature in polar latitudes, although most of the strong wind currents near the arctic tropopause may not qualify as jet streams, according to the usual definition (U.S. Navy Weather Research Facility, 1962). In the analysis of upper level charts, a jet stream is indicated wherever the wind speed is 50 knots or more (25.7 m/sec). Maximum wind speeds over 50 m/sec may occur in the jets north of the Arctic Circle at all times of the year. When the pressure field shows large, blocking ridges near Scandinavia or Alaska, pronounced northward transport of warm air will take place in the high troposphere and extreme jet speeds will be observed in the northern sections of the warm high pressure ridges extending into the central Polar Ocean.

Numerous vertical cross-sections of the wind and temperature fields of the troposphere have been analysed, based on measurements at drifting stations (Gaigerov, 1962). It has been established that the tropospheric jet streams over the Polar Ocean are associated with marked temperature contrasts in the upper troposphere, with fronts reaching up to the tropopause, with warm pressure ridges and with higher elevation of the tropopause on the anticyclonic side of the jet stream and lowering on the cyclonic side. More than half of the jet streams observed at "North Pole" 4 in 1956–1957 were associated with occluded fronts.

The upper limit of the troposphere is the level where the temperature fall with height ceases or reverses sign. This is the tropopause, whose average level varies through the year from around 8 km in February–March to about 9.5 km in July–August. The tropopause rises also in anticyclones, to 12 km or higher, and descends in the centres of cyclones, sometimes to less than 5 km. On occasion the tropopause becomes indistinct, and breaks are found near the jet stream. The variations in height are usually greatest in winter.

Extreme data from some Soviet drifting stations are included in Tables XXIX, XXX.

The layer from the tropopause to about 20 km (50 mbar) may be termed the lower stratosphere. The circulation there is weaker than in the layers of the atmosphere below and above, and there are few disturbances. The tropospheric cyclones and anticyclones are almost completely damped out and have little energy left in this layer. Temperatures range from about −40°C in summer to −60°C in winter, and there is a rapid drop in the water vapour content above the tropopause to values as low as 2% relative humidity. Daily chart and cross-section analyses have lead to certain conclusions (Hare and

TABLE XXIX

MAXIMUM WIND SPEEDS IN THE UPPER ARCTIC TROPOSPHERE
(After Gaigerov, 1962.)

Year	Max. monthly mean wind speed (m/sec)	Month	Height (km)	Mean tropopause height (km)	Max. wind speed (m/sec)	Height (km)	Height of tropopause (km)
1954–1955	24.5	Dec.	7	8.13	63	7	7.71
1955–1956	21.4	Dec.	7	9.40	69	8	9.96
1956–1957	25.3	July	8	8.95	97	7	8.60
1957–1958	19.9	Feb.	8	7.91	78	7	7.64

TABLE XXX

EXTREME HEIGHTS AND TEMPERATURES OF THE ARCTIC TROPOPAUSE
(After GAIGEROV, 1957)

Year	Height				Temperature			
	max.		min.		max.		min.	
	(km)	(month)	(km)	(month)	(°C)	(month)	(°C)	(month)
1950	12.5	July	6.5	July	−37.5	July	−64.2	Oct.
1954–1955	12.4	Oct.	4.7	Feb.	−34.5	July	−70.0	Dec.
1954–1955	11.8	Oct.	4.4	Feb.	−33.0	July	−71.6	Nov.
1955–1956	12.1	Jan.	4.3	Sept.	−33.6	July	−73.0	Dec.

BOVILLE, 1965): the Northern Hemisphere westerly circulation extends from the troposphere through the lower stratosphere to about 20 km (50 mbar). The winds invariably decrease with height above the cores of the jet stream, near the tropopause. The decrease in circulation is caused by the stratospheric temperature distribution, showing a warm belt north of the jet stream and low temperatures in the tropical regions near the 100 mbar level (15 km). This temperature arrangement causes the polar tropospheric cold lows and anticyclones to be damped above the tropopause, and they rarely appear on 50 mbar charts.

Higher levels of the arctic stratosphere are dominated by great, seasonally reversing monsoonlike wind systems—easterlies in summer and westerlies in winter. Near the Pole these currents descend to well below 50 mbar (20 km). The winter 100 mbar maps, north of 60°N, usually show westerly winds which are part of the upper stratospheric system ("polar-night westerlies"). In summer (early April until late in August), light easterly winds descend to well below 20 km in latitudes 70°–75°N.

It can thus be seen that, in the Northern Hemisphere, there are two great sets of zonal winds—westerlies in the troposphere and seasonally reversing currents in the high latitude upper stratosphere. These zonal wind systems are quasi-independent, and a deep barotropic layer normally separates them (HARE and BOVILLE, 1965).

The temperatures of the stratosphere are controlled primarily by the layer of high ozone production, located between 30 and 60 km. Solar ultra-violet energy is absorbed in this layer, creating a temperature maximum (at about 50 km) in much the same way as the bulk of the solar radiation creates a temperature maximum at the ground. The tropopause, in fact, represents the temperature minimum between the two fixed heat sources. Its average height appears to be related to tropospheric convection, and its equatorward slope determines the stratospheric temperature pattern required to eliminate the tropospherically driven westerlies. Much of the uniqueness, and many of the problems, of the polar regions arise from the seasonal setting of the sun and the elimination of the two prime heat sources. Removal of the solar component at the ground tends to create a stable surface inversion and to uncouple the boundary layer of the Arctic from events elsewhere, even though coupling and heat transports may be significant in the troposphere above the inversion. Removal of the solar beam from the stratosphere creates a deep, cold region in the Arctic, and the north–south temperature gradient produces the polar-night westerlies. Due to this radiation effect the temperature minimum near the Pole in winter tends to lie in the 20–30 km layer, where the temperature may be as low as −90°C. At the same time the 8–10 km tropopause, associated with the Ferrel westerlies, is found throughout the peripheral areas and is often observed in the central region. The arctic lower stratosphere and tropopause thus tends to be bimodal, oscillating between the cold of the polar night stratosphere and the relative warmth of the Ferrel warm belt.

Some examples of extreme temperatures observed over Soviet drifting stations are given in Table XXXI.

The polar-night westerlies of the high stratosphere exhibit forced large-scale meanderings, much like the troposphere, and at times they break down spectacularly in sudden "explosive" warmings. Temperatures have risen locally as much as 40°C in a few days. The warmth may spread throughout the lower stratosphere and raise temperatures above summer levels. The result is anticyclogenesis and the slow establishment of the easterly summer wind regime (HARE and BOVILLE, 1965). Such intense, sudden warmings of the arctic stratosphere in winter is a characteristic feature which seems to combine the effects of warm air advection and adiabatic heating due to subsidence. An example of such a case of strong warming in January 1958 is given in Table XXXII (ZUBIAN, 1959).

TABLE XXXI

SOME EXTREME TEMPERATURES IN THE ARCTIC STRATOSPHERE
(After GAIGEROV, 1957)

Year	Maximum temperature			Minimum temperature		
	date	height (km)	temp. (°C)	date	height (km)	temp. (°C)
N.P. 3: 1954–1955	9 July	26.9	−27.0	21 Dec.	8.9	−70.2
N.P. 4: 1954–1955	8 July	27.8	−26.6	13 Nov.	9.5	−71.6
N.P. 4: 1955–1956	7 July	31.5	−21.2	4 Jan.	20.0	−81.2

TABLE XXXII

AN EXAMPLE OF STRONG WARMING

	mbar	km	°C
Jan. 12, 1958 station N.P. 7	100	15.2	−74.2
	50	19.2	−78.8
	30	21.9	−81.2
During 9 days from Jan. 24, the temperature rose as shown:			
	100	15.5	21
	50	19.7	33
	30	23.1	35

The temperatures in the layer from the surface to 300 mbar changed very little during the period of stratospheric warming.
It is at the time of these dramatic warmings of the winter stratosphere that one might look for the most obvious links between the two dynamical domains of the lower atmosphere—that below 20 km and the upper stratospheric domain, which extends to 50 km and perhaps even to much greater heights (HARE, 1962).

Climatic features of the Polar Basin

The first actual measurements of temperature, pressure, wind and other meteorological parameters in the central Polar Basin were made on the earliest Soviet drifting station—"North Pole 1"—from 21 May, 1937 to 19 February, 1938. The station was established on an ice floe near the geographic North Pole and drifted southward along the eastern shore of Greenland, to near Jan Mayen. Before that time, attempts had been made to construct atmospheric circulation schemes and pictures of climatic distributions based on various data obtained on several ship expeditions, drifting on the fringes of the central Polar Ocean.
Many previous suppositions were taken for granted until the era of drifting ice stations and airplane observations in the Arctic. A number of them have been proved inaccurate and contradictory (STEPANOVA, 1965).

Surface temperature

Large and rapid temperature fluctuations are not a common feature over the Polar Ocean. During the greater part of the year the area is covered by a relatively thin layer of cold air, which to a great extent is isolated from the atmosphere above. The air temperature near the surface is primarily dependent on the temperature of the ice surface itself. On occasion, however, warm air advection from the Atlantic may raise the winter temperature near the North Pole by as much as 30°C. After the warm air advection has ceased, radiation control again dominates the surface temperature, which will show an equally drastic drop. Strong winds may also cause higher surface temperatures by breaking down the inversion and mixing the warmer air from above with that near the cold surface. Renewed cooling follows every slackening of wind.
In summer, the prevailing melting of snow and ice holds the surface temperature close to the freezing point. The number of days with maximum temperature slightly above the freezing point is very nearly the same all along the latitude of 75°N—around 40 days. Nearer the Pole, positive temperatures are usually observed in the second half of July—some 10–15 days.
The winter temperatures over the pack ice remain nearly constant for a considerable time. This "flat" minimum represents the temperature at which the loss of heat from the snow and ice surface, by radiation, balances the amount of heat which is conducted to the surface from the water under the ice, and also the heat transported into the Arctic by intense warm air advection in cyclones. In periods of little cyclonic activity, heat transfer to the surface from the atmosphere is small, since the eddy conductivity is very small in the inversion layer. The winter air over the Arctic Ocean is, as a rule, nearly

saturated with water vapour. No heat is involved at the surface in freezing, melting, evaporation or condensation. Surface temperatures are therefore mainly governed by net radiation and conduction of heat to the surface from below.

Minimum air temperatures must occur when the net radiation loss is balanced by transport of heat from the water. This will give a minimum temperature at the surface of about −40°C, or down to −50°C over thick ice. The maximum temperature in overcast, calm weather is reached at about −25°C. The extreme maximum temperatures are nearly linear functions of the wind speed, increasing with increasing wind speed due to the transport of heat from above the inversion.

Fig.30–35 show the surface air temperature distribution over the Arctic, after PRIK (1959), for January, February, April, July, August and October.

The presence, or nearness, of open water is seen in the temperature distribution. All but the summer maps show broad tongues of mild temperatures extending into Barents Sea and Baffin Bay, and across Bering Strait.

The strongest influx of heat by cyclonic activity occurs in the Atlantic section of the Arctic Ocean in January and February, and as a consequence the lowest temperatures are observed in that section in March. In the other regions the lowest winter temperatures usually appear in January or February. From April until June the temperature rises quite rapidly, until the ice begins to melt. The mean summer surface temperatures remain fairly constant, conforming to the nature of the underlying surface. Almost all regions are warmest in July. Temperatures close to the melting point prevail over the pack ice and along the fringes of the Greenland ice cap. Cool temperatures dip southward across Bering Strait and Baffin Bay. Maximum temperatures in the Arctic Basin do not exceed 5°C.

An examination of large departures of monthly mean temperatures from normal values for twenty coastal or island stations, with twenty or more years of observations, showed (STEPANOVA, 1965) that the greatest departures are found in winter (from November to April) and small departures are most probable in July. In winter, the greatest departures were observed in the Kara Sea, the smallest in the East Siberian Sea.

TABLE XXXIII

AVERAGE DEPARTURE FROM THE NORMAL OF MONTHLY MEAN AIR TEMPERATURE, °C
(After PRIK, 1964)

	Jan.	Feb.	Mar.	Apr.	May	June	July	Aug.	Sept.	Oct.	Nov.	Dec.
Wrangel Island	2.4	1.8	2.1	2.1	1.5	0.9	0.5	0.7	1.2	2.0	2.7	1.7
Central region of Arctic Basin:												
Western sector	4.4	3.8	3.0	2.3	1.4	0.9	0.5	0.8	1.8	2.7	3.4	4.2
Eastern sector	2.6	2.3	2.3	2.3	1.3	0.5	0.3	0.7	1.1	2.0	2.8	2.8

In summer, the greatest departures from normal temperatures were observed on the coasts of the peripheral seas and decreasing toward the north. Another study gave the values of temperature departures which are contained in Table XXXIII.

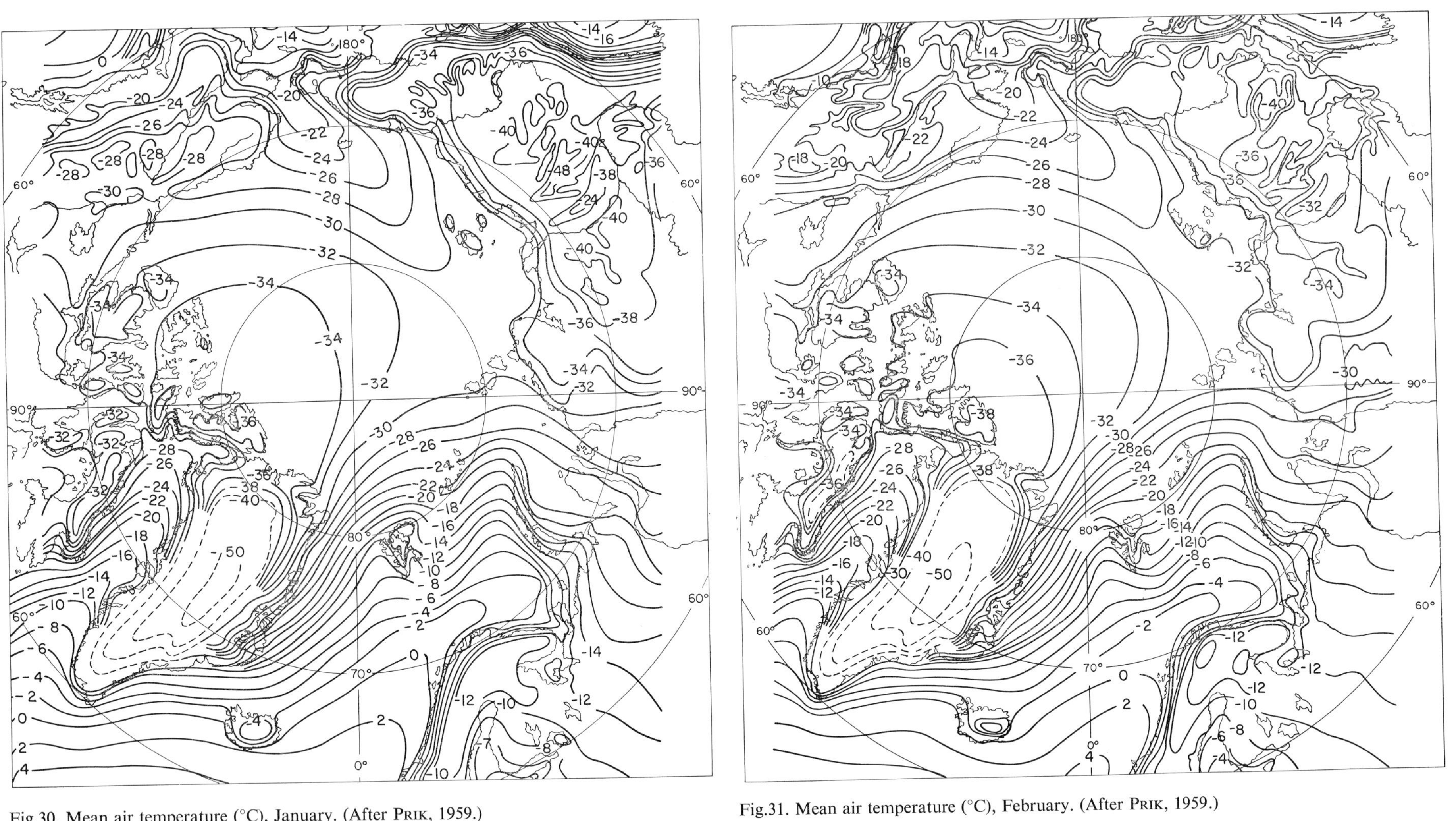

Fig.30. Mean air temperature (°C), January. (After PRIK, 1959.)

Fig.31. Mean air temperature (°C), February. (After PRIK, 1959.)

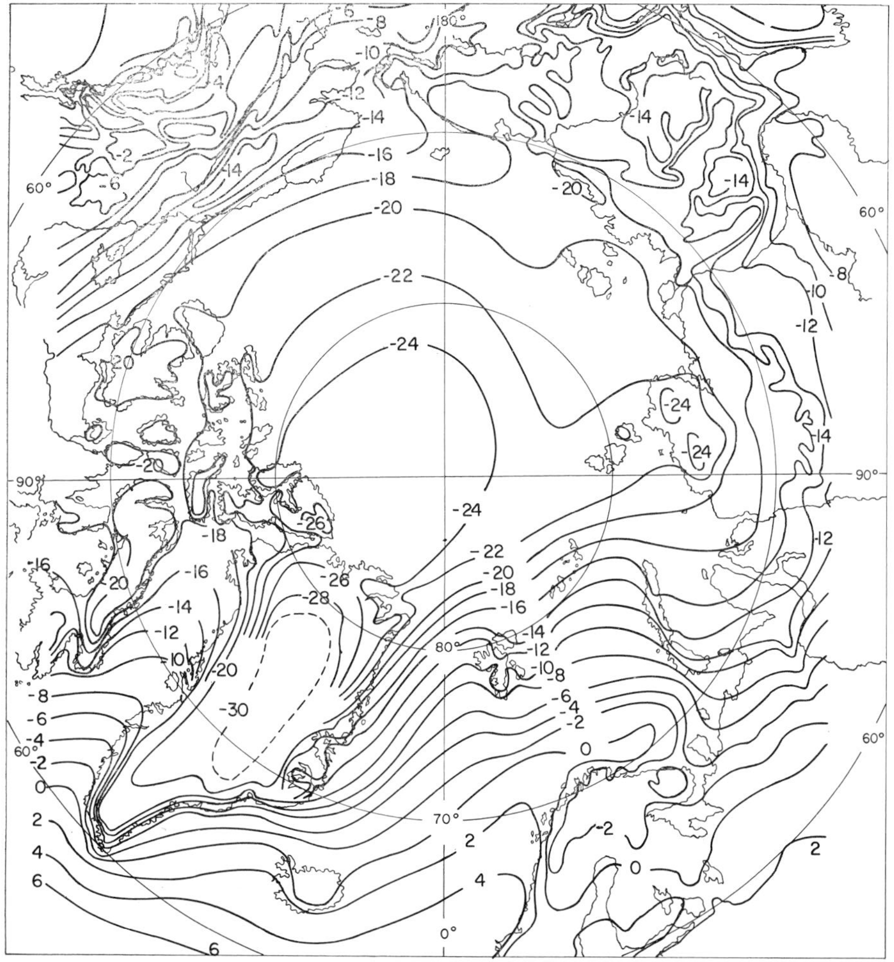

Fig.32. Mean air temperature (°C), April. (After PRIK, 1959.)

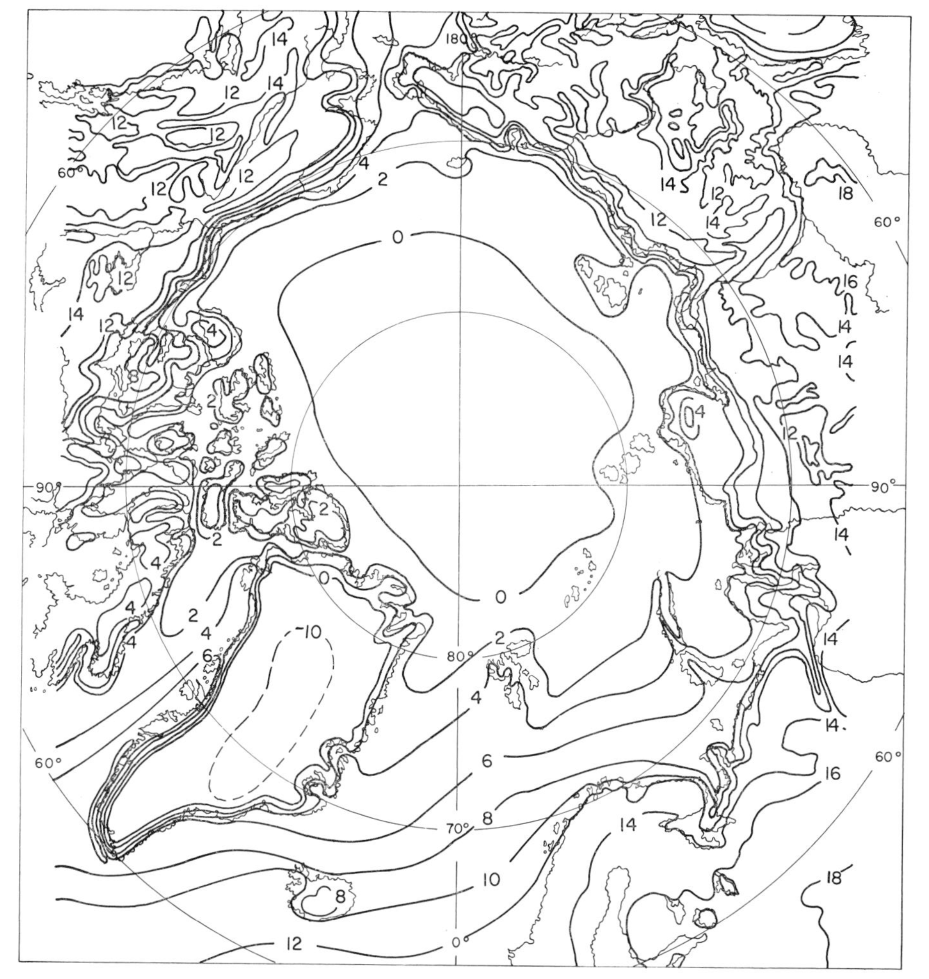

Fig.33. Mean air temperature (°C), July. (After PRIK, 1959.)

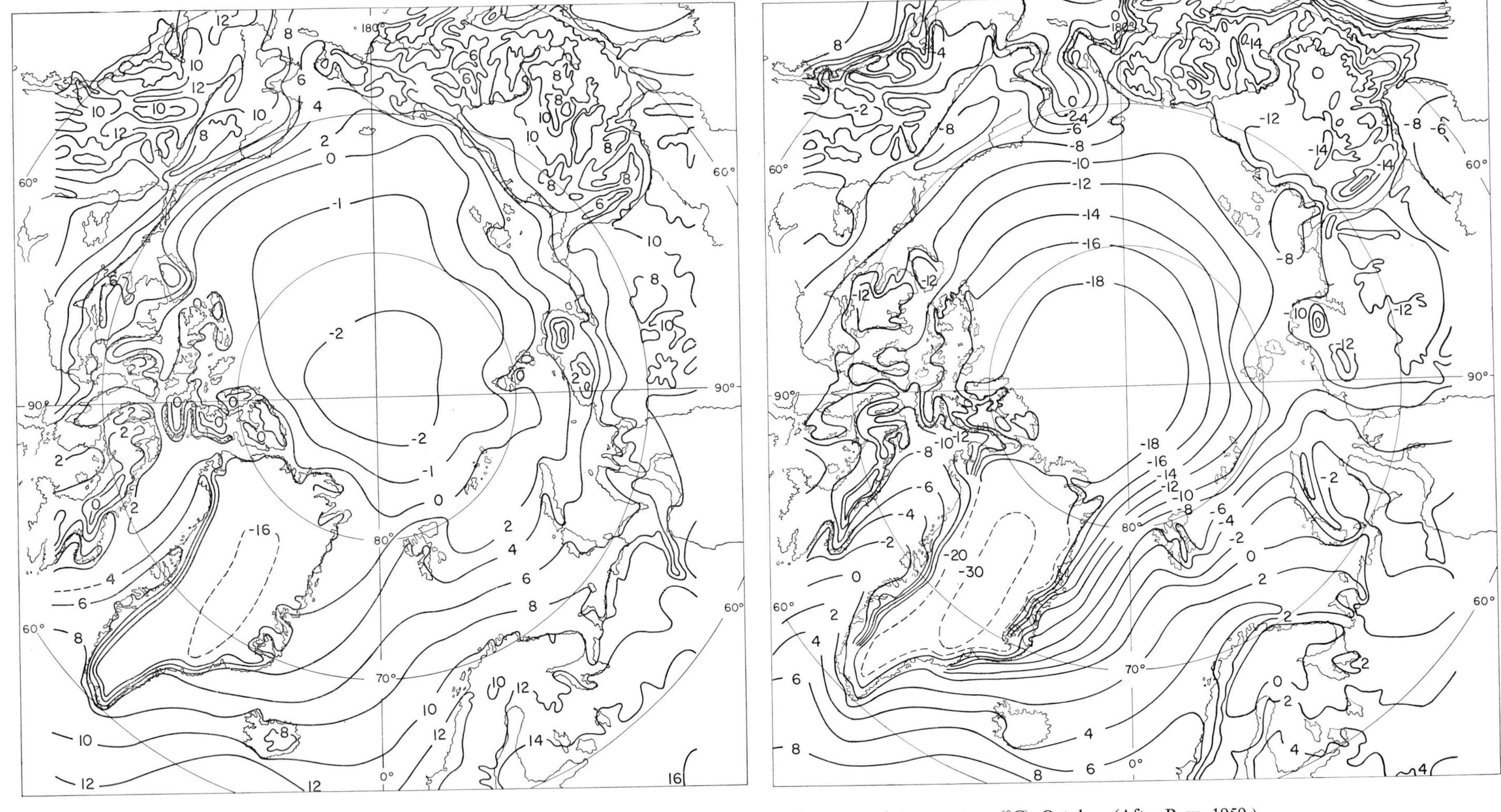

Fig.34. Mean air temperature (°C), August. (After PRIK, 1959.)

Fig.35. Mean air temperature (°C), October. (After PRIK, 1959.)

The temperature inversion

There are two main areas of semi-permanent inversion in the world: the subtropical belt and the polar regions. While the inversions are solely dynamic in the former area and separated from the surface by a highly unstable layer, the polar inversions are generally caused by the energy deficit at the surface. The strongest temperature gradient is therefore most often found near the surface, and the gradients are very much steeper than in the subtropical inversion. Although the main cause of the polar inversion is the negative energy balance at the surface, the phenomenon is not restricted to the surface layers but reaches about 2 km into the atmosphere. The arctic inversion is maintained in its normal position and intensity not only by surface cooling but also by significant subsidence of the air, as well as by warm air advection aloft. The arctic inversion is therefore a complex phenomenon, much more so than the subtropical type. The Polar Ocean is the only region in the Northern Hemisphere where the inversion is dominant practically during the whole year. As there are several causes for the formation of temperature inversions, they can be subdivided into the following groups:
Surface inversions: caused by a higher rate of radiative cooling at the surface than in the air. The gradients are generally very steep. The inversion layer remains relatively shallow.
Subsidence inversions: may be found at any height and are characterized by a sharp decrease of relative humidity in the inversion layer.
Advection inversions: may also occur at any height and are caused by advection of warmer air over an underlying cold air mass. The temperature rise in this inversion usually remains small, the relative humidity remains constant or increases.
It is apparent that reliable moisture observations in the upper air are decisive for a distinction between the different types. The very low air temperatures in the Arctic generally result in unreliable humidity observations which frequently do not reach the critical height. Typical examples of one or the other group are found only rarely. In most cases several causes work together, or they cancel each other, or they follow in such quick succession that the individual aerial sounding shows the result of several causes. This is especially the case in the Arctic, where the detailed structure of the inversion is rather complicated (Belmont, 1958).
Fig.36 shows the frequency distributions of the various stability types in the course of the year (Vowinckel and Orvig, 1967c). The overwhelming importance of inversions over the Polar Ocean is quite apparent: no month of the year shows less than 59% inversions, and in late winter the frequency reaches 100%. By far the majority of cases are inversions starting at the surface: in winter over 80%. Only during summer, from May to September, are upper inversions more frequent. The change from winter to summer conditions is very abrupt in the period April–May. The transition from summer to winter is much more gradual, lasting at least 2–3 months, and longer for surface inversions than for upper inversions. The reason for this longer transition in fall lies in the release of large amounts of energy stored in the ocean during the summer. This release is not terminated by the first ice formation; since the heat conductivity through new ice is quite substantial, the transport is decreased only as the ice gradually thickens.
No relation seems to exist between height and temperature in the inversion. If the surface inversion is regarded as a pure radiative phenomenon, a relation between height and temperature should exist. The surface inversion in the Arctic is not only a phenomenon

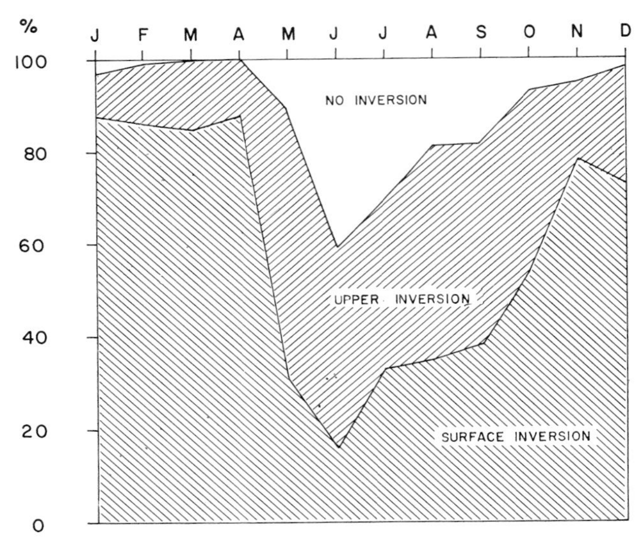

Fig.36. Frequency distribution of different inversion types.

of calm weather, but it also occurs with high wind turbulence, i.e., other than radiative processes play a significant part also in the surface inversion formation.

There is a close relation between intensity of the inversion and cloud amount. The surface temperature under surface inversion conditions is substantially lower with clear skies than in overcast conditions, and the temperature change in the inversion is greater. The inversion under cloudy conditions is less intense and the temperatures at the level of maximum are also higher. This is of greater significance to the surface temperature than the decrease in inversion intensity. The inversion is generally highest under clear sky conditions. On occasion, when a surface radiation inversion merges with an upper subsidence inversion, the system may extend to 4 km and reach an intensity of 25°C.

While the surface inversion is most frequent and has its best development during clear sky periods in winter, the reverse holds for the upper inversion. It is a most pronounced phenomenon in summer and is connected with great cloud amounts. From late September to late in April the frequency of surface inversions (i.e., those lacking a lower mixing layer) is from 70–90%, in summer it is 15–20%, with an average thickness of 1.5–1.9 km. The intensity and thickness of the upper inversion (with a surface mixing layer) are considerably less: 1.2°C and 0.5–0.9 km. Its maximum temperature is, generally, higher than the surface temperature, except for summer and early autumn. This is shown by the following per cent frequencies of occurrences of maximum inversion temperature being higher than the surface temperature (upper inversion):

Winter	Spring	Summer	Autumn
92%	67%	41%	49%

The upper inversion in summer is not a radiative phenomenon but rather one of large scale warm air advection gliding over a cool surface layer, with extended cloud formation in the lower part of the warm air. The advected air mass above does not break through to the surface. The upper warm air forms an inversion with the cool layer. Since the Arctic in summer receives a significant amount of diffuse short wave radiation under the cloud cover, the surface temperature will remain around 0°C, due to the presence of ice. This temperature can only be maintained in the near surface layers by turbulent heat exchange. Without such additional energy gain the air would cool by radiation to the temperature of the lowest cloud layer, which is lower than the surface temperature. This development can last for long periods—as long as the upper warm air advection continues. The stabilization of the surface temperature near 0°C prevents the destruction of the inversion from below. There are, in fact, occasions in summer when this type lasts up to seven consecutive days.

TABLE XXXIV

MEAN DURATION—IN 12-HOUR INTERVALS—OF VARIOUS INVERSION TYPES

	Winter	Spring	Summer	Autumn
Surface	8.1	8.3	2.0	4.5
Upper inversion	1.5	2.3	2.7	2.2
No inversion	1.7	1.8	2.7	1.3

Table XXXIV shows the mean duration of the various inversion types. The greatest variation is shown by the surface inversion with an extremely long mean duration in winter–spring and a normal duration in summer. The variations from season to season remain small for the other types.

The average importance of the inversion is best represented by the mean temperature difference between surface and the 850 mbar level (1.5 km). In summer there is an indication of a slight positive gradient only over the Beaufort Sea, i.e., warming with increasing height. All other areas of the Polar Ocean show slightly negative gradients. The gradients are much less than over the adjoining land masses. The open ocean of the Norwegian–Barents Sea has stronger negative gradients, although these still remain about one half of the continental values which may be as great as 8°C. The Kara and East Siberian seas show gradients intermediate between open ocean and inner polar pack ice area. This can be expected from the ice distribution. The temperature difference between open water and ice is relatively small during summer, and it can therefore be expected that the differences in vertical temperature gradient, induced by the surface, remain small. Both water and ice are cold surfaces with respect to the advected air, and a slightly stable stratification is therefore predominant near the surface. Conditions change drastically towards winter. There are then very unstable gradients over the warm waters of the Norwegian–Barents Sea, and a marked change takes place towards the Polar Ocean, where very strong inversions predominate. The strongest positive vertical gradient is found in the Beaufort Sea and northwest of the Canadian Archipelago (10°C), while the inversion is evidently less strongly developed along the Siberian coast along the main track of the Atlantic cyclones. It is therefore apparent that a belt of less intense inversions is found between the strong inversion area of inner Siberia and the western Polar Ocean. This interruption seems to be absent towards the anticyclonic area of Alaska–Yukon. The differences in intensity of the inversion over the central Polar Ocean remain remarkably small. The vertical gradients may be as steep as 12°C/100 m, while gradients of 1°C/m have been measured close to the snow surface. The highest values of temperature difference between surface and maximum temperature are found just north of the American coast in winter. Considering the relation between cloud amount, type and wind speed, and surface temperature, this distribution of intensity of surface inversion seems quite clear. The greatest intensity is found in the area of the Yukon high and the low cloudiness-low wind speed area north of the Canadian Archipelago.

Very high frequencies of winter surface inversions are found all over the polar area. The frequency does not sink below 70% inside the Polar Ocean. The frequency drops sharply only towards the open, warm water of the Barents Sea (less than 50%). A significant difference seems to exist within the Polar Ocean between the Siberian side, with values generally over 90%, and the American side with values well below 80%. This result is especially noteworthy as the main cyclone track runs from the Barents Sea eastwards just over the area of highest inversion frequency. Moving cyclones found over an area are therefore not sufficient to destroy an existing surface inversion in polar areas. On the contrary, the advection of warm air seems to stabilise the surface condition, and the cold air in the rear is not cold enough to destroy the inversion. However, the tropospheric cold pole is found over the Canadian Archipelago with rather unstable gradients in the upper air. If cyclones move over this area they are not associated to the same degree with warm air advection as along the Siberian coast, and they bring unstable conditions and dissolution of the surface inversion. The drifting stations "North Pole" 6 and 7 recorded certain features of the Polar Ocean inversion (DOLGIN, 1960; GAIGEROV, 1962). The annual mean frequency of inversions in cyclones was found to be 69%.

The mean height of the maximum temperature in the inversion is higher on the Siberian side than on the American side. It thus seems that the advectively determined maximum temperature along the Siberian coast results in higher elevations but less intensity of the inversion than on the American side, where subsidence is the dominating factor.

Fog, precipitation and snow cover

In winter, the relative humidity of the air over the Polar Ocean always remains near 100%. It should, in winter, be calculated as the relative humidity over ice, i.e., with respect to the saturation vapour pressure over ice, which is lower than over water. This permits the air to be slightly supersaturated in respect to ice, while it is not saturated in respect to water. It will cause water droplets to evaporate while water vapour condenses on ice crystals. Supersaturation occurs frequently near the surface in winter, as long as no condensation takes place. Normally, however, hoar frost formation begins when a value slightly over 100% has been reached. The relative humidity over ice has a maximum in midwinter (103%) and a minimum in June (95%), while the relative humidity over water has a maximum in August (96%) and a minimum in midwinter (73%). The actual pressure of the water vapour is very small in midwinter (0.2 mbar), and it has a maximum in July–August (6 mbar).

The amount of hoar frost which is deposited is proportional to the wind speed, and to the difference between actual pressure of the water vapour and the saturation vapour pressure over ice. Hoar frost is formed only when the relative humidity over ice is greater than 100%. The amount of hoar frost deposition is greatest at a temperature of around −29°C, while the probability of hoar frost has a maximum at about −32°C. In July and August there is relatively little hoar frost since air temperatures usually are around 0°C. Otherwise, the smallest amounts occur in April and the greatest in September. In the latter month there is, all over the Polar Ocean, a surplus of water in the air since open lanes are frequent in the sea ice cover. Hoar frost is considerable on vertical surfaces and may amount to 20 mm, about 15% of the annual total precipitation (MALMGREN, 1933). However, the deposition is less on a horizontal surface, and hoar frost is of small importance to the total snow cover in winter.

The melting of the pack ice in summer leads to formation of persistent fog and low cloud. More than 100 days per year experience fog at Polar Ocean stations, most frequently in summer and least in winter. The summer fog is generally caused by advection of relatively warm and moist air over the melting ice or cold water. It is patchy and of fairly short duration. Advection fog is particularly prevalent from June to September, occurring on 10% of all observations in June, 15% in July, 25% in August and 7% in September (U.S. NAVY WEATHER RESEARCH FACILITY, 1962). The fog does not occur at wind speeds above 10 m/sec. During the cold season, small patches of steam fog form over open water leads in the pack ice. This type of fog, sometimes called "arctic sea smoke", develops when very cold air blows over open water, causing rapid discharge of moisture and heat to the air.

The types mentioned above are water fogs. There occurs also in the Arctic winter a phenomenon called "ice fog". This may form where large quantities of water vapour are added to cold air, preferably about −30°C. Light ice particles with small fall velocity remain suspended in the stagnant air near the surface for a considerable period. Such fog occurs locally in the vicinity of human habitation.

The polar atmosphere is fairly uncontaminated by impurities and, apart from fog, only precipitation or blowing snow will reduce visibility. Wind speeds of about 2.5 m/sec will cause unconsolidated snow to drift along the surface. At speeds of 5–8 m/sec the snow is lifted into the air, and when it reaches a height of 1.8 m the term "blowing snow" is used. Coastal stations around the Polar Ocean may experience more than 100 days of blowing snow per year, in winter on more than half the days (U.S. NAVY WEATHER RESEARCH FACILITY, 1962).

In late autumn and winter, falls of ice crystals may be observed when the moisture of advected air condenses over the cold Polar Ocean. Fine ice needles form a slight haze as they settle slowly from an otherwise clear sky. Such falls of ice crystals add little to the snow cover, however. The main precipitation over the Polar Ocean is frontal in nature. Partly for this reason, the annual amounts decrease northward. Along the margins of the Polar Ocean the winter accumulation is for the most part less than 250 mm, the Siberian and Canadian arctic coasts receive as little as 140 mm per year (including rain and the water equivalent of snow). The annual precipitation over the Polar Ocean is meagre. It is mainly in the form of snow and falls during autumn and late spring. The probable average annual water equivalent over the central Polar Ocean is about 135 mm. The minimum precipitation occurs in winter. The cycle, and the paucity of precipitation, is partly caused by the low moisture holding capacity of cold air.

The characteristics of the snow cover, such as thickness and duration, have important climatic effects on the heat and moisture exchange at the surface. Over the central

Polar Ocean, the snow cover becomes established in late August. The thickness may be about 350–400 mm by late spring. Steady snow melt usually begins by the middle of June—caused by solar radiation—and, although there are marked differences from year to year, the ice is usually snow free by the middle of July. Fig.37 shows snow accumulation as measured at the Soviet station "North Pole" 4 (LOSHCHILOV, 1964). The precipitation was very similar during the two winters 1954–1955 and 1955–1956 (curves *2* and *3*). The newly formed ice received a net total of over 300 mm snow from January to April. How much was drifted snow and how much actually fell is not known. Fig.38 shows average snow accumulation and snow density changes throughout the year, based on several years' data for various drifting stations (LOSHCHILOV, 1964). The importance of precipitation in the autumn and spring is seen, as is the onset of rapid melting in early June, and the disappearance of the snow cover in July.

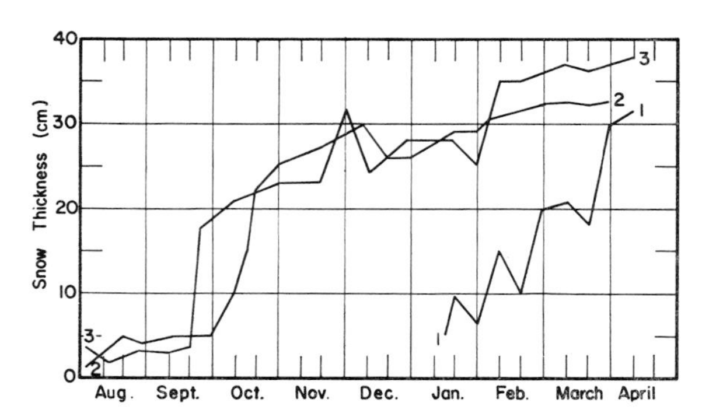

Fig.37. Snow accumulation on station N.P. 4. (After LOSHCHILOV, 1964.) *1* = accumulation on young ice; *2* = accumulation on polar (old) ice during 1954–1955; *3* = accumulation on polar (old) ice during 1955–1956.

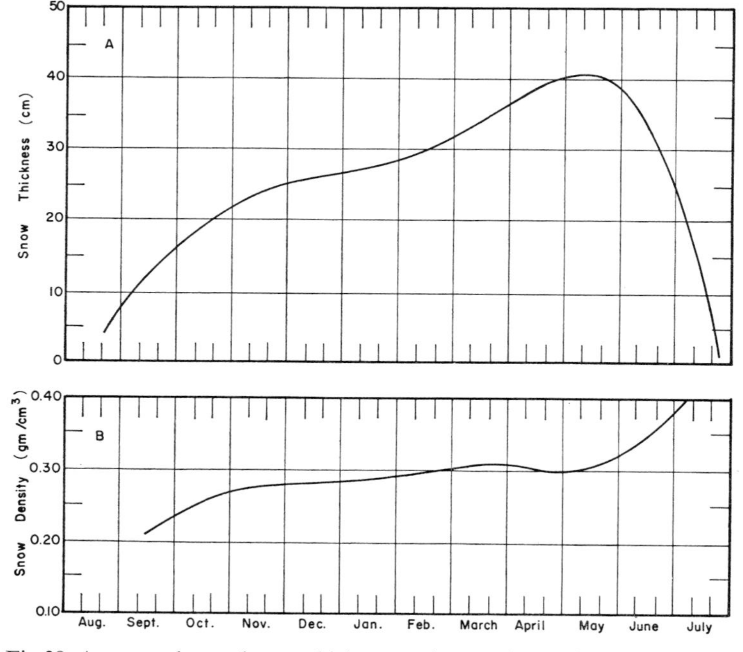

Fig.38. Average observed snow thickness and snow density from the central Polar Ocean. (After LOSHCHILOV, 1964.)

Surface pressure and wind

The meteorological observing network is quite sparse over the Polar Ocean. The main sources of data beyond the continental limits are various island stations and a few drifting ice stations, plus special expeditions and aircraft observation flights. As the surface is flat and uniform, observed winds are generally representative of the pressure field and often used to assist in the surface pressure analysis. Due to frictional influences, however, caution must be used in inferring surface wind from surface pressure patterns. The January map (Fig.39) after PRIK (1959), shows that the sea level circulation is dominated by four large cells, two continental highs and two oceanic lows. The large, semi-permanent Siberian high is joined to a smaller high over the Mackenzie Valley by an elongated ridge across the Polar Ocean. The Icelandic low controls the mean circulation over northern Scandinavia and the sea areas on the Eurasian side of the Pole; especially noteworthy is the elongated trough following open water and running from Iceland north of Norway to Novaya Zemlya and the Chelyuskin Peninsula. Another trough covers Baffin Bay. Both troughs are actually the scene of frequent cyclonic activity. The remaining cell is the smaller Aleutian low. The circulation over the Pole itself

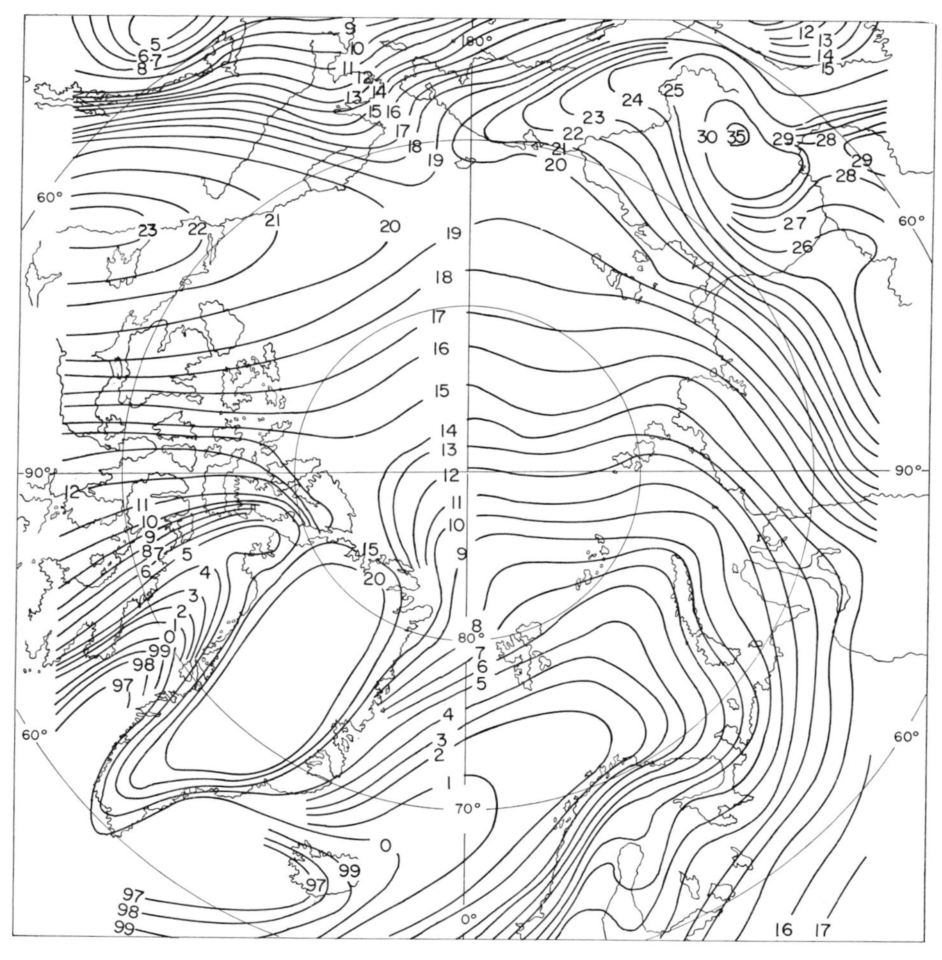

Fig.39. Mean air pressure (mbar), January. (After PRIK, 1959.)

is largely dominated by the cyclonic flow around the Icelandic low. It is noteworthy that there is no "Arctic high" on the January mean map, as stipulated by older theories for the Polar Ocean, although strong anticyclones are common enough over the Basin. The "polar easterlies" exist in the mean flow only over the Norwegian–Barents Sea area and along the north flank of the Aleutian low. Because of the extreme asymmetry and eccentricity of the sea level pressure field, no single latitude circle shows any appreciable integrated easterly flow; hence the polar easterlies are negligible sources of angular momentum for the circulation. Over the central Polar Ocean, the main air stream is directed from the middle and western Siberian coast toward the Pole and thence southward across the Greenland–Spitsbergen area. On the Bering Strait side of the Pole the Siberian air masses cross the Polar Ocean and invade the Canadian Archipelago, mixing with Alaskan–Yukon air. Over regions of elevated topography the pressure gradients are fictitious, as for example over Greenland, where the surface winds are not accurately depicted by the pressure map. Similarly, the strong pressure gradient over the mountain ranges of southeastern Alaska is also spurious.

The January pattern generally persists into March. Thereafter the mean pressure field undergoes rapid change (Fig.40). There is pronounced weakening of the low cells and of the Siberian continental high, and a relative intensification of high pressure over the Polar Ocean, caused partly by the North American high which shifts closer to the Pole. It is in the spring that strong anticyclonic activity is most likely to occur over the central Arctic, and it is at this time that the concept of a "polar anticyclone" is most nearly fulfilled. In the warm season (Fig.41), the central Polar Ocean is under a feeble low pressure area, while weak anticyclones are found over the peripheral areas of the Arctic Ocean, covering the Pacific section, with a ridge over the Beaufort Sea, and the Barents Sea. The Aleutian low fades, in summer, to a trough in the isobars and the Icelandic low is also quite weak. The Siberian high disappears completely and is replaced by a thermal low pressure area over the heated Asiatic land mass. The July mean pressure map thus shows a feeble pattern with a sluggish resultant mean flow. The prevailing air streams over the Eurasian coast are directed from the Polar Basin to the coast, with a marked easterly component, so that on the Siberian coast the winds are generally east-northeast, while on the coast westward of 90°E the wind is prevailing from the northeast. There is a well defined convergence of air streams in the central Polar Ocean in summer.

By October (Fig.42) the pressure patterns show a superficial resemblance to the spring map. Relatively high pressure occurs over the central Polar Ocean, and the sub-polar lows are back in their customary positions. The mean pattern, however, represents a compromise between contrasts of vigorous cyclonic circulation systems and calm periods of anticyclonic control. The Aleutian low has regained its full intensity and the Icelandic low is approaching wintertime strength, the extension over Novaya Zemlya being particularly noticeable.

The Siberian high attains moderate intensity in October. The high over Alaska–Yukon is not established, but the pressure is building over the Canadian Archipelago, to the northeast of its winter-time position.

The annual mean pressure map (Fig.43) shows the main characteristics of the seasonal maps—the sub-polar lows over the Aleutians and Iceland, with its trough to the Taimyr Peninsula, as well as the Baffin Bay pressure trough. Further, the Siberian and North

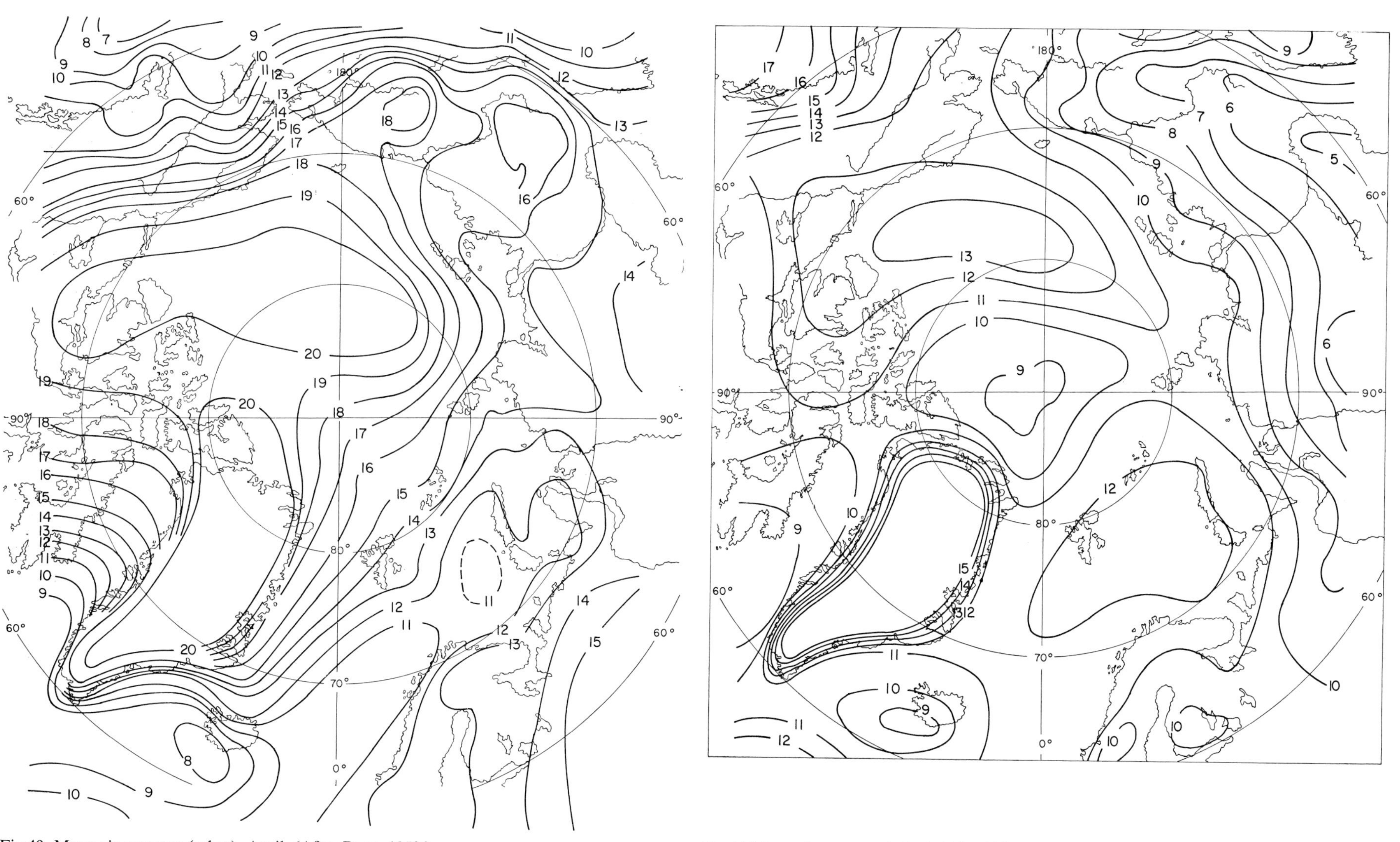

Fig.40. Mean air pressure (mbar), April. (After PRIK, 1959.)

Fig.41 Mean air pressure (mbar), July. (After PRIK, 1959.)

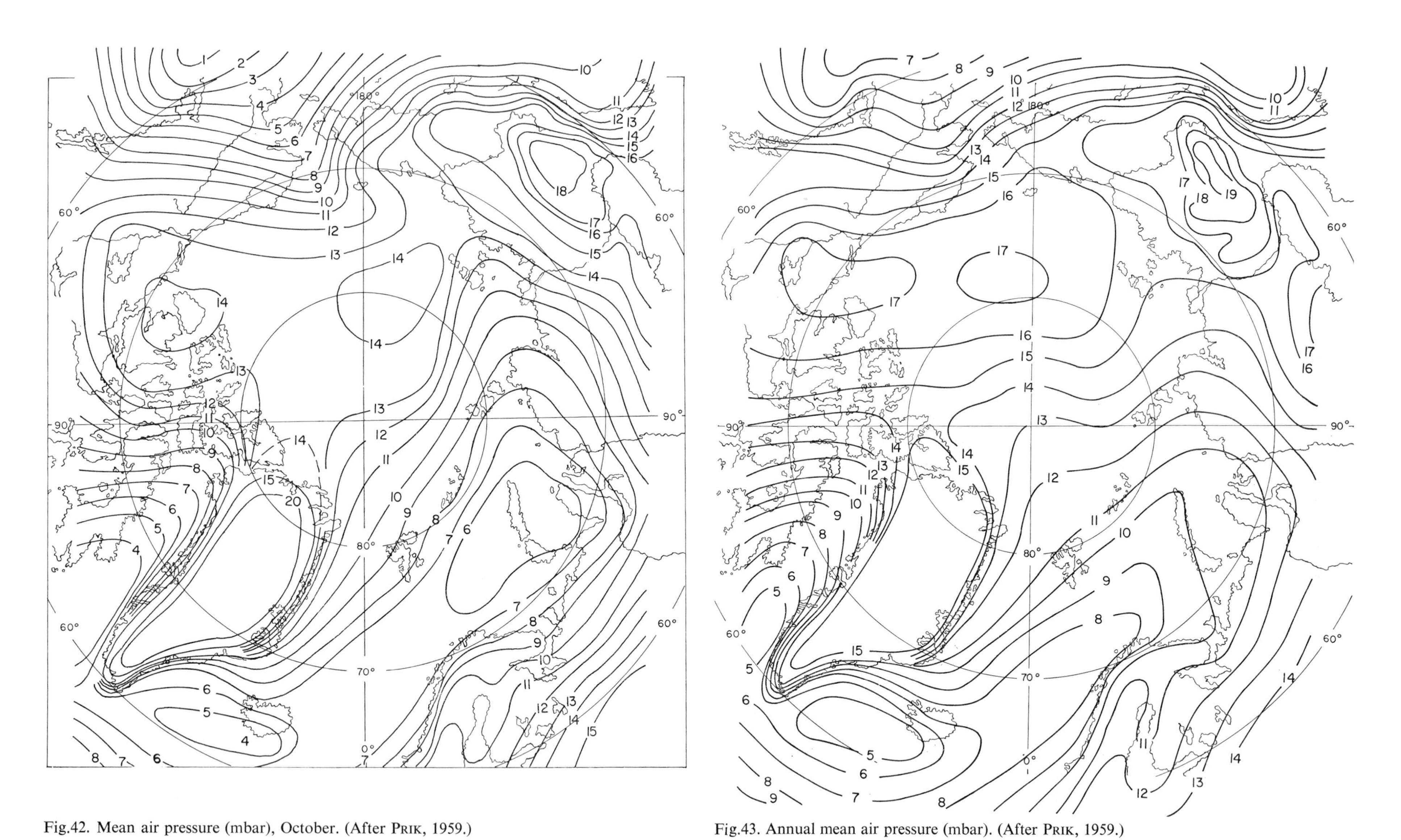

Fig.42. Mean air pressure (mbar), October. (After PRIK, 1959.)

Fig.43. Annual mean air pressure (mbar). (After PRIK, 1959.)

TABLE XXXV

AVERAGE DEPARTURE FROM THE NORMAL OF MONTHLY MEAN AIR PRESSURE (mbar)

	Jan.	Feb.	Mar.	Apr.	May	June	July	Aug.	Sept.	Oct.	Nov.	Dec.
Wrangel Island	3.5	4.6	3.5	2.7	2.2	2.0	1.4	2.5	2.5	3.9	4.4	4.2
Central Region of Arctic Basin:												
Western sector	5.5	8.1	6.8	4.3	4.0	3.5	3.2	3.2	3.4	3.8	4.0	5.7
Eastern sector	5.3	5.7	6.9	4.9	3.2	3.3	2.5	3.6	3.6	4.2	4.8	6.2

American highs, with a ridge extending across the Polar Ocean, and the rather flat, intermediate pressure field over the central Polar Ocean.

The variability of monthly pressure values is shown in Table XXXV.

Surface wind speeds are usually not very high over the Polar Ocean. On the average the force is 4–5 m/sec. When an inversion is present, the surface layer is effectively isolated from the faster moving air above and it is this fact, in combination with the lack of topographic effects, which results in the low number of occurrences of strong winds.

The annual mean wind speeds are greatest at exposed coastal stations near cyclone tracks, and it is in these same locations that gales are most frequent (Jan Mayen, northern Norway to Dickson Island, Bering Strait). A study of wind conditions over the central Polar Ocean, based on data from stations "North Pole" 6 and 7 (ZAVYALOVA and SERGEEVA, 1962) showed that the layer of the atmosphere from the surface to 0.5 km is characterized by a minimum interdiurnal variability of the wind speed near the ice surface (2–3 m/sec), and that the seasonal variability of wind speed is slight.

The interdiurnal variability of wind direction was found to be greatest near the ice surface (50°–70°), and less at greater heights. This is explained by the low surface wind speed

TABLE XXXVI

FREQUENCY DISTRIBUTION OF WIND SPEED OVER CENTRAL POLAR OCEAN (%)

Month	Wind speed (m/sec): 0	1	2	3	4	5	6	7	8	9	10–14	15–19	⩾20	No. of obs.
Jan.	11	7	8	10	13	12	8	8	6	4	10	2	1	564
Feb.	10	6	11	15	17	15	9	5	5	1	6	0	0	548
Mar.	6	6	8	18	22	15	7	6	4	2	5	1	0	585
Apr.	6	5	15	15	17	15	8	7	5	4	3	0	0	479
May	7	5	11	16	16	15	11	7	6	3	3	0	0	744
June	5	5	9	15	13	15	11	9	6	4	7	1	0	669
July	4	3	9	12	13	16	10	9	9	6	8	1	0	609
Aug.	4	4	7	11	11	15	11	11	8	5	11	2	0	570
Sept.	8	4	9	13	15	15	8	10	5	4	7	2	0	545
Oct.	7	5	9	12	12	12	10	9	7	4	11	2	0	586
Nov.	9	7	10	13	16	15	7	6	6	4	6	1	0	607
Dec.	11	10	14	14	17	12	6	5	4	3	4	0	0	607

and the consequent instability of the wind direction. Synoptic observations from stations "North Pole" 4, 6, 7, 8, 9, 10 and 11 have been used to prepare the frequency table (Table XXXVI) of surface wind speed.

The heat budget and stability of the Polar Ocean climate

The heat budget

The heat balance is of prime importance for the understanding of weather development and climatology in all climatic regions. The total budget is of interest, i.e., the sum of positive and negative terms, as are also the individual items. On the expenditure side, for instance, it matters whether the available energy is spent primarily on radiation and sensible heat flux, or on radiation and evaporation. Further, it is important if large amounts of energy are stored from month to month, and if the advective terms are considerable.

Table XXXVII gives the individual terms of the surface energy balance over the central Polar Ocean, and Fig.44 shows this and, in addition, the values for the East Siberian Sea and the Norwegian–Barents Sea, both being areas with characteristic features.

Table XXXVIII gives the individual terms of the surface budget for the whole Polar Ocean.

All radiation terms appear white in the figures and the non-radiative terms in various types of shading. It is apparent that the radiative terms are far greater than all other influences in all areas and months. Of the radiation terms, the long wave components are the greatest in all budgets. This result is not peculiar to polar latitudes, it is found everywhere, but the atmospheric back radiation is far more significant in the Arctic. It is further noteworthy that H plays practically no role in the income side for the surface, even in polar latitudes in winter under inversion conditions. The downward transport of sensible heat is so inefficient, compared to the radiative transfer, that it becomes practically negligible on the income side.

Considering the expenditure side of the surface energy budget, it is apparent from Fig. 44 that all expenditure in winter is radiative over the Polar Ocean, while over the Norwegian–Barents Sea about 20% is non-radiative. This additional loss is rather unfortunate for the surface budget. It will persist as long as the surface temperature remains higher than the air temperature. An effective check of this great loss is introduced by the formation of ice, which acts as a thermal insulator. The relatively small loss from the Polar Ocean is not caused by the exhaustion of the energy supply in the ocean in autumn, but rather by the passing of the critical temperature of $-1.8°C$, resulting in the formation of this insulating layer and the consequent rearrangement in the budget terms. It is further noteworthy that, even over the Polar Ocean, a substantial portion of the incoming radiative energy in summer is put into storage either by direct absorption in the water or by warming and melting of the ice. The storage during summer months amounts to 13,900 cal./cm^2, the total expenditure between September and April to 118,000 cal./cm^2. This indicates that about 12% of the total energy expenditure in winter is covered by the summer storage. It is of interest that the same figures for the Norwegian –Barents Sea are 24,400 and 196,800 respectively, or practically the same 12%. However,

TABLE XXXVII

INDIVIDUAL TERMS OF THE SURFACE ENERGY BALANCE, CENTRAL POLAR OCEAN (cal./cm² month (year) × 10)

	Jan.	Feb.	Mar.	Apr.	May	June	July	Aug.	Sept.	Oct.	Nov.	Dec.	Year
S			59	234	555	771	769	455	120	6			2,969
$L\downarrow$	1,008	902	973	1,095	1,507	1,722	1,916	1,882	1,641	1,448	1,110	1,039	16,243
$L\uparrow$	−1,244	−1,129	−1,271	−1,543	−1,758	−1,980	−2,093	−2,052	−1,812	−1,637	−1,373	−1,299	−19,191
E	9			−7	−50	−44	−16	−66	−69	−9	9		−243
H	12	6	6	−90	−100	−65	4	−56	−100		6	6	−371
O	215	221	233	311	−154	−404	−580	−163	220	192	248	254	593

TABLE XXXVIII

INDIVIDUAL TERMS OF THE SURFACE ENERGY BALANCE, POLAR OCEAN (INCLUDING PERIPHERAL SEAS) (cal./cm² month (year) × 10)

	Jan.	Feb.	Mar.	Apr.	May	June	July	Aug.	Sept.	Oct.	Nov.	Dec.	Year
S		3	67	237	564	788	805	500	148	12			3,124
$L\downarrow$	1,035	927	1,009	1,133	1,530	1,736	1,927	1,896	1,672	1,476	1,148	1,067	16,556
$L\uparrow$	−1,265	−1,153	−1,297	−1,462	−1,793	−1,992	−2,109	−2,080	−1,856	−1,689	−1,408	−1,324	−19,428
E	7	−2	−3	−15	−54	−62	−17	−54	−77	−32	−5		−314
H	13	6	7	−90	−98	−68	7	−40	−93	−34		7	−383
O	210	219	217	197	−149	−402	−613	−222	206	267	265	250	445

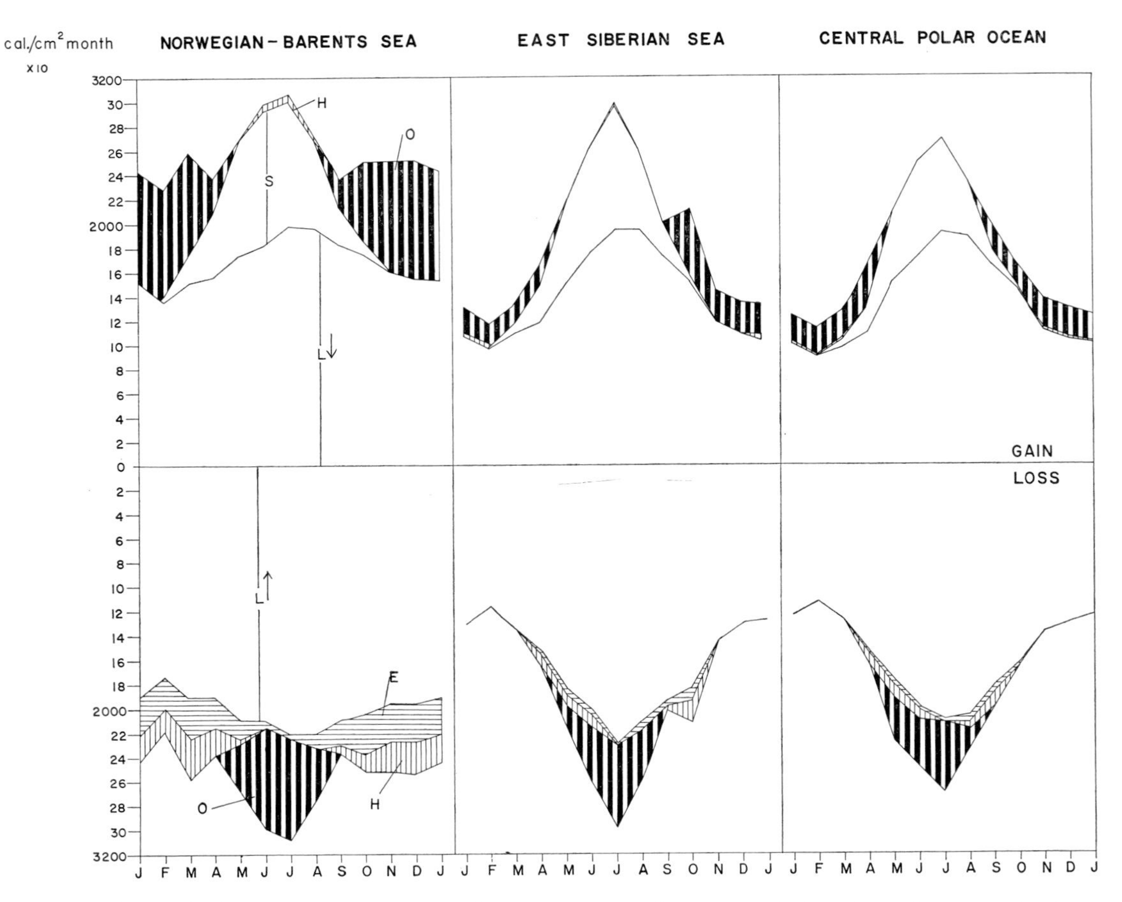

Fig.44. Surface energy budget terms. S = short wave radiation absorbed at the surface; O = heat exchange with deeper layers of ocean; $L\downarrow$ = long wave radiation down (atmospheric back radiation); $L\uparrow$ = long wave radiation from surface (terrestrial radiation); E, H = flux of latent (E) and sensible (H) heat from or to the surface.

considering that the annual total positive radiative flux for the Polar Ocean is 196,800 and for the Norwegian–Barents Sea 252,000 cal./cm², it is apparent that the efficiency of storage is higher in the Polar Ocean than in the Norwegian–Barents Sea. The reason is to be found in the different characteristics of water and ice surfaces. Especially in summer, when the temperatures approach 0°C, a great amount of energy is stored by the melting ice. This keeps the temperature near 0°C, and evaporation and sensible heat flux remain low as does the terrestrial radiation. This favourable effect of the ice is only valid in areas with significant melting in summer. In areas with permanently negative temperatures, as presumably the South Polar Plateau, this effect should not be found. The annual curve of total energy gain for the central Polar Ocean is rather smooth, while the curves for the Eurasian coast show a marked discontinuity in fall, and for the Norwegian–Barents Sea in fall and spring. It can be seen from Fig.44 that this is solely caused by the heat release from the ocean. It can be seen from the expenditure side of the budget that the irregularity is caused by the non-radiative transfer which is affected, to a degree, by factors other than those which govern the radiative transfer. While the radiative loss is solely a function of the surface temperature, and accordingly a smooth curve, H and E are governed by the difference of surface and air parameters which is, to a large extent, independent of the actual value of these parameters. With the onset of temperature fall in autumn, the temperature and moisture differences between air and water rise very rapidly over a short period of time, and hence the non-radiative terms increase sharply until a new equilibrium is reached. Over the Norwegian–Barents Sea, with the readily available energy from the ocean, the rise is able to compensate for the progressively smaller input by radiation, with the result that the energy budget remains extraordinarily stable during the winter. The drop in February is largely fictitious, as it results from the shorter length of that month. These processes decrease again by April, because the temperature differences diminish before the radiation terms can compensate for this drop.

Only the rise in October is apparent over the East Siberian Sea. The rapid formation of an insulating layer of ice actually reduces the high heat release of October to about one half during late fall and winter. Accordingly, the budget curve follows the normal pattern expected from a radiation curve, but at a somewhat higher level. No similarity to the Norwegian–Barents Sea conditions is found in spring, as the whole Siberian Sea area is ice covered by that time. It is clear that no similarities can be expected over the central Polar Ocean.

In Fig.45 a comparison is given of the energy currents at all levels in the annual average for the extreme areas. The illustrations were prepared in a fashion similar to those for the global balance, as used in many text books. The total incoming energy at the top of the atmosphere is taken as 100 units, and all other currents are given accordingly.

Although the general appearance of Fig.45 is rather similar to the global average, there are some noteworthy differences. The short wave radiation actually absorbed at the ground is 47 units in the global average, 37 in the Norwegian–Barents Sea, and 23 in the central Polar Ocean. This means that the available energy is used less efficiently in higher latitudes. A comparison between the Polar Ocean and the Norwegian–Barents Sea shows that part of it is caused by the higher albedo of the snow and ice surfaces. The other part must be ascribed to the higher scattering effect due to the lower elevation of the sun. The contribution of evaporation to the energy turnover in the Polar Ocean is also less than the global average, but in the Norwegian–Barents Sea the difference is insignificant.

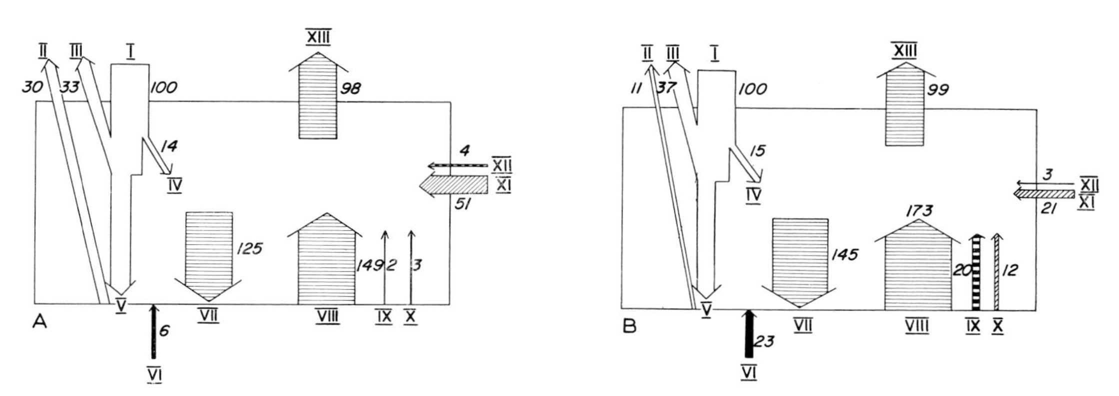

Fig.45. Energy balance. A. Central Polar Ocean. B. Norwegian–Barents Sea. *I* = extra-terrestrial radiation; *II* = reflected solar radiation (from surface); *III* = upward scattered solar radiation; *IV* = absorbed solar radiation (in atmosphere); *V* = global radiation; *VI* = heat flux below surface; *VII* = counter-radiation from atmosphere; *VIII* = terrestrial radiation; *IX* = flux of latent heat; *X* = flux of sensible heat; *XI* = advection of sensible heat; *XII* = advection of latent heat; *XIII* = upward flux of atmospheric radiation.

Climate change

Probably the most interesting and important remaining problem in the North Polar Basin is that of climate change and the future of the ice cover. Observations indicate a reduction in the amount of polar pack ice since the "Fram" expedition, 1893–1896. It is likely that both the ice area and the thickness have decreased. If the ice cover were to disappear completely, a quite different climatic regime would be established. It is interesting to speculate on, and attempt to evaluate, the probable conditions—whether or not the ice pack would re-form and which new climatic characteristics would become established. Speculations have been numerous, attempts at evaluation have been few (Donn and Ewing, 1966; Vowinckel and Orvig, 1967b).

Recent temperature fluctuations in high latitudes have been studied by many investigators, and it is clear that a period of warming began in the 1920's. In Spitsbergen, the average winter temperatures (December–February) were about 8°C lower in the period 1910–1930 than in 1931–1950. The same relationship can be seen in the records from Malye Karmakuly (72°23′N 52°44′E). There, the mean level of winter temperature (December–February) was about 11°C lower in the period 1897–1917 than in the period 1918–1940 (Stepanova, 1965). Other investigations of arctic coastal temperatures indicate that the recent period of warming ended, and a cooling trend began, somewhere in the decade from 1937–1946. Much more dramatic variations have taken place over long periods in the past.

A complete understanding of the climate and its fluctuations and variations can only be obtained if it is known how changes in particular terms in the energy balance equation influence the other terms and, finally, the whole budget. This must be so, because it is apparent that a strong interrelation and balance exists between the individual terms. As an example, it may be of interest to assess the significance of the advective heat transport terms for the temperature conditions in the area of the Polar Ocean.

Only a very crude approximation will be given, in order to show that processes which are numerically rather small may have a profound influence on the temperature level. Short wave radiation is obviously the only radiative heat source for the earth and atmosphere as a whole. Therefore, if advection is excluded, the energy expenditure in the form of long wave radiation would have to balance this income exactly. Under present surface and atmospheric conditions the short wave income would correspond to a certain radiative black body temperature. The present long wave radiation loss to space corresponds to another radiative temperature. The difference between these two temperatures shows the change which would be caused by the termination of advective import in atmosphere and ocean. Calculations of this kind show that the effective radiative temperature would drop by 35°C over the Norwegian–Barents Sea and by about 50°C over the central Polar Ocean. This would certainly not imply a similar temperature fall at the surface, because the stability conditions in the atmosphere would have to be readjusted drastically to obtain the necessary vertical transport, with very sharp cooling in the upper layers and smaller changes in the layers near the ground.

Similar calculations can be carried out, keeping the ocean import constant and omitting only the atmospheric advection. In the Norwegian–Barents Sea this would have a rather significant influence, reducing the temperature drop to 15°C, while in the central Polar Ocean it would still be as much as 36°C.

Of greatest interest and value would be the knowledge of which alterations in the individual heat transport terms are possible and can be realised under the conditions of the world in which we live. As climatic change results from a change in one or several of the terms of the energy balance equation, it can be studied either by determining the changes which have taken place in the various records (geological, botanical, climatological, etc.), or by analysing the present-day energy budget. From such an analysis one can ascertain the climate changes which would occur as a result of different values for the various parameters. There are three different types of influences affecting the energy budget:
(1) Geographical. A region may change its relative location (i.e., continental drift). The distribution of land–sea may change. Mountains may be built or eroded.
(2) Atmospheric. The circulation may change, resulting in changed conditions in certain regions. The composition of the atmosphere may change (e.g., CO_2 content, cloud amount).
(3) Astronomical. The energy output from the sun may change. Solar energy might increase by collision with clouds of interstellar dust. Finally, orbital changes may give rise to changes in the energy budget.
The following considerations assumed that the present-day geographical and astronomical conditions are fixed, and only variations in the atmospheric parameters were supposed. Further, the chemical composition of the atmosphere was regarded as constant. This approach allows quantitative statements to be made. The main difficulty is the close inter-dependency of the terms in the balance equation. At present it is difficult to make reasonable predictions of the complicated interactions subsequent to a change in even one of the terms in the energy budget (VOWINCKEL and ORVIG, 1967b). The terms themselves are uncertain, and it might be argued that it is doubtful if the small changes can be determined, which would result in climatic variations. This would be true especially for the long wave radiation terms, due to their high values. Various methods of determining long wave fluxes will, moreover, give different results. However, this is not particularly serious, if the method used is consistent for all long wave terms. The relative values will remain acceptable, although the absolute values may require adjustments.
The possibility of changes in factors governing long wave radiation is much more likely than for those governing absorbed solar radiation, except for changes in surface albedo which may range from the low values of open water to the high values of dry snow. Changes in the vertical temperature structure of the atmosphere would have a significant influence on the long wave radiation balance, and it would be possible to change the energy budget even without the influence of clouds. The polar atmosphere is at present adjusted in the best possible way for the conservation of energy in winter, by an efficient thermal stratification. The presence of clouds makes the winter inversion most effective. In summer the inversion is much weaker and the clouds accordingly less effective. The most variable factors in the long wave radiation balance are the cloud amount and cloud height. There is therefore a great number of possible variations in the long wave balance, according to the cloud conditions over the Polar Ocean. One can imagine two marked climatic conditions, different from the present—firstly a permanently open Polar Ocean, and secondly an ocean with a solid ice cover, without any snow and ice melting. The individual radiation terms under both conditions can be combined, and various radiation budgets determined. The results for some extreme examples are given

TABLE XXXIX

EXAMPLES OF VARIOUS RADIATION BUDGET POSSIBILITIES 75°–90°N (kcal./cm² per month, season, and year)

		Jan.	Feb.	Mar.	Apr.	May	June	July	Aug.	Sept.	Oct.	Nov.	Dec.	Winter	Summer	Year
Present surface																
cloud free	surface	−3.2	−3.2	−3.1	−1.1	+2.5	+6.3	+8.9	+3.6	−2.8	−4.7	−3.9	−3.6	−24.5	+20.2	−4.3
	earth–atmosphere	−9.6	−8.6	−8.9	−5.9	−1.6	+2.7	+4.4	−1.4	−8.9	−11.2	−10.2	−9.9	−67.3	−1.8	−69.1
10/10 stratus	surface	+1.0	+0.7	+0.9	+1.7	+2.0	+5.0	+5.9	+3.0	+0.2	−0.3	+0.4	+0.8	+3.7	+17.6	+21.3
	earth–atmosphere	−10.0	−9.0	−9.5	−7.4	−4.3	−0.8	−0.9	−5.1	−10.0	−11.2	−10.3	−10.2	−70.2	−18.5	−88.7
Frozen surface																
cloud free	surface	−3.2	−3.2	−3.1	−1.2	+0.4	+1.4	+0.8	−1.3	−3.8	−4.6	−3.9	−3.6	−25.4	+0.1	−25.3
	earth–atmosphere	−9.3	−8.4	−9.0	−5.9	−2.7	−0.9	−2.4	−5.3	−9.1	−10.9	−10.0	−9.9	−66.6	−17.2	−83.8
10/10 stratus	surface	+1.0	+0.7	+0.8	+1.7	+2.7	+3.0	+2.6	+1.4	−0.2	−0.3	+0.4	+0.8	+3.2	+11.4	+14.6
	earth–atmosphere	−10.0	−9.0	−9.5	−7.5	−5.5	−3.9	−5.6	−7.8	−10.4	−11.2	−10.3	−10.2	−70.6	−30.3	−100.9
Open ocean																
cloud free	surface	−6.2	−5.6	−4.6	+3.5	+13.5	+18.1	+15.7	+7.3	−2.1	−5.5	−5.7	−6.2	−35.9	+58.1	+22.2
	earth–atmosphere	−12.5	−11.3	−10.9	−1.7	+9.3	+14.5	+11.6	+2.7	−8.5	−12.7	−12.4	−12.7	−81.0	+36.4	−44.6
10/10 stratus	surface	−2.1	−2.0	−1.4	+3.3	+10.0	+13.0	+10.8	+5.7	0	−1.5	−1.8	−2.2	−11.0	+42.8	+31.8
	earth–atmosphere	−10.7	−10.6	−10.8	−5.3	+2.6	+7.2	+4.3	−2.0	−9.5	−11.9	−11.5	−11.8	−76.8	+6.8	−70.0
Present radiation balance Polar Ocean:																
	surface	−2.3	−2.2	−2.2	−0.9	+3.0	+5.3	+6.2	+3.2	−0.4	−2.0	−2.6	−2.6	−14.3	+16.8	+2.5
	earth–atmosphere	−9.5	−8.5	−8.4	−6.2	−2.4	+0.7	+0.2	−3.9	−8.2	−10.3	−9.9	−9.6	−64.4	−11.6	−76.0
Open Polar Ocean Cloud amount and type as presently at 70°N 0°E																
	earth–atmosphere	−10.7	−10.1	−10.2	−3.6	+5.3	+10.0	+7.1	0	−8.4	−11.4	−11.1	−11.3	−73.2	+18.8	−54.4

in Table XXXIX, which shows monthly, seasonal and yearly values for both surface and earth–atmosphere as a whole. The effects of changes in conditions, especially for the surface, depend on the time of year of occurrence. At present, with relatively high albedo, the summer difference between cloud free and overcast conditions is small, while the difference in winter is very high. The overcast conditions even show a positive winter balance at the surface. This situation cannot come about in reality, as it would require enormous atmospheric heat advection.

If the ocean were permanently frozen, the surface balance would be markedly less positive in summer. It is interesting to note that the annual surface balance with an open ocean and clear sky would be nearly similar to that with overcast sky and present surface conditions. This would be a result of the intense heat loss in winter with clear sky and open ocean, which would compensate for the increased summer heat gain. The optimum surface conditions would doubtless be realized with overcast conditions in winter and clear skies in summer. The following annual balances would result: present conditions: +23.9 kcal./cm^2; frozen surface: +3.3 kcal./cm^2; open ocean: +47.1 kcal./cm^2.

Only for the frozen surface would the value become less than with completely overcast conditions. This is so, because a certain high value of albedo would prevent the gain of heat by short wave radiation from compensating for the long wave loss. These optimum conditions are rather unlikely, however. While an open ocean would probably cause little change in cloud amount in summer, the winter conditions would tend to cause more cloud, perhaps between 70 and 80%, values presently found over the Norwegian Sea.

The radiation budgets for the earth–atmosphere system all show a strongly negative balance in Table XXXIX. In reality, an energy balance must exist. Therefore, the radiational deficit gives the necessary advection in atmosphere and ocean. The present ocean advection amounts to about 5 kcal./year. It is evident that even very large relative changes in this term will remain small compared to the atmospheric advection component.

The last line in Table XXXIX gives the earth–atmosphere radiation budget for an open Polar Ocean with cloud conditions as presently found over the Norwegian Sea. This assumption is probably better than that of clear sky or of completely overcast, but it does involve the supposition that circulation over the Polar Ocean should be similar to that now over the Norwegian Sea. This is, in reality, quite unlikely. It *is* likely, however, that any changes would tend in that direction. Considering the results, and those found for a frozen Polar Ocean, which must require clear winter–cloudy summer (and an annual surface balance near −14 kcal.), it becomes evident that the Polar Ocean is at present in a delicate radiational balance, and relatively minor variations in any term can instigate a gradually intensifying process, ending either in complete freeze-over or in complete melting.

Radiation balance calculations over a year make the tacit assumption that the surpluses in one season can be balanced against deficits in another. Over land surfaces this is certainly impossible. Over ocean surfaces it is more likely, due to the high storage capacity of the water. A second tacit assumption is that no export takes place of radiative energy, i.e., all seasonal surplus goes into storage. This is doubtful. It is only valid if the atmospheric horizontal gradients of temperature and moisture always remain so that wind export is excluded. At present, over the Polar Ocean, this is very nearly the case for monthly means. As soon as present day conditions are altered over the Polar Ocean, while the surrounding areas are kept as before, it becomes highly doubtful and for an open Polar Ocean it would be unlikely.

The results shown in Table XXXIX indicate that, under certain conditions and especially for an open Polar Ocean, the required annual advection decreases significantly. However, it would be erroneous to conclude from the radiation budget that the still large advective term would be possible, and that an energy balance would therefore be achievable. The energy potentially available by advection in a particular area is only partly dependent on the radiation budget of the area. The advected energy comes from other regions, and the upper limit of available energy is given by the difference in total heat content of the air over the area and of the advected air. If the temperature and moisture content of the air over the Polar Ocean are decreased, then the advective term will increase, provided that the circulation remains constant. If the air over the Polar Ocean becomes warmer and more moist, then the advection will decrease. With an open Polar Ocean in winter it is even possible that this term will become negative. This would increase the heat deficit over the Polar Ocean and might lead to reforming of the ice cover.

References

AAGAARD, K. and COACHMAN, L. K., 1968. The East Greenland current north of Denmark Strait. *Arctic*, 21: 181–200 (part I); 267–290 (part II).

ANDREYEVA, N. N. and PYATNENKOV, B. A., 1959. Certain peculiarities of the radiation balance of the central Arctic. *Probl. Arktiki*, 7: 78–89 (Transl.: Am. Meteorol. Soc., 1960, G.R.D.TR. 321).

ANTONOV, N. D., 1936. Amount of heat carried by the rivers into the Kara Sea. *Tr. Vses. Arkticheskii Inst.*, 35: 23–50 (in Russian).

ANTONOV, V. S., 1958. The role of continental drainage in the current regime of the Arctic Ocean. *Probl. Severa*, 1: 55–69 (in Russian).

BADGLEY, F., 1961. Heat balance at the surface of the Arctic Ocean. *Proc. Western Snow Conf., Spokane*, pp.101–104.

BELMONT, A. D., 1958. Lower tropospheric inversions at Ice Island T-3. In: R. C. SUTCLIFFE (Editor), *AGARD Symp. Polar Atmospheres, Oslo, 1956, Meteorol. Sect., Part. I.* Pergamon, London, pp. 215–284.

BRIAZGIN, N. N., 1959. The problem of the albedo on the surface of drifting ice. *Probl. Arktiki*, 1: 33–39 (Transl.: Am. Meteorol. Soc., 1960, G.R.D.TR. 310).

BUDYKO, M. I., 1956. *The Heat Balance of the Earth's Surface.* Gidrometeorologicheskoe Izd., Leningrad, 255 pp. (Transl.: U.S. Weather Bur., 1958, 259 pp.).

CHERNIGOVSKIY, N. T., 1967. Radiation regime of the central Arctic Basin. In: S. ORVIG (Editor), *W.M.O.–S.C.A.R.–I.C.P.M. Symp. Polar Meteorol., Proc., W.M.O. Tech. Note*, 87: 107–115.

COACHMAN, L. K. and BARNES, C. A., 1961. The contribution of Bering Sea water to the Arctic Ocean. *Arctic*, 14: 147–161.

COACHMAN, L. K. and AAGAARD, K., 1966. On the water exchange through Bering Strait. *Limnol. Oceanog.*, 11: 44–59.

COLON, J. A., 1960. On the heat balance of the troposphere and water body of the Caribbean Sea. *U.S. Dept. Comm., Weather Bur., Natl. Hurricane Res. Project*, 41: 65 pp.

DANSK METEOROLOGISK INSTITUT, 1900–1939, 1946–1956. *Isforholdene i de Arktiske Have (The state of the Ice in the Arctic Seas).* Appendices to Nautical-Meteorological Annuals, Charlottenlund, Copenhagen, Yearbooks.

DANSK METEOROLOGISK INSTITUT, 1960–1968. *Isforholdene i de Grönlandske Farvande (The ice conditions in the Greenland Waters).* Charlottenlund, Copenhagen, Yearbooks.

DEPARTMENT OF TRANSPORT, CANADA, METEOROLOGICAL BRANCH, 1961–1968. *Ice thickness data for selected Canadian stations, aerial ice observing and reconnaissance, and charts of ice conditions.* Various circulars. Toronto.

DEUTSCHES HYDROGRAPHISCHES INSTITUT, 1950. *Atlas der Eisverhältnisse des Nordatlantischen Ozeans und Übersichtskarten der Eisverhältnisse des Nord- und Südpolargebietes*, 18 pp., 34 charts.
DOLGIN, I. M., 1960. Arctic aero-climatological studies. *Probl. Arktiki*, 4: 64–75 (in Russian).
DOLGIN, I. M., 1966. Certain results of atmospheric investigations above the Arctic Ocean. In: N. A. OSTENSO (Editor), *Problems of the Arctic and Antarctic. A Collection of Articles*. Arctic Inst. North America, Washington, pp.d1–d14. (Originally published in *Probl. Arktiki*, 11, 1962.)
DONN, W. L. and EWING, M., 1966. A theory of ice ages, III. *Science*, 152: 1706–1712.
DORSEY, H. G., 1951. Arctic meteorology. In: T. F. MALONE (Editor), *Compendium of Meteorology*. Am. Meteorol. Soc., Boston, Mass., pp.942–951.
DROZDOVA, O. A. and POKROVSKAYA, T. V. (Editors), 1959. *Atlas of Maps of Air Pressure, Air Temperature and Atmospheric Precipitation*. Gidrometeorologischeskoe Izd., Leningrad, 38 maps.
DZERDZEEVSKII, B. L., 1945. The circulation of the atmosphere in the central Polar Basin. *Tr. Dreifuiushchei Stantsii "Severnyi Polius"*, 2: 64–200 (in Russian).
FERREL, W., 1889. *A popular Treatise on the Winds*. Wiley, New York, N.Y., 505 pp.
FLETCHER, J. O. (Editor), 1966. *Proceedings of the Symposium on the Arctic Heat Budget and Atmospheric Circulation, Jan. 31–Feb. 4, 1966*. Memo. RM-5233-NSF, The RAND Corp., Santa Monica, Calif., 567 pp.
FLETCHER, J. O., KELLER, B. and OLENICOFF, S. M. (Editors), 1966. *Soviet Data on the Arctic Heat Budget and its climatic Influence*. Memo. RM-5003-PR, The Rand Corp., Santa Monica, Calif., 205 pp.
GAIGEROV, S. S., 1957. Aerological observations made by the drifting station "North-Pole 4" in 1955–1956. *Tr. Tsentr. Aerolog. Observ.*, 18: 49 pp. (in Russian).
GAIGEROV, S. S., 1962. *The Problems of Aerological Structure, Circulation and Climate of the Free Atmosphere in the Central Arctic and Antarctic. Rezul' taty Issledovanii po Programme Mezhduna Rodnogo Geofiziches Kogo Goda. II Razdel Programmy MGG: Meteorologiia, No. 4*, Akad. Nauk S.S.S.R., Moscow, 316 pp. (in Russian).
GAVRILOVA, M. K., 1963. *Radiation Climate of the Arctic*. Gidrometeorologicheskoe Izd., Leningrad, 178 pp. (English translation by Israel Program for Scientific Translation, 1966).
GORDIENKO, P. A., 1958. Arctic ice drift. In: *Arctic Sea Ice, Natl. Acad. Sci., Natl. Res. Council, Publ.*, 598: 210–222.
GORDIENKO, P. A., 1958. Ice drift in the central Arctic Ocean. *Probl. Severa*, 1: 1–30 (in Russian).
GORDIENKO, P. A., 1966. Scientific observations from, and the nature of, drift of the "North-Pole" stations. In: N. A. OSTENSO (Editor), *Problems of the Arctic and Antarctic. A Collection of Articles*. Arctic Inst. North America, Washington, pp.bl–b19. (Originally published in *Probl. Arktiki*, 11, 1962.)
GORDIENKO, P. A. and KARELIN, D. B., 1945. Problems of the movement and distribution of ice in the Arctic Basin. *Probl. Arktiki*, 3: 5–35 (in Russian).
GORDIENKO, P. A. and LAKTIONOV, A. F., 1960. Principal results of the latest oceanographic research in the Arctic Basin. *Izv. Akad. Nauk S.S.S.R.*, 5: 22–33 (Transl.: Defence Res. Board Can. 1961, T 350 R: 12 pp).
HANSON, K. J., 1961. The albedo of sea-ice and ice islands in the Arctic Ocean Basin. *Arctic*, 14: 188–196.
HARE, F. K., 1962. The stratosphere. *Geograph. Rev.*, 52: 525–547.
HARE, F. K. and ORVIG, S., 1958. The Arctic circulation—a preliminary review. *Meteorology (McGill Univ.)*, 12: 211 pp.
HARE, F. K. and BOVILLE, B. W., 1965. The polar circulations. In: *The Circulation in the Stratosphere, Mesosphere and Lower Thermosphere—W.M.O. Tech. Note*, 70: 43–78.
HOUGHTON, H. G., 1954. On the annual heat balance of the Northern Hemisphere. *J. Meteorol.*, 11: 3–9.
INTERNATIONAL HYDROGRAPHIC BUREAU, 1953. Limits of oceans and seas. *Intern. Hydrograph. Bur., Spec. Publ.*, 23: 38 pp.
KEEGAN, T. J., 1958. Arctic synoptic activity in winter. *J. Meteorol.*, 15: 513–521.
LARSSON, P., 1963. The distribution of albedo over arctic surfaces. *Geograph. Rev.*, 53: 572–579.
LONDON, J., 1957. A study of the atmospheric heat balance. *N.Y. Univ., Res. Rept.*, 75 pp.
LOSHCHILOV, V. S., 1964. Snow cover on the ice of the central Arctic. *Probl. Arktiki*, 17: 36–45 (in Russian).
LYON, W., 1961. Ocean and sea-ice research in the Arctic Ocean via submarine. *Trans. N.Y. Acad. Sci., Ser. 2*, 23: 662–674.
MAKSIMOV, I. V., 1944. Determining the relative volume of the annual flow of Pacific water into the Arctic Ocean through Bering Strait. *Probl. Arktiki*, 2: 51–58 (in Russian).

MALMGREN, F., 1933. *On the properties of sea ice. Scientific Results of the Norwegian North Polar Expedition with the "Maud", 1918–1925.* Geofys. Inst., Bergen, 1a (5): 1–67.

MARSHUNOVA, M. S., 1959. Calculation of the longwave radiation balance during an overcast sky in the Arctic. *Tr. Arkticheskogo Nauchn.-Issled. Inst.*, 226: 109–112 (Transl.: Am. Meteorol. Soc., T.R. 304).

MARSHUNOVA, M. S., 1961. Principal characteristics of the radiation balance of the underlying surface and of the atmosphere in the Arctic. *Tr. Arkticheskogo Nauchn.-Issled. Inst.*, 229: 5–53 (Transl. by Rand Corp., Memo. RM-5003-PR).

MARSHUNOVA, M. S. and CHERNIGOVSKII, N. T., 1964. Elements of the radiation regime of the Soviet Arctic during the I.G.Y. and I.G.C. Meteorological conditions in the Arctic during I.G.Y. and I.G.C. *Tr. Arctic Antarctic Inst. Leningrad*, 266: 36–65 (in Russian).

MOSBY, H., 1962. Water, salt and heat balance of the North Polar Sea and of the Norwegian Sea. *Geofys. Publikasjoner, Norske Videnskap-Akad., Oslo*, 24: 289–313.

ORVIG, S. (Editor), 1967. W.M.O.–S.C.A.R.–I.C.P.M. symposium on Polar meteorology. Proceedings. *W.M.O. Tech. Note*, 87: 540 pp.

OSTENSO, N. A. (Editor), 1966. *Problems of the Arctic and Antarctic. A Collection of Arcticles.* Arctic Inst. North America, Washington, 261 pp. (Originally published in *Probl. Arktiki*, 11, 1962.)

PETROV, I. G., 1954–1955. Physical-mechanical properties and thickness of the sea ice cover. In: M. M. SOMOV (Editor), *Observational Data of the Scientific Research Drifting Station of 1950–1951.* Izd. Morskoi Transport, Leningrad, 2 (6): 1–60 (Transl.: Am. Meteorol. Soc., ASTIA DOC. NOS. AD 117132-40).

PRIK, Z. M., 1959. Mean position of surface pressure and temperature distribution in the Arctic. *Tr. Arkticheskogo Nauchn.-Issled. Inst.* 217: 5–34 (in Russian).

PRIK, Z. M., 1960. The main results of meteorological studies of the Arctic. *Probl. Arktiki*, 4: 76–90 (in Russian).

PRIK, Z. M., 1964. Pressure and temperature conditions in the Arctic during the I.G.Y. and I.G.C. *Tr. Arkticheskogo Nauchn.-Issled. Inst.* 266: 11–35 (in Russian).

RAGOZIN, A. I. and CHUKANIN, K. I., 1959. The mean trajectories and speed of movement of baric systems in Eurasian Arctic and sub-Arctic. *Tr. Arkticheskogo Nauchn.-Issled. Inst.*, 217: 35–64 (in Russian).

REED, R. J. and CAMPBELL, W. J., 1960. Theory and observation of the drift of Ice Station Alpha. *Univ. Wash., Res. Rept.*, 255 pp.

REED, R. J. and KUNKEL, B. A., 1960. The Arctic circulation in summer. *J. Meteorol.*, 17: 489–506.

SATER, J. E. (Editor), 1963. *The Arctic Basin.* Arctic Inst. North America, Washington, 319 pp.

SCHWARZACHER, W., 1959. Pack ice studies in the Arctic Ocean. *J. Geophys. Res.*, 64: 2,357–2,367.

SCHWERDTFEGER, P., 1963. The thermal properties of sea ice. *J. Glaciol.*, 4: 789–807.

SHULEIKIN, V. V., 1953. Molecular physics of the sea. *Fizika Moria*, 8 (1–9): 727–786 (Transl.: U. S. Navy Hydrographic Office, 1957).

SMITH, E. H., SOULE, F. M. and MOSBY, O., 1937. Marion and General Green expeditions to Davis Strait and Labrador Sea under the direction of U.S. Coast Guard. *U.S. Coast Guard Bull.*, 19: 200 pp.

SOMOV, M. M. (Editor), 1954–1955. *Observational Data of the Scientific Research Drifting Station of 1950–51.* Izd. Morskoi Transport, Leningrad, 1570 pp. (Transl.: Am. Meteorol. Soc., ASTIA DOC. NOS. AD 117132-40).

STEPANOVA, N. A., 1965. *Some Aspects of Meteorological Conditions in the Central Arctic. A Review of U.S.S.R. Investigations.* U.S. Dept. Comm., Weather Bur., Washington, D.C., 136 pp.

SUTCLIFFE, R. C. (Editor), 1958. *Meteorology Section. Part I, AGARD Polar Atmospheres Symposium, Oslo 1956.* Pergamon Press, London, 341 pp.

SVERDRUP, H. U., 1933. *Meteorology. Part I. Scientific Results of the Norwegian North Polar Expedition with the "Maud", 1918–1925.* Geofys. Inst., Bergen, 2 (1): 331 pp.

SVERDRUP, H. U., 1935. Übersicht über das Klima des Polarmeeres und des Kanadischen Archipels. *Handbuch der Klimatologie*, 2 (K): 3–30.

SVERDRUP, H. U., 1956. Oceanography of the Arctic. In: *The Dynamic North.* U.S. Navy, Washington Op-O3A3, 32 pp.

SVERDRUP, H. U., JOHNSON M. W. and FLEMING, R. H., 1942. *The Oceans.* Prentice-Hall, New York, N.Y., 1087 pp.

SWITHINBANK, C., 1960. *Ice Atlas of Arctic Canada.* Queen's Printer, Ottawa, 67 pp.

SYCHEV, K. A., 1960. The heat content of Atlantic waters and the expenditure of heat in the Arctic Basin. *Probl. Arktiki*, 3: 5–15. (Transl.: Am. Meteorol. Soc., 1960 TR. 336.)

TIMOFEYEV, V. T., 1956. Annual water balance of the Arctic Ocean. *Priroda*, 7: 89–91. (Transl.: Defence Res. Board Can., 1960, T 338 R: 3 pp.)

TIMOFEYEV, V. T., 1957. Atlantic water in the Arctic Basin. *Probl. Arktiki*, 2: 41–51. (Transl.: Am. Meteorol. Soc., TR 196: 13 pp.)

TIMOFEYEV, V. T., 1958. An approximate determination of the heat balance of the Arctic Basin waters. *Probl. Arktiki*, 4: 23–28. (Transl.: Am. Meteorol. Soc., 1958, TR 164: 9 pp.)

TIMOFEYEV, V. T., 1958. On the age of the Atlantic waters in the Arctic Basin. *Probl. Arktiki*, 5: 27–31 (in Russian).

TIMOFEYEV, V. T., 1960. *Water Masses of the Arctic Basin*. Gidrometeorologicheskoe. Izd., Leningrad, 190 pp. (Univ. of Washington transl., 1961).

UNTERSTEINER, N., 1961. On the mass and heat budget of Arctic sea ice. *Arch. Meteorol. Geophys. Bioklimatol.*, 12: 151–182.

UNTERSTEINER, N., 1964. Calculations of temperature regime and heat budget of sea ice in the central Arctic. *J. Geophys. Res.*, 69: 4755–4766.

U.S. NAVY HYDROGRAPHIC OFFICE, 1958. *Oceanographic Atlas of the Polar Seas. 2: Arctic.*, U.S. Navy Hydrographic Office Publ., 705: 149 pp.

U.S. NAVY WEATHER RESEARCH FACILITY, 1962. *Arctic Forecast Guide*. Norfolk, Va., NWRF 16-0462-058: 107 pp.

VOWINCKEL, E., 1962. Cloud amount and type over the Arctic. *Meteorology, (McGill Univ.)*, 51: 63 pp.

VOWINCKEL, E., 1964a. Heat flux through the Polar Ocean ice. *Meteorology, (McGill Univ.)*, 70: 16 pp.

VOWINCKEL, E., 1964b. Ice transport in the East Greenland current and its causes. *Arctic*, 17: 111–119.

VOWINCKEL, E. and S. ORVIG, 1962a. Relation between solar radiation income and cloud type in the Arctic. *J. Appl. Meteorol.*, 1: 552–559.

VOWINCKEL, E. and ORVIG, S., 1962b. Water balance and heat flow of the Arctic Ocean. *Arctic* 15: 205–223.

VOWINCKEL, E. and ORVIG, S., 1964a. Incoming and absorbed solar radiation at the ground in the Arctic. *Arch. Meteorol. Geophys. Bioklimatol.*, 13: 352–377.

VOWINCKEL, E. and ORVIG, S., 1964b. Long wave radiation and total radiation balance at the surface in the Arctic. *Arch. Meteorol. Geophys. Bioklimatol.*, 13: 451–479.

VOWINCKEL, E. and ORVIG, S., 1964c. Radiation balance of the troposphere and of the earth-atmosphere system in the Arctic. *Arch. Meteorol. Geophys. Bioklimatol.*, 13: 480–502.

VOWINCKEL, E. and ORVIG, S., 1966. The heat budget over the Arctic Ocean. *Arch. Meteorol. Geophys. Bioklimatol.*, 14: 303–325.

VOWINCKEL, E. and ORVIG, S., 1967a. The greenhouse effect of the arctic atmosphere. *Probl. Arktiki*, 25: 71–76 (in Russian).

VOWINCKEL, E. and ORVIG, S., 1967b. Climate change over the Polar Ocean, 1. The radiation budget. *Arch. Meteorol. Geophys. Bioklimatol.*, 15: 1–23.

VOWINCKEL, E. and ORVIG, S., 1967c. The inversion over the Polar Ocean. In: S. ORVIG (Editor), *W.M.O. –S.C.A.R.–I.C.P.M. Symp. Polar Meteorol., Proc.—W.M.O. Tech. Note*, 87: 39–59.

VOWINCKEL, E. and TAYLOR, B., 1965. Evaporation and sensible heat flux over the Arctic Ocean. *Arch. Meteorol. Geophys. Bioklimatol.*, 14: 36–52.

YAKOVLEV, G. N., 1955. The thermal regime of the sea ice cover. In: M. M. SOMOV (Editor), *Observational Data of the Scientific Research Drifting Station of 1950–1951*, 2 (7): 1–18. (Transl.: Am. Meteorol. Soc., AD 117138).

YAKOVLEV, G. N., 1958. Heat balance of the ice cover of the central Arctic. *Probl. Arktiki*, 5: 33–45 (in Russian).

ZAVYALOVA, I. N. and SERGEEVA, G. G., 1966. Variability of wind speed and direction above the central Arctic region. In: N. A. OSTENSO (Editor), *Problems of the Arctic and Antarctic. A Collection of Articles*. Arctic Inst. North America, Washington, pp.e1–e16. (Originally published in *Probl. Arktiki*, 11, 1962.)

ZUBIAN, G. D., 1959. On the intralatitudinal exchange of the warm and cold air masses in the stratosphere in winter. *Meteorol. i Gidrol.*, 1: 3–12 (in Russian).

ZUBOV, N. N., 1943. *Arctic Ice*. Izd. Glavsevmorputi, Moscow, 360 pp. (Transl.: Am. Meteorol. Soc. and U.S. Navy Oceanog. Office.)

TABLE XL

CLIMATIC TABLE FOR "NORTH POLE" 6 (MAY 1957–APRIL 1958)
Latitude 75°56′–81°14′N, longitude 170°18′–147°44′E, elevation surface

Month	Mean sta. press. (mbar)	Mean daily temp. (°C)	Mean daily temp. range (°C)	Temp. extremes (°C)		Mean vapour press. (mbar)	Mean precip. (mm)	Max. precip. in 24 h (mm)
				highest	lowest			
Jan.	1024.3	−31.4	8.0	−11.6	−48.7	0.387	10.5	3.6
Feb.	1027.6	−32.4	6.2	−22.6	−41.5	0.759	2.7	0.9
Mar.	1028.5	−30.7	7.0	−13.2	−38.7	0.423	3.4	0.9
Apr.	1023.4	−27.6	7.3	−12.6	−37.0	0.560	2.0	1.1
May	1021.6	−8.5	4.2	0.6	−22.1	2.866	3.4	1.0
June	1010.5	−0.8	2.4	1.5	−4.7	5.358	21.2	6.1
July	1003.3	−0.3	1.8	1.7	−4.5	5.736	33.4	14.4
Aug.	1001.2	−2.0	2.9	0.7	−6.9	4.906	42.5	10.9
Sept.	1016.2	−7.9	6.0	−0.7	−21.3	3.037	16.2	3.7
Oct.	1014.6	−17.4	7.6	−3.7	−32.4	1.392	17.0	6.1
Nov.	1028.4	−25.1	6.2	−12.7	−34.4	0.711	4.0	2.0
Dec.	1020.9	−36.5	5.4	−24.0	−48.1	0.229	1.6	0.8
Annual	1018.4	−18.8	5.4	1.7	−48.7	2.197	157.9	14.4

Month	Relat. hum. (%)	Most freq. wind dir.	Mean wind speed (m/sec)	Max. wind speed (m/sec)
Jan.	87	N	4.5	15
Feb.	84	N	5.1	14
Mar.	89	N	4.5	16
Apr.	88	W	4.2	10
May	89			
June	93			
July	96			
Aug.	93			
Sept.	90			
Oct.	89			
Nov.	89	W	5.6	13
Dec.	85	W	3.9	10
Annual	89		4.6	16

TABLE XLI

CLIMATIC TABLE FOR "NORTH POLE" 6 (MAY 1958–MARCH 1959)
Latitude 81°14′–85°55′N, longitude 147°47′–38°53′E, elevation surface

Month	Mean sta. press. (mbar)	Mean daily temp. (°C)	Mean daily temp. range (°C)	Temp. extremes (°C)		Mean vapour press. (mbar)	Mean precip. (mm)	Max. precip. in 24 h (mm)
				highest	lowest			
Jan.	1024	−33.8	9.3	−18.5	−44.1	0.289	12.8	4.0
Feb.	996	−37.8	8.7	−15.2	−50.2	0.194	12.5	3.9
Mar.	998	−32.6	8.0	−11.1	−45.7	0.325	13.7	3.1
Apr.	—	—	—	—	—	—	—	—
May	1024	−13.0	4.9	−4.5	−19.0	2.071	0.5	0.2
June	1012	−1.9	2.8	0.7	−9.8	4.995	11.0	5.3
July	1013	0.0	1.4	1.3	−2.4	5.985	26.0	13.6
Aug.	1009	−0.6	2.5	1.4	−5.7	5.671	38.9	11.1
Sept.	1008	−9.1	7.5	−0.4	−18.9	2.888	18.1	1.6
Oct.	1009	−21.6	8.8	−9.1	−36.7	0.993	10.9	2.0
Nov.	1020	−30.3	6.1	−16.7	−38.8	0.420	5.4	2.7
Dec.	1024	−28.8	9.4	−14.3	−39.9	0.495	7.6	1.5
Annual	1012	−20.0	6.3	1.4	−50.2	2.211	157.4	13.6

Month	Relat. hum. (%)	Most freq. wind dir.	Mean wind speed (m/sec)	Max. wind speed (m/sec)
Jan.	82	W	5.8	14
Feb.	82	W	5.2	17
Mar.	82	S	5.0	14
Apr.	—	—	—	—
May	92	S	4.0	9
June	94	W	5.0	16
July	98	S	4.7	10
Aug.	97	W	6.9	16
Sept.	94	E	5.8	14
Oct.	91	S	4.8	10
Nov.	85	N	4.7	14
Dec.	87	W	5.6	16
Annual	89		5.2	17

TABLE XLII

CLIMATIC TABLE FOR „NORTH POLE" 7 (MAY 1957–APRIL 1958)
Latitude 82°08′–86°34′N, longitude 164°W–148°25′E, elevation surface

Month	Mean sta. press. (mbar)	Mean daily temp. (°C)	Mean daily temp. range (°C)	Temp. extremes (°C)		Mean vapour press. (mbar)	Mean precip. (mm)	Max. precip. in 24 h (mm)
				highest	lowest			
Jan.	1018.6	−33.0	7.6	−12.4	−46.8	0.248	26.4	10.4
Feb.	1027.3	−37.3	6.4	−25.6	−43.6	0.204	1.1	0.5
Mar.	1027.9	−31.7	6.5	−20.9	−41.7	0.359	5.5	0.9
Apr.	1024.3	−28.8	5.3	−13.7	−40.8	0.478	3.0	1.4
May	1028.9	−5.9	4.7	−1.5	−30.2	3.503	0.4	0.3
June	1016.1	−2.1	3.2	2.6	−8.1	4.765	4.0	1.6
July	1008.1	0.0	1.8	1.6	−3.5	5.802	8.0	3.2
Aug.	1001.8	−2.9	2.9	1.4	−11.9	4.540	26.0	10.6
Sept.	1017.5	−12.3	7.3	−2.1	−30.8	2.096	5.7	1.2
Oct.	1012.3	−15.1	7.4	−2.8	−30.7	1.630	16.3	3.1
Nov.	1031.2	−29.6	4.9	−18.9	−39.0	0.417	0.7	0.6
Dec.	1022.2	−37.1	5.0	−23.6	−49.6	0.213	0.6	0.6
Annual	1019.7	−19.7	5.3	2.6	−49.6	1.961	97.7	10.4

Month	Relat. hum. (%)	Most freq. wind dir.	Mean wind speed (m/sec)	Max. wind speed (m/sec)
Jan.	88	N,E	5.3	20
Feb.	82	E	3.4	14
Mar.	83	N	4.7	12
Apr.	84	E	2.7	
May	89			8
June	91	W	4.8	16
July	95	W	4.9	10
Aug.	92	W	7.8	28
Sept.	88	N	4.7	14
Oct.	86	SE	7.8	24
Nov.	79	E	3.9	16
Dec.	84	N,E	3.0	14
Annual	87		4.8	28

TABLE XLIII

CLIMATIC TABLE FOR "NORTH POLE" 7 (MAY 1958–MARCH 1959)
Latitude 85°14′–86°28′N, longitude 150°02′–34°36′E, elevation surface

Month	Mean sta. press. (mbar)	Mean daily temp. (°C)	Mean daily temp. range (°C)	Temp. extremes (°C)		Mean vapour press. (mbar)	Mean precip. (mm)	Max. precip. in 24 h (mm)
				highest	lowest			
Jan.	1028	−36.3	7.1	−24.6	−46.0	0.195	9.6	1.7
Feb.	1001	−37.5	5.4	−27.5	−49.3	0.183	6.9	0.7
Mar.	1005	−34.2	4.7	−22.4	−44.0	0.261	25.8	4.9
Apr.	—	—	—	—	—	—	—	—
May	1022	−14.2	4.6	−7.6	−20.7	1.674	3.8	1.2
June	1019	−2.4	2.6	0.4	−10.6	4.814	13.1	2.8
July	1015	0.1	1.8	1.0	−1.4	5.967	12.7	3.8
Aug.	1018	−0.6	2.1	1.2	−4.2	5.671	20.9	5.5
Sept.	1006	−11.3	7.3	0.6	−26.8	2.323	27.7	6.4
Oct.	1009	−23.8	8.8	−10.0	−41.1	0.745	22.8	3.8
Nov.	1014	−27.7	7.5	−17.3	−36.2	1.491	24.2	3.8
Dec.	1023	−30.0	8.4	−15.5	−39.3	0.402	13.3	1.7
Annual	1015	−19.8	5.5	1.2	−49.3	1.975	180.8	6.4

Month	Relat. hum. (%)	Most freq. wind dir.	Mean wind speed (m/sec)	Max. wind speed (m/sec)
Jan.	76	E	2.9	7
Feb.	75	E	4.3	15
Mar.	77	E	6.0	16
Apr.	—	—	—	—
May	82	S	3.7	11
June	94	E	4.3	11
July	97	NE	3.4	9
Aug.	97	E	3.8	8
Sept.	90	E	4.6	13
Oct.	83	W	4.1	12
Nov.	78	S	4.5	10
Dec.	79	S	3.9	9
Annual	84		4.1	16

TABLE XLIV

CLIMATIC TABLE FOR "NORTH POLE" 8 (JUNE 1959–APRIL 1960)
Latitude 76°40′–79°22′N, longitude 164°10′W–178°50′E, elevation surface

Month	Mean sta. press. (mbar)	Mean daily temp. (°C)	Temp. extremes (°C) highest	Temp. extremes (°C) lowest	Mean vapour press. (mbar)	Relat. hum. (%)	Most freq. wind dir.	Mean wind speed (m/sec)	Max. wind speed (m/sec)
June 1959	1016	−1.6	1.8	−6.3	5.107	94	N	5.0	14
July	1008	−0.3	2.7	−3.2	5.677	95	S,W	5.2	14
Aug.	1020	−1.2	1.7	−7.0	5.259	94	W	4.6	16
Sept.	1022	−8.2	0.6	−20.2	3.000	91	S	4.8	13
Oct.	1015	−16.3	−0.7	−34.0	1.476	86	E	5.6	16
Nov.	1026	−25.4	−10.6	−32.0	0.622	80	NW	4.6	12
Dec.	1011	−29.9	−14.1	−40.3	0.410	80	E	6.6	16
Jan. 1960	1026	−32.0	−18.6	−45.2	0.332	79	S	4.2	16
Feb.	1024	−28.5	−11.7	−40.9	0.468	80	W	5.9	16
Mar.	1025	−28.6	−16.7	−43.5	0.470	81	E	5.1	14
Apr.	1027	−23.4	−13.0	−35.7	0.754	81	N	5.4	14
Annual	1019	−17.8	2.7	−45.2	2.143	86		5.2	16

TABLE XLV

CLIMATIC TABLE FOR "NORTH POLE" 8 (MAY 1960–APRIL 1961)
Latitude 70°28′–83°57′N, longitude 176°06′E–148°42′W, elevation surface

Month	Mean sta. press. (mbar)	Mean daily temp. (°C)	Mean daily temp. range (°C)	Temp. extremes (°C) highest	Temp. extremes (°C) lowest	Mean vapour press. (mbar)	Mean precip. (mm)	Max. precip. in 24 h (mm)
May 1960	1021.8	−9.6	5.3	−0.1	−19.9	2.540	3.3	0.5
June	1011.9	−2.1	3.9	1.4	−9.2	4.922	10.4	3.6
July	1014.9	0.2	2.1	2.8	−1.9	5.949	11.1	4.7
Aug.	1022.3	−1.9	3.9	1.9	−7.2	4.995	2.3	1.1
Sept.	1017.3	−7.6	6.0	−1.8	−20.2	3.212	16.3	4.7
Oct.	1018.8	−17.0	7.6	−6.3	−26.4	1.359	11.1	6.4
Nov.	1020.4	−25.6	7.0	−15.1	−40.0	0.611	8.3	1.5
Dec.	1020.3	−27.4	7.6	−15.2	−40.1	0.518	9.7	1.3
Jan. 1961	1025.4	−31.1	8.9	−15.8	−40.2	0.362	8.2	1.7
Feb.	1013.5	−35.7	7.3	−18.6	−47.3	0.222	7.0	1.5
Mar.	1036.5	−34.6	3.7	−26.5	−43.7	0.248	2.2	0.6
Apr.	1029.4	−24.8	5.4	−9.4	−34.3	0.624	1.6	0.6
Annual	1021.0	−18.1	5.7	2.8	−47.3	2.324	91.5	6.4

Month	Relat. hum. (%)	Days with precip.	Most freq. wind dir.	Mean wind speed (m/sec)	Max. wind speed (m/sec)
May 1960	86	12	NE	5.1	11
June	94	8	S	5.4	14
July	96	7	E	4.5	11
Aug.	94	3	W	3.5	10
Sept.	93	21	N,E	5.6	16
Oct.	84	23	N	6.0	20
Nov.	80	18	N	5.1	12
Dec.	80	24	NE	5.2	14
Jan. 1961	79	20	E	5.8	12
Feb.	76	17	E	4.8	16
Mar.	76	7	E	3.9	8
Apr.	76	6		—	
Annual	85	166		5.0	20

TABLE XLVI

CLIMATIC TABLE FOR "NORTH POLE" 9 (MAY 1960–MARCH 1961)
Latitude 77°31′–86°36′N, longitude 152°34′E–178°35′W, elevation surface

Month	Mean sta. press. (mbar)	Mean daily temp. (°C)	Mean daily temp. range (°C)	Temp. extremes (°C)		Mean vapour press. (mbar)	Mean precip. (mm)	Max. precip. in 24 h (mm)
				highest	lowest			
May 1960	1022.0	−8.4	6.5	0.8	−21.7	2.888	9.8	6.1
June	1011.3	−1.4	3.3	1.3	−5.8	5.127	6.9	2.9
July	1016.3	0.1	2.5	3.9	−2.5	5.906	8.1	5.7
Aug.	1018.7	−1.1	3.0	2.0	−8.2	5.354	9.8	2.0
Sept.	1014.9	−6.3	5.6	0.7	−21.7	3.512	12.5	4.4
Oct.	1016.2	−17.1	7.7	−6.7	−26.9	1.396	9.3	2.7
Nov.	1019.9	−25.2	7.5	−14.4	−39.0	0.665	2.6	0.7
Dec.	1015.5	−30.0	7.1	−15.1	−42.8	0.396	4.9	1.1
Jan. 1961	1022.9	−31.1	7.3	−19.1	−39.7	0.362	6.3	1.4
Feb.	1011.2	−36.5	5.7	−19.1	−50.8	0.208	3.3	1.2
Mar.	1031.0	−36.3	3.5	−24.2	−46.9	0.217	0.4	0.2
Annual	1018.2	−17.6	5.4	3.9	−50.8	2.367	73.9	6.1

Month	Relat. hum. (%)	Days with precip.	Most freq. wind dir.	Mean wind speed (m/sec)	Max. wind speed (m/sec)
May 1960	89	10			
June	93	8	S	5.0	15
July	96	4	E	4.3	15
Aug.	95	10	W	4.9	12
Sept.	92	14	N	4.4	13
Oct.	87	17	N	5.3	15
Nov.	84	8	NW,E	4.3	13
Dec.	78	9	NE	4.1	12
Jan. 1961	80	12	NE	4.8	12
Feb.	77	9	E	4.0	11
Mar.	79	3			
Annual	86	104		4.6	15

TABLE XLVII

CLIMATIC TABLE FOR DRIFTING STATION "ALPHA" (JULY 1957–NOVEMBER 1958)
Latitude 81°37′–86°25′N, longitude 176°15′–112°30′W, elevation surface

Month	Mean sta. press. (mbar)	Mean daily temp. (°C)	Mean daily temp. range (°C)	Temp.extremes (°C)		Mean precip. (mm)	
				highest	lowest	rain	snow (actual depth)
Jan.	1020	−32.2	6.7	−16.1	−49.4		
Feb.	1028	−36.1	5.6	−25.6	−42.8		
Mar.	1028	−32.2	6.1	−20.6	−46.1		
Apr.	1025	−28.3	5.0	−14.4	−38.9		
May	1020	−13.3	4.4	−2.8	−22.8	0	20.3
June	1017	−1.7	1.7	2.8	−10.6	1.5	170.2
July	1006	0.0	1.1	1.7	−5.0	1.3	125.7
Aug.	1008	−1.4	1.9	2.8	−11.7	4.2	135.9
Sept.	1011	−11.4	5.8	0.6	−30.6	0	54.6
Oct.	1008	−19.2	5.6	−3.9	−35.0	0	92.7
Nov.	1022	−26.4	3.6	−17.2	−29.4	0	12.7
Dec.	1022	−36.7	4.4	−23.3	−49.4		
Annual	1017	−18.3	4.3	2.8	−49.4		

Month	Days with precip.	No. of observ. with fog and ice-fog	Mean cloudiness (oktas)	Most freq. wind dir.	Mean wind speed (m/sec)
Jan.	17		6.0	SW	4.5
Feb.	15		2.9	W,SW	3.3
Mar.	18		5.0	SE	3.5
Apr.	7		3.1	NW	2.5
May	17		5.9	W	3.5
June	21		7.7	S	3.7
July	18	68	7.8	SE	4.2
Aug.	22	43	7.9	S	4.0
Sept.	18	32	7.5	S	3.9
Oct.	23	11	6.7	N	5.0
Nov.	21	26	2.9	W	3.4
Dec.	17	39	4.4	SW	2.9
Annual	214		5.7		3.7

TABLE XLVIII

CLIMATIC TABLE FOR DRIFTING STATION "B" (FLETCHER'S ICE ISLAND, T-3, JULY 1957–DECEMBER 1958)
Latitude 75°24′–82°30′N, longitude 126°30′–98°W, elevation 5.5 m

Month	Mean sta. press. (mbar)	Mean daily temp. (°C)	Mean daily temp. range (°C)	Temp. extremes (°C)		Mean precip. (mm)	
				highest	lowest	rain	snow (actual depth)
Jan.	1014	−30.0	6.1	−7.2	−51.7	0	10.2
Feb.	1020	−37.2	6.1	−21.1	−53.3	0	—
Mar.	1025	−35.0	6.7	−18.9	−46.1	0	—
Apr.	1020	−30.6	6.1	−21.7	−43.9	0	—
May	1019	−12.8	6.1	−2.8	−23.3	0	147.3
June	1020	−1.1	4.4	5.0	−12.8	0.5	25.4
July	1012.5	0.8	2.8	5.6	−2.8	3.9	10.2
Aug.	1011.5	−1.1	2.2	3.3	−9.4	11.0	228.6
Sept.	1010	−9.2	5.6	1.7	−27.2	0.4	61.0
Oct.	1007	−18.1	7.2	−3.9	−37.2	0	53.3
Nov.	1016	−30.6	6.7	−15.6	−45.6	0	10.2
Dec.	1019	−34.7	4.4	−15.0	−43.3	0	3.8
Annual	1014	−18.3	5.4	5.6	−53.3		

Month	Days with precip.	Mean cloud-iness (oktas)	Most freq. wind dir.	Mean wind speed (m/sec)
Jan.	10	5.4	W,NW	7.0
Feb.	4	4.0	W,NW	6.2
Mar.	1	4.0	W	4.1
Apr.	2	2.3	NW	4.5
May	24	5.9	N	4.1
June	14	6.9	NW	3.5
July	13	6.0	N	3.6
Aug.	19	7.7	NE	3.7
Sept.	17	7.8	NE	4.7
Oct.	24	7.3	S	4.0
Nov.	12	4.2	N,NW	4.3
Dec.	9	3.0	N	4.8
Annual	149	5.4		4.5

TABLE XLIX

CLIMATIC TABLE FOR ARLIS II

(Established May 1961 at 73°01′N 156°06′W, abandoned in the East Greenland current in May 1965; elevation 2.5 m)

Month	Mean sta. press. (mbar)	Mean daily temp. (°C)	Mean daily temp. range (°C)	Temp. extr. (°C)		Mean vapour press. (mbar)	Mean precip. (mm)	Max. precip. in 24 h (mm)
				highest	lowest			
Jan.	1030.4	−38.2	9.1	−20.3	−50.5		0.87	0.8
Feb.	1010.5	−37.6	8.9	−21.7	−47.8		0.05	—
Mar.	1012.2	−38.6	6.2	−21.7	−49.2		—	—
Apr.	1015.6	−26.8	6.6	−11.9	−39.4		—	—
May	1016.3	−12.0	5.0	0.6	−20.7		—	—
June	1015.9	−1.8	3.8	3.9	−11.7	4.972	5.16	3.2
July	1013.5	0.2	2.5	5.0	−5.0	5.847	2.52	1.2
Aug.	1013.3	−2.3	4.3	2.8	−8.9	5.198	0.10	0.07
Sept.	1008.2	−11.4	6.2	−2.2	−25.6	2.774	2.07	1.27
Oct.	1014.9	−21.5	8.0	−5.4	−36.7		0.77	1.0
Nov.	1022.1	−25.0	9.2	−5.5	−41.7		7.30	11.2
Dec.	1022.7	−33.2	8.7	−18.2	−46.1		1.08	1.0
Annual	1016.3	−20.7	6.5	5.0	−50.5			11.2

Month	Days with precip.	Mean cloudiness	Most freq. wind dir.	Mean wind speed (m/sec)
Jan.	24	3.0	NE,S	6.3
Feb.	22	4.9	NW,S	4.6
Mar.	19	4.5	E	5.0
Apr.	18	4.2	SW	4.6
May	28	8.4	SE	4.6
June	13	8.0	N	3.7
July	6	6.0	SE	4.9
Aug.	8	7.0	W	4.8
Sept.	10	8.0	SE	5.5
Oct.	18	5.0	W	5.9
Nov.	24	4.0	W	5.3
Dec.	18	2.0	SW	4.5
Annual	208	5.4		5.0

Data based on observations from October 1962 to February 1964.

TABLE L

CLIMATIC TABLE FOR BARROW, ALASKA
Latitude 71°18′N, longitude 156°47′W, elevation 6.7 m

Month	Mean sta. press. (mbar)	Mean daily temp. (°C)	Mean daily temp. range (°C)	Temp. extremes (°C) highest	Temp. extremes (°C) lowest	Mean vapour press. (mbar)	Mean precip. (mm)	Max. precip. in 24 h (mm)
Jan.	1017.6	−26.2	7.3	0.6	−47.2	0.456	4.06	17.78
Feb.	1021.8	−27.7	7.2	0.0	−48.9	0.384	3.81	9.14
Mar.	1018.1	−26.1	7.8	−1.1	−46.7	0.460	3.05	7.11
Apr.	1016.8	−17.9	8.4	5.6	−41.1	1.080	2.54	9.40
May	1018.0	−7.3	6.2	7.2	−27.8	3.040	3.30	7.62
June	1014.4	1.1	5.6	21.1	−13.3	6.084	7.11	20.83
July	1012.0	4.3	7.1	25.6	−5.6	7.596	21.08	21.84
Aug.	1010.0	3.6	5.9	22.8	−6.7	7.350	20.32	21.08
Sept.	1003.3	−0.8	4.0	16.7	−17.2	5.243	13.97	14.22
Oct.	1007.0	−8.3	5.5	6.1	−28.3	2.747	13.21	25.40
Nov.	1015.6	−17.4	6.6	3.9	−40.0	1.173	6.86	10.41
Dec.	1016.8	−23.6	6.7	1.1	−48.3	0.594	5.08	6.60
Annual	1015.1	−12.2	6.5	25.6	−48.9	1.873	104.39	25.4

Month	No. of days with precip.	No. of days with thunder-storms	No. of days with fog	Mean cloud-iness	Most freq. wind dir.	Mean wind speed (m/sec)
Jan.	4	—	2	4	ESE	9.6
Feb.	4	—	1	5	ENE	9.9
Mar.	3	—	1	5	ENE	9.7
Apr.	3	—	3	5	NE	10.0
May	3	—	8	8	ENE	10.2
June	4	—	13	7	E	10.0
July	8	—	13	8	E	10.1
Aug.	10	—	11	8	E	10.1
Sept.	9	—	5	8	E	11.3
Oct.	10	—	4	8	E	12.0
Nov.	6	—	2	7	E	10.9
Dec.	5	—	2	—	ENE	10.0
Annual	69	—	65	6.5	E	10.3

TABLE LI

CLIMATIC TABLE FOR BARTER ISLAND, ALASKA
Latitude 70°07′N, longitude 143°40′W, elevation 15 m

Month	Mean sta. press. (mbar)	Mean daily temp. (°C)	Mean daily temp. range (°C)	Temp. extremes (°C)		Mean vapour press. (mbar)	Mean precip. (mm)	Max. precip. in 24 h (mm)
				highest	lowest			
Jan.	1018.4	−26.4	7.8	2.8	−45.6	0.487	9.91	30.5
Feb.	1023.9	−27.8	7.8	1.1	−50.6	0.422	8.9	31.0
Mar.	1015.6	−25.7	8.7	−0.6	−45.6	0.499	6.6	11.2
Apr.	1016.8	−17.2	1.1	6.1	−38.3	1.162	4.8	6.9
May	1017.4	−6.6	6.8	8.9	−25.6	3.208	5.6	19.3
June	1013.4	1.8	5.7	19.4	−7.8	6.329	10.7	29.2
July	1012.3	4.9	6.9	21.7	−3.3	7.793	30.0	29.7
Aug.	1012.1	4.0	5.8	22.2	−4.4	7.537	30.0	28.2
Sept.	1011.1	−0.4	3.9	17.8	−13.9	5.458	23.1	56.6
Oct.	1005.0	−8.1	5.6	6.1	−26.7	2.857	26.2	50.3
Nov.	1013.0	−17.5	6.8	2.8	−46.1	1.179	15.0	10.9
Dec.	1014.9	−23.9	7.2	−1.1	−46.1	0.620	11.7	14.0
Annual	1014.5	−11.9	6.9	22.2	−50.6	1.924	182.4	56.6

Month	No. of days with			Mean cloud-iness	Most freq. wind dir.	Mean wind speed (m/sec)
	precip.	thunder-storms	fog			
Jan.	6	—	1	6	E	6.1
Feb.	5	—	1	6	W	6.6
Mar.	6	—	2	6	W	5.8
Apr.	7	—	3	6	W	5.5
May	7	—	8	8	E	5.4
June	7	—	11	8	ENE	4.9
July	10	—	15	7	ENE	4.6
Aug.	11	—	15	8	E	5.4
Sept.	10	—	11	8	E	5.6
Oct.	14	—	3	8	E	6.7
Nov.	9	—	3	7	E	6.5
Dec.	6	—	1		W	6.3
Annual	192	—	74	7	E	5.8

TABLE LII

CLIMATIC TABLE FOR SACHS HARBOUR

Latitude 71°57′N, longitude 124°44′W, elevation 84.4 m

Month	Mean sta. press. (mbar)	Mean daily temp. (°C)	Mean daily temp. range (°C)	Temp. extr. (°C)		Mean vapour press. (mbar)	Mean precip. (mm)	Max. precip. in 24 h (mm)
				highest	lowest			
Jan.	1009.4	−29.9	8.7	−5.0	−45.6		2.3	1.0
Feb.	1009.0	−31.5	7.5	−6.1	−47.8		2.3	1.3
Mar.	1010.7	−28.2	7.6	−10.0	−43.9		3.6	4.6
Apr.	1010.1	−19.9	7.3	2.2	−38.9	0.96	2.5	1.0
May	1009.6	−8.7	6.1	7.2	−26.7	2.63	5.6	5.3
June	1006.8	1.8	5.7	17.8	−13.9	5.85	4.6	6.9
July	1002.1	5.6	5.9	16.7	−3.3	7.73	25.4	21.8
Aug.	1002.5	4.4	5.3	15.6	−6.1	7.74	17.0	15.2
Sept.	1002.3	−1.6	3.9	15.6	−13.9	4.89	16.5	12.4
Oct.	1001.9	−12.5	6.4	0.6	−29.4	2.01	11.4	3.8
Nov.	1006.5	−24.3	6.8	−7.8	−36.7	0.64	4.6	3.3
Dec.	1006.1	−27.9	7.0	−10.6	−45.6		3.8	2.5
Annual	1006.4	−14.4	6.6	17.8	−47.8		99.57	21.8

Month	Mean no. of days with			Mean cloudiness	Mean sunshine (h)	Most freq. wind dir.	Mean wind speed (m/sec)
	precip.	thunderstorms	fog				
Jan.	6	—	0	4.4	0	N	5.7
Feb.	4	—	—	3.4	59	E	4.9
Mar.	5	—	1	3.5	187	SE	4.6
Apr.	5	—	3	4.4	260	E,SE	5.7
May	7	—	6	6.6	263	E	5.5
June	4	—	10	6.3	344	N,E	5.5
July	10	—	17	7.3	291	NW	5.6
Aug.	8	—	14	7.5	202	SE	5.8
Sept.	10	—	9	7.9	71	E	6.3
Oct.	15	—	5	7.1	46	E	6.5
Nov.	7	—	1	5.2	3	E	5.6
Dec.	7	—	2	3.7	0	E	5.1
Annual	88		68	5.6	1726		5.6

Data based on records for the period 1955–1960.

TABLE LIII

CLIMATIC TABLE FOR MOULD BAY

Latitude 76°14′N, longitude 119°20′W, elevation 15.2 m

Month	Mean sta. press. (mbar)	Mean daily temp. (°C)	Mean daily temp. range (°C)	Temp. extr. (°C)		Mean vapour press. (mbar)	Mean precip. (mm)	Max. precip. in 24 h (mm)
				highest	lowest			
Jan.	1017.8	−33.1	7.9	−9.4	−48.3		3.6	5.8
Feb.	1015.6	−35.4	7.1	−10.6	−50.0		1.8	2.3
Mar.	1021.5	−32.7	7.4	−10.6	−48.9		3.0	5.6
Apr.	1020.5	−22.7	8.5	−1.7	−41.7		3.3	3.8
May	1019.6	−10.8	6.0	1.7	−28.9	2.20	6.9	7.1
June	1015.3	0.2	4.5	13.3	−13.3	5.26	3.8	1.9
July	1009.9	4.0	4.1	15.6	−3.9	6.91	17.0	14.7
Aug.	1011.1	1.8	4.3	13.9	−10.0	6.26	21.3	47.8
Sept.	1012.9	−6.0	4.6	7.8	−25.0	3.44	10.7	11.4
Oct.	1011.3	−17.1	6.7	0.0	−36.1	1.26	7.6	5.1
Nov.	1014.3	−26.7	7.2	−7.2	−43.3		2.54	1.5
Dec.	1013.5	−31.2	6.8	−9.4	−52.8		3.3	2.0
Annual	1015.3	−17.4	6.4	15.6	−52.8		84.8	47.8

Month	Mean no. of days with			Mean cloudiness	Most freq. wind dir.	Mean wind speed (m/sec)
	precip.	thunderstorms	fog			
Jan.	5	—	—	4.2	NW	4.2
Feb.	4	—	2	4.2	NW	3.6
Mar.	4	—	2	4.0	N	3.4
Apr.	4	—	3	4.7	N,NW	3.6
May	10	—	3	7.5	NW	4.8
June	4	—	3	7.6	NW	6.6
July	9	—	3	7.9	NW	5.2
Aug.	10	—	7	8.4	S,NE	4.8
Sept.	11	—	7	8.5	NW	4.9
Oct.	9	—	2	7.7	NW	4.8
Nov.	4	—	—	5.4	NW	4.3
Dec.	6	—	—	4.5	NW	3.6
Annual	80	—	32	6.2		

Data based on records for the period 1951–1960.

TABLE LIV

CLIMATIC TABLE FOR ISACHSEN

Latitude 78°47′N, longitude 103°32′W, elevation 25.3 m

Month	Mean sta. press. (mbar)	Mean daily temp. (°C)	Mean daily temp. range (°C)	Temp. extr. (°C)		Mean vapour press. (mbar)	Mean precip. (mm)	Max. precip. in 24 h (mm)
				highest	lowest			
Jan.	1013.5	−34.6	7.3	−3.9	−52.8		1.52	1.27
Feb.	1013.3	−36.6	6.8	−20.6	−51.1		1.52	2.03
Mar.	1019.3	−35.0	6.8	−8.3	−53.9		1.02	1.27
Apr.	1018.4	−24.2	7.6	−1.1	−42.2		4.32	4.57
May	1017.0	−11.5	5.8	2.2	−29.4	2.08	7.62	4.83
June	1012.9	−0.2	4.6	16.7	−14.4	5.24	3.30	2.54
July	1007.6	3.7	4.9	18.9	−3.3	6.85	22.10	15.2
Aug.	1009.8	1.4	4.5	14.4	−13.3	6.08	23.10	20.3
Sept.	1011.1	−8.4	5.2	2.8	−27.2	2.93	18.00	19.8
Oct.	1008.3	−18.8	7.1	−1.7	−37.2	1.15	10.20	10.2
Nov.	1011.6	−28.2	7.1	−3.9	−45.6		3.56	6.4
Dec.	1010.1	−32.3	6.9	−9.4	−51.1		1.52	2.0
Annual	1012.8	−18.7	6.1	18.9	−53.9		97.78	20.3

Month	Mean no. of days with			Mean cloudiness	Most freq. wind dir.	Mean wind speed (m/sec)
	precip.	thunderstorms	fog			
Jan.	4	—	1	4.2	N	4.5
Feb.	3	—	2	4.4	N	3.4
Mar.	3	—	5	3.7	N	3.0
Apr.	5	—	5	4.9	N	3.2
May	9	—	3	6.8	N	4.4
June	5	—	3	7.7	N	4.3
July	9	—	6	7.5	NW	4.7
Aug.	9	—	8	8.1	N,SW	4.3
Sept.	12	—	8	8.1	N	4.3
Oct.	8	—	5	7.4	N	4.7
Nov.	4	—	2	4.9	N	3.9
Dec.	3	—	1	4.5	N,NW	4.2
Annual	74	—	49	6.0		4.2

Data based on records for the period 1951–1960.

TABLE LV

CLIMATIC TABLE FOR RESOLUTE

Latitude 74°43′N, longitude 94°59′W, elevation 63.7 m

Month	Mean sta. press. (mbar)	Mean daily temp. (°C)	Mean daily temp. range (°C)	Temp. extr. (°C)		Mean precip. (mm)	Max. precip. in 24 h (mm)	Mean snow-fall (mm)
				highest	lowest			
Jan.	1006.9	−31.8	6.6	−5.0	−47.2	2.5	1.5	25.4
Feb.	1006.5	−33.7	6.5	−13.9	−49.4	2.8	1.8	27.9
Mar.	1011.7	−31.4	6.5	−6.7	−51.7	2.5	2.3	25.4
Apr.	1011.8	−22.1	7.4	−1.1	−40.0	6.1	5.1	61.0
May	1010.8	−10.2	5.6	4.4	−28.9	7.4	2.5	73.7
June	1007.6	0.6	4.3	13.9	−13.3	11.4	19.6	38.1
July	1003.2	4.6	4.9	16.1	−2.2	21.3	16.8	10.2
Aug.	1004.1	2.9	4.2	15.0	−8.3	33.8	25.2	27.9
Sept.	1004.3	−4.4	3.6	8.9	−17.8	15.5	6.4	121.9
Oct.	1000.7	−14.6	5.6	0.0	−34.4	16.3	9.4	162.6
Nov.	1006.1	−24.9	6.2	−2.8	−41.7	5.6	3.6	55.9
Dec.	1003.0	−29.3	6.3	−8.3	−46.1	5.1	2.0	50.8
Annual	1006.4	−16.2	5.6	16.1	−51.7	130.3	25.2	680.8

Month	Mean relat. hum. (%)	Days with precip.	Mean cloud-iness	Mean sun-shine (h)	Most freq. wind dir.	Mean wind speed (m/sec)
Jan.		5	3.9	0	NW	5.3
Feb.		5	4.1	24	E	5.1
Mar.		5	4.0	146	NW	4.6
Apr.	75	7	4.7	264	NW	4.9
May	84	9	6.9	248	NW	5.2
June	86	6	7.2	271	NW	5.6
July	85	9	7.5	284	NW	5.4
Aug.	89	11	7.9	151	NW	5.5
Sept.	90	12	8.2	60	NW	5.6
Oct.	86	16	7.5	16	NW	5.6
Nov.	78	7	4.9	1	NW	4.9
Dec.		7	3.9	0	NW	4.6
Annual		99	5.9	1465		5.2

Data based on records for the period 1951–1960.

TABLE LVI

CLIMATIC TABLE FOR EUREKA
Latitude 80°00′N, longitude 85°56′W, elevation 2.4 m

Month	Mean sta. press. (mbar)	Mean daily temp. (°C)	Mean daily temp. range (°C)	Temp. extr. (°C)		Mean precip. (mm)	Max. precip. in 24 h (mm)	Mean snow-fall (mm)
				highest	lowest			
Jan.	1016.6	−35.9	6.4	−1.1	−51.1	3.1	3.1	30.5
Feb.	1017.1	−37.3	5.8	−12.2	−52.2	1.8	3.6	17.8
Mar.	1023.3	−37.6	5.9	−13.3	−52.8	1.5	5.1	15.2
Apr.	1022.0	−26.8	7.8	−3.3	−45.6	1.8	3.8	17.8
May	1019.6	−9.7	6.3	5.6	−31.1	2.8	5.6	27.9
June	1014.4	2.7	4.8	17.8	−13.3	3.3	14.5	7.6
July	1009.2	5.7	5.6	19.4	−2.2	15.5	13.0	7.6
Aug.	1012.0	3.8	4.5	15.0	−8.3	13.5	41.7	15.2
Sept.	1012.9	−6.7	4.4	5.6	−26.1	10.9	13.7	106.7
Oct.	1011.3	−21.6	6.6	3.9	−41.7	8.9	13.5	88.9
Nov.	1014.9	−30.6	6.6	−1.7	−44.4	2.0	1.5	20.3
Dec.	1013.5	−35.2	5.7	−10.6	−49.4	2.0	2.5	20.3
Annual	1015.6	−19.1	5.9	19.4	−52.8	67.1	41.7	375.8

Month	Mean relat. hum. (%)	Days with precip.	Mean cloud-iness	Most freq. wind dir.	Mean wind speed (m/sec)
Jan.		5	3.8	E	3.2
Feb.		3	3.8	E	3.0
Mar.		2	3.5	E	2.4
Apr.		3	3.9	E	2.6
May	79	3	4.7	NW	3.8
June	79	2	5.7	NW	4.8
July	80	7	6.6	NW	5.1
Aug.	85	6	6.6	NW	4.3
Sept.	86	7	7.4	NE	3.5
Oct.	78	7	5.4	E	3.0
Nov.		4	4.1	E	2.8
Dec.		4	3.2	E	2.4
Annual		53	4.9		3.4

Data based on records for the period 1951–1960.

TABLE LVII

CLIMATIC TABLE FOR ALERT

Latitude 82°30′N, longitude 62°20′W, elevation 62.5 m

Month	Mean sta. press. (mbar)	Mean daily temp. (°C)	Mean daily temp. range (°C)	Temp. extr. (°C)		Mean vapour press. (mbar)	Mean precip. (mm)	Max. precip. in 24 h (mm)
				highest	lowest			
Jan.	1008.8	−31.9	4.0	0.0	−47.8		5.6	2.8
Feb.	1009.3	−33.0	4.1	−1.1	−47.2		5.8	5.1
Mar.	1014.7	−32.9	4.1	−2.2	−47.8		6.4	5.4
Apr.	1014.4	−23.9	4.2	−1.1	−45.6		5.8	3.3
May	1013.7	−11.3	3.3	8.3	−27.2	2.07	9.1	11.4
June	1008.9	−0.1	2.5	17.2	−12.2	4.97	11.7	18.5
July	1003.6	3.9	3.2	20.0	−5.6	6.54	14.7	15.7
Aug.	1006.1	0.8	2.3	15.0	−15.0	5.63	28.4	18.2
Sept.	1006.6	−9.5	2.8	5.6	−26.1	2.52	29.7	15.9
Oct.	1004.4	−19.8	3.5	0.6	−35.6	1.00	16.0	20.3
Nov.	1007.3	−25.8	5.7	−0.6	−40.0		6.1	4.6
Dec.	1005.4	−30.2	3.9	−8.3	−46.1		7.1	6.4
Annual	1008.6	−17.8	3.5	20.0	−47.8		146.4	20.3

Month	Mean no. of days with		Mean cloudiness (oktas)	Most freq. wind dir.	Mean wind speed (m/sec)
	precip.	fog			
Jan.	8	1	4.2	W	2.2
Feb.	6	—	3.8	W	2.2
Mar.	7	1	4.4	W	2.0
Apr.	7	1	4.4	W	2.1
May	8	7	5.6	W,NW	2.2
June	5	7	6.2	NE	2.8
July	10	8	6.7	NE	3.3
Aug.	12	9	7.7	NE	2.6
Sept.	13	6	7.5	W	2.7
Oct.	9	3	6.2	W	3.0
Nov.	8	2	4.9	W	2.5
Dec.	9	1	3.7	W	2.0
Annual	102	46	5.5		2.4

Data based on records for the period 1951–1960

TABLE LVIII

CLIMATIC TABLE FOR ISFJORD RADIO, SVALBARD

Latitude 78°04′N, longitude 13°38′E, elevation 9 m

Month	Mean sta. press. (mbar)	Mean daily temp. (°C)	Mean daily temp. range (°C)	Temp. extr. (°C) highest	Temp. extr. (°C) lowest	Mean rel. hum. (%)	Mean precip. (mm)
Jan.	1006.2	−10.9	5.7	3.5	−30.9	83	29
Feb.	1008.8	−11.2	6.1	4.4	−32.2	83	30
Mar.	1009.6	−12.1	5.9	3.8	−29.0	85	33
Apr.	1011.3	−8.8	4.6	5.6	−28.2	83	17
May	1016.4	−3.3	3.4	13.1	−19.6	83	20
June	1013.3	1.7	3.1	12.5	−8.2	86	24
July	1012.1	4.5	3.2	15.6	−1.3	89	30
Aug.	1012.4	4.2	3.0	14.3	−2.0	87	38
Sept.	1008.6	1.1	3.0	12.0	−9.0	85	38
Oct.	1005.5	−2.7	4.0	8.5	−15.5	82	46
Nov.	1008.9	−6.2	4.7	6.2	−26.9	82	39
Dec.	1006.0	−9.0	5.4	4.1	−28.1	82	34
Annual	1009.9	−4.4	4.3	15.6	−32.2	84	378

Month	Mean no. of days with precip.	Mean no. of days with clear sky	Mean no. of days with overcast sky	Mean cloudiness (oktas)	Most freq. wind dir.	Mean wind speed (m/sec)
Jan.	14.0	4.8	12.0	5.1	NE	8.8
Feb.	13.0	2.4	11.6	5.4	NE	9.1
Mar.	14.1	2.6	13.1	5.3	NE	8.3
Apr.	11.5	4.8	10.5	4.8	NE	7.5
May	10.8	3.7	14.9	5.5	NE	6.3
June	10.5	1.0	17.1	6.2	NE,S	5.1
July	13.3	1.0	19.8	6.5	S,NE	5.3
Aug.	14.4	1.0	18.9	6.3	NE,S	5.3
Sept.	13.8	1.0	17.2	6.3	NE,S	6.5
Oct.	14.5	0.8	18.0	6.3	NE	7.5
Nov.	14.1	2.4	14.5	5.7	NE	8.3
Dec.	13.0	4.2	12.6	5.2	NE	9.6
Annual	157	30	180	5.7		7.3

Data based on records for the periods: Jan.–Aug.: 1947–1965; Sept.–Dec.: 1946–1965.

TABLE LIX

CLIMATIC TABLE FOR BJÖRNÖYA (BEAR ISLAND)
Latitude 74°31′N, longitude 19°01′E, elevation 14 m

Month	Mean sta. press. (mbar)	Mean daily temp. (°C)	Mean daily temp. range (°C)	Temp. extr. (°C) highest	Temp. extr. (°C) lowest	Mean rel. hum. (%)	Mean precip. (mm)
Jan.	1004.5	−6.8	6.0	5.1	−29.8	89	30
Feb.	1006.9	−6.9	6.2	3.5	−29.1	89	25
Mar.	1008.0	−7.5	6.4	6.2	−29.7	89	26
Apr.	1010.2	−5.2	5.2	5.6	−22.6	90	20
May	1015.5	−1.4	3.6	16.5	−17.8	90	20
June	1012.6	1.8	3.9	23.6	−8.4	92	26
July	1012.1	4.2	4.0	19.1	−6.4	93	25
Aug.	1012.2	4.3	3.5	14.4	−1.6	93	37
Sept.	1007.4	3.0	3.2	13.9	−6.3	92	50
Oct.	1004.9	0.5	3.6	10.0	−13.1	89	42
Nov.	1007.5	−2.4	4.4	6.1	−21.0	88	27
Dec.	1003.3	−5.3	5.4	6.4	−25.2	90	29
Annual	1008.8	−1.8	4.6	23.6	−29.8	90	357

Month	Mean no. of days with precip.	clear sky	overcast sky	Mean cloudiness (oktas)	Most. freq. wind dir.	Mean wind speed (m/sec)
Jan.	19.3	1.1	15.0	5.9	NE	8.8
Feb.	18.5	0.8	14.8	6.2	NE	8.8
Mar.	19.8	0.6	16.4	6.1	NE,E	7.7
Apr.	16.8	1.1	14.8	6.0	E	6.7
May	15.8	0.3	21.1	6.8	NE,SE	6.1
June	14.9	0.6	22.1	6.9	E	5.8
July	14.7	0.7	24.0	7.1	E	5.6
Aug.	17.4	0.2	24.5	7.1	E	6.1
Sept.	21.8	0.2	22.7	7.0	E	7.0
Oct.	22.2	0	21.3	6.9	E,SE	7.5
Nov.	20.0	0.2	18.7	6.6	NE	8.3
Dec.	21.0	0.9	15.7	6.1	NE	8.5
Annual	222	7	231	6.6		7.2

Data based on records for the period 1946–1965.

TABLE LX

CLIMATIC TABLE FOR MALYE KARMAKULY (NOVAYA ZEMLYA)
Latitude 72°23′N, longitude 52°44′E, elevation 16 m

Month	Mean sta. press. (mbar)	Mean daily temp. (°C)	Mean daily temp. range (°C)	Temp. extr. (°C) highest	Temp. extr. (°C) lowest	Mean rel. hum. (%)	Mean precip. (mm)	Max. precip. in 24 h (mm)
Jan.	1008.4	−10.6	11.7	1.0	−41.1	88	17.8	5.1
Feb.	1009.2	−13.3	8.3	1.7	−39.4	82	15.2	7.6
Mar.	1008.4	−15.0	6.7	1.1	−43.9	84	12.7	5.1
Apr.	1005.8	−9.4	7.8	5.0	−31.7	82	10.2	5.1
May	1010.6	−3.9	5.6	13.3	−23.3	83	12.7	5.1
June	1013.0	1.7	5.0	21.7	−17.2	80	17.8	10.2
July	1012.8	6.7	11.1	24.4	−10.0	80	25.4	20.3
Aug.	1012.4	6.7	5.0	22.8	−1.1	81	38.1	20.3
Sept.	1009.2	3.3	3.9	18.3	−12.8	84	35.6	22.9
Oct.	1003.3	−1.7	5.6	7.8	−17.8	85	27.4	17.8
Nov.	1008.8	−5.0	15.0	5.0	−33.9	81	15.2	5.1
Dec.	1011.5	−9.4	10.0	2.8	−35.5	84	15.2	5.1
Annual	1009.4	−4.2	8.0	24.4	−43.9	83	243.3	22.9

Month	Mean no. of days with precip.	snow	clear sky	overcast sky	fog	Mean cloudiness	Mean wind speed (m/sec)
Jan.	17	16	2	17	2	8.0	11.6
Feb.	15	14	2	16	3	7.7	11.2
Mar.	13	13	2	15	2	7.1	10.3
Apr.	12	12	2	14	2	7.3	9.4
May	13	12	2	20	4	8.2	8.0
June	12	7	3	16	8	7.6	7.6
July	11	1	3	15	12	7.5	7.2
Aug.	15	1	1	19	11	8.1	5.8
Sept.	18	5	0	21	8	8.4	7.6
Oct.	17	12	1	19	5	8.2	8.9
Nov.	16	14	3	18	3	8.3	9.8
Dec.	14	14	2	15	2	7.5	8.9
Annual	173	121	23	205	62	7.8	8.9

Data based on records, generally, for more than 10 years.

TABLE LXI

CLIMATIC TABLE FOR BUKHTA TIKHAYA (ZEMLYA FRANTSA IOSIFA)
Latitude 80°19′N, longitude 52°48′E, elevation 6 m

Month	Mean daily temp. (°C)	Mean daily temp. range (°C)	Temp. extr. (°C)		Mean precip. (mm)	Mean no. of days with			
			highest	lowest		precip.	snow	blowing snow	clear sky
Jan.	−17.2	8.8	1.7	−38.9	5.1	13	12	9	3
Feb.	−16.7	8.3	0.0	−39.4	7.6	14	11	10	3
Mar.	−13.9	8.4	0.6	−39.4	5.1	11	8	15	6
Apr.	−16.6	7.2	0.6	−33.9	5.1	10	8	13	3
May	−7.8	4.4	5.0	−20.0	5.1	10	9	10	1
June	−1.1	3.9	10.0	−9.4	7.6	9	7	4	1
July	1.1	3.9	10.5	−1.1	17.8	12	5	0	1
Aug.	0.6	7.2	11.7	−5.6	27.9	15	7	0	0
Sept.	−2.2	8.8	7.2	−16.7	17.8	13	11	4	0
Oct.	−7.7	4.4	11.6	−28.3	7.6	16	15	10	0
Nov.	−13.3	6.7	7.2	−38.9	5.1	13	13	8	3
Dec.	−16.6	7.8	3.9	−36.1	5.1	12	11	11	6
Annual	−9.3	6.7	11.7	−39.4	116.9	148	117	94	27

Month	Days with		Mean cloudiness	Mean wind speed (m/sec)
	overcast sky	fog		
Jan.	15	6	7.0	7.2
Feb.	12	7	6.7	6.7
Mar.	12	9	6.1	5.4
Apr.	14	9	7.1	5.4
May	23	9	8.6	4.9
June	22	10	8.5	4.5
July	23	15	8.7	3.6
Aug.	26	16	9.2	4.0
Sept.	25	6	9.2	5.8
Oct.	23	5	9.0	6.7
Nov.	16	4	7.3	6.7
Dec.	13	5	6.4	7.6
Annual	224	101	7.8	5.7

Data based, generally, on records for 10 years.

TABLE LXII

CLIMATIC TABLE FOR OSTROV RUDOLFA

Latitude 81°48′N, longitude 57°57′E, elevation 48 m

Month	Mean daily temp. (°C)	Mean daily temp. range (°C)	Temp. extr. (°C)		Mean rel. hum. (%)	Mean precip. (mm)	Mean no. of days with		
			highest	lowest			precip.	snow	clear sky
Jan.	−20.0	9.4	0.5	−37.2	76	5.1	9	7	4
Feb.	−20.6	8.3	−0.6	−42.8	82	2.5	7	7	5
Mar.	−23.3	7.8	0.0	−37.2	90	2.5	8	7	7
Apr.	−18.3	7.8	−0.5	−35.6	88	2.5	9	8	2
May	−8.9	5.0	0.0	−20.6	90	5.1	12	11	1
June	−1.7	3.3	7.7	−8.3	93	10.2	12	10	1
July	0.6	3.3	11.7	−4.4	97	20.3	13	9	1
Aug.	0.0	3.9	7.2	−6.1	95	15.2	18	13	0
Sept.	−5.0	4.4	5.0	−22.2	93	17.8	18	16	0
Oct.	−10.6	6.1	1.6	−32.8	92	7.6	15	14	0
Nov.	−16.1	7.8	0.5	−36.7	91	5.1	13	10	2
Dec.	−18.9	8.3	−0.5	−35.6	82	2.5	9	8	8
Annual	−11.9	6.2	11.7	−42.8	89	96.4	143	120	31

Month	Days with		Mean cloudiness	Mean wind speed (m/sec)
	overcast sky	fog		
Jan.	11	2	6.4	8.9
Feb.	9	0	5.6	7.6
Mar.	7	4	5.2	6.3
Apr.	16	8	7.6	6.7
May	24	5	8.7	5.8
June	25	13	9.1	4.9
July	25	24	9.1	3.6
Aug.	27	24	9.6	4.5
Sept.	24	14	8.2	5.8
Oct.	19	5	8.5	6.7
Nov.	14	2	6.7	8.0
Dec.	9	2	7.2	8.0
Annual	210	103	7.7	6.4

Data based on 1–4 years' records.

TABLE LXIII

CLIMATIC TABLE FOR OSTROV DOMASHNIY (SEVERNAYA ZEMLYA)
Latitude 79°30′N, longitude 91°08′E, elevation 3 m

Month	Mean daily temp. (°C)	Mean daily temp. range (°C)	Temp. extr. (°C)		Mean precip. (mm)	Mean no. of days with			
			highest	lowest		precip.	snow	blowing snow	clear sky
Jan.	−25.6	10.0	1.7	−46.7	5.1	13	12	9	6
Feb.	−25.0	10.6	−1.7	−47.2	5.1	8	8	6	2
Mar.	−27.8	6.7	−1.6	−45.6	2.5	7	7	11	7
Apr.	−22.2	7.8	−5.5	−40.0	2.5	7	5	8	6
May	−10.0	4.4	0.6	−26.7	5.1	6	6	9	1
June	−1.7	2.8	4.4	−10.6	7.6	8	7	5	2
July	1.1	2.8	6.1	−4.4	27.9	11	4	0	2
Aug.	0.0	2.8	5.6	−8.3	12.7	14	8	0	1
Sept.	−3.3	3.9	5.0	−17.8	15.2	12	9	4	1
Oct.	−10.6	8.3	0.6	−31.1	5.1	10	10	8	1
Nov.	−18.9	7.8	−0.6	−35.6	2.5	9	9	8	6
Dec.	−25.0	9.4	0.0	−41.7	5.1	9	8	8	9
Annual	−14.0	6.4	6.1	−47.2	96.4	114	93	76	44

Month	Days with		Mean cloudiness	Mean wind speed (m/sec)
	overcast sky	fog		
Jan.	9	3	6.0	6.3
Feb.	9	6	6.9	6.3
Mar.	8	7	5.2	4.9
Apr.	9	7	5.6	4.9
May	19	4	8.2	5.8
June	20	6	8.2	5.8
July	22	20	8.4	4.9
Aug.	22	20	8.8	5.4
Sept.	23	13	8.9	7.2
Oct.	20	4	8.3	6.3
Nov.	10	2	6.2	6.3
Dec.	8	1	4.9	5.4
Annual	179	93	7.1	5.8

Data based, generally, on records for 6 years (precipitation records: 24 years).

TABLE LXIV

CLIMATIC TABLE FOR MYS CHELYUSKIN
Latitude 77°43′N, longitude 104°17′E, elevation 13 m

Month	Mean sta. press. (mbar)	Mean daily temp. (°C)	Mean daily temp. range (°C)	Temp. extr. (°C)		Mean rel. hum. (%)	Mean precip. (mm)	Mean snow depth (mm)
				highest	lowest			
Jan.	1005	−27.2	8.3	−2.2	−45.6	86	2.5	178
Feb.	1011	−25.0	8.9	−4.4	−42.8	85	2.5	178
Mar.	1019	−28.3	7.2	−1.1	−42.2	82	2.5	203
Apr.	1019	−21.7	7.8	−3.3	−41.7	84	2.5	203
May	1016	−9.4	5.6	2.2	−25.6	86	2.5	203
June	1017	−1.1	3.3	8.9	−12.2	90	17.8	127
July	1015	1.7	4.4	19.4	−3.3	90	27.9	10
Aug.	1018	1.1	3.3	17.2	−5.0	94	28.0	0
Sept.	1001	−1.7	3.3	13.3	−16.7	90	10.1	25
Oct.	1009	−10.0	5.6	1.7	−29.4	87	7.6	76
Nov.	1014	−18.9	6.7	−0.6	−39.4	84	5.1	127
Dec.	1015	−24.4	7.2	−1.7	−43.3	82	2.5	152
Annual	1013	−13.7	6.0	19.4	−45.6	87	111.5	

Month	Mean no. of days with						Mean cloudiness	Mean wind speed (m/sec)
	precip.	snow	blowing snow	clear sky	overcast sky	fog		
Jan.	8	6	18	7	11	2	5.7	7.2
Feb.	9	8	14	4	9	4	6.0	5.8
Mar.	6	5	19	8	6	7	4.6	5.8
Apr.	4	4	16	5	12	7	6.3	5.4
May	6	6	18	1	22	6	8.6	5.8
June	10	7	9	0	22	11	8.7	6.3
July	12	4	0	1	23	21	8.6	5.8
Aug.	14	6	1	0	26	21	9.3	5.8
Sept.	13	10	7	1	24	11	9.0	6.3
Oct.	13	13	14	1	21	5	8.5	5.8
Nov.	12	11	16	5	12	4	6.4	6.3
Dec.	10	8	18	7	9	1	5.5	6.7
Annual	117	88	150	40	197	100	7.3	6.1

Data based on records for 8 years (precipitation records: 22 years).

TABLE LXV

CLIMATIC TABLE FOR OSTROV KOTELNY (NOVO SIBIRSKIE OSTROVA)
Latitude 76°00′N, longitude 137°54′E, elevation 10 m

Month	Mean sta. press. (mbar)	Mean daily temp. (°C)	Temp. extr. (°C)		Mean rel. hum. (%)	Mean precip. (mm)	Mean cloud-iness
			highest	lowest			
Jan.	1012	−29.3	−9.4	−43.3	86	2.5	3.0
Feb.	1018	−29.6	−12.8	−42.2	87	2.5	4.5
Mar.	1024	−28.9	−11.1	−41.7	88	2.5	3.4
Apr.	1015	−20.6	−3.9	−37.2	88	1.2	6.2
May	1015	−9.1	4.4	−27.8	89	5.1	7.1
June	1014	−0.2	20.6	−7.2	91	10.2	8.5
July	1013	2.8	18.9	−2.8	92	30.5	8.4
Aug.	1015	2.3	17.7	−3.9	92	25.4	9.5
Sept.	1009	−1.2	6.1	−8.9	90	12.7	9.0
Oct.	1011	−9.6	0.0	−20.6	89	5.1	8.1
Nov.	1019	−20.9	−8.3	−31.1	89	2.5	6.6
Dec.	1023	−26.4	−5.6	−40.6	88	5.1	4.4
Annual	1016	−14.2	20.6	−43.3	89	105.3	6.6

Data based on records for 4 years.

TABLE LXVI

CLIMATIC TABLE FOR OSTROV VRANGELYA (WRANGEL ISLAND)
Latitude 70°58′N, longitude 178°32′W, elevation 3 m

Month	Mean daily temp. (°C)	Mean daily temp. range (°C)	Temp. extr. (°C)		Mean precip. (mm)	Mean snow depth (mm)	Mean no. of days with		
			highest	lowest			precip.	snow	clear sky
Jan.	−23.9	7.8	−5.6	−41.7	5.1	203	8	8	6
Feb.	−25.6	7.2	−1.7	−41.6	5.1	305	7	7	4
Mar.	−23.3	8.3	−1.1	−45.6	5.1	305	7	7	6
Apr.	−17.2	7.7	0.6	−36.7	5.1	229	6	6	5
May	−8.3	6.1	9.4	−25.0	5.1	254	8	8	2
June	0.6	4.4	17.2	−10.0	10.2	102	8	4	2
July	2.8	5.6	18.3	−5.0	15.2	0	9	1	1
Aug.	2.2	4.4	16.7	−4.4	22.9	0	12	5	0
Sept.	−1.7	3.3	8.3	−14.4	12.7	0	10	8	1
Oct.	−8.3	4.4	3.3	−23.3	10.1	51	10	8	1
Nov.	−17.2	5.6	1.7	−33.3	2.5	152	7	7	2
Dec.	−21.1	6.7	0.0	−36.7	5.1	76	7	7	4
Annual	−11.7	5.9	18.3	−45.6	104.2		99	76	34

Month	Days with		Mean cloudiness	Mean wind speed (m/sec)
	overcast sky	fog		
Jan.	7	2	5.5	5.4
Feb.	6	2	5.8	4.9
Mar.	7	3	5.0	4.9
Apr.	9	5	5.4	4.0
May	18	7	7.9	4.0
June	15	12	7.4	3.1
July	14	14	7.2	4.0
Aug.	17	14	7.9	3.6
Sept.	15	6	8.1	5.8
Oct.	15	3	7.4	6.7
Nov.	11	4	6.7	6.7
Dec.	10	3	6.4	5.4
Annual	144	75	6.7	4.9

Data based on records for 10 years.

Chapter 4

The Climate of the Antarctic

W. SCHWERDTFEGER

"The wind in its everlasting flight sweeps over these tracks in the desert of snow. Soon all will be blotted out."
Fr. Nansen, Introduction to R. AMUNDSEN'S *South Pole. An Account of the Norwegian Antarctic Expedition in the "Fram", 1910–1912.*

Introduction

The principal geographic characteristics of Antarctica are assumed to be known. The little map, Fig.1, shall serve only as a reminder. It emphasizes the fact that a large part of the surface of the continent, 55%, lies at an elevation of more than 2,000 m, and about 25% at more than 3,000 m above sea level. The highest mountain peaks of Antarctica do not lie in the area of the great icy plateau of East Antarctica; they belong to the Sentinel Range, Ellsworth Mountains, where the summit of the Vinson Massif at 78.6°S 85.4°W, rises to 5,140 m (SILVERSTEIN, 1967).

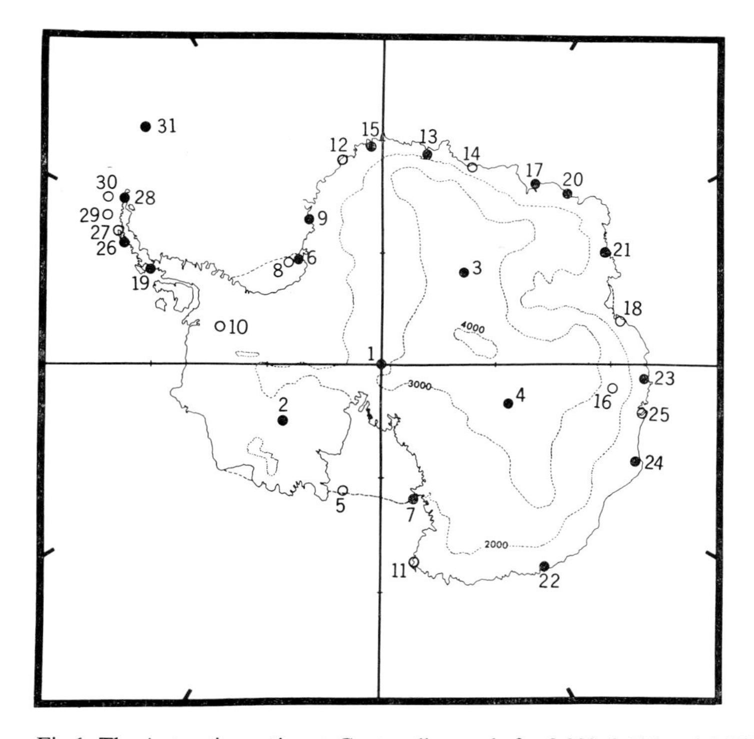

Fig.1. The Antarctic continent. Contourlines only for 2,000, 3,000 and 4,000 m elevation. Full circles = meteorological stations in operation in the winter of 1968. Open circles = stations in operation for two years or more at any time, but not in 1968. Station numbers refer to Table I.

Of the total area of the continent, about $14 \cdot 10^6$ km², less than 3% are estimated to be free of a permanent ice sheet. The map also shows the location of the meteorological stations which have supplied most of the information used in this text. Stations actually in year-round operation at the end of 1968 are marked by solid black circles. Stations in operation at any time for two years or more are listed in Table I.

Climatological considerations are generally based upon several decades of records of observations. No such long records exist for the Antarctic continent itself nor for the immediately adjacent islands or ice shelves. It merits to be mentioned, however, that there is one, and only one, place south of 55°S for which there is an uninterrupted, homogeneous, and really invaluable record since April, 1903. It is the Argentine station Orcadas del Sur, at 60°44′S 44°44′W, north of the Weddell Sea.

Most of the meteorological evidence for Antarctica, and in particular for the interior of the continent, has been obtained since the beginning of the International Geophysical Year (I.G.Y.) in 1957. Nevertheless, an impressive amount of information for the coastal regions of the continent was already available when MEINARDUS in 1938 wrote his comprehensive work on the climate of Antarctica. It gives a full account of the meteorological achievements of the heroic period of Antarctic exploration. For the time between Meinardus' publication and the I.G.Y., extensive descriptions, bibliographies, and climatological tables have been given by PEPPER (1954), by the South African Weather Bureau (VAN ROOY, 1957), and by SCHWERDTFEGER et al. (1959). Also the articles on Antarctica in the *Compendium of Meteorology* (COURT, 1951) and in the journal *Erdkunde* (LOEWE, 1954) should be mentioned.

The outstanding scientific events of the pre-I.G.Y. period were the Byrd Antarctic Expeditions 1928–1930, 1933–1935, 1939–1941 (GRIMMINGER and HAINES, 1939; GRIMMINGER, 1941; COURT, 1949), mainly with respect to the Ross Sea area; the Norwegian–British–Swedish Antarctic Expedition 1949–1952 in the Maudheim area (NORSK POLARINSTITUTT, 1956–1960) and the French Expeditions to Adélie Land 1949–1953 (PRUDHOMME and VALTAT, 1957).

A complete listing, beginning with the year 1772, was compiled by R. J. Venter (in VAN ROOY, 1957), of all expeditions to high southern latitudes as far as they produced meteorological records.

In recent years, detailed climatological statistics for individual stations have been published only by some of the parent national meteorological agencies. Summaries of surface data for a large number of places, including the island stations in the subpolar ocean, were prepared at the International Antarctic Meteorological Research Centre (PHILLPOT, 1967a). Surface and upper air data for selected stations, up to the year 1962, were presented in maps and frequency tables in volume VII of the *Marine Climatic Atlas of the World* (UNITED STATES NAVY, 1965). A selection of climatological charts also constitutes a part of the Soviet *Atlas of Antarctica* (1966). Daily synoptic analyses in the form of surface and 500 mbar maps have been published for several years in the South African journal *Notos*. Preliminary monthly mean values of the basic parameters are regularly included in the *Monthly Climatic Data for the World* (UNITED STATES WEATHER BUREAU and WORLD METEOROLOGICAL ORGANIZATION, 1956–1968). Of the many brief descriptions of weather and climate of Antarctica which have appeared in the literature since the I.G.Y., the concise account of RUBIN (1964) may be singled out as an authoritative source of information.

TABLE I

METEOROLOGICAL STATIONS IN ANTARCTICA WITH A RECORD OF TWO YEARS, AT LEAST[1]

No.	Station	Lat.(S)	Long.	Elev.(m)	In operation
1	× South Pole	90.0°	–	2,800	since Jan. 1957
2	× Byrd	80.0°	120.0°W	1,515	since Jan. 1957
3	× Plateau	79.2°	40.5°E	3,624	Dec. 1965–Jan. 1969
4	× Vostok	78.5°	106.9°E	3,488	1958–1961, and since Feb. 1963
5	× Little America	*78.3°	*163.0°W	* 40	1929–1930, 1934–1935, 1940–1941, 1956–1958
6	× G. Belgrano	78.0°	38.8°W	* 52	since Jan. 1955
7a	Hut Point	77.9°	166.8°E	X	Feb. 1902–Jan. 1904
7b	Cape Evans	77.6°	166.4°E	20	1911 + 1912
7c	× McMurdo	77.9°	166.7°E	24	since March 1956
8	Ellsworth	77.7°	41.1°W	42	March 1957–Dec. 1962
9	× Halley Bay	75.5°	26.6°W	30	since Feb. 1956
10	× Eights	75.2°	77.2°W	421	Dec. 1962–Oct. 1965
11	× Hallett	72.3°	170.3°E	5	Feb. 1956–Feb. 1964, + summers only since 1965
12	Maudheim	71.1°	10.9°W	38	March 1950–Jan. 1952
13	× Novolazarevskaya	70.8°	11.8°E	87	since Feb. 1961
14	Baudouin	70.4°	24.3°E	37	1958–1960, 1964–1966
15a	Norway Station	70.5°	2.5°W	56	April 1957–Jan. 1962
15b	× Sanae	70.3°	2.4°W	52	since April 1962
16	× Pionerskaya	69.7°	95.5°E	*2,740	May 1956–Dec. 1958
17	× Syowa Base	69.0°	39.6°E	15	1957, 1959–1961, since Feb. 1966
18	× Davis	68.6°	78.0°E	12	March 1957–Oct. 1964
19a	Stonington Isl.	68.2°	67.0°W		1946–1949
19b	Marguerite Bay	68.1°	67.1°W	5	1951–1957, incomplete
19c	Horseshoe Isl.	67.8°	67.3°W	9	1955–1960
19d	Adelaide Isl.	67.8°	68.9°W	26	since 1962
20	× Molodezhnaya	67.7°	45.9°E	42	since Feb. 1963
21	× Mawson	67.6°	62.9°E	8	since Feb. 1954
22a	Cape Denison	67.9°	142.7°E	6	1912 + 1913
22b	Port Martin	66.8°	141.4°E	14	1950 + 1951
22c	× Dumont d'Urville	66.7°	140.0°E	41	1952 and since April 1956
23	× Mirny	66.6°	93.0°E	35	since March 1956
24	× Wilkes	66.3°	110.5°E	12	since March 1957
25	Oasis	66.3°	100.7°E	28	Nov. 1956–Oct. 1958
26	× Argentine Isl.	65.3°	64.3°W	11	since 1947
27	× Melchior	64.3°	63.0°W	8	1947–1960
28	× Hope Bay/Esperanza	63.4°	57.0°W	11/7	1945–1960
29	× Deception Isl.	63.0°	60.7°W	8	1944–1967
30	Admir. Bay	62.1°	58.4°W	9	1948–1960
31	× Orcadas	60.7°	44.7°W	4	since April 1903

[1] For the stations marked with a ×, a summary of climatic data is presented at the end of the chapter (climatic tables).

* = approximate, or mean value for various locations.

Several more stations which are not listed here, are and/or have been in operation in the region of the Antarctic Peninsula north of 65°S.

Scientific endeavor in Antarctica distinguishes itself from similar activities in other parts of the world by the auspicious development of friendly and fruitful international co-operation on a broad base. Also in this aspect, beyond the purely scientific achievements, the International Geophysical Year was a unique success. Indeed, in all the years since 1836 (PANZARINI, 1968) international understanding again and again has been an important stimulus as well as a fortunate by-product of scientific work of Antarctica. May it continue to be so.

Radiation and temperature regime

The radiation balance

One of the principal features of the radiation processes over Antarctica is the fact that the snow covered surface of the continent and the surrounding ice fields must reflect about three-quarters of the total incident solar energy flux. The thin atmosphere absorbs only a small fraction of the incoming and reflected radiation. In the long wave radiation band, on the other hand, the snow surface acts very nearly as a black body. In both aspects, the conditions over the continent differ little from those high above the surrounding seas: the average height of the upper surface of the clouds over the polar ocean is comparable to that of the snow surface of the high plateau, the average cloud amount is on the order of eight tenths, and the cloud albedo is not much less than that of snow. However, what happens underneath the average level of the cloud deck over the ocean must also be taken into account. A part of the incident radiation penetrates to the surface. When and where the ocean is free of ice, a large fraction of this incoming radiation is absorbed by the water. When there is ice, even snow covered ice, a greater fraction is reflected. This is one of the circumstances which make the extent and concentration of the pack ice belt around the continent and the seasonal variations of this belt decisive factors for the radiation and heat budget of the subpolar atmosphere (FLETCHER, 1968). The variation of the pack ice belt from fall to spring is large, as Fig.2 clearly suggests, and it must also be borne in mind that the lines in the figure represent average conditions. The variation from year to year, not shown here, is also considerable. Furthermore, referring now to the long wave radiation domain, the downward directed atmospheric radiation ("back radiation") is much larger underneath a thick cloud layer, as is so often found over the polar seas, than in the less cloudy and absolutely drier atmosphere over the continent. This is true at least for the summer months, while in the winter there is relatively strong back radiation over the interior from the comparatively warm layer of air above the strong surface inversion.

Such considerations show the complexity of the radiation processes and can even raise some doubts regarding the often quoted unique role of Antarctica as the "refrigerator of the Southern Hemisphere". Or they may suggest that this role is not due to the radiation conditions alone, but rather to a combination of the radiation effects and the meridional atmospheric circulation, whose development and persistence is favored by the height and the size of the central plateau of the continent. In any case, this line of thought makes it advisable to substitute quantitative arguments for qualitative reasoning.

A summary of the components of the radiation budget is shown in Table II, with average

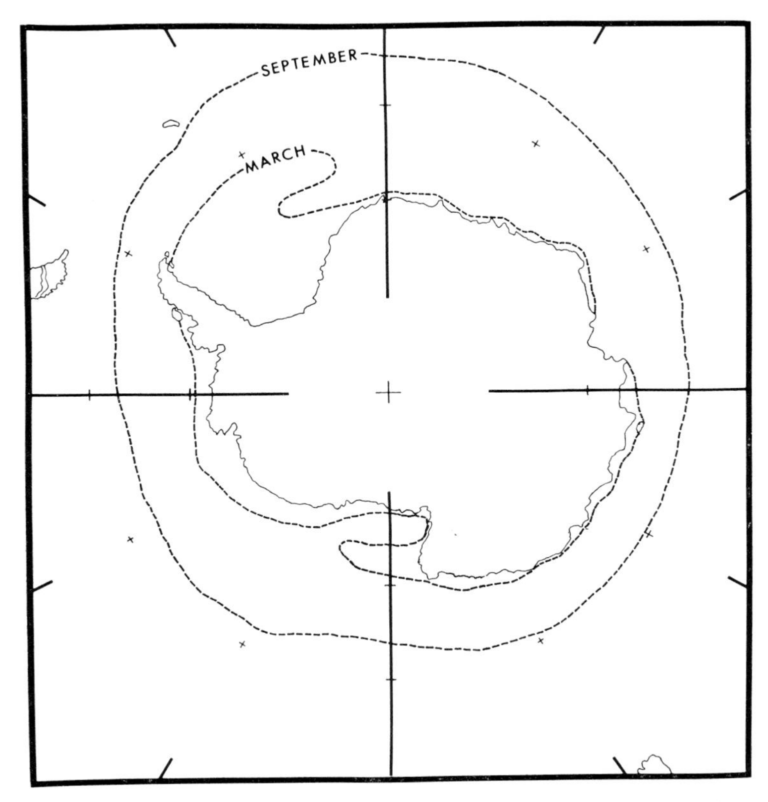

Fig.2. Average extent of the pack ice in March (minimum) and in September (maximum). (After U.S. Navy Hydrographic Office charts and the Soviet Atlas of Antarctica, vol. I.) According to TRESHNIKOV (1967, quoted from BURDECKI, 1968), the area of the region around Antarctica, covered with sea ice in each month of the year, is as follows:

Month	Jan.	Feb.	Mar.	Apr.	May	June	July	Aug.	Sept.	Oct.	Nov.	Dec.
Area in 10^6 km^2	6.8	4.3	2.6	6.4	9.2	11.0	13.9	15.7	18.8	17.8	15.2	11.4

The area of the continent itself is $14 \cdot 10^6$ km^2. This means that the total ice covered surface, sea ice plus continent, at the spring equinox is about twice as large as at the fall equinox.
In the near future, television pictures and infrared radiometer photofacsimiles taken by satellites in polar orbit will provide an almost continuous surveillance of the boundaries and density of the circumpolar pack ice belt. A first step in this direction was reported by PREDOEHL (1966).

values for the two summer and four winter months and the annual sums, for various stations. Detailed statistics for every month of the year as given by RUSIN (1961), for continental stations by DALRYMPLE (1966), and some more recent data are listed in the climatic tables. A critical analysis of the broad scale geographic distribution of the essential items for the continent as well as for the surrounding ocean areas was presented by ZILLMAN (1967). For the duration of polar day and night, see Table III.
Most of the available radiation data are based on measurements of 1–3 years only. Therefore, and in view of the different instrumentation and evaluation procedures, the numerical values are not fully comparable and can only be considered as tentative (HANSON, 1960).
The values for the effective short wave radiation ($R_S = (1 - \alpha)R_G$; α = albedo, R_G = global radiation) in the summer, and consequently the annual values, are strongly

TABLE II

SUMMARY OF ITEMS IN THE RADIATION BUDGET AT THE SURFACE; AVERAGE MONTHLY AND ANNUAL AMOUNTS, IN 10^3 Ly

	Vostok*	S. Pole	Pion.	Maudh.	Mirny**	Pt. Mart.	Oasis
Summer (XII + I)/2							
Global radiation	26.8	27.2	24.0	20.0	21.3	16.6	17.9
Albedo	0.81	0.84	0.84	0.83	0.79	0.48	0.18
Eff. shortwave rad.	5.1	4.4	3.8	3.5	4.5	8.6	14.7
Eff. longwave rad.	−4.4	−4.4	−2.2	−2.9	−2.2	−3.7	−4.1
Net radiation	0.7	0.0	1.6	0.6	2.3	4.9	10.6
Winter (V to VIII)/4							
Global radiation	0.0	0.0	0.2	0.2	0.6	0.4	0.5
Albedo	–	–	0.9	0.9	0.9	0.8	0.44
Eff. shortwave rad.	–	–	0.0	0.0	0.1	0.1	0.3
Eff. longwave rad.	−0.9	−0.9	−1.4	−1.4	−2.0	−2.8	−2.0
Net radiation	−0.9	−0.9	−1.4	−1.4	−1.9	−2.7	−1.7
Year (I to XII)							
Global radiation	108.4	103.3	109.0	93.7	103.7	78.4	89.1
Eff. shortwave rad.	19.5	16.5	16.5	15.3	17.6	32.9	68.6
Eff. longwave rad.	−25.6	−23.4	−24.0	−24.3	−22.6	−38.2	−31.0
Net radiation	−6.1	−6.9	−7.5	−9.0	−5.0	−5.3	+37.6

* Global radiation and albedo values for the period 1963–1965 (courtesy M. Rubin); for 1958, RUSIN (1961) gives considerably higher values for the global radiation.
** Global radiation and albedo values for 1956–1959 and 1963–1965 (M. Rubin).

TABLE III

SIGNIFICANT DATES ON THE DURATION OF DAYLIGHT AND TWILIGHT AT VARIOUS LATITUDES[1]

	S. Pole[2]	85°S	80°S	75°S	70°S
Sun stays above horizon until	III 23	III 10	II 25	II 12	I 26
Sun remains below horizon from	III 23	IV 4	IV 19	V 5	V 25
No civil twilight from	IV 6	IV 20	V 5	V 27	–
No astronom. twilight from	V 11	VI 6	–	–	–
Astronom. twilight returns at	VIII 1	VII 4	–	–	–
Civil twilight returns at	IX 8	VIII 26	VIII 10	VII 20	–
Sun returns at	IX 21	IX 8	VIII 25	VIII 9	VII 19
Sun stays above horizon from	IX 21	X 5	X 18	XI 2	XI 19
Approx. duration of astronom. twilight at winter solstice	–	–	6 h	7 h	10 h
Approx. duration of civil twilight at winter solstice	–	–	–	–	4 h

[1] Unobstructed horizon and normal refraction conditions are assumed.
[2] At the South Pole, the polar day has 183, the polar night 182 days; at the North Pole, it is 189 and 176 days.

affected by the uncertainty of the albedo determinations. This is the case especially in the coastal areas and the pack ice belt where the reflectivity of the surface varies with changing weather conditions: bare ground or snow, broken or closed and old or new ice cover, presence or absence of polynyas. Results of airborne measurements over various types of ice in the western Ross Sea and a comparison with such measurements made near Mirny were given by PREDOEHL and SPANO (1965) and SPANO (1965). Over the perennial snow and ice fields of the interior, on the other hand, a useful value of 0.8 as the lower limit of surface albedo appears to be certain.

Some published data include values up to 0.98—much larger than the figures in Table II suggest. It is difficult to say whether this is due to different local and temporal conditions as, for instance, a smoother versus a more sculptured snow surface, a difference in the angle of incidence at the time of the observation, or to different measurement techniques. Furthermore, it appears questionable whether the values for the (all-wave) net radiation, R_{net}, in the summer at Vostok and the South Pole, as given in Table II, will stand up when more and better measurements become available. The value for the South Pole, the result of continuous measurements by a C.S.I.R.O. net radiometer, refers to December 1965 and January 1966. The series of daily data indicates that positive values of R_{net} prevailed from November 15 until December 5 and December 26 until January 1, while most of the remaining December and January days had small negative values. Thus, the sum for the two summer months is very slightly negative. The correlation of the daily values with observed sky conditions is good; cloudy days at the high plateau have little or no decrease of global radiation (in comparison to clear days) but have more back radiation and therefore tend to have a positive (all-wave) budget. The two summer months 1965/1966 had a smaller than normal number of cloudy days. There is some reason, however, to assume that in the summer the net radiation as measured at the South Pole station is systematically biased in favor of negative values, because a slightly elevated rim around the upper sensor plate of the C.S.I.R.O. net radiometer causes some shading. Furthermore, the fact that there is a pronounced diurnal variation (around 00h00 G.M.T. more positive, around 12h00 G.M.T. more negative values) raises doubts whether the average of 24 hourly data really is the most representative value for an entire day. On the other hand, it appears possible that the average net radiation in the summer at the South Pole is less than at Vostok, where the incident radiation must be stronger and the surface temperature is lower. In any case, one must realize that an average of 0.016 Ly/min, adding up to 700 Ly/month, is a rather small energy flux and of the same order of magnitude as the possible systematic errors of the instruments and/or computations used.

Nevertheless, the main features of the radiation regime are clearly represented by the values shown. Very large amounts of solar energy reach the surface of the Antarctic Plateau in the summer. At two Soviet stations, values slightly over $3 \cdot 10^4$ Ly/month have been measured in December 1958. RUSIN (1961) makes the interesting point that "at the latitudes of the 80's where the coldest point on earth is situated, the point of maximum monthly amounts of solar energy is also situated". The factors which combine to make this possible are astronomical (earth nearest to the sun in the southern summer), geographical (high elevation of the plateau), and meteorological (purity and minimal H_2O content of the air). The high surface albedo and the character of the clouds enhance the role of the diffuse sky radiation, so that the global radiation on cloudy days differs little

TABLE IV

LONG WAVE RADIATION REGIME IN SUMMER AND WINTER (Ly/h)

	Vostok	S. Pole	Pion.	Maudh.	Mirny	Oasis
Summer (XII, I)						
Average sfc. temp.	240	245	250	268	271	275°K
Average emitted rad.	16.2	17.6	19.1	25.3	26.4	27.9 (Ly/h)
Average eff. radiation	5.9	5.9	3.3	3.9	2.4	5.7 (Ly/h)
Ångstrom ratio	0.37	0.34	0.17	0.15	0.09	0.21
Average cloudiness	3.7	5.2	6.6	6.9	7.0	6.6/10
Aver. temp. inversion	1	1	0	–	–	–°C
Winter (V–VIII)						
Average sfc. temp.	206	215	227	247	256	257°K
Average emitted rad.	8.4	10.4	13.0	18.2	21.0	21.3 (Ly/h)
Average eff. radiation	1.2	1.2	2.2	2.0	2.7	2.7 (Ly/h)
Ångstrom ratio	0.14	0.12	0.17	0.11	0.13	0.13
Average cloudiness	3.5	3.5	5.7	6.7	6.4	7.0/10
Aver. temp. inversion	23	20	*	5	1	3°C

* = no data

from that on clear days; as the direct radiation diminishes when clouds move in, the diffuse radiation increases.

It may be noted that the latter is true also for "clouds" of volcanic dust which produced a remarkable diminution in the intensity of the direct solar radiation at the South Pole in late 1963. The dust cloud in the stratosphere could be traced back to the eruption of Mount Agung, Bali. In their analysis of the relevant measurements, VIEBROCK and FLOWERS (1968) state: "Even though the direct component shows a large decrease, this is almost completely compensated by an increase in the amount of diffuse radiation".

As far as the long wave radiation alone is concerned, a comparison of summer and winter conditions is quite revealing. For three stations near sea level and three at high elevation, Table IV lists the mean temperature at surface T_0 and the corresponding emitted radiation ($R_E = \varepsilon\sigma T_0^4$), the effective long wave radiation[1] R_L, the so-called Ångstrom ratio $A_0 = R_L/R_E$, the mean cloudiness, and the average strength of the temperature inversion in the lowest 500 m of the atmosphere. A small value of the ratio A_0 indicates that the downward directed atmospheric energy flux (the "back radiation") amounts to a large fraction of the energy flux going upward from the surface. The pronounced increase of A_0 from the coast to the high plateau, in the *summer* months, corresponds to the decrease of cloudiness and of the water vapor content of the atmosphere, the latter from about 5 to 0.5 mm liquid equivalent ("precipitable water"). In the *winter*, nevertheless, when again cloudiness and water vapor content decrease toward the interior (the latter approximately from 2 to 0.25 mm water equivalent), A_0 changes very little, an indication that the back radiation is relatively strong. The explanation lies in the presence of a deep layer of comparatively warm air above the surface inversion, see last line in Table IV. This feature and its implications for the temperature regime will be discussed further below.

[1] Effective long wave radiation R_L is upward directed emitted (black body) radiation R_E, minus downward directed atmospheric ("back") radiation R_B.

The seasonal variation of the net radiation (all-wave radiation balance) over the continent and the adjacent ocean can best be summarized in the words of ZILLMAN (1967) who describes the conditions for the four mid-season months as follows: "In *January*, the entire region is seen to be undergoing a net radiative heat gain, although values approaching zero are found over the greater part of the plateau. By *April*, all the continent and the ocean southward of 55°S are experiencing a net loss of radiation. A belt of maximum loss surrounds the continent where the upward emission from the still warm ocean considerably exceeds the counter-radiation from the cold air flowing off the continent and the very small short wave income. In *July*, the region of net radiative loss from the surface has expanded northward to 45°S with largest values in the 60°–65°S latitude band. Despite the low cloudiness over the plateau, the intense surface inversion, now fully developed, suppresses the long wave loss from the surface and a minimum in the pattern is located over the interior of East Antarctica. By *October*, the return of the sun brings a net radiative heat gain to parts of the Ross Sea and eastern escarpment of the plateau extending as far south as 85°. Over the remainder of the continent, the absorbed short wave radiation has not yet compensated the long wave loss, and the all-wave balance is still negative, with the exception, of course, of individual small areas with low surface albedo." From Zillman's fig.10, latitudinal mean values can be derived (Table V).

TABLE V

ESTIMATES OF LATITUDINAL AVERAGES OF ALL-WAVE NET RADIATION AT THE SURFACE ($1 \cdot 10^3$ Ly/month)

Latitude (°S)	Jan.	Apr.	July	Oct.
80	1.2	−1.7	−1.4	−0.8
70	3.1	−1.8	−1.9	−0.1
60	4.8	−0.8	−2.3	1.1

Naturally, such averages as given in Table V can only serve to describe the broad scale conditions. Characteristic values for any given place depend critically upon the physical properties of the material below the interface and the structure of the atmosphere above it. Several micrometeorological field programs have been carried through with the aim to appraise all components of the heat budget at the snow surface, including radiation absorption and heat conduction in the ice or snow and the eddy heat flux in the atmospheric surface layer. These studies refer to a variety of glaciological conditions: Maudheim (LILJEQUIST, 1956), Mirny (KOPANYEV, 1960; RUSIN, 1961), Little America (HOINKES, 1961), South Pole (HANSON and RUBIN, 1962; DALRYMPLE et al., 1963; DALRYMPLE, 1966), as well as sea ice and blue ice (continental ice bare of snow) at low elevation in the Mawson area (WELLER and SCHWERDTFEGER, 1967; WELLER, 1968).

Average temperature conditions and meridional temperature gradients

The average temperature conditions near the surface in January (summer) and July (winter) are shown in two maps, Fig.3 and 4, for the continent itself and the surrounding

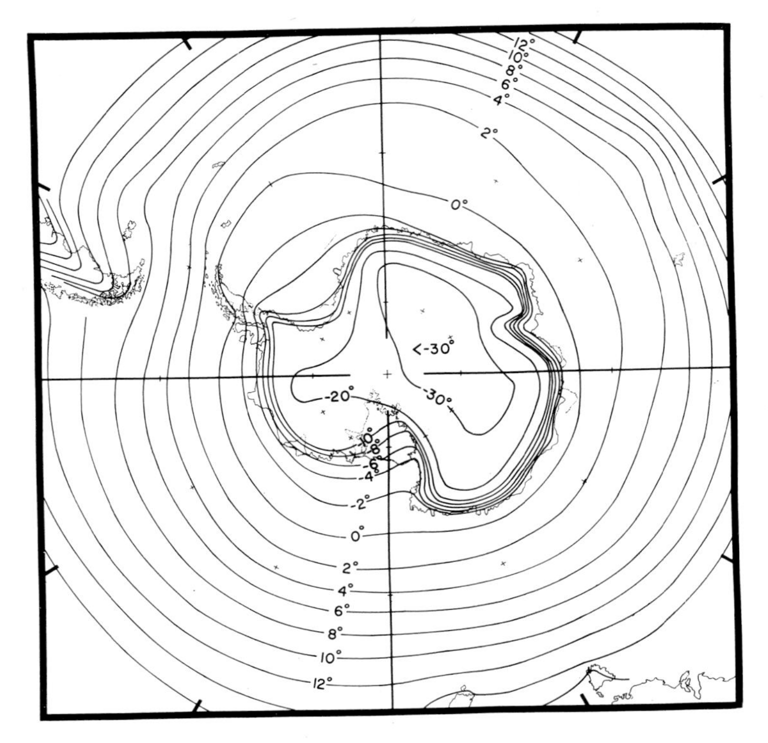

Fig.3. Mean isotherms at surface, January.

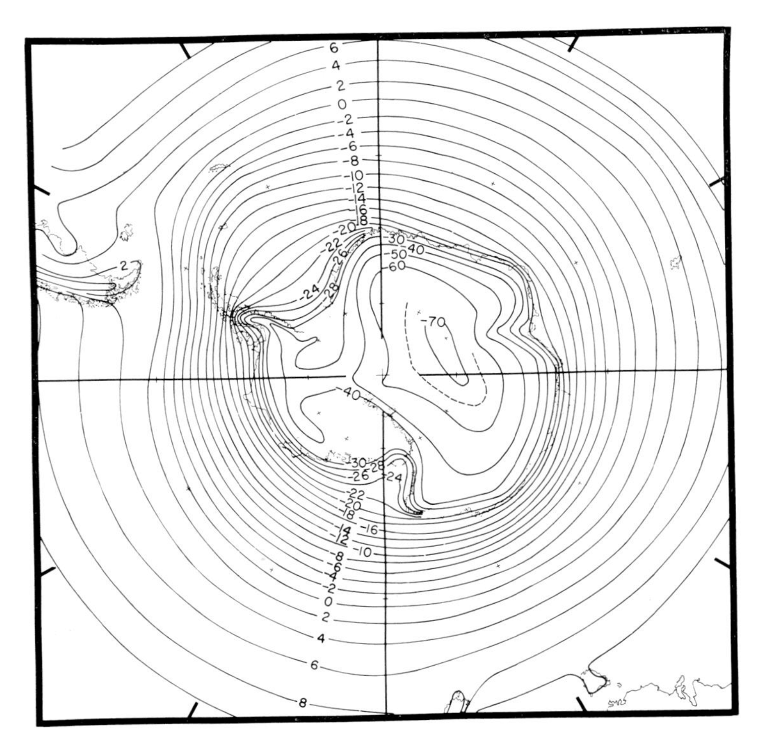

Fig.4. Mean isotherms at surface, July.

TABLE VI

LATITUDINAL AVERAGES OF THE TEMPERATURE NEAR SEA LEVEL AND PRESSURE AT SEA LEVEL

Lat. °S	January		July		$\bar{t}_{I} - \bar{t}_{VII}$ (°C)
	temp.(°C)	press.(mbar)	temp.(°C)	press.(mbar)	
50	7.9	1,002.2	4.3	1,002.8	3.6
55	5.2	994.5	0.9*	995.8	4.3
60	2.5	988.8	−4.0*	989.5	6.5
65	0.2	987.4	−11.6*	986.2	11.8

* It may be of interest that these values, derived from recently elaborated maps (TALJAARD et al., 1969) are 3 to 5 degrees higher than earlier estimates (MEINARDUS, 1938).

ocean (TALJAARD et al., 1969). The isotherms are drawn with 2° intervals over the ocean, with 10° intervals over the continent. Multi-annual average values for each month and for the year are listed in the climatic tables for various stations at the end of the chapter. Latitudinal averages of the air temperature over the sub-Antarctic ocean have been determined for January and July, Table VI; these data may also be of interest for a comparison with Northern Hemisphere values. For the over-all picture of tropospheric temperature conditions, reference to the 500 mbar level is more meaningful; this is the lowest standard level which can be considered to represent the "free atmosphere" for the high southern latitudes including the Antarctic Plateau (Fig. 5 and 6, Table VII).
The mean meridional temperature gradients in January and July can immediately be derived from the values in Tables VI and VII. However, such average values for selected months can be misleading. The curves in Fig.7 clearly show that there are significant differences between various sectors. They also demonstrate a remarkable characteristic of the annual variation of some meteorological parameters: in the first instance the temperature in the lower layers of the troposphere. In Antarctica, there are only two months of summer, December and January, but six months of winter, April to September. (In this context, the reader may disregard the thin curve *f* in Fig.7; it refers to the special

TABLE VII

LATITUDINAL AVERAGES OF TEMPERATURE AND HEIGHT (g.p.m.) OF THE 500-MBAR SURFACE

Lat. °S	January		July		$\bar{t}_{I} - \bar{t}_{VII}$ (°C)
	temp.(°C)	height(g.p.m.)	temp.(°C)	height(g.p.m.)	
50	−20.4	5,452	−28.2	5,333	7.8
55	−23.7	5,320	−31.1	5,210	7.4
60	−26.7	5,221	−33.8	5,094	7.1
65	−28.7	5,164	−36.4	5,004	7.7
70	−30.3	5,138	−38.9	4,945	8.6
75	−32.6	5,118	−41.1	4,908	8.5
80	−34.3	5,095	−43.0	4,880	8.7
85	−35.8	5,072	−44.5	4,857	8.7
90	−36.0	5,063	−45.0	4,848	9.0

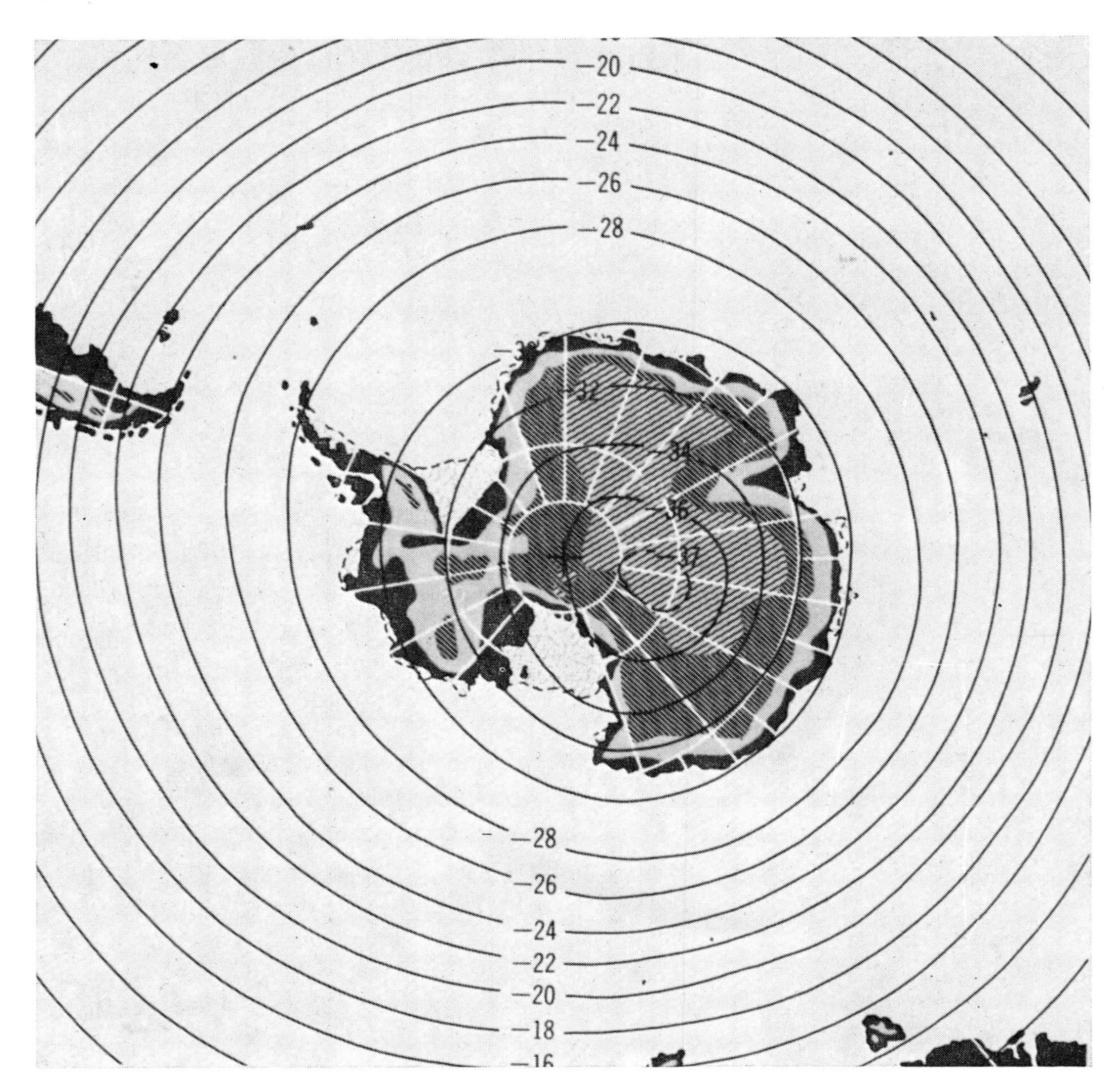

Fig.5. Mean isotherms at the 500 mbar level, January.

conditions of the eastern Ross Sea, and the McMurdo Sound in particular, which will be discussed later.) In terms of an analysis of periodic variations, this implies that in addition to a simple annual oscillation there is a remarkably strong semi-annual variation of the meridional temperature gradient, its maxima appearing around the equinoxes. Accordingly, the annual march of the meridional pressure gradient, the strength of the tropospheric westerlies, and other meteorological parameters in subpolar and polar latitudes are characterized by equinoctial maxima. This will be discussed in a later section.

The short summer and long winter clearly define the temperature regime of the Antarctic continent, and in view of the relatively short observational record of most of the Antarctic stations, many features of the regime can best be shown by presenting average conditions for these two "seasons". This is done in Table VIII for the regions poleward from the coast where the steep rise of the continent and the increase in latitude combine their effects to

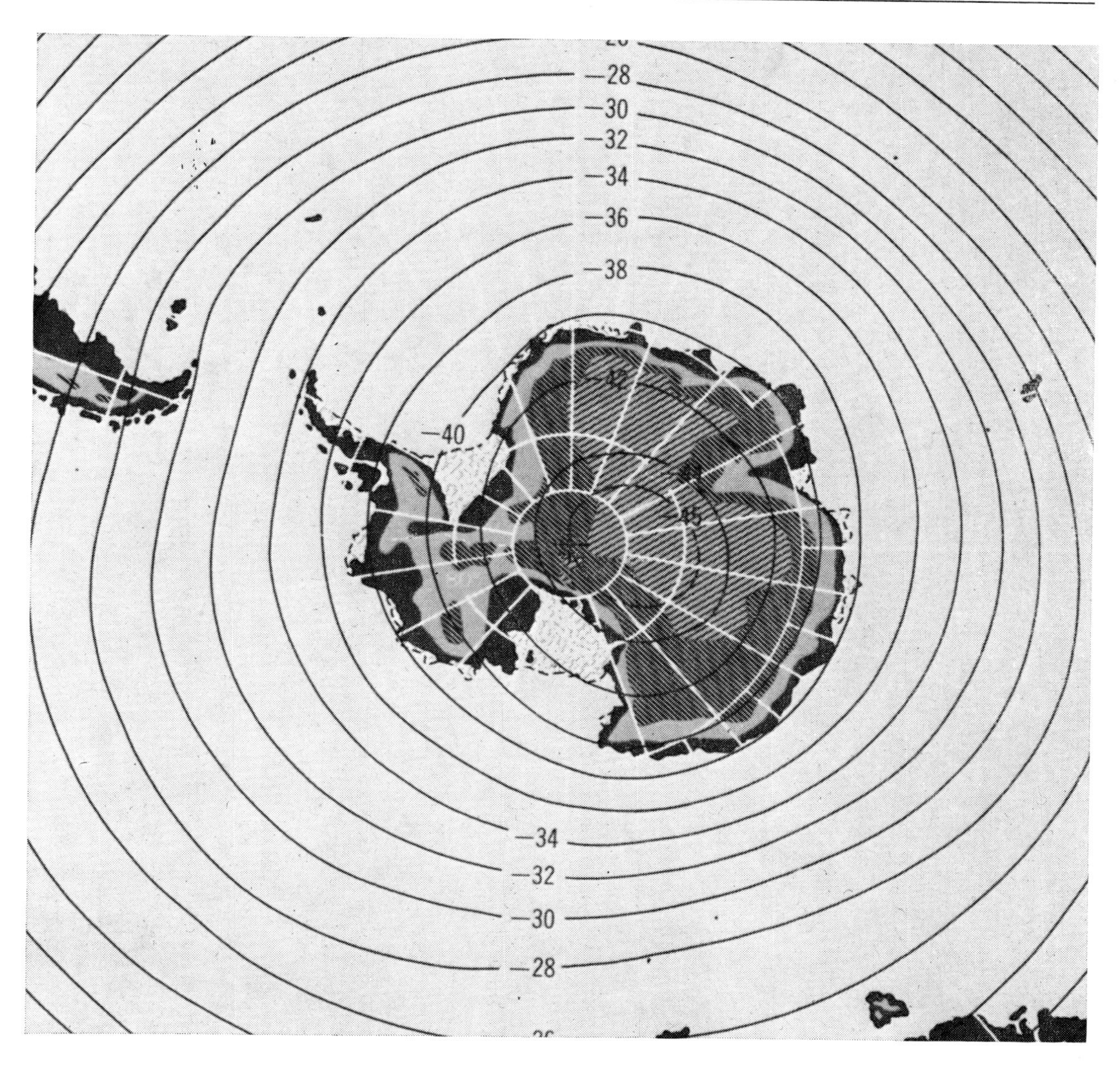

Fig.6. Mean isotherms at the 500 mbar level, July.

produce a dramatic change of the temperature in the surface layer. The first two pairs of stations can be considered typical for the escarpment of the East Antarctic Plateau. In the winter months, the mean temperature differences are more than dry adiabatic; air moving downslope under adiabatic conditions would arrive colder than the air it is going to replace. In individual cases, temperature differences twice as large as the winter mean have been recorded. The third and fourth pair of stations in Table VIII show that the conditions are quite different to the southeast of the two deep indentations of the nearly circular Antarctic shoreline; the Ross and Weddell seas, about 11° farther south. Warm air advection in the troposphere, though not in the surface layer itself, is the prevailing feature in these regions, as can be derived from the four maps (Fig.5 with 19 and Fig.6 with 20), and has been shown for the Byrd Station area in a detailed synoptic study by KUTZBACH and SCHWERDTFEGER (1967). Consequently, and also due to the considerably lower temperatures over the ice shelves than at the coast of East Antarctica,

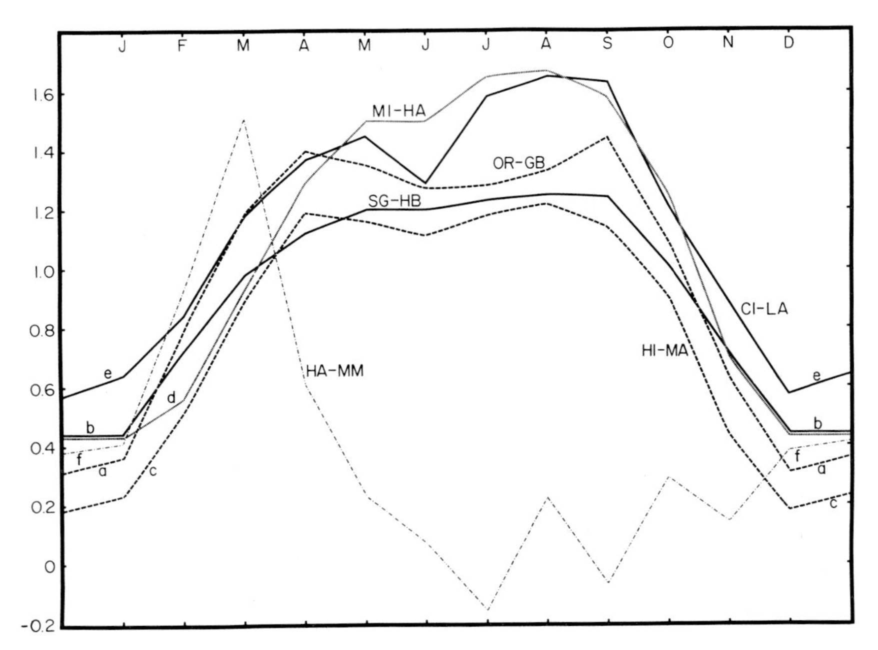

Fig.7. Meridional temperature gradient (°C/degree latitude) between sub-Antarctic islands and Antarctic stations near sea level.

Pair of stations	Diff. in latitude (°)	Av. of longitudes
a = Orcadas–G. Belgrano	60.7–78.0	32°W
b = S. Georgia–Halley Bay	54.3–75.5	38°W
c = Heard Isl.–Mawson	53.0–67.6	68°E
d = Macquarie–Hallett	54.5–72.3	165°E
e = Campbell–Little America	52.6–78.4	177°W
Two Antarctic stations		
f = Hallett–McMurdo	72.3–77.9	168°E

the temperature gradients are much smaller than in Queen Mary Land and Adélie Land. The last two pairs of stations show the temperature decrease poleward over terrain where the effect of increasing height is negligible in comparison to the effect of increasing latitude. Of particular interest here is the record of Admiral Byrd's advanced winter base south of Little America II, Bolling Base, the setting of one of the truly great epics of Antarctic literature (Byrd, 1938). The large temperature difference of almost 7°C over a distance of less than 200 km clearly indicates the importance of the factor "distance from the coast" even during winter when the water, in this case the southern Ross Sea, is mostly ice covered. Nevertheless, in other parts of the big ice shelves (where no adequate observations exist) the poleward temperature gradient may even be larger, as is suggested by the fact that Little America is the coldest of three comparable places at about 78°S near sea level: the multi-annual six winter months mean of McMurdo in the southwestern Ross Sea is −24.3°C, of G. Belgrano in the southeastern Weddell Sea −31.0°C, against −33.2°C for Little America/Framheim.

Mean temperature differences over the interior of the continent are listed in Table IX. Only in the summer of the central area, between South Pole, Vostok and Plateau stations

TABLE VIII

TEMPERATURE DIFFERENCES BETWEEN COASTAL AND NEIGHBORING INLAND STATIONS, IN SUMMER (DEC., JAN.) AND WINTER (APR.–SEPT.)

Station	Lat.(S)	Long.	Height (m)	t_{XII+I}	t_{IV-IX}	Record
Mirny	66°33′	93°01′E	35	−1.7	−15.7	
Pionerskaya	69°44′	95°30′E	2,700	−23.2	−44.3	
Difference	360 km		2,665	21.5	28.6	2 years
Dumont d'Urville	66°42′	140°01′E	40	−1.0	−16.2	
Charcot	69°22′	139°01′E	2,400	−22.4	−44.5	1 summer
Difference	300 km		2,360	21.4	28.3	2 winters
Little America	78°23′	163°46′W	42	−6.8	−33.2	7 years
Byrd	80°00′	120°00′W	1,512	−14.8	−34.1	10 years
Difference	890 km		1,470	8.0	0.9	
Shackleton	77°59′	37°10′W	58	−4.4*	−28.6	
South Ice	81°57′	28°50′W	1,350	−15.3**	−37.8	
Difference	470 km		1,292	10.9	9.2	1 winter
Bahia Marg.	68°08′	67°08′W	21	+0.3	−11.4	12 years
Eights	75°14′	77°10′W	421	−10.6	−33.8	4 sum., 3 wint.
Difference	850 km		400	10.9	22.4	
Little America II	78°34′	163°56′W	40	−6.3	−32.5	
Bolling Base	80°08′	163°55′W	80	–	−39.1	
Difference	174 km		40	–	6.6	1 winter

* Dec. 1956 only.
** Dec. 1957 only.

TABLE IX

TEMPERATURE DIFFERENCES BETWEEN INLAND STATIONS, IN SUMMER (DEC., JAN.) AND WINTER (APR.–SEPT.)

Station	Lat.(S)	Long.	Height (m)	t_{XII+I}	t_{IV-IX}	Record
Byrd	80°00′	120°00′W	1,512	−14.8	−34.1	
S. Pole	90°	–	2,800	−28.5	−58.3	
Difference	1,117 km		1,288	13.7	24.2	10 years
Pionerskaya	69°44′	95°30′E	2,700	−23.2	−44.3	
Komsomolskaya	74°05′	97°29′E	3,420	−31.8	−61.7	
Difference	500 km		720	8.6	17.4	2 years
S. Pole	90°	–	2,800	−28.5	−58.3	
Vostok	78°28′	106°48′E	3,488	−33.1	−66.4	
Difference	1,288 km		688	4.6	8.1	9 years
S. Pole	90°	–	2,800	−27.2	−59.3	
Plateau	79°15′	40°30′E	3,624	−32.3	−67.6	
Difference	1,200 km		824	5.1	8.3	3 years

do the differences correspond to a less than adiabatic lapse rate. Again, the relatively mild temperature regime of West Antarctica is evident, represented by Byrd Station. The annual temperature conditions over the interior in their relationship to the elevation of the terrain are graphically shown in Fig.8. As explained in the legend, the steeper slope of the upper line (compared to the lower one) between Pionerskaya and the higher part of the plateau (Plateau Station, Sovietskaya and Vostok) indicates a decrease of the temperature with height at a rate of more than 1°C/100 m.

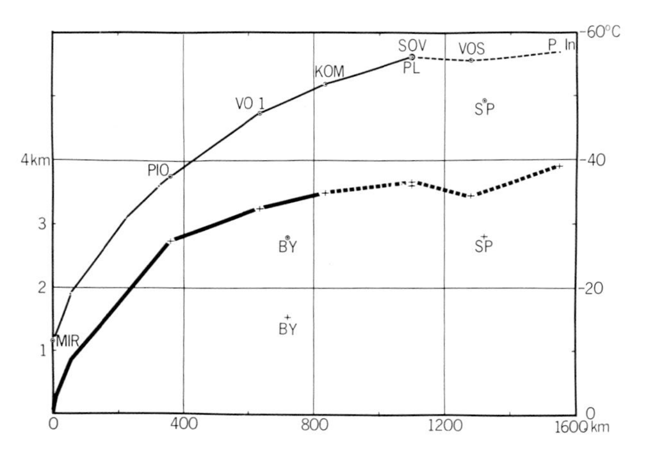

Fig.8. Profiles of height above sea level (lower curve, left side scale) and mean annual temperature (upper curve, right side scale) as function of distance from the nearest coast (abscissa). The two solid lines join the data points of U.S.S.R. stations in East Antarctica, close to the 90°E meridian. The double circles refer to temperature values of meteorological stations, the single circles to snow pit measurements of the temperature several meters below surface. For the latter places, the elevation values are somewhat uncertain. The temperature- and height-scales are such that parallelism of the two lines corresponds to the adiabatic lapse rate, 1°C temperature decrease for 100 m increase in height. *BY* = Byrd Station, *PL* = Plateau Station, *SP* = South Pole.

Aperiodic and periodic temperature variations

Monthly values of the mean daily temperature range (M.D.T.R.), that is, the multi-annual mean difference between the monthly average of the daily maximum and the daily minimum temperature (in older texts often called the aperiodic daily variation), are listed in the climatic tables at the end of the chapter for all stations for which such data were available. The general annual trend is characterized by low values, mostly between 3° and 5°C, in the warmest months December to March, and higher values in winter and/or spring. This is true for most of the coastal stations as well as for the two places in the interior for which a ten-year record of this parameter has been published, Byrd and South Pole. For the stations on the East Antarctic Plateau, Pionerskaya, Vostok and Plateau Station itself, there are only shorter records, but all existing evidence suggests that there the opposite annual variation of the M.D.T.R. is the rule, the highest values appearing in the months when the sun is above the horizon. It becomes evident that the conditions at the South Pole can certainly not be considered typical for most of the interior area; the pole is, of course, the only place where there is no diurnal variation, in any month, of the solar radiation incoming at the top of the atmosphere, and no significant diurnal variation of cloudiness. This circumstance must also affect the annual average of the M.D.T.R., but Table X reveals some other surprising features.

TABLE X

ANNUAL AVERAGES OF THE DAILY MEAN TEMPERATURE RANGE (°C)

Station	Temp. range	Station	Temp. range
Little America I	10.6	Little America II	10.6
Little America III	10.6	Little America V	7.9
Halley Bay	8.1	Belgrano	9.2
Sanae	9.7	South Pole	4.8
Pionerskaya	6.5	Byrd	7.6
Vostok	8.6	Plateau	10.7

Twelve coastal stations have ranges between 5.2° and 7.0°C.

The peculiar annual march of the M.D.T.R. is shown in the climatic tables (see pp.331–355). Varying height of the sun and variable cloud conditions (back radiation) certainly are contributing factors, but a full, quantitative explanation of the differences between various areas has not yet been given. Again it becomes apparent that the stations on the east coast of the Weddell Sea differ significantly from other places near sea level. In this case, it might be suggested that the frequent occurrence of open water areas, polynyas, in the proximity of the stations has an important role in producing the large M.D.T.R. and the largest values at the time of the year when the temperature at the continent is lowest.

Only a few of the many temperature records which have been accumulated in recent years have been evaluated with regard to the *periodic* temperature variation. HISDAL's (1960) analysis of the Maudheim record is a fine example. He also explained that a strange phenomenon which had puzzled polar meteorologists in years past, that is, systematic diurnal temperature variations found for special types of days (clear vs. cloudy, with high vs. low winds, etc.) during the polar night, is the result of a statistical bias rather than of physical processes.

As a simple substitute for determining the amplitudes of the periodic diurnal variations by harmonic analysis of the series of hourly mean values, Table XI shows the monthly mean temperature differences between two hours which can be assumed to be close to the time of occurrence of the maximum and the minimum, respectively. For Byrd Station, 16h00 and 04h00 instead of 15h00 and 03h00 L.M.T. have been chosen only because they are the hours of the synoptic radio soundings. It is interesting to note that in this series the highest values appear when the sun is permanently above the horizon (see Table III); of course, its elevation angle varies by 20°, on October 18 and February 25 from 0° to 20°, at the summer solstice from 13.5° to 33.5°. The Maudheim data permit a comparison of temperature differences for the selected hours with the amplitude of the first harmonic component of the diurnal variation. The low summer values of the Orcadas series indicate the maritime character of this station when skies are almost permanently cloud covered and the waters around the island often free of ice. Opposite climatic conditions are shown in Fig.9. Hourly temperature data from the high plateau give evidence of how shallow the layer affected by the diurnal heating really is, when a strong inversion exists in the surface layer. The full amplitude of the periodic diurnal variation at Plateau Station

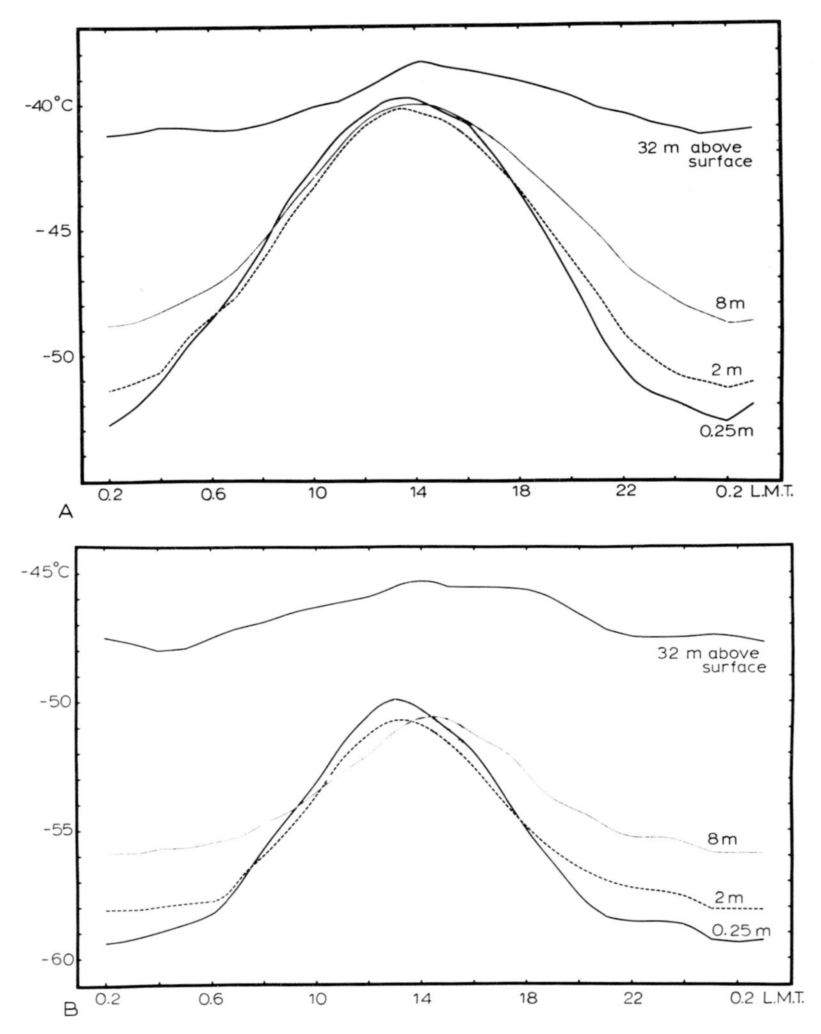

Fig.9. The diurnal variation of the temperature at various levels above the surface at Plateau Station (3,625 m). The hourly value of each day at each level of the micrometeorological tower is the average of 10–20 individual measurements recorded in the first 30 min after the full hour, Greenwich time. 00h00 G.M.T. = 02h42 L.M.T. The temperatures are all negative, −°C. A. Hourly means for 25 days of February, 1967. B. Hourly means for 22 days of March, 1967.

in the month of February, for instance, amounts to 13°C at 0.25 m above surface, while at only 32 m it has already diminished to about 3°C.

The extreme shallowness of the layer of strongest temperature increase with height explains the relatively large variability of the temperatures at screen height, about 1.5 m, which is observed at all high plateau stations. A slight divergence in the local wind field with the resulting vertical advection of heat, as well as temporarily enhanced vertical mixing due to minor changes in the wind vector profile, must lead to a pronounced temperature rise; an analysis of such a case was given by LETTAU (1966). With the cessation of the heating processes, the strong inversion re-establishes itself under the influence of the overall radiation conditions and by horizontal advection of air from neighboring areas in which the above phenomena happen to be weaker, or absent. Strong variations of this kind are restricted to the lowest layers.

TABLE XI

ANNUAL VARIATION OF THE DIFFERENCE OF THE MEAN TEMPERATURES (°C) AT HOURS CLOSE TO THE TIME OF AVERAGE OCCURRENCE OF THE DAILY MAXIMUM AND DAILY MINIMUM

Month	Station: Byrd 80°S Time: 16–04h L.M.T. Record: 10 years Parameter: Δt	Maudheim 71°S 15–03h L.M.T. 2 years		Orcadas 60.7°S 15–03h L.M.T. 47 years Δt
		Δt	$2A_1$**	
Jan.	3.8	5.2	5.1	1.1
Feb.	3.9*	4.8	4.7	0.9
Mar.	2.0*	3.5	3.5	0.5
Apr.	0.3	1.5	1.6	0.2
May	0.1	1.0	0.7	0.0
June	−0.5	1.8	1.3	0.0
July	−0.4	0.0	0.4	0.1
Aug.	0.0	0.9	0.9	0.6
Sept.	2.0	2.1	2.0	1.1
Oct.	3.9	5.4	5.8	1.2
Nov.	4.5	7.2	7.0	1.2
Dec.	2.8	4.8	5.1	1.2

* For a comparison with values for a station at similar latitude but much higher elevation, see Fig.9.
** In the case of Maudheim, the *full* amplitude of the first harmonic component of the periodic diurnal variation has been added under the heading "$2A_1$". For this first harmonic, the mean time of occurrence of the maximum is 14.3 L.M.T. (HISDAL, 1960).

Extreme values and their possible errors

Absolute temperature extremes are included in the climatic tables at the end of the chapter. Fig.10 shows a comparison of extreme values recorded at two stations located at equal latitude but very different elevation. The comparability of such values from station to station is limited. One reason is the difference in length of the observational records, though this problem could be taken care of by statistical procedures. Another reason, regarding the absolute maxima as well as the mean temperatures of the summer months, is the effect of solar radiation on values measured in meteorological shelters (screens); it can lead to errors of the order of 2°C or more, on days with little wind. Finally, the character of the strong winter surface inversion makes the absolute minima and the mean temperature of the winter months a function of the exact height of the thermometer above surface. This is immediately evident from the temperature curve in Fig.11, as well as from Fig. 9.

The possible radiation effect on the temperature measurements depends upon local conditions, like the albedo of the ground underneath the screen, and upon the specific instrumentation. At Maudheim in 1950 and 1951 (LILJEQUIST, 1956) as well as at Norway Station during the I.G.Y., a ventilated Assmann psychrometer was used, outside the screen, at the time of the 3-hourly routine readings of the thermometers in the screen. The difference "inside minus outside" can be called the radiation error. The results of Norway Station (VINJE, 1965) are summarized in Table XII; its upper part refers to observations when the global radiation was >0.45 Ly/min, the lower part to smaller values.

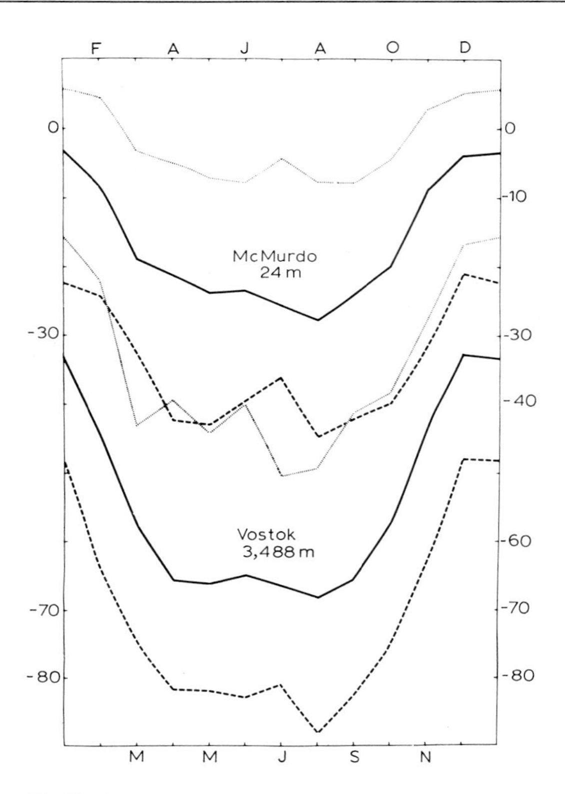

Fig.10. Average and absolute extreme temperatures for each month, at two stations of approximately equal latitude.

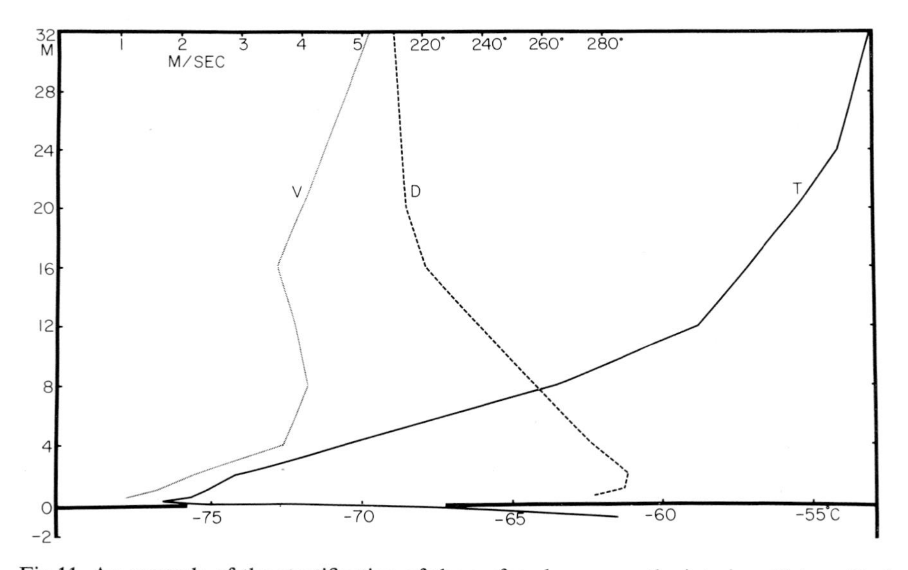

Fig.11. An example of the stratification of the surface layer over the interior; Plateau Station, April 9, 1967, average of 49 runs, 19h30–19h54 G.M.T. T = temperature from −1 to 32 m; D and V = wind direction and speed, respectively, from 0.5 to 32 m above surface.

Vinje points out that even with winds of considerable strength the radiation error is not zero. Therefore, the mean temperatures can also be seriously affected, particularly in the months with permanent daylight. Liljequist cites the maximum radiation error found at Maudheim: 5.4°C (screen −2.2°C, Assmann −7.6°C), 15h00 L.M.T., 19 February, 1951, bright sunshine and 6/8 cloud cover (optimum global radiation conditions).
Nevertheless, there is no doubt that at all Antarctic stations near sea level, including the stations on the ice shelves at about 78°S, temperatures considerably above the freezing point have been observed. Even at Byrd, 700 km from the nearest coast and at an elevation of 1,500 m, the absolute maximum of −0.6°C comes close to that limit. Only a few of the stations near sea level north of 70°S have observed absolute minima below −40°C, but on the ice shelves the limit is close to −60°C. In the interior, −80°C has been passed once in 10 years at the South Pole, and is frequently passed at the higher elevations of the central plateau. The observed absolute minimum stands at −88.3°C, Vostok, August 24, 1960. The absolute range amounts to 45°–55°C for the coastal stations, and lies between 64° and 68°C for the ice shelves and the interior. That is not a particularly large range, when compared with continental areas in middle latitudes of the Northern Hemisphere.

TABLE XII

THE AVERAGE RADIATION ERROR OF TEMPERATURE MEASUREMENTS IN A METEOROLOGICAL INSTRUMENT SCREEN AT NORWAY STATION, AS A FUNCTION OF WINDSPEED
(After VINJE, 1965)

Mean wind speed (m/sec)	Mean radn. error (°C)	Mean global radn. (Ly/min)	No. of observations
$R_G > 0.45$ Ly/min			
1.5	1.95	0.86	21
3.6	1.17	0.78	48
7.1	0.49	0.82	57
13.3	0.37	0.80	24
$R_G < 0.45$ Ly/min			
1.5	0.48	0.12	10
3.9	0.50	0.16	26
6.7	0.25	0.15	44
11.6	0.16	0.24	11

The surface inversion

The most conspicuous characteristic of the Antarctic temperature regime, besides the low temperatures of the interior, is the surface inversion. Excepting the two summer months, it is an ever present feature over the high plateau, and a frequent one over the rest of the continent, the coastal areas included. The average strength of the inversion over various stations is given in Table XIII, average winter conditions over the entire continent are shown in Fig.12. Regarding its diurnal variation, the data for Byrd Station are of particular interest; Byrd is located at 120°W where the times of the two daily radio-soundings happen to be close to the hours of the average occurrence of the daily extremes. The difference between the temperature of the 750 mbar level (400–600 m above ground) and the surface temperature has been computed separately for the two daily soundings. As the average 12-hourly temperature difference at 750 mbar does not exceed 0.5°C in any month and therefore is negligible, a comparison of the two columns, for 04h00 and 16h00 M.L.T., directly shows the effect of diurnal variation of the temperature near the

TABLE XIII

AVERAGE TEMPERATURE INCREASE WITH HEIGHT IN THE LOWER TROPOSPHERE

	Summer (XII+I)	Winter (IV–IX)
S. Pole, sfc. to 650 mbar*	0.8°C	18.1°C
S. Pole, sfc. to 600 mbar*	−0.6°C	20.5°C
Vostok, sfc. to 4 km*	1.0°C	22.7°C
Byrd, sfc. to 750 mbar, 04 L.M.T.*	0.0°C	9.5°C
Byrd, sfc. to 750 mbar, 16 L.M.T.*	−3.3°C	9.2°C
Little Am., sfc. to 950 mbar	−0.8°C	8.6°C
Little Am., sfc. to 900 mbar	−2.9°C	10.8°C
Mirny, sfc. to 500 m*	−2.0°C	1.0°C

* For monthly values and number of soundings see the climatic tables at the end of the chapter.

surface (as it is listed in Table XI) on the inversion. It becomes apparent that the surface inversion undergoes strong diurnal changes from about mid-October to March, and that a diurnal change from inversion to lapse conditions is the prevailing feature from November through February.

Fig.13 shows average summer and winter temperature distributions at Vostok, the only station in the higher part of the plateau with a long record of radiosoundings. An individual temperature curve for a winter day with more than normal strength of the inversion has been added. It is necessary to note that almost always there is a slow, gradual decrease of dT/dz, the change of temperature with height, so that the determination of the top of the inversion is quite arbitrary, depending upon a few tenths of a degree, the density of evaluated points of the radiosoundings, and the definition of "top". This explains the widely scattering values for the "depth" of the surface inversion, anywhere between 100 and 1,000 m, to be found in the literature. It appears, however, that there is a real annual variation of the height above ground of the average tropospheric temperature maximum, with the lowest values ($<$ 300 m) in the summer, the highest ($>$500 m) in the winter (Phillpot and Zillman, 1969). In effect, a conventional radiosonde rising 250–350 m/min, is not designed for a detailed study of a strong surface inversion. On a linear temperature–height diagram, a straight line drawn from surface to the next data point measured, cannot give a true picture of the inversion. As Fig.11 shows for an individual case, the increase of temperature in the lowest few meters above ground is much stronger than in the rest of the inversion layer. This is the normal,

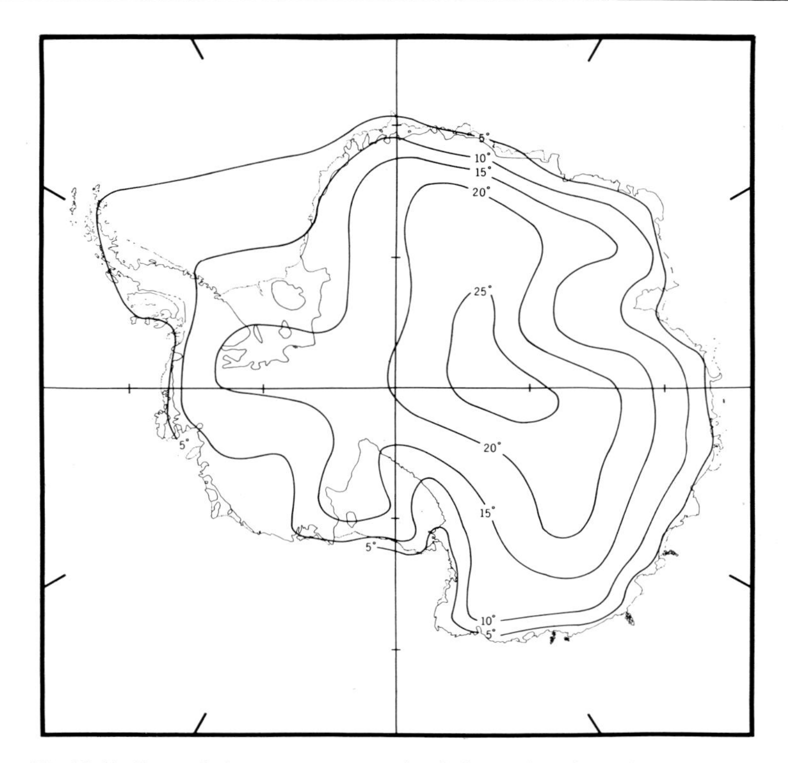

Fig.12. Isolines of the average strength of the surface inversion in the winter (June–August). (After PHILLPOT and ZILLMAN (1969) with slight modifications.)

typical structure of the surface inversion over the continent, as proved by measurements on masts (KUHN, 1969) and by special, slow rising radiosondes (SPONHOLZ, 1968). A detailed statistical study of the inversion conditions in the lowest 15 m, over one of the large ice shelves, was made by HOINKES (1967). Still in the process of elaboration, in 1969, is the most detailed analysis of the great inversion, from data obtained since January 1967

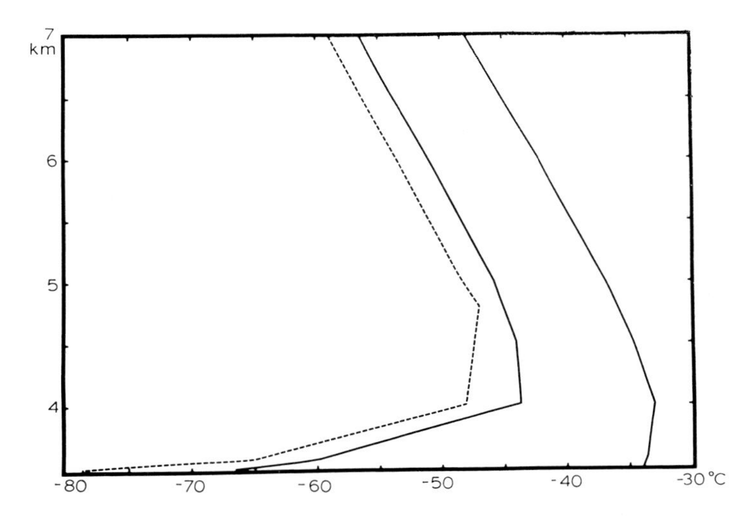

Fig.13. Vostok. Average temperature between surface and 7,000 m in the summer (December, January) and winter (April–September), and an example of an extremely strong inversion (June 2, 1960).

by an almost continuous series of measurements of temperature and wind at ten levels between surface and 32 m, along the micrometeorological tower erected at Plateau Station. These data promise a wealth of information of which Fig.9 and 11 give only a preliminary example.

The maintenance of the inversion and the coreless winter

There is no doubt that the strong surface inversion is brought about through the intense radiative heat loss from the surface under the geographical and atmospheric conditions of Antarctica. However, the facts that the inversion never comes to exceed a certain limit, $< 40°C$, and that at the end of the long winter night it is not stronger than at the beginning, deserve some comment. When the skies become clear after a period of cloudy weather and relatively high temperatures in the surface layer, there is at first a considerable heat loss from the surface because the emitted long wave radiation is much larger than the atmospheric back radiation; large values of the Ångstrom ratio A_0 result. With the corresponding marked decrease of the surface temperature, the outgoing heat flux diminishes, while the temperature in the free atmosphere above the inversion layer, and hence the back radiation, change little. Consequently, even under excellent "radiation conditions", i.e., clear skies and negligible moisture content of the air, an equilibrium

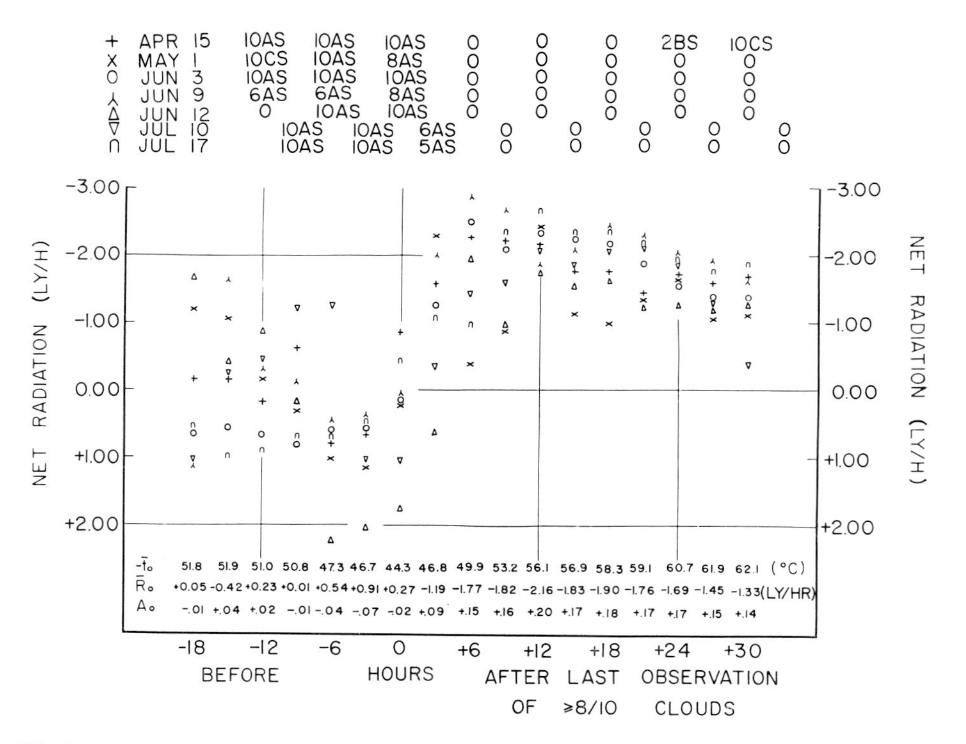

Fig.14. Variation of the net radiation with time, in seven typical cases of clearing skies, South Pole, winter 1965. Top of the graph: dates of the seven days of the analysis, and symbols used for each day; amount (in tenths) and type of the clouds observed at the times indicated at the bottom of the graph; AS = altostratus; CS = cirrostratus; BS = blowing snow. Underneath the +2.00 Ly/hour line, $\bar{t}_0$ and $\bar{R}_0$ stand for the mean temperature and the mean net radiation at the times shown farther below; Å$_0$ is the Ångstrom ratio as explained in the text.

state can be established in which the heat loss by emitted (black body) radiation is balanced by the positive components of the heat budget. These are atmospheric back radiation, vertical eddy flux of heat, heat conduction from the ground, and heat of sublimation (deposition of H_2O from the atmosphere). The data presented in Fig.14 (SCHWERDTFEGER, 1968) support this interpretation of the role of the two radiation fluxes. The figure shows the relevant parameters for seven cases of rapid clearing at the South Pole, in 1965. It appears that about 12 hours after disappearance of the cloud deck, the net outgoing radiation R_n reaches a maximum; so does the strength of the surface inversion. Later, R_n diminishes sufficiently to make an approximate heat balance possible.

On a large time scale, the same evidence and reasoning can be used to explain why the surface temperature in Antarctica and particularly over the Antarctic Plateau decreases very little during the winter night, from April to August or early September, presenting the so-called coreless winter phenomenon (POLLOG, 1924; MEINARDUS, 1938). When the effective short wave radiation (global radiation minus reflected) becomes negligible and the surface temperature still is relatively high, the emitted long wave radiation is considerably larger than the atmospheric back radiation. With the ensuing temperature decrease at surface and much less cooling of the free atmosphere (see Fig.13 and 15), the net radiation (outgoing) decreases and the above explained equilibrium of the heat budget components will be approached very soon. A definite change of the radiation regime imposes itself only after the return of the sun. The short time variations of the surface temperature during the winter night can be understood as caused by changes of sky conditions (clouds or no clouds), by advection of air masses of different moisture content mainly in the free atmosphere above the inversion layer, and by vertical motion and mixing in the surface layer itself, as discussed earlier under aperiodic and periodic temperature variations.

The role of warm air advection

The above explanation of the coreless winter refers to the surface temperature. It is incomplete in so far as it does not include any reasoning why the same phenomenon exists in the troposphere also above the inversion. This reason can best be stated in WEXLER's words (1959a): "As the sun sets, the temperature drops rapidly over the continent but less so over the surrounding oceans which are only partly ice-covered. The increasing meridional temperature gradient brings about the release of baroclinic instability in the troposphere which initiates the formation of numerous intense cyclones. These cyclones move vast quantities of warm marine air southward in the lower troposphere, effectively 'ventilating' large portions of Antarctica above a thin surface layer of cold air and slowing and even sometimes reversing the normal seasonal decline in surface temperature".

The notion that synoptic scale perturbations from the subpolar belt of strongest cyclonic activity, instrumental for the meridional exchange of air masses, reach far into the continental areas is confirmed by the daily weather maps (Weather Bureau of South Africa, 1962, for the I.G.Y.; in the journal *Notos* for later years). The climatological parameter which best reflects such cyclonic activity is the average magnitude of the interdiurnal pressure variation, $\overline{\Delta p} = |\overline{p_2 - p_1}|$, shown in Table XIV. Of course, one must refer to

TABLE XIV

INTERDIURNAL PRESSURE VARIATION (mbar) AT VARIOUS LATITUDES AND ELEVATIONS. SELECTION OF TWO-YEAR PERIODS ACCORDING TO AVAILABILITY OF DATA

Station		Season			
		XII–II	III–V	VI–VIII	IX–XI
Orcadas	$\overline{\Delta p}$	5.0	6.4	7.8	6.3
1941 + 1942					
60.7°S	$\bar{p}_s$	989.3	990.6	993.1	993.1
4 m	$\overline{\Delta p}/\bar{p}_s$	5.0‰	6.5‰	7.8‰	6.4‰
Syowa	$\overline{\Delta p}$	3.7	5.5	6.5	6.2
1960 + 1961					
69.0°S	$\bar{p}_s$	985.7	983.6	983.5	982.8
15 m	$\overline{\Delta p}/\bar{p}_s$	3.8‰	5.6‰	6.6‰	6.3‰
S. Pole	$\overline{\Delta p}$	2.3	3.4	4.2	3.3
1964 + 1965					
90°S	$\bar{p}_s$	691.2	677.8	682.3	676.7
2,800 m	$\overline{\Delta p}/\bar{p}_s$	3.0‰	5.1‰	6.1‰	4.9‰
Plateau	$\overline{\Delta p}$	1.9	2.8	2.9	2.5
1967 + 1968					
79.3°S	$\bar{p}_s$	618.0	609.6	603.0	606.4
3,625 m	$\overline{\Delta p}/\bar{p}_s$	3.1‰	4.6‰	4.8‰	4.1‰

For explanation see text.

the relative variation $\overline{\Delta p}/\bar{p}_s$, where $\bar{p}_s$ is the average *surface* pressure, in order to make a meaningful comparison of variations observed at low and at high elevation.

The three curves in Fig.15 clearly show in which layers of the atmosphere the advection process and meridional exchange of airmasses must be most pronounced. The range between mean winter and summer temperatures above the surface layer of the high plateau is practically the same as on the coast. Finally, to make the model complete, one should add that over central Antarctica also seasonal variations of vertical "advection", more downward motion in winter than in summer, contribute to keep the annual temperature range in the troposphere relatively small.

The importance of warm air advection and prevailing sinking motion for the temperature regime of the continent, particularly in the troposphere above the inversion, is shown by direct measurements of long wave radiation fluxes. The results of radiometer soundings made at the South Pole, shown in Fig.16, indicate a radiative cooling rate of about 4°C/day in the warmest layer above the surface inversion. The actual mean local decrease of temperature with time is more than an order of magnitude less. The only processes to counterbalance the radiative cooling during the winter night are sinking motion in a stable atmosphere and advection. Furthermore, an evaluation of satellite data (RASCHKE, 1968) showed that the average outgoing radiative energy flux amounted to 10.3 Ly/h at the top of the atmosphere over the South Pole area during two mid-winter months of 1966. This value is in good agreement with the results of direct radiometer soundings which for the four polar winter nights 1959–1962 indicated an average outward radiative flux at the 50 mbar level of 10.5 and 12.9 Ly/h, at the South Pole and Byrd Station, respectively

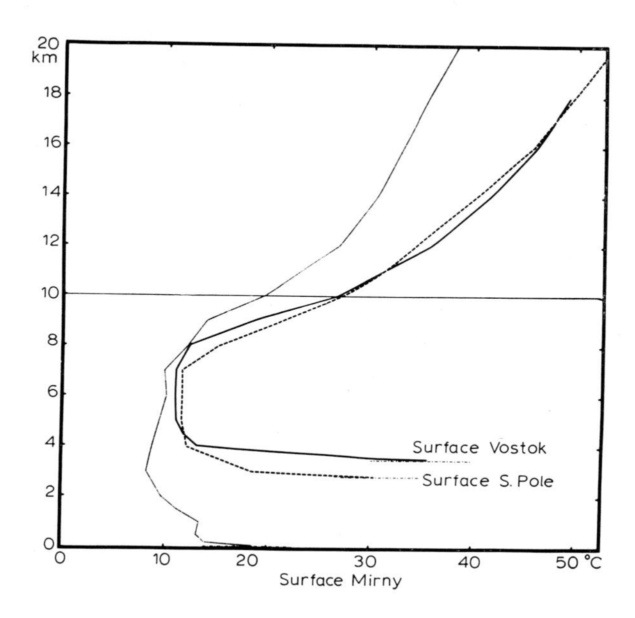

Fig.15. Average range of temperature, $\overline{T}$(January) $-\overline{T}$(July), as function of height, at 66°, 78°, and 90°S.

(WHITE and BRYSON, 1967). Reliable measurements made at the South Pole meteorological station indicate that at the same time of the year (in 1965 and 1966) the outgoing net radiation at the surface was about 1.2 Ly/h (SCHWERDTFEGER, 1968). The difference, flux at the top minus flux at the bottom, represents the radiative energy loss of the South Pole atmosphere, the flux divergence. The corresponding temperature decrease, the "radiative cooling rate", is about 1.3°C/day, or 40°C/month. The actually observed decrease of the vertically averaged temperature, according to radiosonde measurements up to 50 mbar and an extrapolation for the highest layers of the atmosphere, is not more than 3°C/month; in fact, it is less in the troposphere, somewhat more in the stratosphere.

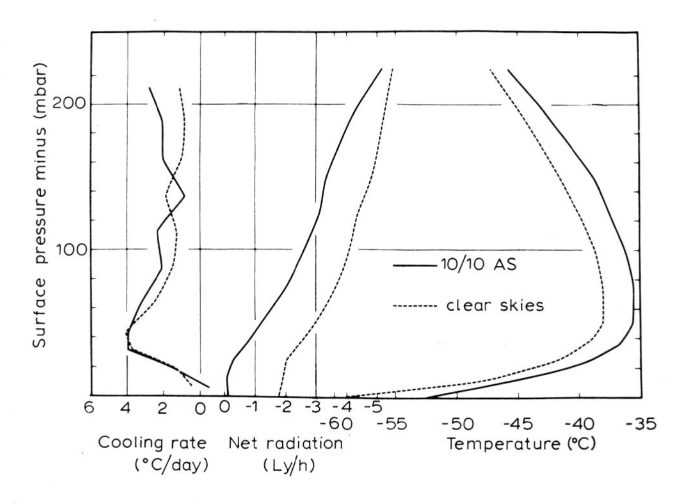

Fig.16. Cooling rate, net radiation, and temperature in the lower troposphere over the South Pole; averages of ten cases of overcast and 56 cases of clear skies, winter 1965.

This discrepancy, 40°C/month against 3°C, clearly illustrates the main factors which determine the climate of Antarctica for the larger part of the year: heat loss through radiation, but heat gain through horizontal advection of air from lower, warmer latitudes, and through prevailing downward vertical motion. In the surface layer itself, however, there is a predominant outflow from the continent. For this layer, therefore, the back radiation from the warmer layers above the inversion is the decisive process which prevents a more intense cooling during the winter night.

The "coreless" annual temperature curve carries an interesting implication: the average temperature of the coldest month of the year, $\overline{T}_{mi}$, differs considerably less from the annual mean than that of the warmest month. $(\overline{T}_{ma} + \overline{T}_{mi})/2 > \overline{T}_{year}$. If one knows the temperature of December or January from a summer expedition, and the annual average from a measurement a few meters below the surface, one cannot deduce the temperature of the coldest month without more information on the annual march. This led to a pleasant surprise for the members of the first South Pole wintering party in 1957 who had expected to suffer a much colder polar night (SIPLE, 1959). In the cases of the South Pole and Vostok, for instance, the disregard of this phenomenon would lead to an error of more than 10°C.

The important process of advection refers, of course, both to sensible and latent heat. The relatively warm air coming from the ocean areas surrounding the continent also transports moisture, mainly in the form of water vapor; the liquid or solid H_2O content of clouds at high latitudes is rather small. A large part of the moisture accumulates through precipitation and deposition (negative sublimation) and nourishes the persistent flow back into the ocean of ice in glaciers and shelves, and of blowing snow.

Pressure and wind

Average conditions, circulation pattern, and meridional gradients

The sea level pressure field for the months January and July is shown in Fig.17 and 18, after TALJAARD et al. (1969); no isobars are drawn over the major part of the continent because of its elevation. Indeed, the average *surface* pressure over wide areas of East Antarctica is less than 700 mbar, over the highest parts of the central plateau even less than 600 mbar. Therefore, the first standard pressure level which can represent the free atmosphere over the entire continent is the 500 mbar level. Its absolute topography is shown in Fig.19 and 20.

The most remarkable features of these maps are the significant asymmetry of the circumpolar vortex (SCHWERDTFEGER, 1967), and the fact that its centre does not coincide with the centre of the approximately circular isotherm pattern shown in Fig.5 and 6. At the 200 mbar level, not shown, the situation is similar in more pronounced form still. Average conditions of this kind only suggest, but an analysis of daily upper air data proves, that in the upper troposphere and lower stratosphere warm and cold air advection prevails over different sectors of the continent, respectively (KUTZBACH and SCHWERDTFEGER, 1967). Specifically, it is West Antarctica where warm air advection and positive vertical motion are frequently found, as a lapse rate analysis shows. Opposite conditions, cold advection and subsidence, predominate over East Antarctica, in particular in the sector between 120° and 170°E. In the latitudinal belt 75°–80°S at the 200 mbar level, for instance, Ellsworth (77.7°S) and Halley Bay (75.5°S) have in the winter months lower temperatures than Little America (78.3°S) and even Byrd (80°S).

Oversimplifying a bit, one can visualize relatively warm and moist air masses moving toward West Antarctica from the northwest sector, rising and cooling while travelling across this part of the continent, then east of the Weddell Sea slowly sinking and gradual-

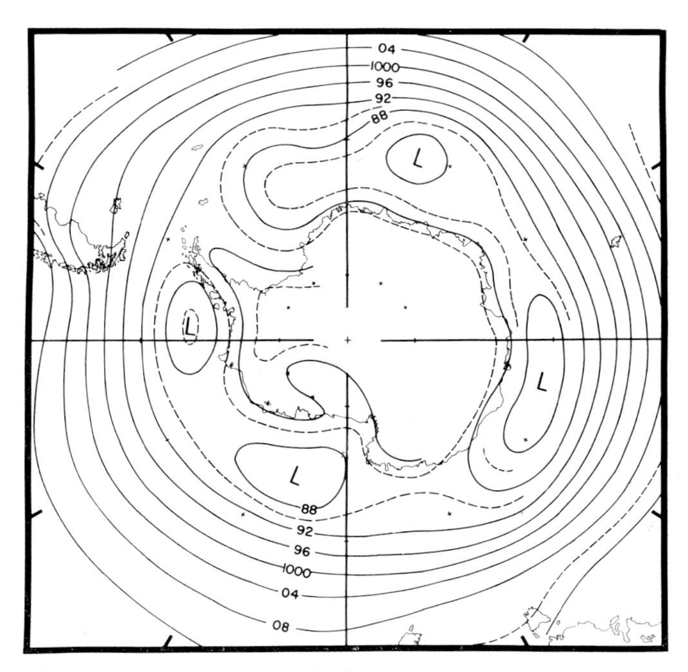

Fig.17. Mean pressure at sea level, January.

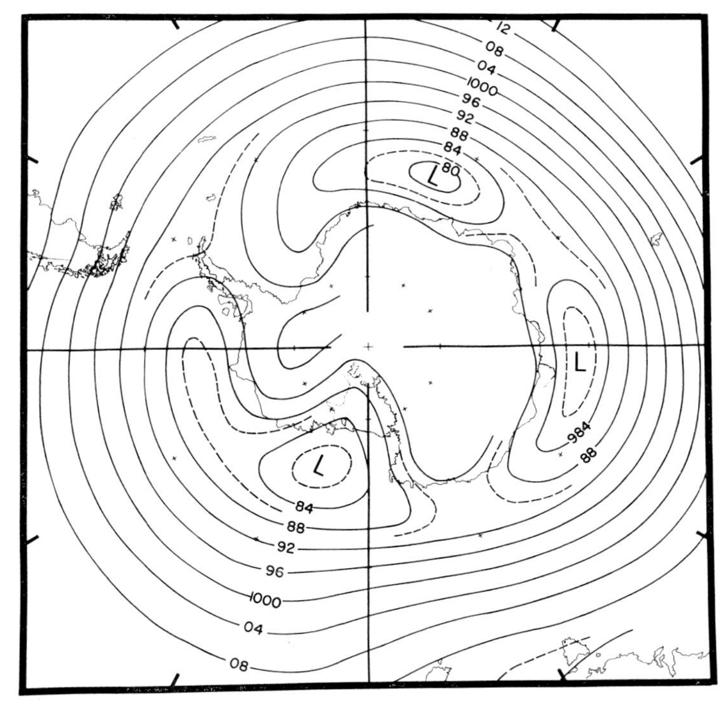

Fig.18. Mean pressure at sea level, July.

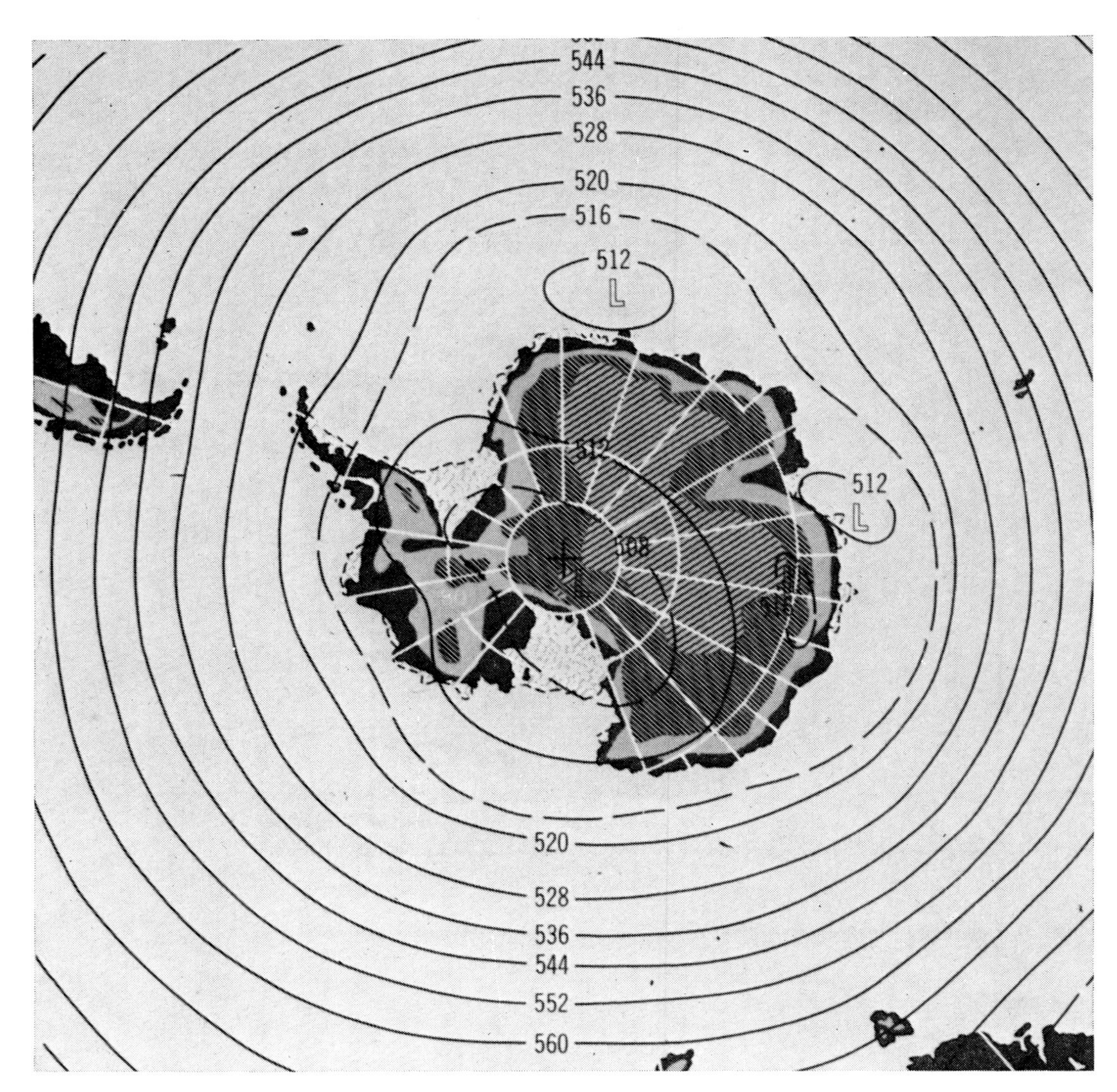

Fig.19. Mean height of the 500 mbar level, January.

ly incorporating themselves into the lower tropospheric circulation over the high plateau. Naturally, the development of synoptic systems like ridges and troughs will temporarily modify such a flow pattern. The daily synoptic charts for the Southern Hemisphere south of 20°S, surface and 500 mbar, elaborated by the Weather Bureau of South Africa and published as part III of the I.G.Y. World Weather Maps series, as well as the continuation of this comprehensive analysis for later years published in the South African journal *Notos*, contain a wealth of information. Of course, the lack of daily routine observations in the vast areas of the Southern Ocean is still a serious handicap. This gap can now be filled, for the 200 mbar level (between 10 and 11 km) at least, by the Global Horizontal Sounding Technique (GHOST)—balloons flown by the U.S. National Center for Atmospheric Research (N.C.A.R.) from New Zealand (SOLOT, 1968). A project of this kind was originally proposed to the World Meteorological Organization's Commission for Aerology, Toronto session 1956, by the Argentine representative. The movement of

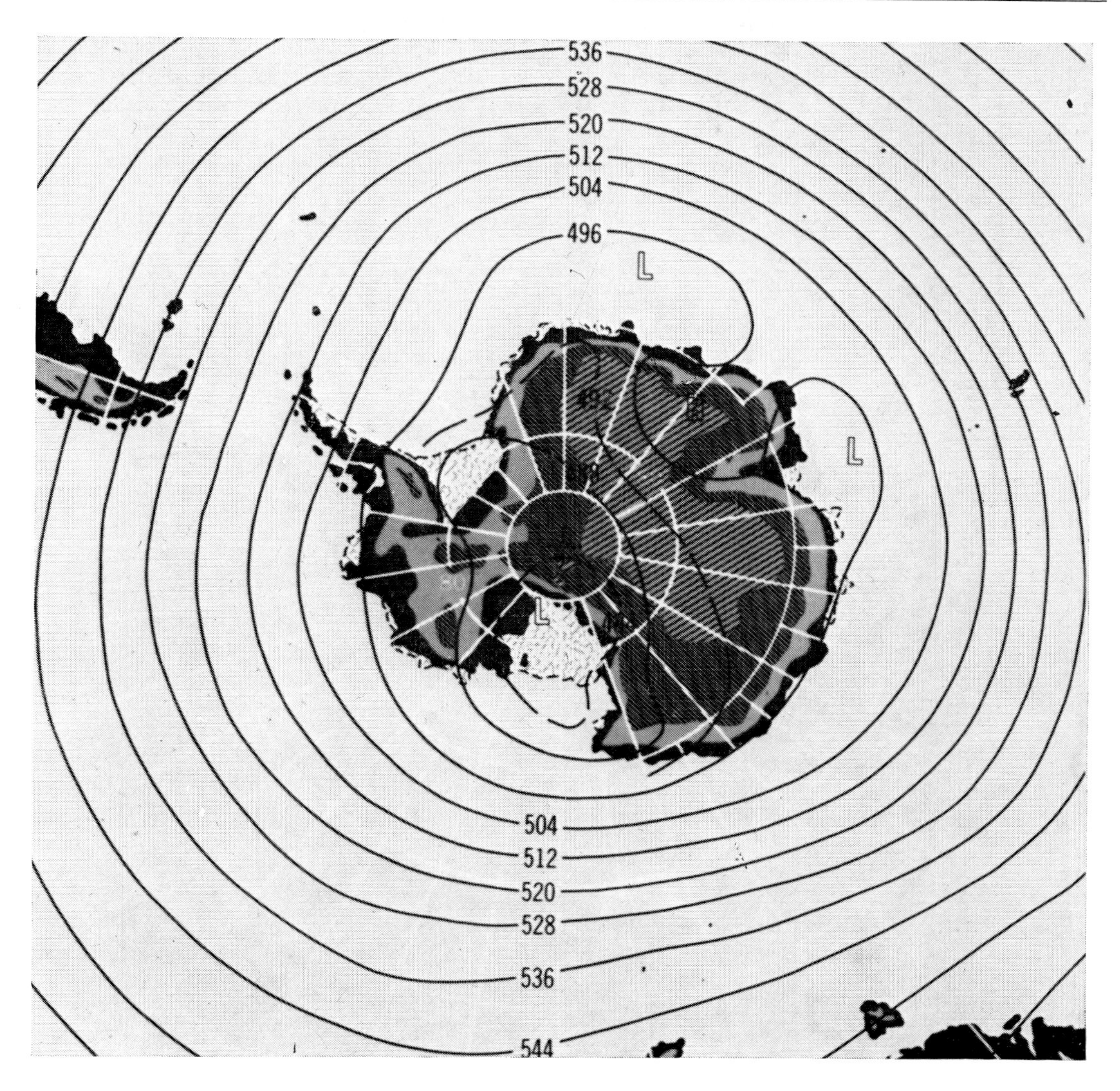

Fig.20. Mean height of the 500 mbar level, July.

N.C.A.R.'s superpressurized, constant volume balloons, floating at predetermined constant density levels, clearly depicts the main characteristics of the circumpolar vortex and its large scale wave pattern, as the balloon trajectories reproduced in Fig.21 exemplify.

In all layers, certain seasonal variations are superimposed on the principal vortex field which in the troposphere and lower stratosphere is always cyclonic, but which above about 20 km reverses to a warm, anticyclonic vortex during the summer months. Especially, it must be borne in mind that the July maps, Fig.4, 6, 18 and 20, do not represent extreme monthly conditions in the troposphere. The annual curve of mean pressure in the 40°–50°S belt has two maxima, in the equinoctial months, whereas in the polar regions, near 70°, maxima appear at the time of the solstices. This leads to a pronounced semi-annual variation of the meridional pressure gradient (SCHWERDTFEGER and PROHASKA, 1955, 1956; HOFMEYR, 1957; SCHWERDTFEGER, 1960, 1962a), as it is shown in Fig.22, the

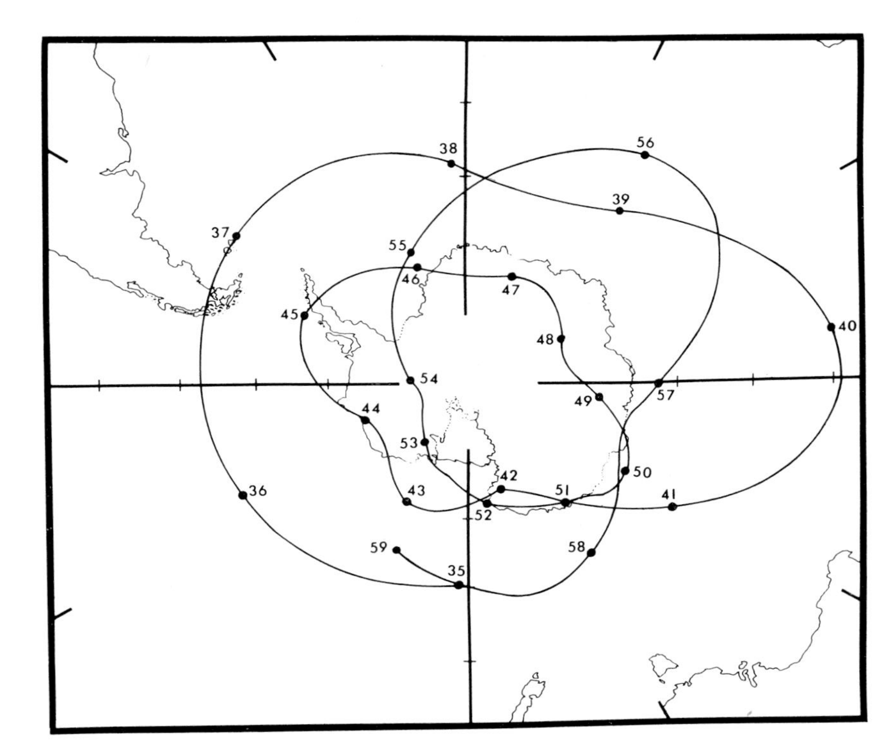

Fig.21. Trajectory of a GHOST balloon floating at 200 mbar, crossing the Antarctic continent various times. The numbers indicate the 135th to 159th day since the start of the balloon at Christchurch, New Zealand, 25 May 1966. (After SOLOT, 1967.)

pressure gradients having been translated into average zonal components of the geostrophic wind. These are latitudinal averages, resulting from a comprehensive new analysis of all information available up to 1967 (TALJAARD et al., 1969).

A statistically adequate number of direct surface wind observations made on board ships in high southern latitudes in all parts of the year and over many years, exists only for one sector, the area south of Cape Horn. These data, more than 77,000 observations between 55° and 60°S, clearly confirm the fall and spring maxima of the westerly wind component, the "equinoctial storms" according to old mariner's lore (SCHWERDTFEGER, 1962b). This semi-annual variation of the strength of the subpolar westerlies is not restricted to the surface layer, it appears rather conspicuously at the 500 mbar level (Fig.23), and characterizes the atmospheric flow in the troposphere. The phenomenon is related, of course, to the tropospheric meridional temperature gradient caused by differential heating. Indeed, as a consequence of the geometry of the sun–earth system, the annual march of the difference between the heat budgets of the troposphere at 50° and at 80°S shows well defined equinoctial maxima.

The curves in Fig.22 also contain the answer to the often discussed question of the seasonal variation of the mean latitudinal position of the circumpolar belt of lowest sea level pressure. Obviously, at any time of the year this zone of transition from prevailing westerlies to prevailing easterlies lies between 60° and 70°S. More precisely, however, one can now state that it is north of 65°S in the solstitial months, while it is south of that parallel in fall and spring, most clearly in March, September and October.

The average wind speed (independent of direction) is illustrated for the four mid-season months in Fig.24. In these four small maps the isolines are drawn for multiples of 2 knots, which approximately equals 1 m/sec.
It becomes evident that low wind speeds are typical for the interior, and that the highest values are found in the coastal belt, particularly in the areas of, and close to, the steep escarpments of the Antarctic Plateau. The surface wind of the interior will be discussed first.

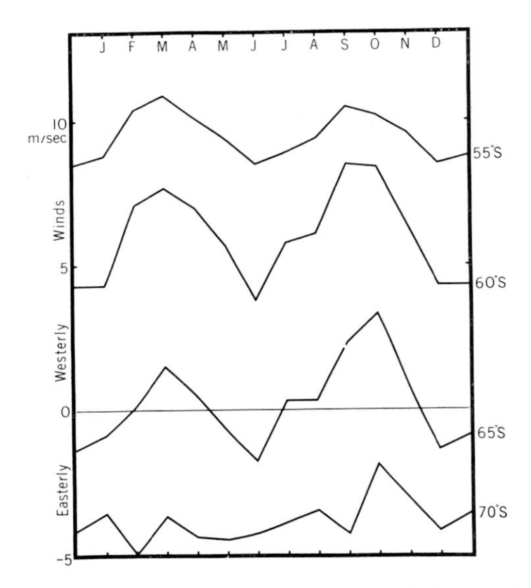

Fig.22. Monthly latitudinal averages of the zonal component (west–east = +) of the geostrophic wind corresponding to the sea level pressure field. (After data from VAN LOON et al., 1969.)

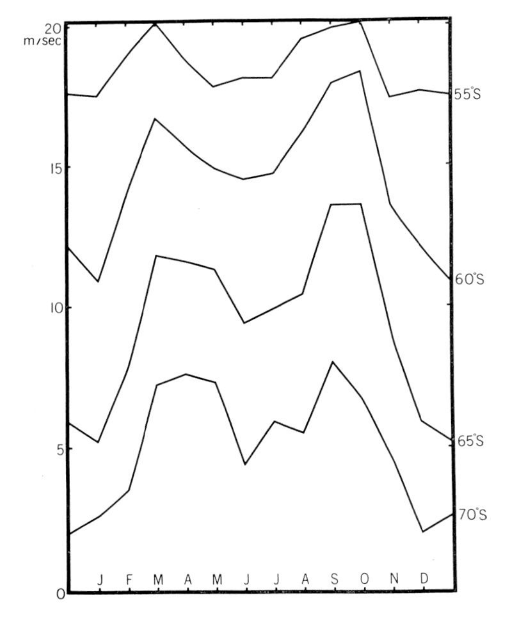

Fig.23. Monthly latitudinal averages of the zonal component of the geostrophic wind at the 500 mbar level. (After data from VAN LOON et al., 1969.)

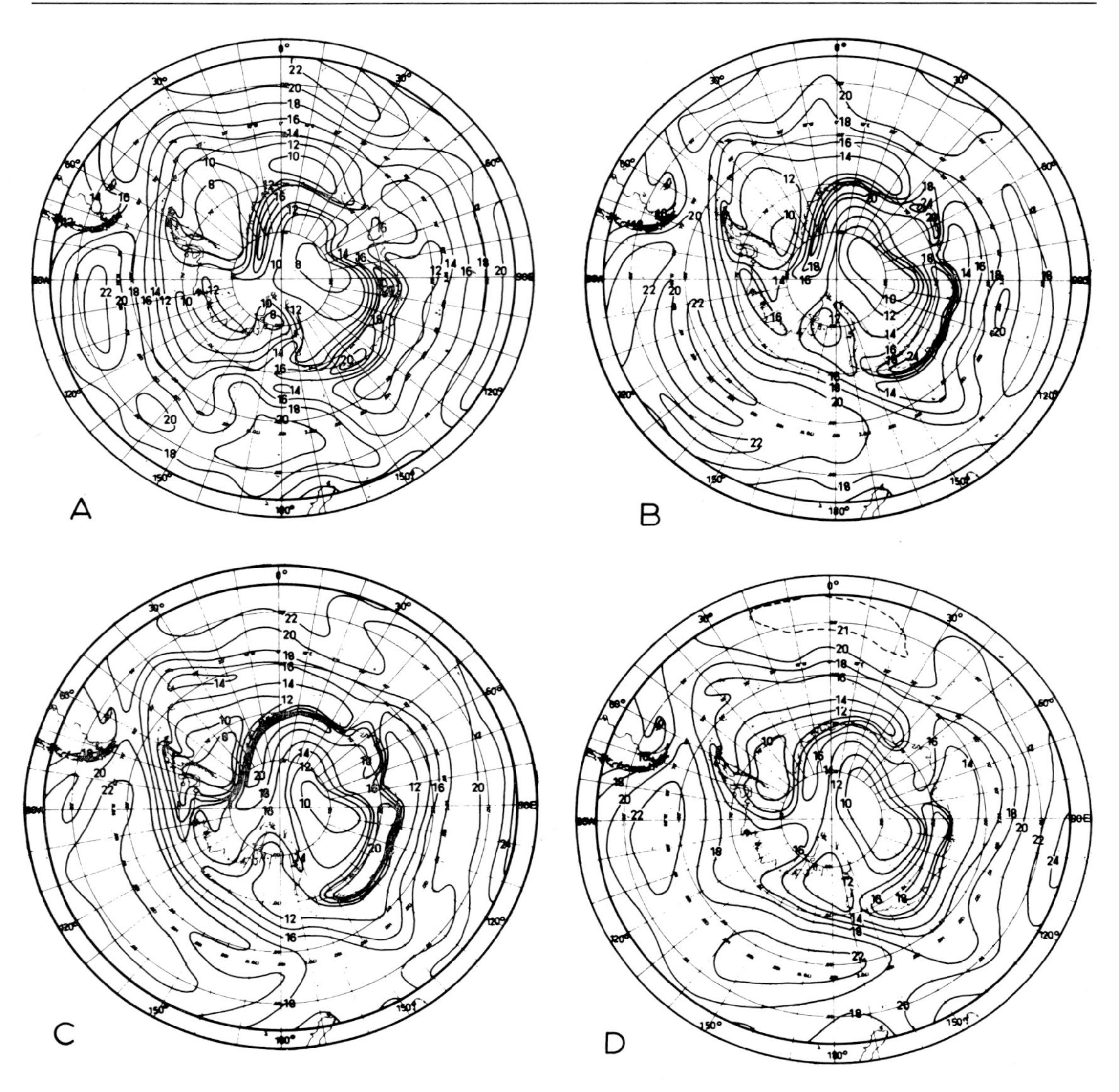

Fig.24. Average scalar wind speed (knots) in the four mid-season months. A. January; B. April; C. July; D. October. (Published with permission of J. Zillman.)

Surface winds over the Antarctic Plateau in relation to the temperature inversion

An overall picture of the flow of air near the surface was composed by MATHER and MILLER (1967) and is reproduced in Fig.25. The internationally recommended reference level for routine wind observations is 10 m above ground, though this is not fully respected at all stations. In addition to the records of the few permanent inland stations, information about the prevailing wind direction also presents itself in the orientation of the sastrugi. Therefore, the reports of the many traverse-parties which have crossed the high plateau and other parts of the continent in recent years have proved most valuable. From all this evidence emerges the clear result that over the gentle slopes of the plateau the directional constancy of the surface wind at a given place is remarkably great, the vector standard deviation small. The predominant direction itself is clearly related to the

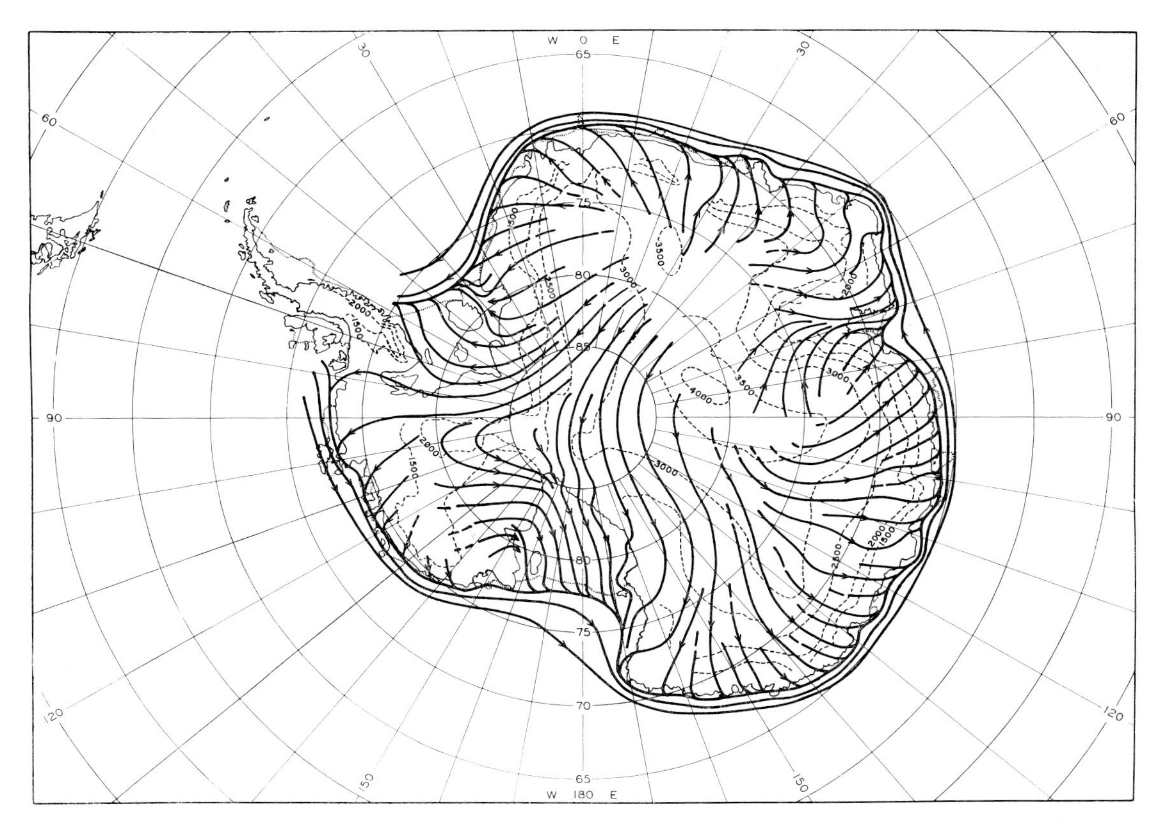

Fig.25. Average pattern of the surface wind flow, inferred from predominant wind frequencies at the stations and from traverse records. (After MATHER and MILLER, 1967.)

TABLE XV

SOME RESULTS OF THE THERMAL WIND EVALUATION

	Vostok 4 yrs. $\Delta T_i \geqslant 20°C$	South Pole 5 yrs. $\Delta T_i \geqslant 20°C$		Byrd 9 yrs. $\Delta T_i > 15°C$
		a	b	
Upper level used	4,000 (± 200) m	650 mbar	600 mbar	750 mbar
Mean dist. from sfc.	500 m	290 m	830 m	510 m
Number of cases	504	426	593	323
Average ΔT_i*	26°C	22°C	23°C	19°C
Upper result. wind	216° 5.1 m/sec	025° 4.4 m/sec	354° 2.1 m/sec	018° 3.0 m/sec
Upper constancy	55%	59%	27%	41%
Sfc. result. wind	256° 4.1 m/sec	063° 5.1 m/sec	062° 5.0 m/sec	020° 6.9 m/sec
Sfc. constancy	86%	80%	80%	91%
Thermal wind	008° 7.2 m/sec	205° 5.6 m/sec	190° 7.6 m/sec	163° 17.5 m/sec
Vect. st. dev.	1.4 m/sec	1.9 m/sec	2.7 m/sec	2.4 m/sec
Average angle α_0	57° (2.3)	44° (5.7)	55° (6.9)	36° (5.3)
Average ratio r_0	0.33 (0.02)	0.55 (0.05)	0.51 (0.07)	0.34 (0.02)
Comp. slope of inv.	100 m/100 km	90 m/100 km	130 m/100 km	370 m/100 km

* Average of the specified inversion class.

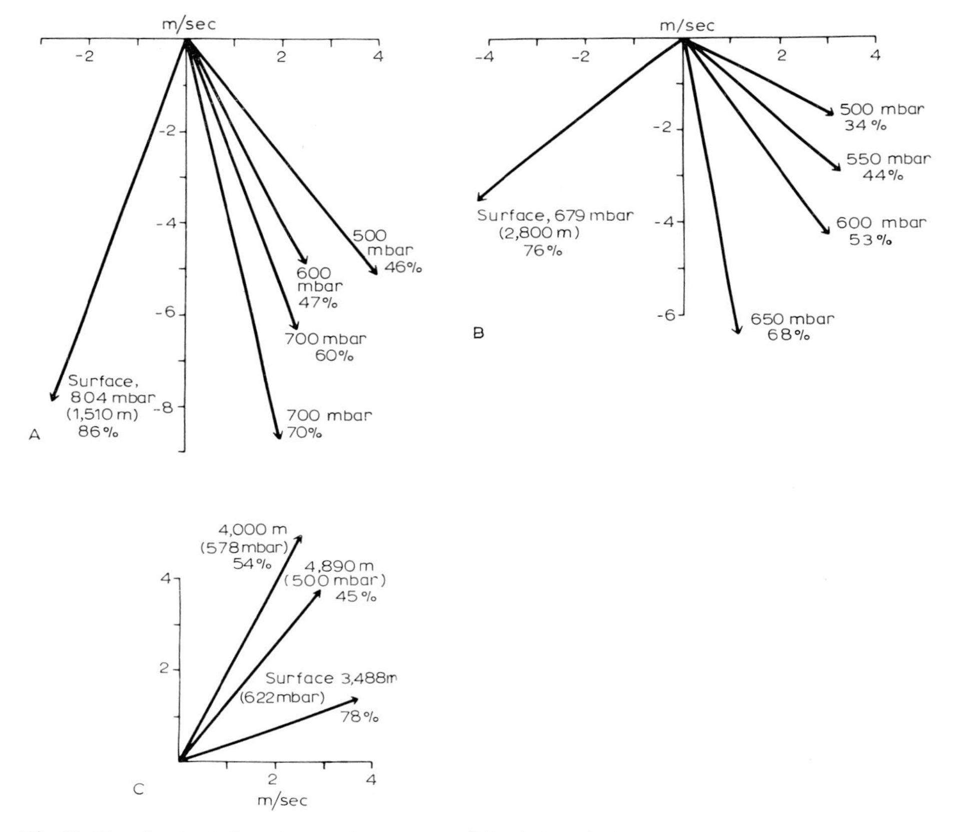

Fig.26. Resultant wind vectors and constancy (%) of the winds at surface and various levels up to 500 mbar. Vector averages of daily winds, April through September, at: A. Byrd Station (8 years); B. South Pole (5); and C. Vostok (4). At the South Pole, facing any direction one is looking northward; "north" has arbitrarily been defined as the direction of the Greenwich meridian.

orientation of the fall line of the sloped terrain, in a sense that the air moves downslope crossing the contour lines at an angle of approximately 45° to the left of the fall line. Two other significant features, unmistakably established by the long series of upper wind soundings of Byrd, South Pole, and Vostok, are that in the lowest few hundred meters the wind turns with height so as to blow parallel to the contour lines of the terrain, and that the constancy of the winds decreases significantly with height. A few numerical values are listed in Table XV, and resultant wind vectors at various levels are shown in Fig.26.

This predominant pattern of atmospheric motion in the lowest few hundred meters, which is most pronounced in the months with a strong temperature inversion, March to October (see the climatic tables at the end of the chapter for South Pole, Byrd and Vostok), can be explained in the following way (DALRYMPLE et al., 1963; DALRYMPLE, 1966; LETTAU and SCHWERDTFEGER, 1967; SCHWERDTFEGER and MAHRT, 1968). The presence of a cold air layer of approximately constant thickness over sloped terrain implies the existence of a horizontal temperature gradient through the inversion layer, as sketched at the bottom of Fig.27. Such a horizontal temperature gradient is equivalent to a thermal wind, blowing into the paper in the case of the sketch of the figure. This thermal wind, which must be proportional to the slope of the terrain and the strength of

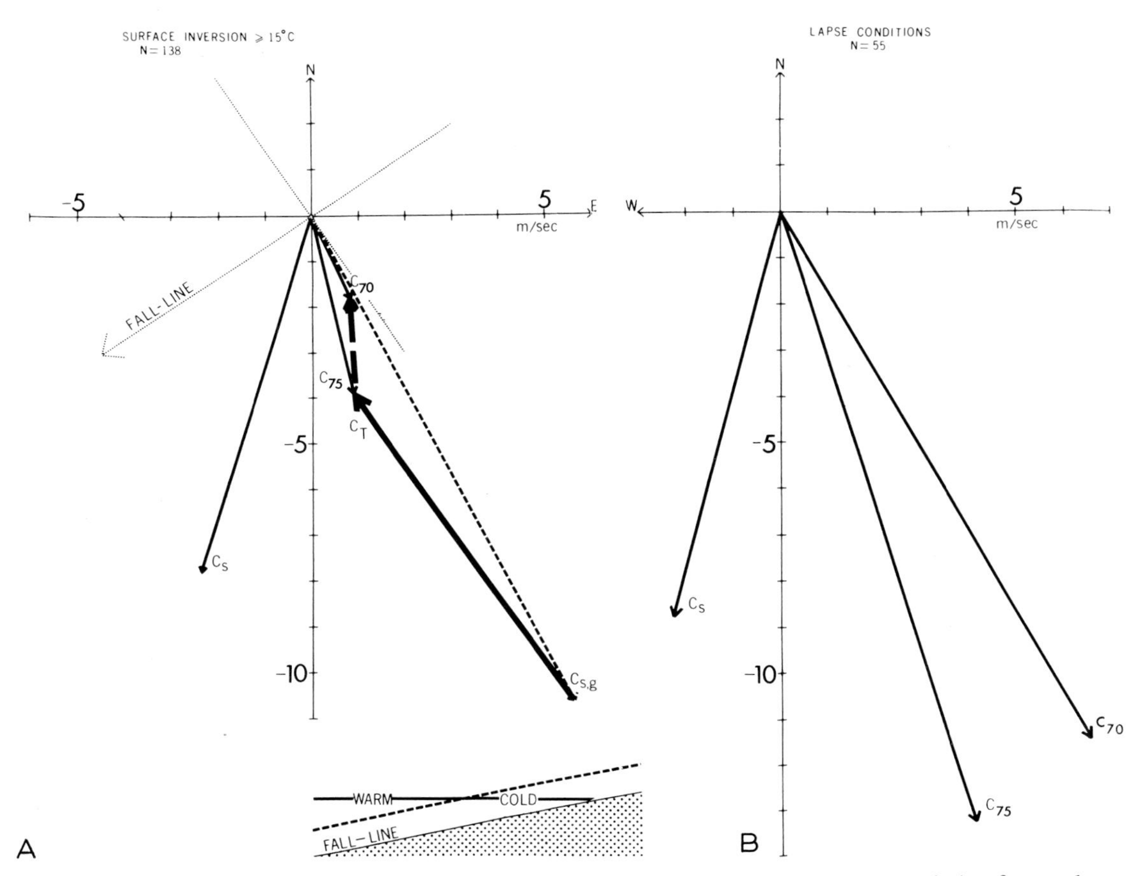

Fig.27. Resultant wind vectors at Byrd Station, April–September, 1961–1965. Graph A refers to days with a strong surface inversion; graph B to days with lapse conditions. c_s = observed surface wind; $c_{s,g}$ = geostrophic wind at the surface level, derived; c_{75} and c_{70} = observed wind at the 750 and 700 mbar level, respectively.

the temperature inversion, can also be interpreted as the wind shear between the geostrophic wind at surface level and geostrophic wind in the upper level. Under the conditions of the Antarctic Plateau, the possible deviations from geostrophic equilibrium can be assumed to be negligible when the vector average of the upper winds of a large number of soundings is considered. With this assumption, the average thermal wind vector can be determined and the fictitious geostrophic surface wind vector be derived (SCHWERDTFEGER and MAHRT, 1968; MAHRT, 1969). Comparing this vector with the resultant of the observed surface winds, one obtains two parameters which characterize the surface friction effect; that is, the mean difference in direction between surface wind and geostrophic surface wind, α_0, and the ratio of the magnitude of these two winds, r_0. Numerical results of the evaluation of a large number of soundings made at the three continental stations with long series of daily wind and temperature soundings are shown in Table XV. Considering how smooth the surface of the Antarctic Plateau is in comparison to many land surfaces, the high values of the frictional deviation angle α_0 and the low values of the ratio r_0 are attributable to the extreme stability of the vertical structure of the surface layer. There are, however, some differences from station to station. At Byrd, where the mean and the mode of the geostrophic surface wind is greater than at the two other stations, the frictional deviation angle is considerably smaller.

Fig.28 shows the relationship between the contour lines of the terrain and the direction and magnitude of the thermal wind of the cold, sloped inversion layer. In this map, estimates of the thermal wind derived from observations made at Sovietskaya and Vostok I during one I.G.Y. winter only, have been added. The three stations in East Antarctica happen to be located on three sides of the main ridge of the plateau, and therefore illustrate the slope effect very clearly. The main point is that the thermal wind tends to be directed parallel to the contour lines of the terrain, with the higher ground to the right hand side. The Sovietskaya data, though only for 121 winter days, are interesting also because the station lies not more than 150 km west of the crest of the aforementioned East Antarctic Ridge (NUDELMAN, 1965, p.103). This gives at least some indication regarding the size of the area of approximately uniform slope, necessary to produce the inversion wind equilibrium.

The term "katabatic wind" has been avoided intentionally, when referring to the surface wind regime over the Antarctic Plateau. Indeed, as suggested by H. H. Lettau in 1963 (in DALRYMPLE et al., 1963), a clear distinction should be made, from the point of view of atmospheric dynamics, between katabatic flow over and at the foot of steeply inclined surfaces, and the equilibrium flow ("inversion wind") over the much less inclined snow and ice fields of the plateau. Typical katabatic flow is characterized by a relatively short duration of very strong surface winds over a wide area, or by limited horizontal extent (in a glacier valley, for instance) if the winds last for a longer period. This is necessarily so because of the limited cold air reserves upstream which cannot be replenished fast enough to maintain a persistent cold air flow. Therefore, stations haunted by strong katabatic storms like the famous Cape Denison, Mawson or Mirny, also experience typical "sudden lulls", abrupt cessations of the wind, followed by extremely sharp onsets of a new blast of cold air with high directional constancy, see STRETEN (1963) and MATHER and MILLER (1967); the latter publication carries a comprehensive bibliography.

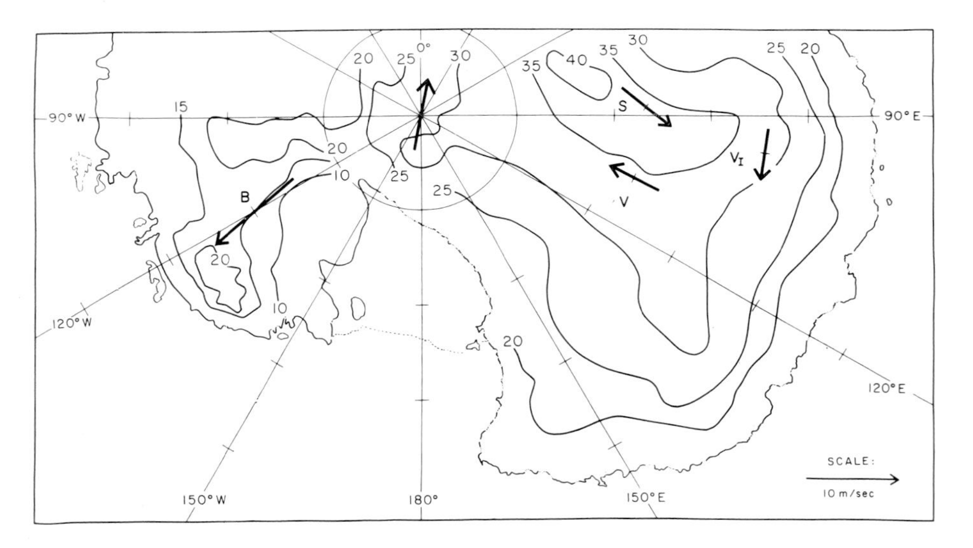

Fig.28. Terrain contour lines of the Antarctic continent (elevation in hundreds of meters) and the thermal wind due to the sloped inversion between "surface" (10 m) and approximately 1000 m above ground. The values for Sovietskaya (*S*) and Vostok I (V_I) are estimates based upon observations of one I.G.Y. winter only. *B* = Byrd Station; *V* = Vostok, present location.

In essence, the ample observational evidence is not compatible with the assumption (BALL, 1960) of equilibrium flow for pure katabatic winds.
In contrast, the typical inversion wind which also is characterized by pronounced directional constancy, can persist over much larger time intervals on the order of weeks, and exhibits much less range in speed. As long as the radiation conditions remain favorable, a near-equilibrium flow establishes itself, with only slight variations imposed by changing pressure patterns in the free atmosphere above the inversion. The rather large frictional deviation angle given in the tables rapidly decreases with height. Hence, the total cross contour outflow rate is relatively small, so that the low temperatures in the surface layer can persist through radiational and conductive cooling of warmer air sinking from the upper layers. Therefore, the inversion wind over the plateau can really be treated as an equilibrium flow.
From a practical point of view, it is interesting and of potential value for the planning of Antarctic field programmes to state the following: the prevailing surface wind direction and strength is so closely related to the direction and steepness of the slope of the terrain, that the former two values can well be estimated if the topography is known, and vice versa. This statement is also supported by the comprehensive, descriptive surface wind analysis of MATHER and MILLER (1967).

The climatic significance of the surface wind regime

From a more general point of view, and referring to the problem of maintenance and persistence of the Antarctic ice cap through the ages, the role of the inversion and the inversion winds appears to be an important one. There is no doubt that the winds over the plateau are rather inefficient agents of downslope transport of cold air and drifting snow. If true katabatic storms were frequent phenomena in the interior of the continent, its ice cap would not persist, would not have built up in the first place. The prevailing surface wind field, due to the existence of the sloped inversion and, of course, the Coriolis force, must be considered as necessary for the maintenance of the ice dome. Certainly, due to surface friction, the wind in the lowest layers has a downslope component as shown in Table XV and Fig.25. But the snow transport due to this relatively weak component is small, so that an approximate balance can be maintained with the similarly small amounts of precipitation on the plateau. Minor deviations, positive or negative, appear equally possible on a regional scale, depending upon the frequency, intensity and trajectory of cyclonic disturbances.

Katabatic winds and blizzards

The meaning of the term "katabatic" wind as used in meteorology is somewhat ambiguous. The *Glossary of Meteorology* (HUSCHKE, 1959) cites two definitions: (*1*) "any wind blowing down an incline"; and (*2*) "a 'gravity wind', that is, a wind directed down the slope of an incline *and* caused by greater air density near the slope than at the same levels some distance horizontally from the slope". In the sense of the first definition one might distinguish three different types of wind, all three characterized by high directional constancy:
(*a*) Those winds which blow with relatively small variation of speed for weeks or months

over extended ice or snow fields of slight or moderate slope and which do not produce a marked temperature change at a given location. These winds can be understood, from the dynamical point of view, as a balanced flow with pressure gradient, Coriolis, and frictional forces keeping an (approximate) equilibrium. They may be called "inversion winds" because they are essentially controlled by the thermal wind due to the existence of a sloped inversion, as it is found during the major part of the year over the Antarctic Plateau.

(*b*) Foehn winds, which are warmer than the air they are displacing when they arrive at the foot of a slope or the bottom of a valley; their occurrence depends upon the presence of a favorable synoptic situation.

(*c*) Winds of the bora type, which are colder than the air they are going to displace.

The two latter types show extremely large variations in speed and therefore cannot be adequately represented as an equilibrium flow. For instance, STRETEN (1962) writes: "Katabatic winds onset with characteristic suddenness, the wind speed often jumping instantaneously from calm to 30–40 knots, and may be likened to the sudden rush of water from a lock gate in a stream". And equally sudden cessations ("lulls") appear. Both types, with temperature increase and decrease, respectively, can appear frequently in the coastal regions of Antarctica, though many explorers have used the term "katabatic" to describe only the more vehement phenomena of this kind.

The frequency and intensity of katabatic winds naturally depends upon the configuration and the average angle of the incline of the coastal escarpment of the continent and upon the distance of the observing station from the slope. There are pronounced annual and, in the months with daylight, also diurnal variations of frequency and intensity of these winds, parallel to the temperature contrast between the foot of the escarpment and the rim of the plateau. It is obvious that the development of katabatic storms can be favored by the synoptic pressure distribution. A depression whose centre is located over the Antarctic Ocean provides in its southwest sector the appropriate pressure gradient over a west–east oriented coast. Hence, the displacement of such a depression along the coast causes considerable temporal variations of the speed of katabatic winds while their direction, determined mainly by the orientation of the fall line and the configuration of the terrain, changes little. Flow lines can be estimated for different types of topography (BALL, 1960; MATHER and MILLER, 1966) and such a qualitative analysis indicates that comparatively minor terrain features are responsible for noticeable spatial variations in the intensity of katabatic winds. On a west–east coast, the eastern flanks of promontories and the western sides of inlets are the places where the strongest winds should occur.

Reports and comments on katabatic storms are numerous, in travelogues as well as in meteorological publications. For a comprehensive bibliography, see MATHER and MILLER (1967). Here it may suffice to describe briefly the absolutely extreme conditions of the infamous Cape Denison, conditions which challenge the imagination of anybody who has not yet flown through snow showers in a small aircraft without windshield. When MAWSON (1915) first reported the observations, measurements and physical experiences of the members of the Australasian Antarctic Expedition, his account was received with much skepticism or even incredulity. But a recalibration of the anemometer after the expedition returned to Melbourne, and the painstaking elaboration and detailed publication of all observations by MADIGAN (1929) brought full vindication. Furthermore, the French Antarctic Expedition 1949–1952 found at Port Martin, about 60 km west-northwest of Cape Denison, similar though perhaps not quite so extreme storm conditions (LOEWE, 1954; PRUDHOMME and VALTAT, 1957). After the base at Port Martin was destroyed by fire in January 1952, the station was moved to Point Géologie, 65 km to the west; here, the average wind speed amounted to only about half as much as at the former place.

The numbers given in Table XVI, extracted from MADIGAN's publication (1929), describe the Cape Denison winds without further comment. The specific local topography was analyzed by MATHER and MILLER (1966).

The question of how far off the coast the extreme conditions of a katabatic storm extend over the ocean has aroused much interest; its importance for ships and aircraft operating in the coastal areas is obvious. Various sets of observations indicate that at a few kilometers from the coast the windfield changes drastically, provided it is a "true" katabatic wind situation, without much support of the flow pattern by the synoptic pressure field. Captain J. K. Davis commanding the Aurora, in the ship's story in *The Home of the Blizzard* (MAWSON, 1915) states that by keeping three miles from the shore he seemed to be beyond the reach of the more violent gusts (KIDSON, 1946). A detailed study carried out by meteorologists of the Soviet Antarctic Expedition 1956 in the Mirny area, with simultaneous observations at various mobile stations on the slope, at its foot, and on the sea ice 14 km off the coast, was described by TAUBER (1960) and RUSIN (1961). For similar observations in the Mawson area, see SHAW (1957) and WELLER (1969).

In the band parallel to the coast where the force of a katabatic storm abates, a pronounced convergence and forced upward motion must exist; conditions which lead to the appearance of almost vertical walls of cloud and blowing snow as they have been observed from aboard ships farther off shore. Another strange phenomenon are Mawson's "whirlies", strong vortices with vertical axis and diameters "a few to a hundred yards" similar to

TABLE XVI

WIND SPEED AT CAPE DENISON (MADIGAN, 1929). TWELVE MONTHS PERIOD MARCH 1912 THROUGH FEBRUARY 1913. NUMBER AND RELATIVE FREQUENCY OF DAYS ON WHICH THE 24-h AVERAGE WIND SPEED WAS BELOW AND ABOVE VARIOUS LIMITS

Limit	Number of days		Percentage	
< 4.5 m/sec	1		0.3	
< 8.9 m/sec	25		6.9	
< 13.4 m/sec	71		19.5	
⩽ 17.9 m/sec	130	365	35.7	100
> 17.9 m/sec	235		64.3	
> 22.4 m/sec	143		39.1	
> 26.8 m/sec	63		17.2	
> 31.3 m/sec	6		1.6	
> 35.8 m/sec	0		0	

Total record: 1 February, 1912–15 December, 1913.
Annual mean wind speed: 19.4 m/sec.
Mean of the quietest month: 11.7 m/sec, February 1912.
Mean of the stormiest month: 24.9 m/sec, July 1913.
Mean of the stormiest day: 36.0 m/sec, 16 August 1913.
Mean of the stormiest hour: 42.9 m/sec, 5/6 July 1913.

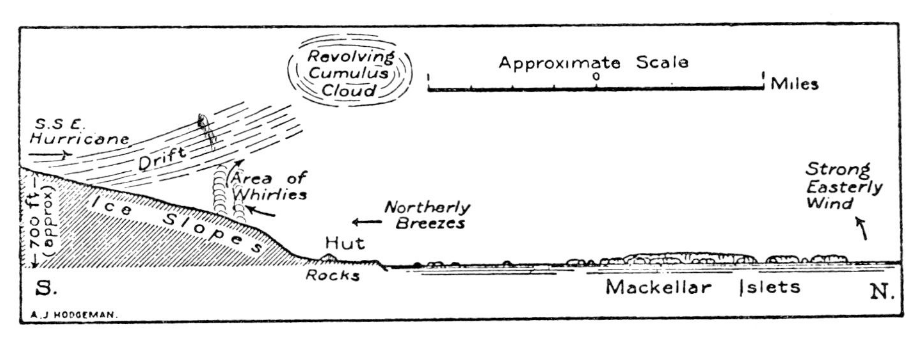

Fig.29. The meteorological conditions at Cape Denison, noon, September 6, 1913. (After MAWSON, 1915.)

dust-devils, which appear in the surface layer when a katabatic storm temporarily does not sweep through to the foot of the slopes. Fig.29, a reproduction of an original sketch drawn by A. J. Hodgeman, published in MAWSON's book (1915) and discussed also by BALL (1957), describes such a situation better than one could do in words.

Blizzards, severe windstorms laden with snow lifted from the ground, falling from clouds, or both, naturally do occur also as symptoms of specific synoptic situations without participation of katabatic flow. They do not differ essentially from the blizzards experienced in winter-cold continental areas of the Northern Hemisphere. The fact that very few case studies of such larger-scale weather situations have been made for Antarctica is simply the consequence of the great distances between the reporting stations, a circumstance which makes a detailed synoptic analysis impossible. Nevertheless, a synoptic study of a blizzard situation by ALVAREZ and LIESKE (1960), made when the International Antarctic Analysis Center was in operation at Little America V during the I.G.Y., is quite instructive. In the middle part of May, 1957, when a blocking ridge extended from the southeastern Pacific Ocean more than half way across the Antarctic continent, a series of deep cyclonic disturbances over the Ross Sea brought heavy snow fall and very strong surface winds to McMurdo, Little America and Byrd stations. The remarkable feature in this case is the extension and intensity of the poleward advection of warm and correspondingly moist air: the temperature rose at Little America to −1°C (multi-annual average for May −31.3°C), at Byrd to −8°C (−33.1°C), and at the South Pole where −73°C had been recorded four days earlier, to −34°C (−57.4°C). For the 700 mbar level, Byrd Station reported −10°C while at Hobart, Tasmania, 37° of latitude closer to the equator, the temperature at this level was −9°C.

Hydrometeors

With reference to Antarctica, the term "cryometeors" might be more appropriate than "hydrometeors", but it is not used.

Moisture conditions

The problematic nature of routine measurements of the moisture content of the air in the polar regions, and at low temperatures in general, is well known. For the stations near the coast, of course, the difficulties, limitations and errors are not greater than for any continental mid-latitude area in winter. The real problem begins with the rise to the Antarctic Plateau. The very low temperatures are only one of the two principal reasons. The other is the fact that the air in the surface layer frequently is in the state of supersaturation (with respect to ice), and the interesting yet unanswered question is: "how much?" If it is small, one can use the much more reliable temperature measurements to estimate, by means of the relationship between temperature and saturation vapor pressure, the corresponding moisture content of the air. But this may turn out to be a rough approximation only. Ice crystals are frequently observed in the surface layer at all inland stations under conditions which ascertain that these particles have formed in loco: too little wind to whirl them up from below, cloudless sky and warmer air aloft. Hoar frost observations confirm this notion. For instance, a report from KUHN (1969) for Plateau Station says: "Hoarfrost growth on the tower scaffolds has been observed during the daily maintenance of the tower instrumentation. The deposit on the crane on top of the tower (33 m above surface) was not removed during the winter (1967); although occasionally diminished in thickness by stronger winds, it had reached by September a radius of 40 cm around a base of 10 cm in diameter".

In any case, humidity measurements at very low temperatures is a wide open field for practical research with modern instrumentation. The present, unsatisfactory state of the art is illustrated in Fig.30. It shows two series of relative humidity measurements (by convention always referring to a plane water surface) plotted against the respective temperature values. Both series are the most recent daily values available to the author for the respective part of the year. The four months for Vostok were selected to include temperatures above and below —40°C. The four months for Byrd were chosen at random; the U.S. stations do not report humidity parameters when the dew point temperature drops below —40°C. Under the circumstances, this might be a wise regulation. The Vostok data in the scatter diagram clearly suggest that a reliable method of moisture

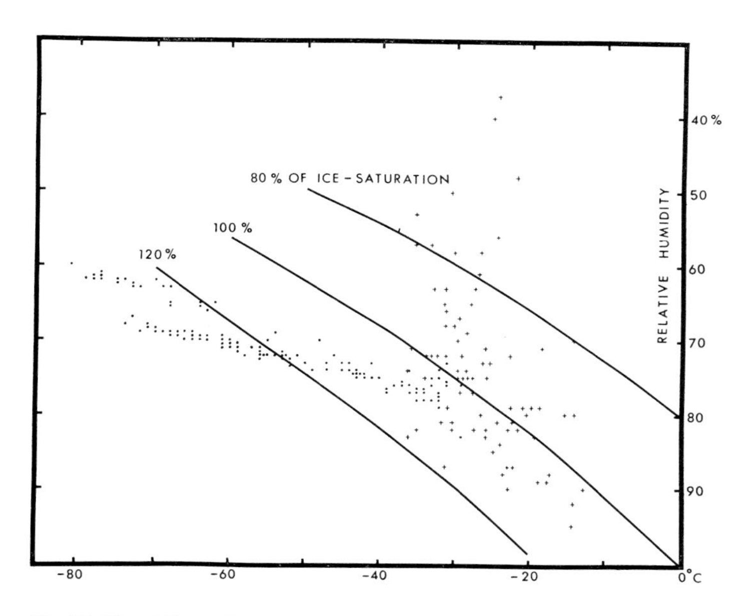

Fig.30. Humidity values measured in the surface layer, at low temperatures. One measurement daily. Crosses = Byrd Station, April–July, 1965; dots = Vostok, January–April, 1964.

measurements at very low temperatures, simple enough to be used in the daily observational praxis, was not available.

Nevertheless, there is no reason to question the reality of the occurrence of supersaturation with regard to ice, and the Byrd data *may* give a first estimate of its frequency and magnitude at temperatures between −20° and −40°C.

Average values of moisture parameters are listed in the climatic tables in a regrettably unsystematic manner according to the form in which they were available in the climatological statistics of the stations. These averages are presented with all necessary reservation. Because of the non-linear relationship between temperature and saturation vapor pressure, neither "mean relative humidity" nor "mean dew point" values permit the derivation of the average moisture content of the air at a given place. The example in Table XVII of an average of values for two different air masses, both at 1,000 mbar, suggests that the error can be substantial even if the majority of individual data in a large series lies closer to the average. Consequently, mixing ratio or specific humidity must be computed from individual soundings in order to obtain unbiased monthly or seasonal averages. Daily data of these parameters have been published in recent years only for the Australian aerological stations (AUSTRALIAN NATIONAL ANTARCTIC RESEARCH EXPEDITIONS, 1957–1965).

TABLE XVII

AN EXAMPLE OF MISLEADING COMPUTATIONS OF AVERAGE VALUES OF HUMIDITY PARAMETERS

Air mass	Measured		Derived for individual data	
	T(°C)	U(%)	dew point (°C)	mixing ratio (g/kg)
A	−10	20	−28.5	0.36
B	6	80	2.7	4.72
Aver.	−2	50	−12.9	2.54

Mean mixing ratio derived from average T and U = 1.66 g/kg.
Mean mixing ratio derived from average dew point = 1.42 g/kg.

An analysis of mean mixing ratio and mean liquid equivalent (precipitable water) in the layer between surface and 500 mbar, based on the daily radiosonde data of stations south of 50°S, has been made, to the knowledge of the author, only for the month of January, i.e., when the temperatures are relatively high and the uncertainty of the humidity measurements somewhat less than in other months (BRYAN, 1966). The results of this study are shown in Fig.31 and 32. Longitudinal differences which undoubtedly exist have been disregarded in this summary presentation. Nevertheless, the change from maritime to continental regime at the average latitude of the coastline is quite distinct. What limit of error has to be attributed to the numerical values shown in the figures, is difficult to decide. At 78°S, BRYAN's (1966) values are almost twice as large as those derived by BROWN and PYBUS (1964) from a small series of special soundings with a dew point hygrometer at McMurdo, December 1960–February 1961.

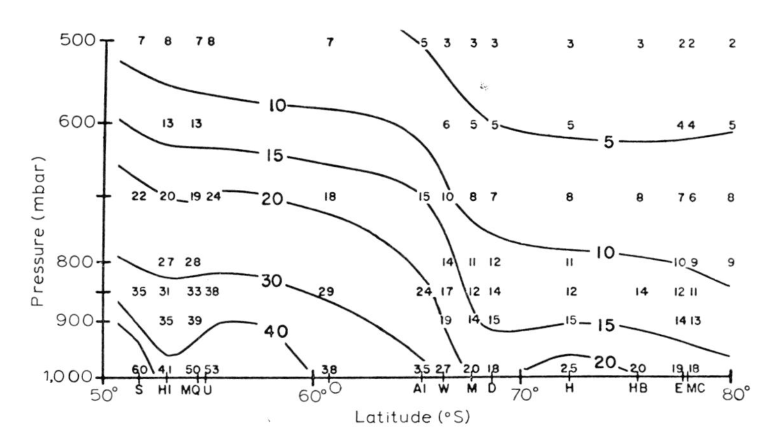

Fig.31. Mean meridional cross section of the mixing ratio, January. Numbers are in decigrams per kilogram.

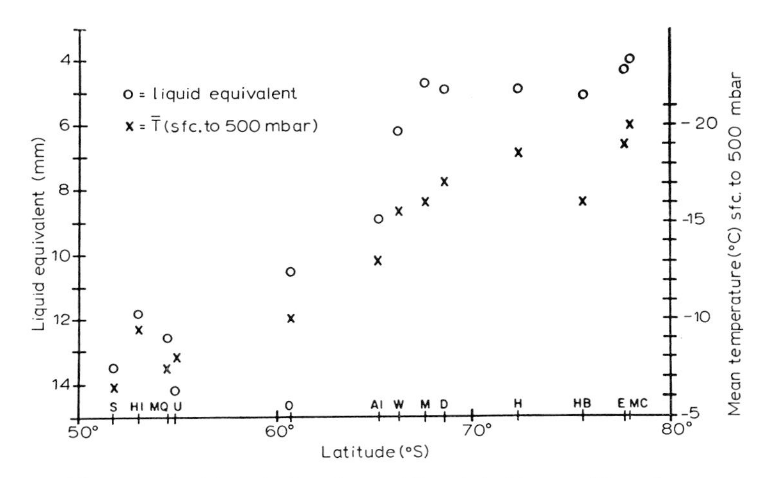

Fig.32. Mean liquid equivalent (mm of "precipitable water"), and mean temperature of the layer from surface to 500 mbar, January. The letters at the bottom line stand at the respective latitude for the meteorological stations whose records have been used: *S* = Stanley/Falkland Islands; *HI* = Heard Island; *MQ* = Macquarie Island; *U* = Ushuaia/Tierra del Fuego; *O* = Orcadas; *AI* = Argentine Islands; *W* = Wilkes; *M* = Mawson; *D* = Davis; *H* = Hallett; *HB* = Halley Bay; *E* = Ellsworth; *MC* = McMurdo.

As far as the moisture conditions in the surface layer are concerned, no further discussion of regional or local regimes is intended. It is obvious that the air rushing down from the snow or ice covered inland has low absolute and relative humidity, that maritime air is comparatively moist, and that the relative humidity of such air still increases when it moves over colder water, pack ice, shelf ice, or snow covered ground.

The frequency of the occurrence of fog at the coastal and subantarctic island stations, and more so still over the surrounding ocean itself, is closely related to the advection of relatively warm maritime air over a colder surface. The conditions in the region of the Antarctic Peninsula where every summer and fall a fair number of ships is in operation, have been described in detail by SCHWERDTFEGER et al. (1959).

TABLE XVIII

RELATIVE FREQUENCY OF FOG DURATION[1]

Station	Period	Fog persists				Total no. fog sit.
		4 h	4–8 h	8–12 h	12 h	
Orcadas	Nov.–Apr.	37	29	19	15	127
(1951–1956)	May–Oct.	55	30	8	7	60
Decepción	Nov.–Apr.	64	24	7	5	107
(1948–1956)	May–Oct.	62	26	5	7	109
Melchior	Nov.–Apr.	36	36	9	19	69
(1947–1956)	May–Oct.	34	39	13	14	113

[1] Expressed in % as: $\frac{\text{number of fog situations of specified duration}}{\text{number of all fog situations}}$

The customary statistics of the average monthly number of days with fog have not been included in the climatic tables in view of the relative uselessness of such numbers as long as nothing can be said about the duration of the phenomenon. For many practical purposes it is more interesting to ask: once the visibility has diminished to less than 1 km (or any other significant limit), what is the climatological probability of various time intervals of fog duration? Table XVIII gives an example.

Cloudiness

"The difficulties of cloud observation are especially high in Antarctic latitudes, where one constantly observes a thin haze all over the sky, which is often so thin as hardly to be noticeable. The difficulty has then to be faced, whether the cloud amount is 0 or 10 tenths. A factor which frequently leaves no room for graduated estimate, but must be classed as either in the minimum or maximum class, is obviously unsuited for statistical investigation ..." (G. C. Simpson, 1919, quoted from Vowinckel, 1957). To this pessimistic observation which refers essentially to fair weather conditions one may add the problems arising from blowing snow and from lack of light in moonless winter nights in order to realize the problematic character of cloud cover statistics for months or even entire years. Notwithstanding, such statistics belong to the conventional repertoire of climatology and are included in the climatic tables.

The major part of this information is summarized in Table XIX, combining the data of various stations in accordance with latitude, location, and season. On a monthly as well as on a seasonal basis the circumpolar band of maximum cloud amount is clearly north of the belt of lowest pressure at sea level and south of the belt of strongest westerly winds.

Another point of interest is the type of the clouds. As an example only, Table XX shows the relative frequency of the occurrence of various cloud forms over a station at relatively high latitude at which blowing snow, the greatest obstacle to such observations, is a rare phenomenon.

Over the Antarctic Plateau, i.e., over most of the continent, the interpretation of the conventional names and definitions becomes problematic, and the observers at different stations do not always follow the same guidelines. In the South Pole records for instance,

TABLE XIX

SEASONAL AND ANNUAL AVERAGES OF TOTAL CLOUD COVER (tenths) FOR VARIOUS CLIMATIC REGIMES

Latitude (°S)	Stations	Record (years)	Summer	Fall	Winter	Spring	Year
50–55	six islands	10–20	7.9	7.4	7.3	7.5	7.5
60.7	Orcadas	55	9.3	8.9	7.9	8.7	8.7
63–65	three stations west coast Antarct. Penins.	10–15	8.5	8.0	7.5	8.3	8.1
66–69	six stations coast of E. Antarct.	7–10	6.9	6.9	6.4	6.7	6.7
70–75	three stations near sea level	5–10	6.8	6.2	5.6	6.2	6.2
78	three ice shelf stations	7–11	6.9	6.3	5.1	6.4	6.2
69.5	two plateau stations 2,400–2,700 m	2	5.5	6.0	4.6	5.0	5.3
79	two plateau stations 3,500–3,600 m	3–9	3.8	3.4	3.3	3.9	3.6

TABLE XX

RELATIVE FREQUENCY (%) OF DIFFERENT CLOUD TYPES AT SCOTT-BASE, ROSS ISLAND (77.8°S 166.8°E); FOUR OBSERVATIONS DAILY, MARCH 1957–JANUARY 1959
(After THOMPSON and MACDONALD, 1962)

Most extensive of all cloud layers present, covering ⩾2/8	XII–II	III–V	VI–VIII	IX–XI	Year
Ci, Cs, Cc	8	8	28	11	14
As	16	14	8	9	12
Ac	10	2	1	5	5
Ns, St	22	28	21	21	22
Cu, Sc	15	18	5	22	15
No clouds ⩾ 2/8	29	30	37	32	32

the most frequently used cloud name is altostratus, a term applied even when the disappearance of the radiosonde balloon in the clouds indicates that their base is less than 1,000 m above surface. An annual summary of cloud type observations made at U.S.S.R. stations on the high plateau and given by RUSIN (1961) did not do so. Therefore, a comparison is not possible, but it is interesting to note that even at Plateau Station where the surface temperature never exceeds −18°C, cumulus mediocris can be found in the summer "around noontime on calm and otherwise clear days. In all cases it started with extended bands that indicate convection along a line of convergence. These bands developed into cloud fields covering the sky to various degrees. All cumulus clouds observed displayed brilliant colors at their edges whenever they moved into the vicinity of the sun" (KUHN, 1969).

Precipitation and accumulation

The measurement of precipitation with rain gauges, snow bins, ombrometers, or any other kind of collector is a rather hopeless undertaking most of the time, over most of the continent. Only in the coastal regions is liquid precipitation found occasionally. Consequently, in Antarctica the accent is on determining accumulation instead of precipitation. This is entirely appropriate for one of the most intriguing problems of the seventh continent, the mass budget of its ice shield. For purposes of synoptic meteorology, of course, the amount of precipitation produced in an atmospheric process would be the more desirable quantity, but that is an elusive goal. The net accumulation is to be understood as the end result of solid precipitation proper, evaporation, hoarfrost formation by negative sublimation, better called deposition (MCDONALD, 1958), and the effects of snow drift (LOEWE, 1962). The relative importance of these different processes shall be briefly discussed.

The conditions are relatively easy to assess for the interior of the continent. The uniformity of the vast snow surface and the almost total absence of stormy winds suggest that advective changes of accumulation, of either sign, must be small. They will nearly cancel each other when the average for a certain area is considered, for instance, the average of 100 snow stakes on an area of 1 km^2. Hence, the validity of the assumption that in the interior of the continent accumulation equals precipitation, can be impaired only by possible losses through evaporation and gains by deposition.

Evaporation has been estimated on the order of 1 mm/month or less for the two summer months, December and January, a much smaller rate in the transition months, and practically zero in the winter (LOEWE, 1962). Even in the two summer months the elevation of the sun is low, the albedo of the snow surface very high and, most important, the air near the surface is close to saturation with respect to ice. For the eight months, March–October, in effect, it must be concluded that "normal" and most frequent conditions over the plateau are characterized by an increase of vapor pressure with height in the layer of the strong surface inversion. The reasoning—reliable measurements are not available—is the following: the saturation vapor pressure with respect to ice at the very low temperatures observed at and near the surface, is only a small fraction of that at the higher temperatures found at the top of the inversion. Table XXI lists a few relevant values.

TABLE XXI

SATURATION PRESSURE WITH RESPECT TO ICE, AND THE RATIO $e_{s(ice)}/e_{s(water)}$ AS A FUNCTION OF TEMPERATURE

Temperature (°C):	0	−20	−40	−60	−80
$e_{s(ice)}$ (mbar)	6.11	1.03	0.13	0.01	0.00055
$e_{s(ice)}/e_{s(water)}$	1	0.82	0.68	–	–

Numerically, the values given for e_s in millibars are equal to the saturation specific humidity (or saturation mixing ratio) expressed in g/kg at the pressure level of 622 mbar. This happens to be close to the average pressure at stations like Vostok, Komsomolskaya, and Plateau.

All aerological measurements of the moisture content of the air at temperatures below $-40°C$ must be considered inexact, to say the least. In the middle troposphere there is, over the interior of Antarctica, a prevailing inflow of comparatively moister air from the sub-Antarctic regions, so it appears safe to assume (and is also in agreement with published data) that in most cases the relative humidity of the air above the inversion is not less than 20–50% in cloudless air, and higher, of course, when there are clouds.

On a cloudless day over the plateau, the temperature in the isothermal layer above the inversion is about $-40°$ to $-45°C$, with relatively small variations from day to day; the vapor pressure, correspondingly, is between 0.03 and 0.1 mbar. With a surface inversion of about 20°–25°C, it follows that an increase in moisture content of the air with height must be a frequent state of affairs, even if there is some super-saturation (ice) in the air near the surface. Under such conditions there must exist a vertical transport of water vapor downward by means of eddy diffusion (weak but not zero in a very stable layer with considerable wind shear) as well as by slow (steady) sinking motion which is due to the slight divergence in the prevailing surface wind pattern over the central part of the plateau. This downward transport of moisture leads to the presence of ice particles in the lower part of the inversion layer, as described under "moisture conditions"; it also leads to the deposition of hoarfrost on the snow surface. LOEWE (1962) has estimated a downward transport which would produce an accumulation by deposition in the amount of 2.8 mm/year. This could be an underestimate, not only because he has taken into account only the eddy, not the steady, transport, but also in the light of some observations made at two Russian stations during the I.G.Y. (RUSIN, 1961). At Sovietskaya (for 10 months) and at Komsomolskaya (for a full year), two stations with a mean wind speed less than 4 m/sec and infrequent occurrence of snow drift, the observers have tried to distinguish the occurrence of precipitation originating in visible clouds, from precipitation due to the settling of ice crystals formed in the inversion layer, and to hoarfrost or rime formation. Their results, as tabulated by RUSIN (1961, table 60), lead to the impression that no-cloud precipitation is an everyday phenomenon on the high plateau. This notion is not completely confirmed by all meteorologists who spent a full year in the area. One of them writes: "No-cloud precipitation is definitely a real phenomenon and was observed during all months of the year. However, on the basis of my own witness and the somewhat contradictory statistics of Rusin, I could not believe it to be an almost continuous phenomenon on the high Antarctic Plateau" (M. Sponholz, personal communication, 1969). On the other hand, SIPLE (1959, p.290) reports from the South Pole: "In the winter when you turned your flashlight upward, you saw ice crystals falling continually." This appears to be confirmed also by the observational records of Plateau Station for the last two years which tend to agree with those of the U.S.S.R. I.G.Y.-stations. In the years 1967 and 1968, respectively, there were 40 and 62 days with observed snowfall (never more than "traces"), but 317 and 314 days with ice crystals floating in the air. It may be noted that such observations can be made very reliably when there is sunshine, and during the polar night, with the help of a torchlight. Monthly mean values are listed in the climatic table of Plateau Station (p. 333). In the two years 1967 and 1968 there were five winter months for which no snowfall was reported. In these, ice crystals were observed on 23–29 days/month and the mean net accumulation change, read from 49 snow stakes, amounted to 1.2 cm/month (real height change values, *not* converted to water equivalent), to be compared with a net change of 0.9 cm/month for the full two years.

One or two years of observation at three places is hardly enough for any categorical statement, but it is interesting to note that in these regions, probably, the major part of the snow and ice deposit is not brought about by precipitation as this term is conventionally understood in meteorology. As far as the net accumulation itself is concerned, of course, it is irrelevant by which process it is achieved, whether by snowfall proper, or by water vapor transport downward with subsequent settling of ice particles formed in the lowest layers in the atmosphere, plus hoarfrost at the surface itself.

For the coastal regions and the lower and steeper parts of the glacial slopes, the relation between precipitation and accumulation is much more uncertain and can vary from place to place. The temperature at the surface can rise above the freezing point, so that melting and run-off must be taken into account. Over slopes with strong katabatic winds and over the downward adjacent areas there is a considerable net transport of mass in form of drifting and blowing snow, which can amount to a sizeable portion of the total precipitation falling upon a certain region. Indeed, blowing snow is the phenomenon which defeats any exact measurement of precipitation. Practically all statistics of amounts as well as of frequency of occurrences must be looked at with reservation and eventually must be interpreted with regard to the possible effects of wind, local terrain conditions and surface characteristics.

On the other hand, measurements of the net annual accumulation based upon stratigraphic determinations in excavations can be and have been made on many spots in the surroundings of permanent stations and during traverses. Wide areas have thus been covered,

TABLE XXII

SOME RECENT DATA ON ACCUMULATION MEASUREMENTS

On the route from Byrd Station to Mount Chapman, (83°34′S 105°55′W), average accumulation between summer 1962/1963 and November/December 1965, for the first 300 km was 16.1 g/cm^2 year; for the remaining 60 km it was 21.3 g/cm^2 year. (After Brecher, 1966.) The author suggests that stratigraphic studies have lead to underestimates by 4–28%.
On a traverse from Plateau Station to 75.9°S 07.2°E to 78.7°S 06.9°W, by excavation of 168 shallow pits, one about every 8 km. In the first 370 km accumulation 5 g/cm^2 year; in the last 420 km, values between 9 and 11 g/cm^2 year. (After Rundle, 1968.)
At Plateau Station itself, from a network of 99 stakes (5 km × 7 km) and from snow pit stratigraphy: 1966. 2.5 g/cm^2 year (stakes), 2.6 g/cm^2 year (50 pits), (after Koerner et al., 1967); 1967. 3.4 g/cm^2 year (pits), (after Rundle, 1968); Feb. 1966–Jan. 1969. Total three year accumulation, average of the height changes of 99 poles, 23.63 cm; standard deviation σ = 6.94 cm, σ_m = 0.70 cm (T. Frostman, personal communication, 1969); Average for the past 128 years (one pit, 10 m deep), 2.8 g/cm^2 year. (After Koerner et al., 1967; Orheim, 1968.) At the "Pole of Relative Inaccessibility", 82.1°S 55.1°E, 3,720 m, from five mutually independent methods, multi-annual average 3.0 g/cm^2 year. (After Picciotto et al., 1968.)

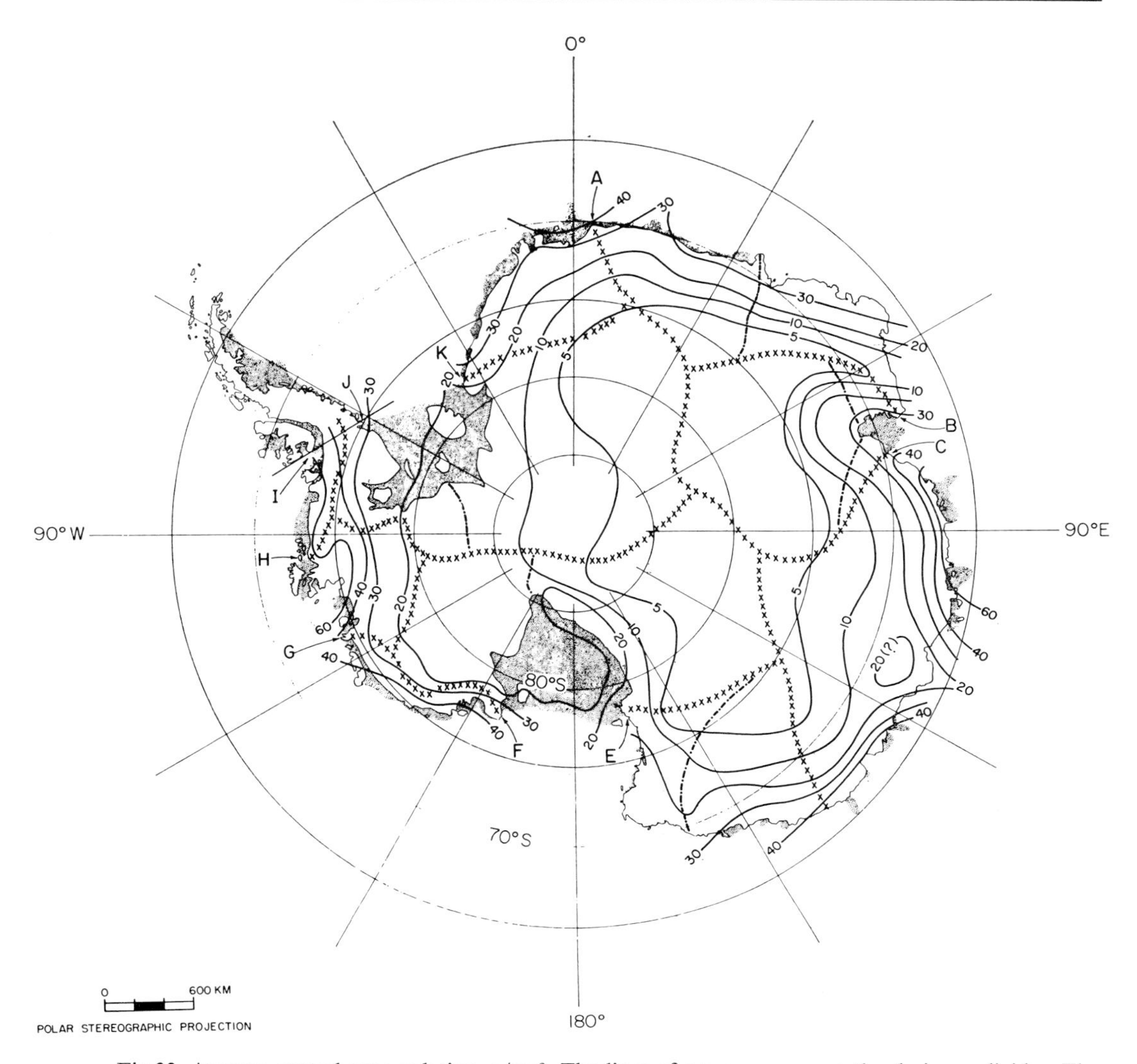

Fig.33. Average annual accumulation, g/cm². The lines of crosses represent the drainage divides. The capital letters A through K refer to the limits of the drainage systems as defined by GIOVINETTO (1964).

and at some places long-term records have been obtained. As a result of these activities, the regional distribution of the average annual net accumulation of mass is now relatively well established. Fig.33 shows the main features according to GIOVINETTO (1964, 1968). Some additional data published since Giovinetto's compilation are listed in Table XXII. The information obtained from snow pits along traverse routes also has made it possible to interpret the accumulation pattern of parts of the interior in terms of topographic and synoptic-meteorological factors (RUBIN and GIOVINETTO, 1962; VICKERS, 1966).

Drifting and blowing snow

When the wind blowing over an extended snow field reaches sufficient speed and thus the horizontal stress becomes strong enough, snow particles disengage from the surface, are displaced in the general direction of the wind, and are raised to a certain height in the turbulent flow of air ("negative precipitation").

The snow in the air can diminish the horizontal visibility appreciably. If the height up to which that happens is less than that of an upright standing man, the term "drifting snow" is generally used. The more intense phenomenon, then, which in extreme cases can extend to a height of several hundred meters above the ground, is "blowing snow". It is a serious impediment to the measurement of precipitation. It also causes serious problems to improperly designed above-surface constructions and can present, in its extreme form, a very real danger for surface travel and for landing aircraft. It is, in the case of Antarctica as well as Greenland, an ablation process which has to be considered in the overall mass budget of an ice cap. With the strongly prevailing off-shore component of the surface wind all along the coast of the frozen continent, huge amounts of snow are blown into the Southern Ocean.

Average monthly and annual frequency values for the occurrence of drifting or blowing snow are given for various stations in the climatic tables. More detailed and more revealing statistics, even though for shorter periods of observation, can be quoted from LOEWE (1956) for an area where the frequency of storms of truly amazing intensity exceeds that recorded at any other spot on earth, Cape Denison (MAWSON, 1915; MADIGAN, 1929; KIDSON, 1946) and Port Martin about 60 km to the west (LOEWE, 1954), (Table XXIII).

TABLE XXIII

RELATIVE FREQUENCY (% of total time) OF THE OCCURRENCE OF DRIFTING OR BLOWING SNOW IN THE COASTAL REGION WHERE THIS PHENOMENON REACHES EXCEPTIONAL DIMENSIONS

	Spring	Summer	Fall	Winter	Year
Port Martin (1951):					
Drifting snow, light or mod.*	13	7	16	12	12
Drifting snow, strong*	6	4	16	9	9
Blowing snow, light or mod.*	22	3	17	26	17
Blowing snow, strong*	10	4	23	21	15
Total	51	18	72	68	53
Cape Denison:					
Near surface only	6	1	6	2	3
Light	19	6	24	27	19
Moderate	16	2	35	24	18
Strong	13	2	24	20	15
Sky invisible	4	8	43	8	16
Total	54	11	89	73	55

* Synoptic code ww numbers 36–39.

Recently, a comprehensive study of the problems of drifting and blowing snow has been published by BUDD et al., 1966). The authors consider the physical theory of the phenomenon as well as the results of extensive measurements made in 1962 by W. R. J. Dingle at Byrd Station (see BUDD et al. (1966). They also give an ample bibliography from Andrée, 1886 (cited after BUDD et al., 1966), to the reports on measurements made during the I.G.Y. Observations and data obtained at the Russian I.G.Y.-stations are discussed in some detail by RUSIN (1961).

Different types of snow traps at several levels above surface have been employed at various stations, but the results are not necessarily comparable in an absolute sense because the efficiency factors of snow traps are not know with any exactness. Even for the same wind speed at a chosen reference level, the drift density can vary considerably from day to day, depending upon the surface conditions, freshly fallen snow versus an older crusty layer, "dry" powdery snow at low temperatures versus wet, heavier snow in warmer air, changing roughness characteristics of the surface, ripples, sastrugi, etc.; and all this over an undetermined, possibly quite large, distance upwind. These modifying factors are particularly critical for the coastal stations where the weather and surface conditions are much more variable than in the interior, and it is just the rim of the continent where reliable measurements are most needed for the mass budget problem. Nevertheless, the main features and the magnitude of the aeolian snow transport as a function of the wind in the surface layer have been established by the measurements, though the agreement with the existing theory is not complete.

Horizontal snow transport under normal Antarctic conditions starts near the surface when the mean wind speed at the 10 m anemometer level exceeds 8 m/sec (LILJEQUIST, 1956, referring to Maudheim), or when the wind at 5 m reaches 6–10 m/sec (RUSIN, 1961, referring to Mirny); LOEWE (1956) states that drifting of snow begins with a wind speed of 6–7 m/sec, reaches a height of 20 cm with 10 m/sec and more than 3 m with 18 m/sec. Frequency data of the occurrence of drifting or blowing snow as function of wind speed at two inland stations are given in Table XXIV. BUDD et al. (1966) determined drift densities at

TABLE XXIV

RELATIVE FREQUENCY (%) OF OBSERVATIONS OF DRIFTING OR BLOWING SNOW FOR VARIOUS CLASSES OF WIND SPEED, AT THE SOUTH POLE (ANEMOMETER AT 9 m ABOVE SURFACE) AND BYRD STATION (10 m); FOUR OBSERVATIONS DAILY, YEARS 1964 AND 1965

Station	Wind speed			
	⩽ 6 m/sec	7–9 m/sec	10–12 m/sec	⩾ 13 m/sec
S. Pole	1	34	91	100
Byrd	0.3	18	74	100

seven levels between 3 cm and 4 m above the ground together with detailed wind profiles at Byrd Station. For two of these levels their values are shown in Table XXV. At higher wind speeds than those listed in the table, with strong turbulence in the surface layer, the visibility reduces to values seldom found at other latitudes, even in thick fog. In Port Martin, 1951, a full 10% of all weather observations indicated a visibility of 10 m or less, all of them due to strong blowing snow. Under such conditions, the blowing snow can be carried up to considerable heights, but only sporadic observations are available. It is well known, however, that in katabatic storms in the coastal area the wind speed decreases notably between 100 and 300 m above ground.

BUDD et al. (1966) have suggested a numerical relationship for the aeolian snow transport

TABLE XXV

MEAN DRIFT-SNOW DATA, GROUPED ACCORDING TO WIND SPEED, FOR BYRD STATION

Wind speed at 10 m (m/sec)	Drift density at		Approx. visibility at 2 m (m)
	25 cm (g/m³)	2 m (g/m³)	
11.7	1.2	0.2	500
13.2	2.5	0.4	300
14.5	4.7	0.6	200
16.8	10	1.1	100
18.5	15	1.5	70
22.1	30	4.0	30

Q between 1 mm and 300 m above ground (practically the total transport to be expected), across a 1 m line at right angle to the prevailing wind direction at 10 m, as a function of the wind speed at this level. The formula is:

$$\log_{10} Q = 1.1812 + 0.0887 \cdot v_{10},$$

where Q is given in g/m sec, and v_{10} the wind speed at the 10 m reference level, in m/sec. This formula represents a theoretically justifiable best fit curve for the grouped averages of the measurements made at Byrd Station. The apparent exactness of such a nice mathematical formula is, of course, fictitious. The measurements have been made in one year only and another year can, and most likely will, bring different prevailing snow characteristics and/or surface roughness conditions in the Byrd area. Moreover, at other stations, somewhat different numerical values for the constants in the equation can be expected. Nevertheless, it is interesting to see to what transport values such an exponential relationship leads (Table XXVI).

TABLE XXVI

DRIFT-SNOW TRANSPORT OVER BYRD STATION IN THE LAYER FROM SURFACE TO 300 m HEIGHT, AS A FUNCTION OF THE WIND SPEED AT THE 10-m ANEMOMETER LEVEL

Wind speed (m/sec)	Transport in the direction of the prevailing wind (kg/m h)
10	400
15	1,200
20	3,300
25	9,000

The magnitude of the transport brought about by high wind speeds is certainly impressive, but the Byrd Station measurements tend to confirm earlier estimates. For a wind speed of 35 m/sec over Adélie Land, LOEWE (1954) arrived at an approximate value of 30 metric tons/m h.

TABLE XXVII

ANNUAL SNOW-DRIFT TRANSPORT ESTIMATES
(After LOEWE, 1954; RUSIN, 1961; BUDD et al., 1966)

Station	Transport ($1 \cdot 10^9$ g/m year)[1]	Remarks
Wilkes	2.1	
Byrd	3.2	
S 2 (66.5°S 112.3°E)	3.5	
Mirny	3–5	probably underestimated
Mawson	20	probably overestimated
Port Martin	50	
Cape Denison	60	

[1] The unit of $1 \cdot 10^9$ g/m year is equivalent to $1 \cdot 10^6$ ton/km year.

With a realistic appraisal of all the problems and questions involved, it becomes obvious that any estimate of the total annual off-shore transport of snow over one coastal station cannot be more than an educated guess. For the entire coastline of the continent, then, it can only be a rather wild guess. Of the many thousands of kilometers of shoreline, only for a few isolated spots is the frequency distribution of the off-shore wind components at various levels known, and only for a still smaller number of spots have some measurements of the aeolian transport been carried through. Table XXVII lists such values.
As will be seen later, though, even with generous assumptions regarding the outflow over the entire coastline of the continent, the total loss of mass through wind transport can only be a small fraction of the ice discharge by calving from glaciers and ice shelves.
Where storms of blowing snow are a frequent phenomenon, wind erosion must be expected. When Mawson revisited, in 1931, the hut at Cape Denison in which the members of his earlier expedition had wintered twenty years before, he found: "Remarkable effects of snow blast erosion were everywhere evidenced on the exposed timbers. In many places the planks had thus been reduced in thickness by more than half an inch" (MAWSON, 1932, p.115).

Annual budgets and related problems

The ice mass budget

A question of considerable importance for various branches of the earth sciences refers to the mass budget of Antarctica. In simple words: does the total ice mass accumulated in Antarctica decrease with time, remain constant, or increase, and can we estimate the rate of change if there is any?
The first to approach this problem in a detailed and realistic manner was LOEWE (1956). He came to the conclusion that at present the budget is positive; the mass of ice increases with time, at least in the last decades. This result was at first received with much scepticism. However, the literature, which, since Loewe's paper has also grown at an increasing

TABLE XXVIII

ESTIMATES OF THE ANTARCTIC MASS BUDGET ($1 \cdot 10^{17}$ g/year)

	M. Giovinetto	Others
Net accumulation	21	19
Subglacial freezing	1	1
Subglacial melting	−2	−3
Ice discharge, calving	−10	−11
Net mass budget	10	6

rate, and in particular the most recent, comprehensive study of the problem by GIOVINETTO (1968), essentially confirm Loewe's result. Table XXVIII lists the various items to be taken into account and gives the corresponding estimates: in the first column Giovinetto's estimates; in the second column the averages of values obtained by various other investigators. These values of "others" are the results of eleven different studies of which ten indicate a net mass gain. Two more analyses had to be disregarded because they a priori assumed the budget to be in equilibrium.

The most critical term in all budget estimates for Antarctica is not really the total net accumulation, but rather the mass output given under the heading "ice discharge, calving". Giovinetto points out that, paradoxically, the use of only the more reliable observational data can introduce a serious bias. He writes (GIOVINETTO, 1968, p.141): "This bias may be explained considering that the logistic limitations impose the establishment of permanent and semi-permanent bases in atypical areas of the coastal regions; it is generally near the main base camps that mass output estimates are reliable because the field data are more accurate, being both abundant in number and covering relatively long periods."

This is not the place to discuss the probable errors of the estimates of the individual terms and of the net result, but it can be stated that they are not large enough to lightly dismiss the notion of a positive net mass budget. Indeed, the larger value given by Giovinetto is based upon more and better evidence. Nevertheless, Giovinetto himself points out (GIOVINETTO, 1968; and personal communication, 1968) that a value of 10^{18} g/year is most likely too large. That becomes obvious when one computes the average annual mass increase per unit area. With the total area of Antarctica being about $14 \cdot 10^6$ km^2, a net value of 10^{18} g would lead to 7 cm water equivalent, i.e., more than the total annual accumulation over large parts of the interior. Furthermore, there is glaciological evidence which suggests that the net mass budget in the marginal areas of the continent is close to steady state or even slightly negative. This asks for a correspondingly larger contribution to the total budget from the interior, and again speaks in favor of a smaller net value than the 6 and $10 \cdot 10^{17}$ g/year, respectively, given in Table XXVIII.

The objection has sometimes been raised (HOLLIN, 1962a,b) that the rise of the level of the oceans (0.5–1 mm/year), observed during past decades, is not compatible with an increase of the mass of H_2O held in solid form in Antarctica. However, it must be borne in mind that there can be other circumstances which lead to a variation of the average level of the world's oceans (GUTENBERG, 1941). One of these circumstances would be a slight change in the volume of the ocean basins, a possibility which leaves much room for speculation. Another one might be a very small increase of the mean temperature of the oceans (HEINSHEIMER, 1958; LOEWE, 1961). Indeed, a sea level rise of 1 mm can be produced by a (non-measurable) increase of the average water temperature by 0.0025°C, or 0.1°C in 40 years. Furthermore, it is not yet certain whether the sea level rise really has continued since the late 1950's. On the other hand, the major part of the observations in favor of a positive ice mass budget refer to the years since the beginning of the I.G.Y.

If one accepts the idea of a positive mass budget for the interior of the continent and a slightly negative one for its marginal regions, it follows that the Antarctic ice dome is, at present, building up, its slopes steepening at a minute rate. Of course, this can be only one phase in the life of the dome. When the accumulated weight of the snow forces the continental ice crust to yield outward, there will be an increased mass discharge by the glaciers and consequently by the ice shelves, resulting in a decrease of the slopes of the flanks of the dome. This change of slope, as well as the altered ice conditions of the Southern Ocean, would again affect the atmospheric circulation over the continent and probably lead to a modification of the precipitation pattern, and so on, ad infinitum.

One must bear in mind, however, that any such changes, if they happen, must be very slow ones. At present, the annual accumulation on the continent amounts to approximately $2 \cdot 10^{12}$ tons (Table XXVIII). The total amount of ice in Antarctica has been estimated to $2 \cdot 10^{16}$ tons. Even a comparison of only these two figures makes it clear (RUBIN, 1964) that, ruling out a catastrophe of astronomical proportions, any major change in the topography of the continent through atmospheric processes could only occur on a geological time scale.

The heat budget of the Antarctic atmosphere

As a sound basis for the discussion of the heat budget it can be assumed that year to year changes of the total heat content of the atmosphere, and here of the Antarctic atmosphere in particular, can be disregarded. Any such changes should be negligible indeed in comparison to the magnitude of the two main items in the budget, the radiative loss of heat and the gain through poleward transport of sensible and latent heat across an imaginary boundary as, for instance, an idealized circular (but not necessarily pole-centered) coastline. The question then arises whether the various atmospheric processes which make a balance possible can adequately be accounted for.

For that purpose it is useful to distinguish between two different kinds of mechanism by which the transport of sensible and latent heat is carried out: the mean meridional circulation and, superimposed on it, large horizontal eddies in the form of cyclonic and anticyclonic circulations.

There can be no doubt that in the lower layers of the troposphere a net outflow of air from the continent prevails, see Fig.25. Based on a detailed analysis of wind components in various layers at ten appropriately located stations in Antarctica, 1958, RUBIN and WEYANT (1963) estimated that the zero level for the mean meridional wind component lies at about 600 mbar. Observations show, and simple continuity considerations require, that above such a level there is a mean inflow of equal efficiency. Otherwise, a systematic change of the area-mean annual pressure would exist over the polar regions. The accent on *annual* mean is essential, as will be shown in the next section (p.311). Thus there is an influx of relatively warm air in the upper layers, a heat loss by net outgoing radiation (see Fig.16) and an outflux of relatively cold air in the lower layers where additional cooling is provided by contact with the cold snow surface. This notion implies, of course, that there is a prevailing sinking motion over the continent, and this is important indeed for the understanding of climate and weather in Antarctica.

On a shorter time scale, the action of transient cyclonic and anticyclonic eddies or troughs and ridges is quite similar: on one side the inflow of relatively warm and moist air with northerly wind components, on the other side outflow of air which has suffered the cooling process. All this establishes the chain of events which is meant when one says that Antarctica acts as a "heat sink" for the atmosphere of its hemisphere. An indirect proof of the contention that the tropospheric meridional exchange of air really reaches far into the interior of the continent, may be seen in Fig.15 which shows the summer to winter temperature range as a function of height, at 66°, 78°, and 90°S.

The concept of the two kinds of transport, steady and eddy, applies also to the latent heat of the water vapor contained in the inflowing air masses. The amount of latent heat becoming available can directly be determined from the total amount of precipitation on the continent, provided one assumes that the amount of H_2O transported across the imaginary boundary in form of cloud droplets or ice crystals is negligible in comparison to the amount in the gaseous phase. In this context, precipitation is understood in the widest sense of the word, including deposition of hoarfrost, rime, and slow settling of ice crystals formed in the cold surface layer, not necessarily in clouds proper.

Numerical values tentatively derived by RUBIN (1962) for the year 1958 are, when referred to the entire area south of 70°S ($15.5 \cdot 10^6$ km²), as follows:

Influx of sensible heat	$13.2 \cdot 10^{21}$ cal./year
Influx of latent heat	$1.8 \cdot 10^{21}$ cal./year
Total advective heat gain	$15.0 \cdot 10^{21}$ cal./year

With heat losses through melting and evaporation being found negligible against the two influx figures, $15 \cdot 10^{21}$ cal. should be the estimated annual radiative heat loss through the top of the atmosphere over the continent. An earlier estimate by GABITES (1960) had been $16 \cdot 10^{21}$ cal./year.

These values can now be compared with the results of direct measurements made by means of satellites and evaluated by VONDERHAAR (1968; also personal communication, 1969). Monthly mean values of the net outgoing radiation (derived for the period July 1964–November 1965) lead to an *annual* mean of 6.84 Ly/h for the area south of 80°S, 6.30 Ly/h for the belt 70°–80°S. From this, taking into account the different size of the two reference areas, one obtains as heat loss from the entire area south of 70°S, the value of $8.7 \cdot 10^{21}$ cal./year, with an error on the order of $\pm$ 10%.

There can be no doubt that a much wider range of errors is unavoidable for an estimate of the heat flux toward Antarctica based on aerological data from ten unevenly distributed stations, at more than 1,000 km distance apart. Consequently, with upper air records for ten years or more now available for several places, a new analysis of the meridional circulation and the eddy transport of heat and H_2O south of the belt of surrounding oceans and strongest westerly winds appears highly desirable.

Periodic variations of atmospheric mass over the Antarctic continent

For all stations with a record of several years situated south of 65°S, on the coast as well as in the interior, an analysis of the series of monthly mean pressure values indicates the occurrence of solstitial maxima and equinoctial minima. Consequently, by properly weighting and combining the records of stations in different parts of Antarctica, one can produce a series which represents the area-mean monthly deviations from the annual average of pressure over the entire continent (SCHWERDTFEGER, 1967). This means that one can determine the seasonal changes of total atmospheric mass weighing upon the continent.

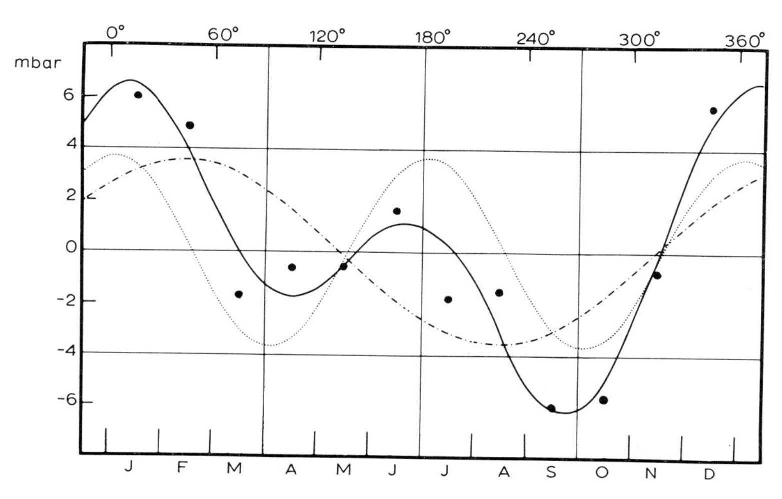

Fig.34. Average monthly deviations from the annual mean atmospheric pressure over Antarctica (black circles); first and second harmonic computed from these twelve values, and sum of the two (curves). 1 mbar ≈ $12 \cdot 10^{16}$ grams change of mass over the continent.

The results of such an analysis, summarized in Fig.34, show that the annual and the semi-annual oscillations are significant and of about equal amplitude. The sum of these two variations accounts for about 92% of the total variance of the mean annual series. The maximum of the combined curve occurs on January 11 with +6.6 mbar, the main minimum on September 24 with −6.2 mbar. The range of 12.8 mbar, taken as representative for the total area of the Antarctic continent, corresponds to a change of atmospheric mass of the order of $1.6 \cdot 10^{18}$ g, about 2% of the total mass over this area. These variations, besides their very slight effect upon the moment of inertia of the earth, can be interpreted as the result of a net meridional transport of air across, say, the polar circle. The increase from October to December and, to a lesser degree, from April to June means an enhanced inflow of comparatively warm air from the southern ocean toward the cold continent. On a yearly basis, it is a mechanism of exchange of antarctic and subpolar air masses with a transport of sensible and latent heat poleward, additional to the macroturbulent exchange that operates on a much shorter time scale. This concept is supported by a study of VAN LOON (1967a, b) who presented other evidence for a meridional mass inflow into antarctic regions from spring to summer and from autumn to winter, and corresponding outflow at the end of summer and winter.

Climate change since the beginning of Southern Polar exploration

The subantarctic ocean

The reports of the early expeditions into the Southern Ocean include some information about the ice conditions in times long before any regular measurements of meteorological elements could be made. The cause of a change in ice conditions is not necessarily of a meteorological nature. For instance, subglacial volcanic activity in Antarctica could affect the rate of glacier outflow, or great tsunamis originated by earthquakes in lower latitudes could lead to a break up of parts of the large ice shelves.

However, once icebergs are advancing into previously free waters in sufficient quantity and size, repercussions on the climate must be expected in the entire subpolar region and possibly far beyond it. Therefore, a critical study of old reports and logbooks can reveal facts of climatological significance, as the following lines quoted from LAMB and JOHNSON (1961, p.391) suggest: "The farthest south positions reached by the early voyagers between Cook's expedition in 1773 and Biscoe's in 1831 in various sectors were all 1°–2° south of the normal limit of pack ice at the end of the melting season in recent years. When the frailty of the wooden ships of those times is allowed for, it seems clear that the southern sea ice tended to be then somewhat less extensive than since. A deterioration followed, the ice increasing on the whole until 1900 or rather later, with some recession after 1907. There were several groups of extraordinarily bad years, 1832–1834, 1840, 1844, 1854, 1888–1890, 1892–1896, 1898–1899, 1904–1907, and perhaps 1929–1931, with large numbers of great icebergs in temperate latitudes in many sectors—in extreme cases reaching the River Plate and to near the Cape of Good Hope. Taking into account some circumstantial evidence as, for instance, the air temperatures at Punta Arenas, it seems probable that most of the years 1888–1907 were bad ice years around Cape Horn and that reports are sometimes lacking because ships avoided the risk by keeping to Magellan Strait. Accounts are known to exist of ships about the turn of the century finding only a hundred miles of sailing room south of the Horn, and such reports probably spread as a warning amongst mariners."

In 1903, with the wintering of W. S. Bruce's Scottish National Antarctic Expedition 1902–1904 (BRUCE, 1907; MOSSMAN, 1907, 1909) the first permanent meteorological station south of 60°S began its activity which fortunately has been continued uninterrupted since. Its temperature trends and the record of the duration of the ice cover in Bahía Escocia are shown in Fig.35; mean values and standard deviations are listed in Table XXIX. The temperature curves indicate that in late summer and early fall, when the islands of the South Orkneys group normally are not enclosed by ice, the temperature regime is typically maritime with very small variations from year to year. In the winter the waters around the islands, in particular toward the south, are ice covered. It must be assumed that frequently, though not always, an almost continuous layer of pack ice extends from the Weddell Sea that far northward. In accordance with the prevailing winds, strong temperature variations, interdiurnal as well as from month to month and year to year, characterize the regime in the winter half-year (PROHASKA, 1951, 1954). Consequently, the annual temperatures are essentially determined by the winter conditions and clearly related to the duration of the presence of ice (last two lines of Table

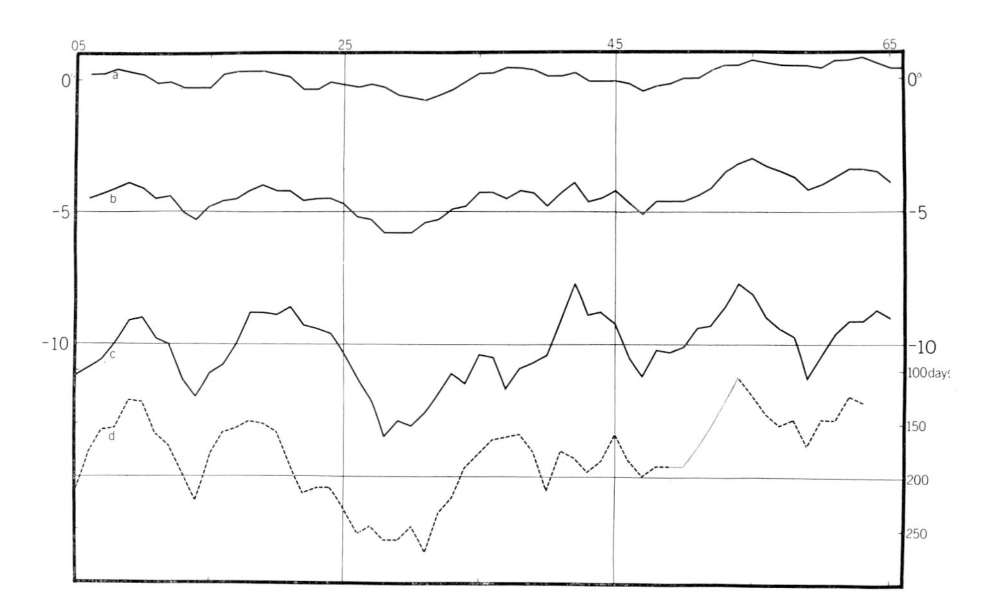

Fig.35. The temperature and ice-regime at Orcadas. Five year running means of: (*a*) temperature of the three months of pronounced oceanic regime, January–March; (*b*) annual temperature; (*c*) temperature of the three coldest months, June–August; (*d*) duration of the ice cover in Bahía Escocia, southeast of the station; data for 1951 and 1952 missing.

XXIX), as it must be expected from radiation and heat budget considerations (FLETCHER, 1968).

An analysis of possible climatic change in the present century in the Orcadas area has been made by PROHASKA (1951) with regard to temperature and by SCHWERDTFEGER et al. (1959, fig.79) with regard to other elements. Four different precipitation parameters—number of days with snowfall, number of days with snow on the ground, average and maximum snow accumulation—show notable differences between the two periods 1904–1934 and 1935–1957, concurring in the sense of lesser precipitation activity, which

TABLE XXIX

AVERAGE TEMPERATURE VALUES (°C) FOR ORCADAS (60.7°S 44.7°W) AND DURATION OF ICE COVER IN BAHÍA ESCOCIA, 1903–1966

Parameter	Period	Average	Stand. dev.
t (year)	1904–67	− 4.4	1.1
t (year)	1904–35	− 4.8	1.0
t (year)	1936–67	− 4.0	1.2
t (Jan.–Mar.)	1904–68	+ 0.1	0.7
t (June–Aug.)	1903–67	−10.2	2.7
Duration ice	1903–65*	180 days	69 days

Correlation t (May–Sept.) × duration ice = −0.69.
Correlation t (year) × duration ice = −0.77**.

* Ice data for 1951 and 1952 missing.

** There is a significant, positive correlation between the annual temperatures at Orcadas and at Argentine Islands, $r = 0.76$, for the 21 years 1947–1967.

may be interpreted as less intense cyclonic activity, since 1935. A statistical test indicates with a probability of about 95%, that these differences are not due to chance alone; this is not much when one also considers that the two periods were arbitrarily selected in favor of a maximum effect. A comparison of the multi-annual mean temperatures prior to, and after, 1935 can be extended to include the records up to 1967. The result is shown in Table XXIX. The difference of 0.8°C is statistically not significant. Hence, the conclusion to be drawn is that the only station in the far south with an adequate observational record does not give convincing evidence of a warming trend in the first half of the century, nor of a reversal in the more recent years.

The continent

At the beginning of the I.G.Y. there was some expectation that records of the temperature at various depths in deep drill holes in the continental ice sheet would yield relevant information about multi-annual trends of the temperature at the surface in years long past. Up to the present, however, one must say that the interpretation of such records remains ambiguous because it depends upon assumptions regarding the motion of the ice and other factors not exactly known (WEXLER, 1961). Therefore, no reliable results on long-time temperature trends have become available. No continuous record of direct temperature measurements exists for higher latitudes, with a period long enough to permit any statement regarding secular change. WEXLER (1959a, b) and MELLOR (1960) have used the data of Amundsen's Framheim base 1911–1912 and the various Little America stations of the Byrd expeditions and the I.G.Y., to suggest the probable existence of a warming trend between the first and the last year of occupancy of the ice shelf near the Bay of Whales, on the order of 2°C per 40 years. However, Wexler and Mellor did not take into account the fact that the geographic position of the various stations changed in several steps from Framheim at 78°38′S to Little America V at 78°11′S. The meridional temperature gradient in this area can be estimated from measurements of the temperature at a depth of 8 m in the ice near Little America V and at a point 40 miles to the southeast, as listed by WEXLER (1959b, in his table 2). These values, which can be interpreted as multi-annual averages, give a meridional temperature gradient of 4°C per degree latitude in the ice south and southeast of the rim of the shelf. Such an estimate agrees satisfactorily with another that can be derived for the winter half-year only from the simultaneous observations at Little America III and Byrd's advanced "Bolling Base", 1°33′ to the south and at only about 65 m higher elevation. The mean air temperature difference between these two places is 6.7°C, or 4.3°C per degree latitude. This is the temperature gradient poleward from Little America. The temperature increase in the opposite direction can only be estimated from isotherm maps like Fig.3 and 4. It is not likely to be smaller, and it must be of lesser importance for the problem because the prevailing winds at Little America are from the southeast. Consequently, the temperature change from the Framheim year to the three years of the I.G.Y. activities at Little America V, with intermediate values at intermediate positions, can easily be explained by the change of nearly half a degree of latitude. In conclusion, then, it must be said that the existing temperature records do not support the hypothesis of any significant trends. The situation is different when one looks at the records of snow accumulation. Snow pits or snow mines and deep bore holes have been used successfully to determine the annual accumulation by stratigraphic analysis as far back as a clear distinction of the annual layers is possible. A fair number of such studies has been carried through since the beginning of the I.G.Y. and more and longer series can be produced in the future. The most ambitious field work of this kind has just recently come to a dramatic conclusion. The U.S. Cold Regions Research and Engineering Laboratory (C.R.R.E.L.) completed in January 1968, the drilling to the bottom of the Antarctic ice at Byrd Station (elevation 1,510 m). The thickness of the ice sheet was found to be 2,164 m (GOW, 1968; UEDA and GARFIELD, 1968). The analysis of the ice cores promises a wealth of information regarding ice density, signs of melting, crystal structure, enclosed air, dirt layers, etc.; the results are not yet available at the time of this writing.

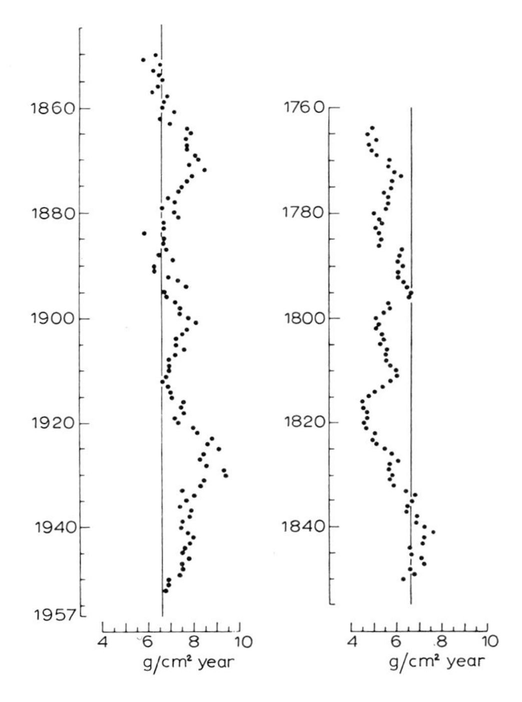

Fig.36. Ten-year running means of snow accumulation at the South Pole in relation to the average for the entire series 1760–1957.

As an example of earlier stratigraphic investigations and their climatological interpretation, Fig.36 shows in the form of ten-year running means the 198-year series of annual accumulation at the South Pole (GIOVINETTO and SCHWERDTFEGER, 1966). For approximately 200 more years (GIOVINETTO, 1960) the annual layers, at a depth from 30 to 48 m, could not be identified unequivocally. The average values for the three 66-year periods from 1760 to 1957 give evidence of a pronounced increase of the rate of accumulation: 5.4 for the first, 6.8 for the second, and 7.5 g/cm^2 year for the most recent interval. A variance analysis of the total series indicates a very high probability ($>$ 99.9%) for the assumption that the increase is real. The coefficient of linear increase with time amounts to 0.15 g/cm^2 per 10 years. Of course, there is no "a priori" reason why the increase should be linear. A line defined by no increase in the first part of the series, a pronounced rise within a limited number of years (1820–1870), and no change for the remaining part would give a bit better approximation of the observed trend. From the meteorological point of view, taking into account that a significant change of annual precipitation or annual moisture transport toward the continent must be assumed to be related to a slight modification in the prevailing pattern of the atmospheric circulation, a non-linear relationship has at least as much merit as a linear one. However, the variability of the accumulation data makes it impossible to decide by statistical means what simple accumulation versus time relationship might be the most likely one. Therefore, the only conclusion to be reached is that the increase in the rate of annual accumulation at the South Pole is significant. The analysis does not permit one to say whether the increase is still continuing through the last decades or occurred only during some time interval in the 19th century.

No similar change of the rate of accumulation has been found at two other places for which comparable series, extending over more than 100 years, have been elaborated: Little America (Gow, 1963) and Wilkes (Cameron et al., 1959). Both series show a very small and definitely not significant increase. Of course, the distance of these two stations from the Pole is great, 1,300 and 2,600 km respectively, and it is easy to conjecture that the effect of a minor variation of the general atmospheric circulation is different in various parts of the continent. Closer to the Pole there is at least one bit of evidence which could support the notion of an increase of accumulation in the interior. At Mount Chapman, 82.6°S 105.9°W, Koerner (1964) found that corrosion features on the face of a granitic nunatak extended downward below the present snow level; but the time of the obvious changes is unknown. On the other hand, at the Gaussberg (66.8°S 89.2°E, 170 km west of Mirny) "the thickness of the ice sheet has decreased by an average of 8.1 m from 1902 to 1957; the minimum decrease among the 15 points surveyed is 5.8 m, and the maximum is 21.7 m" (Feodosyev, in Dolgushin et al., 1962, quoted from Giovinetto, 1968). And in the region of Wilkes Station, on the Windmill Islands, Cameron (1964) found indications of a thinning of the ice cover. These latter two observations, both at a latitude close to the polar circle, are in accordance with the prevailing decrease of glacierization in the middle latitudes of the Southern Hemisphere.

Altogether, then, there is some evidence for an increase of snow accumulation in the interior of the continent and for a decrease of the ice in the coastal regions. It remains uncertain, however, whether one should conceive of this as a general, continent-wide phenomenon and hence as a temporary steepening of the ice domes. More observations and critical analyses are needed, and for any notion regarding the related variations of the atmospheric circulation it would be particularly rewarding to see whether the same changes have occurred in East and in West Antarctica.

Brief comments and bibliographic references on various climatological topics

Variable atmospheric constituents

The total atmospheric ozone content has been measured with a Dobson spectrophotometer at Halley Bay since 1956 (MacDowall, 1960; Dobson, 1966). Shorter series of this kind have been obtained at the South Pole station, Byrd (Aldaz, 1965), and Argentine Islands. The most remarkable result of these measurements is that the annual variation of total atmospheric ozone in the Antarctic regions differs significantly from that in the Northern Hemisphere and in lower southern latitudes. In the Arctic, the minimum occurs in the fall (September or October), the maximum in early spring (generally the month of March) with the most pronounced rise in January. The same annual march, only with reduced amplitude, is found in the mid-latitude and subtropical regions of the Northern and Southern Hemispheres, allowing for the six months phase shift, at least up to the latitude of Tasmania (41°S). In Antarctica, on the other hand, the ozone content remains low from fall through winter into October, then rises abruptly to the maximum in November. This change is very pronounced in high latitudes, as documented by the data from the South Pole, Byrd and Halley Bay (75.5°S), less so and with a wider scatter of the daily values at Argentine Islands (65.3°S), and cannot be found anymore (DOBSON, 1966) in a one year record of Macquarie Island (54.5°S).

The different annual variation in the two polar regions must be related to the different behavior of the two great circumpolar vortices and the vertical motions in their core, in the stratosphere. The Arctic vortex is the less persistent of the two; it can break down temporarily with the well known sudden stratospheric warmings as early as in January. The southern vortex south of about 50°S is more stable, reaches its maximum intensity in August or September when the temperature contrast between middle and polar latitudes is largest, and loses intensity in October with the regular spring warming of the stratosphere. The so-called breakdown of the vortex is equivalent to an inflow of air from lower latitudes, illustrated in Fig.34 by the rise of pressure over the continent, and is accompanied by sinking motion and an increase of ozone in the stratosphere. This increase is strongest in the 15–20 km layer, as the ozone soundings made at Halley Bay (MACDOWALL, 1960) and the South Pole (WEYANT, 1967) have shown. At the latter station, the height of the maximum ozone density lies all year long at the relatively low height of 15–16 km. The ozone concentration in the *surface* air shows a characteristic annual variation which is not parallel to that of the total content of the atmosphere. Measurements made during the I.G.Y. at Little America have been used by WEXLER et al. (1960) to estimate the outflow of ozone from the continent in the lowest layers during the winter months and to suggest a scheme of the meridional circulation of the atmosphere in the high southern latitudes. A re-examination of Wexler's model in the light of the observations made at various stations for a longer period and with improved analytical methods would be highly desirable.

The concentration of *carbon dioxide* (CO_2) in the atmosphere has been measured by means of continuously operating gas analysers at Little America in 1958, and at the South Pole from 1957 through 1963, with some interruptions (KEELING, 1960; BROWN and KEELING, 1965). Since there is no vegetation which could cause seasonal variations, nor industrial installations to produce local and temporal irregularities, Antarctica is the ideal terrain to determine the natural level of atmospheric CO_2 and its possible multi-annual trend. Mount Erebus, the only known active volcano, at a distance of about 700 km from Little America V and 1,400 km from the Pole, can be a source of volcanic CO_2, but its effect all over Antarctica is considered to be negligible. The only sources of artificial contamination on the entire continent are the few scientific stations themselves, and their CO_2 production is much too small to be of any consequence.

BROWN and KEELING (1965) have determined the average value of the period 1958–1963, adjusted to the constant datum January 1960, for Antarctica to be 313.1 p.p.m. The

annual increase for the same period amounts to 0.7 p.p.m. per year, in close agreement with the increase found at the Mauna Loa Observatory, Hawaii. The fact that the above annual mean is about 0.5 p.p.m. less than the corresponding value for Mauna Loa must be considered as a confirmation of the positive trend of the CO_2 content of the atmosphere. The principal cause of the increase is the release of large amounts of CO_2 into the atmosphere as a by-product of the combustion of fossil fuels. This locates the main source which disturbs the natural equilibrium clearly in the Northern Hemisphere, excludes the higher southern latitudes entirely, and thus requires a time-delay for the increase in Antarctica. Eventually, the oceans are capable of absorbing most of the artificially produced CO_2, but they may not be able to keep pace with the release which has become greater and greater during the past 60 or 70 years (PALES and KEELING, 1965). The delayed increase in an area far from the source region suggests that CO_2 can be used as a tracer substance and makes it possible to deduce quantitatively the efficiency of the large scale meridional mixing in the atmosphere (BOLIN and KEELING, 1963).
This line of reasoning is supported by the observation that even at the South Pole there is a small seasonal variation of the CO_2 content of the atmosphere, composed of an annual oscillation with 80% of the total variance and a semi-annual one with 18%. The dates of the occurrence of the maxima (October 23, and May and November 8, respectively) are quite compatible with the notion of a net influx of air into the Antarctic region late in spring and to a lesser amount late in fall, as shown on p.311. The amplitudes, 0.57 and 0.27 p.p.m., are about one quarter of the corresponding values derived for Hawaii.
Determinations of the concentration of various elements, Na, Cl, K, Ca, Mg, Ni, in the snow have also been made in Antarctica (MATVEYEV, 1961, 1963 (quoted after RUBIN, 1964); BROCAS and DELWICHE, 1963; BROCAS and PICCIOTTO, 1967; PICCIOTTO, 1967). The finding of relatively high concentrations of sulphur particles collected in the air was discussed by CADLE et al. (1968).
Results of measurements of atmospheric radioactivity have been reported by LOCKHART (1960), and LOCKHART et al. (1966). Analyses of particles for indications of extraterrestrial origin have also been carried out (WRIGHT et al., 1963; HODGE et al., 1964, 1967; BROCAS and PICCIOTTO, 1967; HODGE and WRIGHT, 1968). HODGE and WRIGHT (1968) re-evaluate earlier studies in the light of recent results of satellite experiments.

The spring warming of the stratosphere[1]

It was not intended to present a general comparison of the climatic regimes of the northern and the southern polar regions, mainly because the differences are so obvious in view of the different geographic conditions of the two regions. A comparison had to be made, however, in the discussion of the ozone concentration in the polar atmosphere, for the sake of clarity. It must now be continued for the description of a phenomenon intimately related to the ozone content, the spring warming of the Antarctic stratosphere. The so-called "sudden" warmings in the northern polar stratosphere are well known (SCHERHAG, 1952; LABITZKE, 1962; WILSON and GODSON, 1962; GODSON, 1963; REED, 1963); one of the distinguishing characteristics is the irregularity of their occurrence, in

[1] The question of an adequate terminology has been discussed by JULIAN (1967).

some years so deep in the winter that any direct causal relationship with the absorption of solar radiation in the ozone layer cannot be inferred. It appears that the northern circumpolar vortex in the stratosphere can become unstable at any time of the winter half-year. When that happens, a mass flux toward the vortex centre occurs, and the subsequent sinking motion leads to a pronounced temperature rise in the stratosphere. The main point to be made is that up to the present, that is in twelve successive winter seasons, an unseasonable temperature increase of comparable magnitude has *not* been observed in the Antarctic stratosphere (Godson, 1963; Schwerdtfeger, 1963; Phillpot, 1964, 1967b; Berson, 1966, 1967; Julian, 1967). The accent is on "comparable magnitude", because minor day to day temperature variations on the order of 10°C do occur; they can be understood as phenomena which accompany the development, passage or decay of synoptic systems whose amplitude can be considerable even over the heart of the continent, as documented in Table XIV. The accent is also on the word "unseasonable", because once the time is ripe, that is, the sun is high enough in the sky, in the month of October there is a fast rise of stratospheric temperatures. Indeed, in several years the highest temperatures at the 30 mbar level (about 23 km, the highest level reached by an adequate number of radiosoundings) have been recorded in late November or early December with values of about −30°C, which means warmer air than found at the same level over the Arctic; and this after the lowest temperatures, about −90°C, had appeared in August. In every year the annual temperature curve is highly asymmetrical, with a fast rise and slower decrease; it contains a strong semi-annual component. Naturally, the stratospheric flow pattern reflects this state of affairs. The circumpolar vortex, with its centre closer to the geographic pole in the stratosphere than in the lower layers, proves to be highly stable through the winter half-year. Only when the meridional temperature gradient between middle and polar latitudes in the stratosphere reverses itself will the westerlies decrease with height. Above the 20 km level, approximately, the polar summer anticyclone establishes itself, with easterly winds dominating the upper half of the stratosphere over subpolar and middle latitudes (Schwerdtfeger and Martin, 1964). In this aspect, then, there is no essential difference between southern and northern regions.

It shall not be implied by this brief description, of course, that every aspect of the change from winter to summer regime in the Antarctic stratosphere is fully known and understood. The number of radiosoundings in the winter or early spring which reach sufficiently high levels is much too small, and the rocket soundings made at McMurdo are too few and sporadic (Quiroz, 1966; Julian, 1967). There are differences of the order of two to three weeks in the onset of the warming over various sectors of the southern polar regions (Godson, 1963) as well as from year to year at any given station (Phillpot, 1967b). It is still an open question whether this should be interpreted as only due to different timing and location of synoptic disturbances. Another question refers to the relationship between the spring warming and the quasi-biennial upper-stratospheric oscillation of lower latitudes. Furthermore, it has been suggested that the Indian Ocean sector, represented by two stations with relatively long records, Mirny and Wilkes, is the preferred region of first appearance of the warming; however, with large parts of the continent and the sub-Antarctic ocean void of high reaching upper air observations, also this point needs further confirmation. It can be expected that stratospheric temperature data, deduced from spectral radiation measurements made by satellites, will soon become available on a broad scale (Nordberg et al., 1965; Phillpot, 1967b; Belmont et al.,

1968; SHEN et al., 1968). It will then be possible to carry out complete case studies of the spring warming and to monitor the southern polar stratosphere for any earlier impairment of the circumpolar vortex.

Cooling power as a climatic index

The discussion of one or several climate classifications, almost a standard ingredient of climatological texts, has been intentionally omitted in the present. If such a classification aims at a comparison of the relevant meteorological and biometeorological factors in different parts of the inhabitable world, for agricultural or biological purposes for instance, there is little room for application indeed. As far as precipitation is concerned, most of the continent is a desert. For a summary description of the severity of the hostile atmospheric environment of the Antarctic, from the physiological point of view, the concept of "cooling power" appears more appropriate. CONRAD (1936) describes the history, back to 1826, of attempts to measure the combined effect of various elements, temperature, wind, and moisture conditions, upon the heat loss from a sensor kept at the approximate temperature of the human body. The cooling effect may be expressed, for instance, as the loss of body heat in calories per hour per square metre of skin surface. The interest in such a parameter, now called wind-chill index or wind-chill factor, and its application to the extreme conditions of Antarctica revived when Siple reported on experiments conducted in Little America in 1939 and suggested an adequate formula for the cooling power as function of wind speed and air temperature (SIPLE and PASSEL, 1945; FALCONER, 1968). A critical commentary on the representativeness of this wind-chill index for different persons, clothing and activities was given by COURT (1948).

The Antarctic Peninsula as a climatic divide

Descriptions of the climate of Antarctica written prior to the I.G.Y. have always given much attention to the Antarctic Peninsula, at that time variously known as Palmer Peninsula, Graham Land, or Tierra de San Martín. This was so mainly because many more meteorological observations had been made in this region than in the rest of the continent. In recent years the pendulum has swung to the other side, the peninsula frequently being excluded from general climatological considerations.

Projecting toward the northeast about four degrees of latitude beyond the polar circle and separating the Bellinghausen from the Weddell Sea, this mountainous land is a climatic divide of the first order. The waters and islands on its west side can easily be reached by ships every summer and fall, poleward to about the southern tip of Adelaide Island, 67.8°S. This is the area which during the past hundred years has seen the full series of human endeavor in polar regions, from early whaling activities to daring geographical exploration, proprietary ambitions, all kinds of scientific investigation, and finally the first steps toward Antarctic tourism (SCHULZE, 1959; CADWALADER, 1968). The east side on the other hand, hemmed in by ice shelves south of 64.5°S and blocked by pack ice drifting from the south, is one of the least accessible and least explored portions of the entire coast of the frozen continent. Only in the northernmost part of this region have meteorological observations been obtained for a full year or more, and here the exemplarily elaborated record of Snow-Hill (64.4°S 57.0°W), the wintering station of O.

Nordenskjoeld's famous and ill-fated expedition 1901–1903 (Bodman, 1910) must be named in the first place. One degree of latitude farther north, an Argentine and a British station have been in operation intermittently since 1945 on the shores of the Hope Bay (see Table I and the climatic tables at the end of the chapter). South of Snow-Hill, only sporadic observations have become known from traverses, going southward from Hope Bay as well as across the mountains from Marguerite Bay (Pepper, 1954; Tapper and Scholten, 1964). In recent years, an Argentine station has been in operation at 65.0°S 60.0°W, station Teniente Matienzo on one of the Seal Nunatacs, but unfortunately no observations or results of any kind have been published to date, to the knowledge of the author. A comparison of the records of this place with simultaneous observations of the British station Argentine Islands at 65.3°S 64.3°W, and of the new U.S. station at Anvers Island, Palmer Station (64.8°S 64.1°W) would be rewarding.
The climate contrast between the two sides of the Antarctic Peninsula is great. From the scanty evidence it can be estimated that on the multi-annual average the west side is about 4–6°C warmer than the east side. And the same appears to be true for the wider areas of the Bellinghausen Sea in comparison to the Weddell Sea, as far as one can conclude from the observations made during the Belgica drift 1898–1899 (average position 70.6°S 86.4°W) against those of the Deutschland drift 1911–1912 (69.1°S 38.4°W) and the Endurance drift 1915 (72.7°S 44.4°W); for details and bibliography, see Meinardus (1938) or Van Rooy (1957). Also in other parameters like cloudiness and temperature ranges, the west side shows certain maritime characteristics while on the other side of the mountain chain, generally at less than 100 km to the east, the regime is as continental as it is on the larger ice shelves south of Weddell and Ross Sea.
Typical for the two stations on the east side, Snow-Hill and Hope Bay, are the strong, gusty, and cold winds from the southwest sector. From synoptic analysis in the wider peninsula area south of 60°S, where in some years up to eighteen meteorological stations (Argentine, British and Chilean ones) have been in full year operation and several more stations and ships during summer and fall, one can ascertain that these winds frequently do not reflect the overall sea level pressure field. Rather, they impose themselves when cold, stable air masses are flowing from the east or southeast over the ice covered Weddell Sea towards, and are dammed up at the foot of, the mountain barrier (Schwerdtfeger et al., 1959, pp.200–204). Disregard of this peculiar type of wind can lead, and has led, to substantial errors in synoptic analysis. A similar phenomenon, though perhaps in a less pronounced form, can be found in the area around southern Greenland (Rodewald, 1955).
As far as precipitation is concerned, a comparison of the west against the east side of the peninsula is not possible because of the total lack of data for the coast of the Weddell Sea. However, the conditions on the Bellinghausen side alone are interesting enough to merit a brief discussion, especially as in this area precipitation and accumulation can by no means be equated. The orientation of the chain of mountains on the peninsula and adjacent islands, with peaks of more than 2,500 m above sea level, is such that relatively warm and moist air masses from the northwest quadrant find them as a formidable obstacle on their way toward the Weddell Sea. Consequently, frequency, duration, and amount of precipitation in some limited areas on the west coast significantly exceed the respective values observed or conjectured in any other part of Antarctica. For instance, an average annual sum of considerably more than 1,000 mm can be assured for Melchior (see Table I) and Almirante Brown (64.9°S 62.9°W). A high correlation exists between

the occurrence and intensity of northwesterly flow and the amount of precipitation in this region (SCHWERDTFEGER et al., 1959). The year 1956 brought extremely high temperature averages to the entire area, and a record value of the average sea level pressure gradient between Orcadas and Argentine Islands (see Fig.1) was recorded; in that year the annual sum of precipitation at Melchior amounted to 2,300 mm. As all precipitation measurements in cold and windy climates are doubtful, to say the least, such a value would not be mentioned in this context were there not some supporting evidence by a different kind of measurement from another period with close to normal conditions. At Anvers Island, a team of glaciologists operating from Palmer Station has recently determined the snow accumulation in a study area of 275 km^2 located at an unbroken ice cap between 60 and 850 m elevation on the western side of the island, about 70 km south-southwest from Melchior, for the three years February 1965 to January 1968. One of the results of the comprehensive investigation is that "over the study area a total average annual mass income of $380 \cdot 10^6$ metric tons has been calculated" (RUNDLE, 1969). This is equivalent to 140 g/cm^2 year, or, disregarding the probably very small effects of evaporation, ice melt, and runoff, roughly a mean annual precipitation of 1,400 mm.

Other regional climatic peculiarities

The climatic regime of the so-called dry valleys in southern Victoria Land, and of the "oases" in other regions of low elevation near the coast, must here be mentioned in the first place. In these areas with ice free ground, bare of snow during the major part of the year, the surface radiation budget and the heat budget of the overlying air differ radically from the normal Antarctic conditions. As an example, radiation data for the Russian I.G.Y. station Oasis have been included in Table II. Other observational data and comments on the oases, with detailed comparisons of albedo values, temperature, moisture, and wind conditions are presented in Rusin's text (RUSIN, 1961, part I, chapter 6). The summer climate of the dry valleys has been described by BULL (1966). More information on the microclimatology of this region will in due time become available from a yearlong meteorological field programme which was begun in 1969 in the Wright Valley and which includes observations for a meso-analysis of the particular flow pattern prevailing in the lower layers of the valley and on the surrounding slopes.
Since the beginning of the I.G.Y., several summertime expeditions have explored the dry valleys with the interest focussed on the microbiology of the soils (CAMERON and CONROW, 1969, with bibliography), on the suitability of the valleys as testing ground for detection devices of plant and animal life on other planets (*Antarctic Journal*, 2(2), 1967), on the permanently ice covered Lake Vanda whose temperature at a depth of 60 m rises to +25°C (RAGOTZKIE and LIKENS, 1964), and on the geology of the region. The latter has produced results of great climatological interest: in the middle Taylor Valley... "clear evidence was found to support and strengthen the hypothesis that only a few thousand years ago the climate of the area of the dry valleys and the nearby coasts was sufficiently warm and humid to permit vigorous action by expanded glaciers of the temperate or warm type" (DORT, 1967). Studies of the glacial history of Antarctica on a much larger time scale have found important evidence in the dry valleys (HOLLIN, 1962a; WILSON, 1964; UGOLINI and BULL, 1965). In a recent report on the results of Potassium–Argon dating of three samples of basaltic scoria, ARMSTRONG et al. (1968) come to the

conclusion that major glaciation in the Taylor Valley is at least 2.7 million years old. Another point of interest is the relatively mild climate of McMurdo Sound. The southwest corner of the Ross Sea is a remarkable area indeed, not only because there is found the only known active volcano and one of the largest and oldest[1] human settlements of Antarctica, but also because of its climatic peculiarities. For the period May through October, McMurdo Station has an average temperature slightly higher than Hallett which is located about 600 km closer to the equator; see curve *f* in Fig.7. In the same six months, McMurdo is about 10°C warmer than Little America which lies less than 100 km closer to the Pole and about 700 km to the east. The older observations in the McMurdo Sound area, a region of complex orographic features, 1902–1904, 1908–1909, 1911–1912, were already discussed by MEINARDUS (1938). The weather conditions were described, from the synoptician's point of view, by MIRABITO (1960). The main reason for the warm winter and spring must be sought in the atmospheric motion in the lower troposphere. Cyclonic circulation centered over the Ross Sea characterizes the predominant synoptic situation. Additionally, the surface wind regime, affected by the configuration of the terrain, can prevent the formation, or favor the destruction, of a strong surface temperature inversion more frequently than at other stations in comparable latitude. It may be of historical interest that the first statistics of the prevailing wind direction in the middle troposphere (from west-southwest) have been obtained in the McMurdo Sound area, by observations of the volcanic cloud emanating from Mount Erebus (3,745 m); a summary of these data was given by MEINARDUS (1938).

Acknowledgements

Climatological statistics and information from recent years for the climatic tables were kindly supplied by various National Meteorological Services. Beyond that, the author had the valuable cooperation of Messrs. H. R. Phillpot and J. Zillman of the International Antarctic Meteorological Research Centre at Melbourne, H. van Loon at the National Center for Atmospheric Research, Boulder, Colorado, T. Taljaard at the Weather Bureau of South Africa, and H. Crutcher at the U.S. Environmental Science Services Administration, who graciously permitted the inclusion of maps and other unpublished results of their own work in this text. Conversations with a colleague of great first-hand knowledge of the Antarctic environment, Dr. M. Giovinetto, and with the Plateau Station meteorologists M. Sponholz, M. Kuhn, and T. Frostman, helped the writer to overcome his own lack of personal experience in Antarctica. For the last two decades, the author has been engaged in studies of the meteorological problems of the high southern latitudes, and during the last six years he has enjoyed the support of his work by the National Science Foundation, Washington, D.C. All this generous help is gratefully acknowledged.

[1] First meteorological station maintained on the continent through an Antarctic winter: Cape Adare (71.3°S 170.2°E, 6 m) in the year 1899, by Borchgrevink's "Southern Cross" Expedition 1898–1900 (BORCHGREVINK, 1901).

References

ALDAZ, L., 1965. Atmospheric ozone in Antarctica. *J. Geophys. Res.*, 70: 1767–1773.

ALVAREZ, J. A. and LIESKE, B. J., 1960. The Little America blizzard of May 1957. *Proc. Symp. Antarctic Meteorol., Melbourne, 1959*, pp.115–127.

ARMSTRONG, R. L., HAMILTON, W. and DENTON, G. H., 1968. Glaciation in Taylor Valley, older than 2.7 million years. *Science*, 159 (3811): 187–189.

ASTAPENKO, P. D., 1960. *Atmospheric Processes in the High Latitudes of the Southern Hemisphere. Section 2 of the I.G.Y. Program (Meteorology), 3.* Academy of Sciences of the U.S.S.R. (Translated by Israel Program for Scientific Translations, Jerusalem, 286 pp.)

AUSTRALIAN NATIONAL ANTARCTIC RESEARCH EXPEDITIONS, 1957–1965. *ANARE Data Reports. Series D, Meteorology* (annual volumes).

BALL, F. K., 1956. The theory of strong katabatic winds. *Australian J. Phys.*, 9 (3): 373–386.

BALL, F. K., 1957. The katabatic winds of Adélie Land and King George V Land. *Tellus*, 9: 201–208.

BALL, F. K., 1960. Winds on the ice slopes of Antarctica. *Proc. Symp. Antarctic Meteorol., Melbourne, 1959*, pp. 9–16.

BELMONT, A. D., NICHOLAS, G. W. and SHEN, W. C., 1968. Comparison of 15μ TIROS VII data with radiosonde temperatures. *J. Appl. Meteorol.*, 7: 284–289.

BERSON, F. A., 1966. The mass and heat budget of the atmosphere over the Antarctic Plateau, with particular reference to spring. *Intern. Antarctic Meteorol. Res. Centre, Tech. Rept.*, 6: 51 pp.

BERSON, F. A., 1967. Spring warming transfer processes in the lower Antarctic stratopshere. *Tellus*, 19 (1): 161–173.

BODMAN, G., 1910. Meteorologische Ergebnisse der Schwedischen Südpolar-Expedition. In: *Wissenschaftliche Ergebnisse der Schwedischen Südpolar-Expedition 1901–1903.* Lithographisches Institut des Generalstabs, Stockholm, 2: 282 pp.

BOLIN, B. and KEELING, C. D., 1963. Large-scale atmospheric mixing as deduced from the seasonal and meridional variations of carbon dioxide. *J. Geophys. Res.*, 68: 3899–3920.

BORCHGREVINK, C. E., 1901. *First on the Antarctic Continent; Being an Account of the British Antarctic Expedition 1898–1900.* Newnes, London, 333 pp.

BRECHER, H. H., 1966. Measurement of ice surface movement by aerial triangulation. *Antarctic J.*, 1: 139.

BROCAS, J. and DELWICHE, R., 1963. Cl, K and Na concentrations in Antarctic snow and ice. *J. Geophys. Res.*, 68: 3999–4000.

BROCAS, J. and PICCIOTTO, E., 1967. Nickel content of Antarctic snow: implications of the influx rate of extraterrestrial dust. *J. Geophys. Res.*, 72: 2229–2236.

BROWN, C. W. and KEELING, C. D., 1965. The concentration of atmospheric carbon dioxide in Antarctica. *J. Geophys. Res.*, 70: 6077–6085.

BROWN Jr., J. A. and PYBUS, E. J., 1964. Stratospheric water vapor soundings at McMurdo Sound, Antarctica, December 1960–February 1961. *J. Atmos. Sci.*, 21 (6): 597–602.

BRUCE, W. S., 1907. *The Scottish National Antarctic Expedition 1902–1904.* The Scottish Oceanographical Laboratory, Edinburgh, (7 vols.).

BRYAN, T. E., 1966. *An Analysis of Moisture Measurements Made at Stations South of 50°S Latitude in January.* Thesis, University of Wisconsin, Madison, Wisc., 33 pp.

BUDD, W. F., DINGLE, W. R. T. and RADOK, U., 1966. The Byrd snow drift project: outline and basic results. In: M. J. RUBIN (Editor), *Studies in Antarctic Meteorology—Am. Geophys. Union, Antarctic Res. Ser.*, 9: 71–134.

BULL, C., 1966. Climatological observations in ice-free areas of southern Victoria Land. In: M. J. RUBIN (Editor), *Studies in Antarctic Meteorology—Am. Geophys. Union, Antarctic Res. Ser.*, 9 (9): 177–194.

BURDECKI, F., 1957. Climate of the Graham Land region. In: M. P. VAN ROOY (Editor), *Meteorology of the Antarctic.* Weather Bureau of South Africa, Government Printer, Pretoria, pp.153–172.

BURDECKI, F., 1968. Climatic regions in the Antarctic. *S. African Weather Bur., Newsletter*, 233: 133–140.

BYRD, R. E., 1938. *Alone.* Putnam's, New York, N.Y., 296 pp.

CADLE, R. D., FISCHER, W. H., FRANK, E. R. and LODGE JR., J. P., 1968. Particles in the Antarctic atmosphere. *J. Atmos. Sci.*, 25 (1): 100–103.

CADWALADER, J., 1968. Antarctic Peninsula tourism. *Antarctic J.*, 3 (4): 149–150.

CAMERON, R. L., 1964. Glaciological studies at Wilkes Station, Budd Coast, Antarctica. In: *Antarctic Snow and Ice Studies—Am. Geophys. Union, Antarctic Res. Ser.*, 2: 1–36.

CAMERON, R. L. and CONROW, H. P., 1969. Soil moisture, relative humidity, and microbial abundance in dry valleys of southern Victoria Land. *Antarctic J.*, 4 (1): 23–28.

CAMERON, R. L., LOKEN, O. H. and MOLHOM, J. R. T., 1959. Wilkes Station glaciological data. *Ohio State Univ. Res. Found., Rept.*, 825-1, 1: 170 pp.

CONRAD, V., 1936. Die Klimatologischen Elemente und ihre Abhängigkeit von terrestrischen Einflüssen. In: W. KOEPPEN und R. GEIGER (Herausgeber), *Handbuch der Klimatologie*. Bornträger, Berlin, 1 (B): 556 pp.

COURT, A., 1948. Windchill. *Bull. Am. Meteorol. Soc.*, 29 (10): 487–493.

COURT, A., 1949. Meteorological data for Little America III, 1939–1941. *Monthly Weather Rev., Suppl.*, 48: 150 pp.

COURT, A., 1951. Antarctic atmospheric circulation. In: F. T. MALONE (Editor), *Compendium of Meteorology*. Am. Meteorol. Soc., Boston, Mass., pp.917–941.

DALRYMPLE, P. C., 1966. A physical climatology of the Antarctic Plateau. In: M. J. RUBIN (Editor), *Studies in Antarctic Meteorology—Am. Geophys. Union, Antarctic Res. Ser.*, 9: 195–231.

DALRYMPLE, P., LETTAU, H. and WOLLASTON, S., 1963. South Pole micrometeorology program, 2. Data analysis. *Quartermaster Res. Eng. Center, Earth Sci. Div., Tech. Rept.*, ES-7: 93 pp. (Also published in: M. J. RUBIN (Editor), *Studies in Antarctic Meteorology—Am. Geophys. Union, Antarctic Res. Ser.*, 9: 13–58.)

DOBSON, G. M. B., 1966. Annual variation of ozone in Antarctica. *Quart. J. Roy. Meteorol. Soc.*, 92: 549–552.

DOLGUSHIN, L. D., EVTEEV, S. A. and KOTLIAKOV, V. M., 1962. Current changes in the Antarctic ice sheet. *Intern. Assoc. Sci. Hydrol., Publ.*, 58: 286–294.

DORT JR., W., 1967. Geomorphic studies in southern Victoria Land. *Antarctic J.*, 2 (4): 113.

FALCONER, R., 1968. Windchill, a useful wintertime weather variable. *Weatherwise*, 21 (6): 227–229, 255.

FLETCHER, J. O., 1968. *The Polar Oceans and World Climate*. RAND Corporation, Santa Monica, Calif., P-3801: 60 pp.

FUNK, J. P., 1959. Improved polythene-shielded net radiometer. *J. Sci. Instr.*, 36: 267–270.

GABITES, J. F., 1960. The heat balance of the Antarctic through the year. *Proc. Symp. Antarctic Meteorol., Melbourne, 1959*, pp.370–377.

GAIGEROV, S. S., 1964. *Aerology of the Polar Regions*. Central Aerological Observatory of the U.S.S.R., Arctic and Antarctic Scientific Research Institute. (Translated by the Israel Program for Scientific Translations, Jerusalem, 280 pp.)

GIOVINETTO, M. B., 1960. Glaciological report for 1958, South Pole Station. *Ohio State Univ. Res. Found. Rept.*, 825-2, 4: 104 pp.

GIOVINETTO, M. B., 1964. The drainage systems of Antarctica: accumulation. In: M. MELLOR (Editor), *Antarctic Snow and Ice Studies—Am. Geophys. Union, Antarctic Res. Ser.*, 2: 127–155.

GIOVINETTO, M. B., 1968. *Glacier Landforms of the Antarctic Coast, and the Regimen of the Inland Ice*. Thesis, University of Wisconsin, Madison, Wisc., 164 pp.

GIOVINETTO, M. B. and SCHWERDTFEGER, W., 1966. Analysis of a 200 year snow accumulation series from the South Pole. *Arch. Meteorol. Geophys. Bioklimatol., Ser. A*, 15 (2): 228–250.

GODSON, W. L., 1963. A comparison of middle-stratosphere behavior in the Arctic and Antarctic, with special reference to final warmings. *Proc. Intern. Symp. Stratospheric Mesospheric Circulation, Berlin, 1962—Meteorol. Abhandl.*, 36: 161–206.

GOW, A. J., 1963. The inner structure of the Ross Ice Shelf at Little America V, Antarctica, as revealed by deep-core drilling. *Intern. Assoc. Sci. Hydrol., Publ.*, 61: 272–284.

GOW, A. J., 1968. Preliminary analysis of ice cores from Byrd Station. *Antarctic J.*, 3 (4): 113–114.

GRIMMINGER, G., 1941. Meteorological results of the Byrd Antarctic Expeditions 1928–1930, 1933–1935. Summaries of data. *Monthly Weather Rev., Suppl.*, 42: 146 pp.

GRIMMINGER, G. and HAINES, W. C., 1939. Meteorological results of the Byrd Antarctic Expeditions 1928–1930, 1933–1935. Tables. *Monthly Weather Rev., Suppl.*, 41: 377 pp.

GUTENBERG, B., 1941. Changes in sea level, postglacial uplift and mobility of the earth's interior. *Bull. Geol. Soc. Am.*, 52: 721–772.

HANSON, K. J., 1960. Radiation measurements on the Antarctic snow field, a preliminary report. *J. Geophys. Res.*, 65: 935–946.

HANSON, K. J. and RUBIN, M. J., 1962. Heat exchange of the snow–air interface at the South Pole. *J. Geophys. Res.*, 67 (9): 3415–3424.

HEINSHEIMER, G., 1958. Zur Geophysik der eustatischen Schwankungen des Meeresspiegels. *Arch. Meteorol. Geophys. Bioklimatol., Ser. A*, 10: 242–256.

HISDAL, V., 1960. *Norwegian-British-Swedish Antarctic Expedition 1949–1952. Scientific Results: Temperature*. Norsk Polarinstitutt, Oslo, 1 (2): 125–181.

HODGE, P. W. and WRIGHT, F. W., 1968. Studies of particles for extraterrestrial origin, 6. Comparisons of previous influx estimates and present satellite flux data. *J. Geophys. Res.*, 73 (24): 7589–7592.

HODGE, P. W., WRIGHT, F. W. and LANGWAY, C. C., 1964. Studies of particles for extraterrestrial origin, 3. Analyses of dust particles from polar ice deposits. *J. Geophys. Res.*, 69 (14): 2919–2931.

Hodge, P. W., Wright, F. W. and Langway, C. C., 1967. Studies of particles for extraterrestrial origin, 5. Compositions of the interiors of spherules from Arctic and Antarctic ice deposits. *J. Geophys. Res.*, 72 (4): 1404–1406.

Hofmeyr, W. L., 1957. Atmospheric sea-level pressure over the Antarctic. In: M. P. Van Rooy (Editor), *Meteorology of the Antarctic.* Weather Bureau of South Africa, Government Printer, Pretoria, pp.51–70.

Hoinkes, H. C., 1961. Studies in glacial meteorology at Little America V, Antarctica. *Symp. Antarctic Glaciol., Intern. Union Geodesy Geophys.—Intern. Assoc. Sci. Hydrol., Publ.*, 55: 29–48.

Hoinkes, H. C., 1967. Low-level inversions at Little America V, 1957. *Proc. Symp. Polar Meteorol., Geneva 1966—World Meteorol. Organ. Tech. Note*, 87: 60–79.

Hollin, J. T., 1962a. On the glacial history of Antarctica. *J. Glaciol.*, 4: 173–195.

Hollin, J. T., 1962b. Some problems of the Antarctic mass budget. *J. Glaciol.*, 4: 312–313.

Huschke, R. E. (Editor), 1959. *Glossary of Meteorology.* Am. Meteorol. Soc., Boston, Mass., 638 pp.

Julian, P. R., 1967. Mid-winter stratospheric warmings in the Southern Hemisphere. General remarks and a case study. *J. Appl. Meteorol.*, 6 (3): 557–563.

Keeling, C. D., 1960. The concentration and isotopic abundances of carbon dioxide in the atmosphere. *Tellus*, 12: 200–203.

Kidson, E., 1946. Discussion of observations at Adélie Land, Queen Mary Land and Macquarie Island. *Australasian Antarctic Expedition 1911–1914, Sci. Rept., Ser. B*, 6: 121 pp.

Koerner, R. M., 1964. Firn stratigraphy studies on the Byrd–Whitmore Mountains traverse, 1962–1963. In: *Antarctic Snow and Ice Studies—Am. Geophys. Union, Antarctic Res. Ser.*, 2: 219–236.

Koerner, R. M. and Scott Kane, H., 1967. Glaciological studies at Plateau Station. *Antarctic J.*, 2 (4): 122–123.

Kopanyev, I. D., 1960. *Snow Cover Over the Antarctic.* Chief Administration of the Hydrometeorological Service, Leningrad, 143 pp.

Kuhn, M., 1969. *Preliminary Report on Meteorological Studies at Plateau Station Antarctica.* Meteorology Department, University of Melbourne, Melbourne, 20 pp.

Kuhn, P. M., Stearns, L. P. and Stremikis, J. R., 1967. Atmospheric infrared radiation over the Antarctic. *Environ. Sci. Serv. Admin., Tech. Rept.*, IER 55-IAS 2: 75 pp.

Kutzbach, G. and Schwerdtfeger, W., 1967. Temperature variations and vertical motion in the free atmosphere over Antarctica in the winter. *Proc. Symp. Polar Meteorol., Geneva 1966,—World Meteorol. Organ., Tech. Note*, 87: 225–248.

Labitzke, K., 1962. Beiträge zur Synoptik der Hochstratosphäre. *Meteorol. Abhandl.*, 28 (1): 93 pp.

Lamb, H. H., 1967. On climatic variations affecting the far south. *Proc. Symp. Polar Meteorol., Geneva 1966—World Meteorol. Organ., Tech. Note*, 87: 428–453.

Lamb, H. H. and Johnson, A. I., 1961. Climatic variation and observed changes in the general circulation, 3. *Geograf. Ann.*, 43: 363–400.

Lettau, H. H., 1966. A case study of katabatic flow on the South Polar Plateau. In: M. J. Rubin (Editor), *Studies in Antarctic Meteorology—Am. Geophys. Union, Antarctic Res. Ser.*, 9: 1–12.

Lettau, H. H. and Schwerdtfeger, W., 1967. Dynamics of the surface-wind regime over the interior of Antarctica. *Antarctic J.*, 2 (5): 155–158.

Liljequist, G. H., 1956. Energy exchange of an Antarctic snow-field. In: *Norwegian–British–Swedish Antarctic Expedition 1949–1952. Scientific Results.* Norsk Polarinstitutt, Oslo, 2 (1): 1–298.

Lockhart, L. B., 1960. Atmospheric radioactivity in South America and Antarctica. *J. Geophys. Res.*, 65 (12): 3999–4005.

Lockhart, L. B., Patterson Jr., R. L. and Saunders Jr., A. W., 1966. Airborne radioactivity in Antarctica. *J. Geophys. Res.*, 71 (8): 1985–1991.

Loewe, F., 1950. A note on katabatic winds at the coasts of Adélie Land and King George V Land. *Geofis. Pura Appl.*, 16 (3/4): 1–4.

Loewe, F., 1954. Beiträge zur Kenntnis der Antarktis. *Erdkunde*, 8 (1): 1–15.

Loewe, F., 1956. Etudes de glaciologie en Terre Adélie, 1951–1952. In: B. Valtat (Rédacteur), *Expéditions Polaires Françaises.* Hermann, Paris, 1247: 159 pp.

Loewe, F., 1960. Notes concerning the mass budget of the Antarctic inland ice. *Proc. Symp. Antarctic Meteorol., Melbourne 1959*, pp.361–369.

Loewe, F., 1961. Beiträge zum Massenhaushalt des Antarktischen Inlandeises. *Petermanns Geograph. Mitt.*, 1961: 269–274.

Loewe, F., 1962. On the mass economy of the interior of the Antarctic ice cap. *J. Geophys. Res.*, 67 (13): 5171–5177.

Lorius, C., 1963. *Le Deutérium, Possibilités d'Application aux Problèmes de Recherches concernant la Neige, la Névé et la Glace dans l'Antarctique.* Comité National Français des Recherches Antarctiques, Paris, 8: 102 pp.

MacDowall, T., 1960. Some observations at Halley Bay in seismology, glaciology and meteorology. *Proc. Roy. Soc., London, Ser. A*, 256: 149–192.

MacDowall, T., 1962. Meteorology. In: D. Brunt (General Editor), *The Royal Society I.G.Y. Expedition Halley Bay, 1955–1959*. Royal Society, London, pp.49–109.

Madigan, C. T., 1929. Meteorology of the Cape Denison station. *Australasian Antarctic Expedition 1911–1914, Sci. Rept., Ser. B*, 4: 286 pp.

Mahrt, L. J., 1969. *The Wind Regime of the Sloped Inversion Friction Layer in the Interior of Antarctica*. Thesis, University of Wisconsin, Madison, Wisc., 65 pp.

Martin, D. W., 1968. Satellite studies of cyclonic developments over the Southern Ocean. *Intern. Antarctic Meteorol. Res. Centre, Tech. Rept.*, 9: 64 pp.

Mather, K. B. and Miller, G. S., 1966. The problem of the katabatic winds on the coast of Terre Adélie. *Polar Record*, 13 (85): 425–432.

Mather, K. B. and Miller, G. S., 1967. Notes on topographic factors affecting the surface wind in Antarctica, with special reference to katabatic winds; and bibliography. *Univ. Alaska, Tech. Rept.*, UAG-R-189: 125 pp.

Matveyev, A. A., 1961. Chemical balance of the atmospheric precipitation salts in Antarctica. In: *The Antarctic, Committee Reports, 1961*. Academy of Sciences of the U.S.S.R., Moscow, pp.18–26.

Matveyev, A. A., 1963. *The Chemical Content of Atmospheric Precipitation According to the Observations at Mirny. Section 2 of the I.G.Y. Program, (Meteorology)*, 5. Academy of Sciences of the U.S.S.R., Moscow, pp.100–107.

Mawson, D., 1915. *The Home of the Blizzard, Being the Story of the Australian Antarctic Expedition 1911–1914*. Heinemann, London, 1: 349 pp., 2: 338 pp.

Mawson, D., 1932. The B.A.N.Z. Antarctic Research Expedition, 1929–1931. *Geograph. J.*, 80: 101–131.

McDonald, J. E., 1958. "Deposition"—a proposed antonym for 'sublimation'. *J. Meterol.*, 15 (2): 245–247.

Meinardus, W., 1938. Klimakunde der Antarktis. In: W. Koeppen und R. Geiger (Herausgeber), *Handbuch der Klimatologie*. Bornträger, Berlin, 4 (U): 133 pp.

Mellor, M., 1960. Temperature gradients in the Antarctic ice sheet. *J. Glaciol.*, 4: 773–782.

Mirabito, J. A., 1960. Notes on Antarctic weather analysis and forecasting. *U.S. Navy Weather Res. Facility, Publ.*, 16-1260-038: 71 pp.

Mossman, R. C., 1907. Meteorology. In: William S. Bruce (Editor), *Reports on the Scientific Results of the Voyage S.Y. Scotia 1902–1904, 2 (1). Physics*. Edinburgh, V, pp.1–306.

Mossman, R. C., 1909. The meteorology of the Weddell quadrant and adjacent seas. *Trans. Roy. Soc. Edinburgh*, 47 (5): 103–136.

Nitschke, P., 1966. Zur Bearbeitung von Luftdruckmessungen an Ost-Antarktischen Inlandstationen und zum Problem der glazialen Antizyklone. *Z. Meteorol.*, 18 (5/7): 223–233.

Nordberg, W., Bandeen, W. R., Warnecke, G. and Kunda, V., 1965. Stratospheric temperature patterns based on radiometric measurements from the TIROS VII satellite. *NASA (Natl. Aeron. Space Admin.) Tech. Note*, D-2798: 12–36.

Norsk Polarinstitutt, 1956–1960. *Norwegian–British–Swedish Antarctic Expedition 1949–1952. Scientific Results*. Oslo, 1 (2): 181 pp., 2 (1): 298 pp.

Nudelman, A. V., 1965. *Soviet Antarctic Expeditions 1961–1963*. Academy of Sciences of the U.S.S.R., Moscow. (Translated by Israel Program for Scientific Translations, Jerusalem, 1968, 220 pp.)

Orheim, O., 1968. Studies of surface snow at Plateau Station. *Antarctic J.*, 3 (2): 37.

Pales, J. C. and Keeling, C. D., 1965. The concentration of atmospheric carbon dioxide in Hawaii. *J. Geophys. Res.*, 70 (24): 6053–6076.

Panzarini, R. N. M., 1968. Official organizations for scientific cooperation in the Antarctic before the I.G.Y., 1957–1958. *SCAR-Bull.*, 30. (Also, *Polar Record*, 14 (90): 438–440.)

Pepper, I., 1954. *The Meteorology of the Falkland Islands and Dependencies 1944–1950*. Falkland Islands Dependencies Survey, London, 250 pp.

Phillpot, H. R., 1964. The springtime accelerated warming phenomenon in the Antarctic stratosphere. *Intern. Antarctic Meteorol. Res. Centre, Tech. Rept.*, 3: 87 pp.

Phillpot, H. R., 1967a. *Selected Surface Climatic Data for Antarctic Stations*. Australian Bureau of Meteorology, Melbourne, 113 pp.

Phillpot, H. R., 1967b. Some further observations on Antarctic stratospheric warming. *Proc. Symp. Polar Meteorol., Geneva 1966—World Meteorol. Organ., Tech. Note*, 87: 379–406.

Phillpot, H. R. and Zillman, J. W., 1969. *The Surface Temperature Inversion over the Antarctic Continent*. Commonwealth Bureau of Meteorology, Melbourne. In press.

Picciotto, E. E., 1967. Geochemical investigations of snow and firn samples from East Antarctica. *Antarctic J.*, 2 (6): 236–240.

Picciotto, E., Cameron, R., Crozaz, G., Deutsch, S. and Wilgain, W., 1968. Determination of the rate of snow accumulation at the Pole of relative inaccessibility, eastern Antarctica. A comparison of glaciological and isotopic methods. *J. Glaciol.*, 7 (50): 273–287.

Pollog, C. H., 1924. Untersuchung von jährlichen Temperaturkurven zur Charakteristik und Definition des Polarklimas. *Mitt. Geogr. Ges. München*, 17 (2): 165–253.

Predoehl, M. C., 1966. Antarctic pack ice boundaries established from NIMBUS I pictures. *Science*, 153 (3738): 861–863.

Predoehl, M. and Spano, A. F., 1965. Airborne albedo measurements over the Ross Sea, October–November 1962. *Monthly Weather Rev.*, 93 (11): 687–696.

Prohaska, F., 1951. Zur Frage der Klimaänderung in der Polarzone des Südatlantiks. *Arch. Meteorol., Geophys. Bioklimatol. Ser. B*, 3: 72–81.

Prohaska, F., 1954. Bemerkungen zum säkularen Gang der Temperatur im Südpolargebiet. *Arch. Meteorol. Geophys. Bioklimatol., Ser. B*, 5: 327–330.

Prudhomme, A. and Valtat, B., 1957. Les observations météorologiques en Terre Adélie 1950–1952, analyse critique. In: *Expéditions Polaires Françaises, No. S V.* Paris, 179 pp.

Pybus, J. A. and Brown Jr., J. A., 1967. Polar atmospheric water vapor. *Proc. Symp. Polar Meteorol., Geneva 1966—World Meteorol. Organ., Tech. Note*, 87: 3–14. (Also in: *J. Atmos. Sci.*, 21: 597–602.)

Quiroz, R. S., 1966. Midwinter stratospheric warming in the Antarctic revealed by rocket data. *J. Appl. Meteorol.*, 5 (1): 126–128.

Radok, U., Schwerdtfeger, P. and Weller, G., 1968. Surface and subsurface meteorological conditions at Plateau Station. *Antarctic J.*, 3 (6): 257–258.

Ragotzkie, R. A. and Likens, G. E., 1964. The heat balance of two Antarctic lakes. *Limnol. Oceanog.*, 9 (3): 412–425.

Raschke, E., 1968. The radiation balance of the earth–atmosphere system from radiation measurements of the NIMBUS II meteorological satellite. *NASA (Natl. Aeron. Space Admin.) Tech. Note*, TN D-4589: 81 pp.

Reed, R. J., 1963. On the cause of the stratospheric sudden warming phenomenon. *Proc. Intern. Symp. Stratospheric Mesospheric Circulation, Berlin 1962—Meteorol. Abhandl.*, 36: 315–334.

Rodewald, M., 1955. *Beiträge zum Wettergeschehen in den nordeuropäischen Gewässern, 3. Klima und Wetter der Fischereigebiete West- und Süd-Grönlands.* Veröffentlichung des Seewetteramtes Hamburg, 126 pp.

Rubin, M. J., 1960. Advection across the Antarctic boundary. *Proc. Symp. Antarctic Meteorol., Melbourne 1959*, pp.378–393.

Rubin, M. J., 1962. Atmospheric advection and the Antarctic mass and heat budget. *Antarctic Res., Geophys. Monograph, 7,—Am. Geophys. Union, Publ.*, 1036: 149–159.

Rubin, M. J., 1964. Antarctic weather and climate. In: *Research in Geophysics, 2. Solid Earth and Interface Phenomena.* Mass. Inst. Technol., Boston, Mass., pp.461–478.

Rubin, M. J. and Giovinetto, M. B., 1962. Snow accumulation in central West Antarctica as related to atmospheric and topographic factors. *J. Geophys. Res.*, 67 (13): 5163–5170.

Rubin, M. J. and Weyant, W. S., 1963. The mass and heat budget of the Antarctic atmosphere. *Monthly Weather Rev.*, 91 (10/12): 487–493.

Rundle, A. S., 1968. Snow stratigraphy and accumulation SPQMLT III. *Antarctic J.*, 3 (4): 95–96.

Rundle, A. S., 1969. Snow accumulation and ice movement on the Anvers Island ice cap, Antarctica. A study of mass balance. *Proc. Intern. Symp. Antarctic Glaciological Exploration, Dartmouth, N.H., 1968* (in press).

Rusin, N. P., 1961. *Meteorological and Radiational Regime of Antarctica.* Leningrad. (Translated by Israel Program for Scientific Translations, Jerusalem 1964, 355 pp.)

Scherhag, R., 1952. Die explosionsartigen Stratosphärenerwärmungen des Spätwinters 1951–1952. *Ber. Deut. Wetterdienst. U.S. Zone*, 6: 51–63.

Schulze, A. A. R., 1959. La campaña Antártica 1958–1959. *Contrib. Inst. Antártico Argentino*, 44: 13 pp.

Schwerdtfeger, W., 1960. The seasonal variation of the strength of the southern circumpolar vortex. *Monthly Weather Rev.*, 88: 203–208.

Schwerdtfeger, W., 1962a. Die halbjährige Periode des meridionalen Temperaturgradienten in der Troposphäre und des Luftdrucks am Boden in Südpolargebiet, ihre Erscheinungsform und kausalen Zusammenhänge. *Beitr. Phys. Atmosphäre*, 35: 234–244.

Schwerdtfeger, W., 1962b. Meteorología del área del Pasaje Drake. *Serv. Hidrografía Naval, República Argentina, Publ.*, H. 410: 78 pp.

Schwerdtfeger, W., 1963. The southern circumpolar vortex and the spring warming of the polar stratosphere. *Proc. Intern. Symp. Stratospheric Mesospheric Circulation, Berlin 1962—Meteorol. Abhandl.*, 36: 207–224.

SCHWERDTFEGER, W., 1967. On the asymmetry of the southern circumpolar vortex, in the winter. *Antarctic J.*, 2 (5): 198.

SCHWERDTFEGER, W., 1968. New data on the winter radiation balance at the South Pole. *Antarctic J.*, 3 (5): 193–194.

SCHWERDTFEGER, W. and MAHRT, L. J., 1968. The relation between terrain features, thermal wind, and surface wind over Antarctica. *Antarctic J.*, 3 (5): 190–191.

SCHWERDTFEGER, W. and MAHRT, L. J., 1969. The relation between the temperature inversion in the surface layer and its wind regime, over Antarctica. *Proc. Intern. Symp. Antarctic Glaciological Exploration, Dartmouth, N.H., 1968*, (in press).

SCHWERDTFEGER, W. and MARTIN, D. W., 1964. The zonal flow of the free atmosphere between 10°N and 80°S, in the South American sector. *J. Appl. Meteorol.*, 3 (6): 726–733.

SCHWERDTFEGER, W. and PROHASKA, F., 1955. Análisis de la marcha anual de la presión y sus relaciones con la circulación atmosférica en Sud América austral y la Antártida. *Meteoros*, 5: 223–237.

SCHWERDTFEGER, W. and PROHASKA, F., 1956. Der Jahresgang des Luftdrucks auf der Erde und seine halbjährige Komponente. *Meteorol. Rundschau*, 9: 33–43, 186–187.

SCHWERDTFEGER, W., DE LA CANAL, L. M. and SCHOLTEN, J., 1959. Meteorología descriptiva del sector antártico sudamericano. *Inst. Antártico Argentino, Publ.*, 7: 425 pp.

SHAW, P. J. F., 1957. The climate of Mawson during 1955. *Australian Meteorol. Mag.*, 18: 1–20.

SHEN, W. C., NICHOLAS, G. W. and BELMONT, A. D., 1968. Antarctic stratospheric warmings during 1963 revealed by 15 TIROS VII data. *J. Appl. Meteorol.*, 7: 268–283.

SILVERSTEIN, S. C., 1967. The American Antarctic Mountaineering Expedition. *Antarctic J.*, 2 (2): 48–50.

SIPLE, P., 1959. *90°South; the Story of the American South Pole Conquest.* Putnam, New York, N.Y., 384 pp.

SIPLE, P. A. and PASSEL, C. F., 1945. Measurement of dry atmospheric cooling in subfreezing temperatures. *Proc. Am. Phil. Soc.*, 89 (1): 177–199.

SOLOT, S. B., 1967. GHOST atlas of the Southern Hemisphere. *Natl. Center Atmos. Res., Publ.*, 10 pp.

SOLOT, S. B., 1968. GHOST balloon data. *Natl. Center Atmos. Res., Tech. Note*, 34: 1729 pp. (11 volumes).

SPANO, A. F., 1965. Results of an airborne albedo program in Antarctica. *Monthly Weather Rev.*, 93 (11): 697–703.

SPONHOLZ, M. P., 1968. Meteorological studies on the Antarctic Plateau. *Antarctic J.*, 3 (5): 189–190.

STRETEN, N. A., 1962. Note on weather conditions in Antarctica. *Australian Meteorol. Mag.*, 37: 1–20.

STRETEN, N. A., 1963. Some observations of Antarctic katabatic winds. *Australian Meteorol. Mag.*, 42: 1–23.

STRETEN, N. A., 1968. Some features of mean annual wind speed data for coastal East Antarctica. *Polar Record*, 14 (90): 315–322.

TALJAARD, J. J., VAN LOON, H., CRUTCHER, H. L. and JEUNE, R. L., 1969. *Climate of the Upper Air;* Part 1. *Southern Hemisphere, 1. Sea Level Pressures and Selected Heights, Temperatures and Dew-points.* U.S. Government Printing Office, Washington, D.C., NAVAIR 50-16-55 (in press).

TAPPER, G. A. G. and SCHOLTEN, J., 1964. Expedición Terrestre Invernal Antártica entre Bahía Esperanza y Bahía Margarita en 1962. *Contrib. Inst. Antártico Argentino*, 83: 44 pp.

TAUBER, G. M., 1960. Characteristics of Antarctic katabatic winds. In: *Proc. Symp. Antarctic Meteorology, Melbourne 1959*, pp.52–64.

THOMPSON, D. C. and MACDONALD, W. T. P., 1962. Radiation measurements at Scott Base. *New Zealand J. Geol. Geophys.*, 5: 874–909.

TRESHNIKOV, A. F., 1967. The ice in the Southern Ocean. *Proc. Symp. Pacific–Antarctic Sci., Pacific Sci. Congr., 11th, Tokyo 1966*, pp.113–123.

UEDA, H. T. and GARFIELD, D. E., 1968. Deep-core drilling program at Byrd station (1967–1968). *Antarctic J.*, 3 (4): 111–112.

UGOLINI, F. C. and BULL, C., 1965. Soil development and glacial events in Antarctica. *Quaternaria*, 7: 251–269.

U.S. NAVY HYDROGRAPHIC OFFICE, 1957. *Oceanographic atlas of the Polar Seas*, 1. *Antarctica.* U.S. Navy, Hydrographic Office Publ., 705: 70 pp.

U.S. NAVY, 1965. *Marine Climatic Atlas of the World*, 7. *Antarctica.* U.S. Government Printing Office, Washington, D.C., NAVWEPS 50-1C-50: 361 pp.

U.S. WEATHER BUREAU and WORLD METEOROLOGICAL ORGANIZATION, 1956–1968. *Monthly Climatic Data for the World.* Vol. 9–21.

U.S. WEATHER BUREAU, 1962–1968. *Climatological Data for Antarctic Stations* (9 volumes).

U.S.S.R., 1966. *Atlas of Antarctica.* Moscow, 225 plates.

VAN LOON, H., 1966. On the annual temperature range over the Southern Oceans. *Geograph. Rev.*, 56 (4): 495–515.

Van Loon, H., 1967a. On the half-yearly oscillation of the sea level pressure in middle and high southern latitudes. *Proc. Symp. Polar Meteorol., Geneva 1966,—World Meteorol. Organ., Tech. Note*, 87: 419–427.

Van Loon, H., 1967b. The half-yearly oscillations in middle and high southern latitudes and the coreless winter. *J. Atmos. Sci.*, 24 (5): 472–486.

Van Loon, H., Taljaard, J. J., Jeune, R. L. and Crutcher, H. L., 1969. *Climate of the Upper Air: I. Southern Hemisphere; 2. Average Zonal and Meridional Geostrophic Winds for the Southern Hemisphere.* U.S. Government Printing Office, Washington, D.C., NAVAIR 50-iC-55 (in press).

Van Rooy, M. P. (Editor), 1957. *Meteorology of the Antarctic.* Weather Bureau of South Africa, Government Printer, Pretoria, 240 pp.

Vickers, W. W., 1966. A study of ice accumulation and tropospheric circulation in western Antarctica. In: M. J. Rubin (Editor), *Studies in Antarctic Meteorology.—Am. Geophys. Union, Antarctic Res. Ser.*, 9: 135–176.

Viebrock, H. J. and Flowers, E. C., 1968. Comments on the recent decrease in solar radiation at the South Pole. *Tellus*, 20 (3): 400–411.

Vinje, T. E., 1965. Climatological tables for Norway Station, instruments and surface observations. *Norsk Polarinstitutt, Arbok 1963*, pp.181–183.

VonderHaar, T., 1968. Variations of the earth's radiation budget. *Dept. Meteorol., Univ. Wisc., Res. Rept.*, 118 pp.

Vowinckel, E., 1957. Climate of the Antarctic coast. In: M. P. Van Rooy (Editor), *Meteorology of the Antarctic.* Weather Bureau of South Africa, Government Printer, Pretoria, pp.137–152.

Weller, G. E., 1968. The heat budget and heat transfer processes in Antarctic plateau ice and sea ice. *ANARE (Australian Natl. Antarctic Res. Expeditions) Sci. Rept., Ser. A (4), Glaciology, Publ.*, 102: 155 pp.

Weller, G. E., 1969. A meridional surface wind speed profile in MacRobertson Land, Antarctica. *Pure Appl. Geophys.* (in press).

Weller, G. and Schwerdtfeger, P., 1967. Radiation penetration in Antarctic Plateau and Sea Ice. *Proc. Symp. Polar Meteorol., Geneva 1966—World Meteorol. Organ., Tech. Note*, 87: 120–141.

Wexler, H., 1959a. Seasonal and other temperature changes in the Antarctic atmosphere. *Quart. J. Roy. Meteorol. Soc.*, 85: 196–208.

Wexler, H., 1959b. A warming trend at Little America, Antarctica. *Weather*, 14 (6): 191–197.

Wexler, H., 1961. Growth and thermal structure of the deep ice in Byrd Land, Antarctica. *J. Glaciol.*, 3 (30): 1075–1088.

Wexler, H., Moreland, W. B. and Weyant, W. S., 1960. A preliminary report on ozone observations at Little America, Antarctica. *Monthly Weather Rev.*, 88: 43–54.

Weyant, W. S., 1967. Interpretation of ozone measurements at U.S. Antarctic stations. *Proc. Symp. Polar Meteorol., Geneva 1966—World Meteorol. Organ., Tech. Note*, 87: 29–36.

White, F. D. and Bryson, R. A., 1967. The radiative factor in the mean meridional circulation of the Antarctic atmosphere during the polar night. *Proc. Symp. Polar Meteorol., Geneva 1966—World Meteorol. Organ., Tech. Note*, 87: 199–224.

Wilson, A. T., 1964. Origin of ice ages; an ice shelf theory for Pleistocene glaciation. *Nature*, 201 (4915): 147–149.

Wilson, C. V. and Godson, W. L., 1962. The stratospheric temperature field at high latitudes. *Arctic Meteorol. Res. Group, Publ. Meteorol.*, 46: 191 pp.

Wright, F. W., Hodge, P. W. and Langway, C. C., 1963. Studies of particles for extraterrestrial origin. 1. Chemical analyses of 118 particles. *J. Geophys. Res.*, 68 (19): 5575–5587.

Zillman, J. W., 1967. The surface radiation balance in high southern latitudes. *Proc. Symp. Polar Meteorol., Geneva 1966—World Meteorol. Organ., Tech. Note*, 87: 142–174.

TABLE XXX

CLIMATIC TABLE FOR SOUTH POLE
Latitude 90°00'S, elevation 2,800 m

Month	Mean sta. press.[1] (mbar)	Temperature[1] (°C) daily mean	mean daily range	extremes max.	extremes min.	Inversion strength[2] surf. to 650 mbar	surf. to 600 mbar	Number of soundings[2]	Global radiat.[3]	Number of months[3]
Jan.	687.8	−28.8	2.9	−15.0	−40.6	1.3	−0.5	471	25.9	7
Feb.	686.8	−40.1	3.8	−22.2	−56.1	9.3	8.9	334	13.4	6
Mar.	680.2	−54.4	5.0	−28.9	−70.0	17.4	18.9	394	2.6	6
Apr.	679.6	−58.5	5.7	−31.7	−72.2	19.3	21.4	378	–	–
May	680.5	−57.4	6.3	−35.0	−73.3	20.0	22.1	450	–	–
June	682.1	−56.5	6.7	−29.4	−76.1	17.8	19.3	425	–	–
July	677.4	−59.2	6.5	−35.6	−80.6	17.4	20.2	477	–	–
Aug.	676.9	−58.9	6.2	−32.8	−77.2	17.1	20.0	462	–	–
Sept.	674.8	−59.0	6.2	−37.8	−77.2	17.2	20.3	447	0.4	8
Oct.	675.4	−51.3	4.1	−30.0	−67.2	12.9	14.7	486	9.4	8
Nov.	681.8	−38.9	2.6	−19.4	−53.9	6.2	5.8	519	22.0	6
Dec.	687.6	−28.1	2.0	−18.9	−38.3	0.2	−1.8	579	29.6	8
Annual	680.9	−49.3	4.8	−15.0	−80.6	13.0	14.1	5422	103.3	–

Month	Blowing snow[4] (%)	Total nr. of observ.[4]	Resultant wind vector[1] dir. (°)	speed (m/sec)	Wind constancy[1]	Mean cloudiness[1] (%) 0–3/10	4–7/10	8–10/10	Mean wind speed[1] (m/sec)	Net radiation[5] (Ly/h)
Jan.	5	2,362	031	3.1	0.74	44	13	43	4.2	
Feb.	11	2,188	040	4.1	0.78	45	14	41	5.2	
Mar.	24	2,235	043	5.0	0.80	46	17	37	6.3	
Apr.	18	2,206	054	4.8	0.75	68	13	19	6.4	−1.15
May	26	2,338	035	5.4	0.78	69	11	20	6.9	−1.19
June	28	2,277	035	5.8	0.79	65	10	25	7.3	−1.11
July	28	2,350	037	5.8	0.79	66	14	20	7.4	−1.28
Aug.	21	2,342	039	5.6	0.77	63	12	25	7.3	−1.28
Sept.	28	2,285	039	5.6	0.78	47	16	37	7.3	−1.30
Oct.	24	2,405	033	5.8	0.85	41	14	45	6.8	
Nov.	9	2,394	039	4.1	0.80	55	13	32	5.1	
Dec.	5	2,479	026	3.1	0.70	47	11	42	4.4	
Annual	19	27,861	038	4.8	0.77	55	13	32	6.2	

[1] I/57–XII/66.
[2] Average temperature increase from surface to 650 mbar and from surface to 600 mbar I/57–XII/66.
[3] Average monthly sum of global radiation (10^3 Ly); *not* reduced to equal number of days per month. Series IX/57–III/66 incomplete; number of months used for averages given in next column.
[4] Relative frequency of occurrence, % of all weather observations; I/57–XII/66.
[5] Net radiation in Ly/h, from C.S.I.R.O. net radiometer. Months IV–IX for 1965 and 1966.

TABLE XXXI

CLIMATIC TABLE FOR BYRD

Latitude 80°01′S, 80°S[1], longitude 119°31′W, 120°W[1], elevation 1,533 m, 1,511 m[1]

Month	Mean sta. press.[2] (mbar)	Temperature[2] (°C)				Inversion[3]			
		daily mean	mean daily range	extremes		mean strength at 04h (L.M.T.)	number of soundings	mean strength at 16h (L.M.T.)	number of soundings
				max.	min.				
Jan.	813.3	−15.1	5.5	−0.6	−29.4	0.4	262	−3.4	259
Feb.	812.6	−19.9	6.7	−3.3	−38.9	1.6	198	−2.3	199
Mar.	806.4	−28.2	7.5	−9.4	−52.8	6.1	144	4.5	145
Apr.	806.7	−29.3	9.0	−5.6	−58.9	8.4	144	7.6	144
May	804.8	−33.1	8.7	−10.6	−60.6	9.2	165	8.9	174
June	807.0	−33.5	8.8	−10.6	−59.4	9.9	167	10.4	166
July	802.7	−35.7	8.8	−12.2	−62.8	9.4	170	9.6	170
Aug.	801.8	−37.0	8.5	−13.9	−62.2	10.2	163	10.4	166
Sept.	796.7	−36.2	8.1	−10.0	−61.1	9.9	161	8.4	165
Oct.	798.4	−30.2	8.5	−11.1	−50.0	8.5	193	4.5	196
Nov.	805.3	−20.9	6.2	− 6.7	−41.1	3.1	276	−1.0	272
Dec.	813.9	−14.6	5.2	− 2.8	−29.4	−0.5	275	−3.2	277
Annual	805.8	−27.9	7.6	− 0.6	−62.8	6.4	2,318	4.5	2,333

Month	Blowing snow[4] (%)	Total nr. of weather observ.[4]	Days with snow-fall[2]	Mean cloudiness[2] (%)			Wind			
				0–3/10	4–7/10	8–10/10	mean speed[2] (m/sec)	result. direct. (°)	vect. speed[5] (m/sec)	con-stancy[5]
Jan.	9	2,407	14	26	15	59	6.0	010	4.6	0.78
Feb.	20	2,219	16	21	12	67	7.2	011	5.9	0.82
Mar.	25	2,192	17	23	15	62	7.9	001	6.7	0.84
Apr.	31	2,139	12	33	15	52	8.9	012	8.3	0.88
May	39	2,201	13	50	13	37	9.5	008	9.1	0.92
June	30	2,129	14	48	13	39	9.8	019	9.2	0.82
July	38	2,200	16	45	16	39	9.9	016	9.7	0.90
Aug.	40	2,200	13	44	14	42	10.1	003	8.8	0.85
Sept.	40	2,146	9	38	15	47	10.2	014	8.4	0.83
Oct.	33	2,321	13	24	16	60	9.3	013	8.7	0.85
Nov.	24	2,397	12	28	15	57	8.1	010	6.9	0.82
Dec.	13	2,480	16	24	14	62	6.4	357	5.0	0.81
Annual	29	27,031	165	34	14	52	8.6	010	7.5	0.83

[1] Second figure applies to period before February 18, 1962.
[2] I/57–XII/66, mean station pressure refers to 1,530 m above sea level.
[3] Average temperature increase from surface to the 750-mbar level; summer 1958–66; winter 1957–62.
[4] Relative frequency of occurrence, % of all weather observations.
[5] III/62–XII/66.

TABLE XXXII

CLIMATIC TABLE FOR PLATEAU STATION
Latitude 79°15′S, longitude 40°30′E, elevation 3,625 m

Month	Mean sta. press.[1] (mbar)	Temperature[1] (°C)				Number of days with[3]		
		daily mean	daily range	extremes		drifting snow	snowfall	ice crystals
				max.	min.			
Jan.	619.0	−33.9	10.4	−18.5	−48.9	4	8	20
Feb.	615.7	−44.4	11.8	−24.9	−60.8	8	10	28
Mar.	612.7	−57.2	9.9	−35.9	−75.3	8	8	28
Apr.	606.7	−65.8	8.1	−42.7	−78.0	6	3	22
May	609.3	−66.4	9.4	−38.9	−80.6	6	1	25
June	606.3	−69.0	9.0	−32.8	−82.2	5	—	26
July	605.3	−68.0	9.1	−43.9	−86.2	9	2	30
Aug.	597.3	−71.4	9.1	−41.2	−85.0	11	—	28
Sept.	602.0	−65.0	12.0	−37.8	−84.4	11	2	27
Oct.	604.0	−59.5	15.4	−37.1	−80.0	5	5	25
Nov.	613.3	−44.4	12.9	−26.7	−66.1	10	6	29
Dec.	619.3	−32.3	13.6	−20.6	−47.8	7	6	28
Annual	609.2[2]	−56.4	10.9	−18.5	−86.2	90	51	316

Month	Mean cloudiness[1] (%)			Mean wind speed[1] (m/sec)	Most frequ. wind dir.[3]	Peak gust[4] (m/sec)
	0–3/10	4–7/10	8–10/10			
Jan.	52.5	32.4	15.1	3.0	N	9
Feb.	35.8	43.4	20.8	4.2	NW	12
Mar.	47.3	43.0	9.7	5.0	N	16
Apr.	64.2	28.9	6.9	5.2	NNE	18
May	74.5	20.2	5.3	5.4	N	16
June	71.9	15.6	12.5	5.0	N	22
July	70.9	21.8	7.3	5.8	N	13
Aug.	60.2	28.7	11.1	5.9	NNW	16
Sept.	33.4	43.7	22.9	5.7	NNW	20
Oct.	61.5	27.2	11.4	5.1	NNW	13
Nov.	59.6	23.5	16.9	4.6	N	25
Dec.	47.6	32.9	19.5	3.8	NNW	12
Annual	56.6	30.1	13.3	4.9	N	25

[1] XII/65–XII/68.
[2] Boiling point of water at average station pressure is 86.3° C.
[3] I/67–XII/68.
[4] III/66–XII/68, except XII/66.

TABLE XXXIII

CLIMATIC TABLE FOR VOSTOK

Latitude 78°28′S, longitude 106°48′E, elevation 3,488 m

Month	Mean sta. press.[1] (mbar)	Temperature[1] (°C) daily mean	extremes max.	min.	Mean relat. humid.[1] (%)	Inversion[2] strength	number of soundings
Jan.	632.7 (7)	−33.4 (9)	−22.3 (9)	−48.3 (9)	75 (6)	1.0	118
Feb.	630.7 (7)	−44.2 (9)	−24.3 (9)	−64.0 (9)	72 (6)	10.2	113
Mar.	624.5 (7)	−57.4 (9)	−32.5 (9)	−75.0 (9)	71 (6)	17.2	124
Apr.	622.1 (7)	−65.7 (9)	−42.4 (9)	−81.8 (9)	68 (6)	22.4	99
May	622.7 (7)	−66.2 (9)	−43.0 (9)	−82.0 (9)	68 (6)	23.0	120
June	623.2 (7)	−66.0 (9)	−39.5 (9)	−83.0 (9)	68 (6)	22.4	121
July	624.4 (8)	−66.7 (9)	−36.1 (9)	−81.1 (9)	69 (7)	22.6	103
Aug.	620.3 (8)	−68.4 (9)	−44.9 (9)	−88.3 (9)	69 (7)	24.6	115
Sept.	617.2 (7)	−65.6 (9)	−42.1 (8)	−82.8 (8)	69 (7)	21.1	124
Oct.	617.8 (7)	−57.4 (9)	−39.9 (8)	−75.7 (8)	71 (7)	15.7	134
Nov.	623.8 (7)	−43.6 (9)	−31.7 (8)	−63.1 (8)	72 (7)	7.2	129
Dec.	631.5 (7)	−32.7 (9)	−21.0 (8)	−48.0 (8)	74 (6)	1.1	133
Annual	624.2	−55.6	−21.0	−88.3	70	15.7	1.433

Month	Number of days with[3] T̄< −40° C	drifting snow	Mean cloudiness[1] (tenths)	Wind preval. dir.[3]	mean speed[1] (m/sec)	Global rad.[4]	Albedo[4] (%)
Jan.	0	1	3.6 (6)	SW	4.4 (8)	25.7	80
Feb.	23	1	3.5 (6)	SW	4.5 (9)	13.8	82
Mar.	31	5	4.0 (6)	SW	5.4 (9)	4.9	81
Apr.	30	4	3.1 (7)	SW	5.3 (9)	0.3	80
May	31	6	3.6 (6)	SW	5.4 (9)	—	—
June	30	8	3.3 (6)	SW	5.3 (9)	—	—
July	31	7	3.4 (8)	SW	5.3 (9)	—	—
Aug.	31	7	3.6 (8)	W	5.2 (9)	0.0	97
Sept.	30	6	4.0 (7)	SW	5.5 (8)	2.2	85
Oct.	31	5	4.6 (6)	SW	5.3 (8)	11.0	86
Nov.	26	2	3.5 (6)	SW	5.0 (8)	22.7	83
Dec.	0	1	3.0 (6)	SW	4.7 (8)	27.8	82
Annual	294	53	3.6	SW	5.1	108.4	84

[1] Number of months used for computation of mean values between brackets; I/58–I/68; no data for I/62–I/63.

[2] Average temperature increase from surface to the 4 km level, 512 m above ground; I/58–XII/61.

[3] I/58–XII/61 and I/63–XII/65.

[4] Monthly and annual global radiation (in 10^3 Ly) and albedo: II/63–XII/65.

TABLE XXXIV

CLIMATIC TABLE FOR LITTLE AMERICA

Latitude 78°18′S[1], longitude 163°00′W[1], elevation 40 m (approx.)[1]

Month	Mean sta. press.[2] (mbar)	Temperature[2] (°C)				Snow[3] (cm)	Number of days with snowfall[3]	Mean cloudiness[4] (tenths)
		daily mean	daily range	extremes				
				max.	min.			
Jan.	989.5	−6.6	6.1	5.8	−22.2	15	17	7.1
Feb.	987.6	−12.9	7.8	1.7	−33.6	34	18	7.7
Mar.	984.8	−21.8	9.8	−1.0	−48.5	46	20	7.4
Apr.	983.9	−28.3	9.4	−2.2	−50.0	18	16	6.8
May	981.2	−25.8	10.9	−1.7	−57.4	25	18	5.7
June	983.9	−24.6	10.2	−5.0	−52.9	26	20	6.2
July	979.9	−29.6	11.6	−5.2	−57.9	18	19	4.4
Aug.	977.3	−30.7	11.2	−6.1	−60.6	16	16	5.4
Sept.	972.1	−30.2	10.7	−6.1	−59.4	19	18	5.5
Oct.	972.5	−21.1	10.1	−1.7	−48.1	17	21	7.2
Nov.	979.6	−13.2	4.7	−1.7	−37.4	11	18	6.5
Dec.	989.7	−6.4	6.3	5.9	−18.9	20	15	6.8
Annual	981.8	−20.9	9.1	5.9	−60.6	265	216	6.4

Month	Wind	
	preval. direct.[5]	mean speed[2] (m/sec)
Jan.	SE	4.8
Feb.	SE	5.0
Mar.	SE	6.1
Apr.	SE	5.3
May	SE	5.6
June	SE	5.8
July	SE	5.4
Aug.	SE	5.4
Sept.	SE	5.2
Oct.	SE	5.5
Nov.	SE	4.5
Dec.	SE	4.6
Annual	SE	5.3

[1] Mean values for various locations (see COURT, 1949; U.S. WEATHER BUREAU, 1962).
[2] I/29–I/30; II/34–I/35; II/40–I/41; II/56–XII/58.
[3] I/29–I/30; III/34–I/35; II/56–XII/58, except I/57.
[4] I/29–I/30; II/34–I/35; I/57–XII/58.
[5] II/56–XII/58.

TABLE XXXV

CLIMATIC TABLE FOR BASE GENERAL BELGRANO
Latitude 77°58′S, longitude 38°48′W, elevation 50 m (approx.)

Month	Mean sta. press.[1] (mbar)	Temperature (°C)				Number of days with		
		daily mean[2]	daily range[3]	extremes		winds[4] >15 m/sec	precip.[5]	blowing snow[6]
				max.[3]	min.[2]			
Jan.	988.8	−6.0	7.9	6.8	−26.6	1	10	6
Feb.	991.2	−13.2	9.8	6.6	−36.0	1	11	9
Mar.	984.5	−21.0	9.6	−1.1	−42.3	2	10	17
Apr.	984.8	−27.1	11.4	−3.0	−55.0	2	9	16
May	987.7	−30.0	11.3	−2.6	−52.8	4	10	18
June	988.3	−31.7	9.2	−9.0	−52.8	4	8	17
July	989.0	−32.7	10.9	−7.2	−53.3	4	9	20
Aug.	985.3	−32.9	10.5	−6.7	−57.2	6	8	19
Sept.	983.8	−31.4	10.3	−6.6	−56.0	5	9	20
Oct.	979.2	−22.4	9.5	−1.3	−48.6	4	10	16
Nov.	983.1	−12.9	8.5	0.6	−37.1	4	10	15
Dec.	988.2	−6.0	7.5	5.7	−21.5	1	8	7
Annual	986.2	−22.3	9.7	6.8	−57.2	38	112	180

Month	Number of days		Mean cloudiness[2] (tenths)	Wind[2]	
	clear[7]	cloudy[7]		preval. direct.	mean speed (m/sec)
Jan.	7	15	6.7	S	3.6
Feb.	6	14	6.8	S	4.2
Mar.	7	16	6.7	S	5.9
Apr.	9	11	5.9	S	5.4
May	7	12	5.2	S	6.5
June	11	9	4.4	S	6.3
July	10	10	4.6	S	6.5
Aug.	9	10	5.0	S	6.6
Sept.	6	12	5.9	S	6.7
Oct.	4	18	6.9	S	5.6
Nov.	6	16	6.5	S	4.9
Dec.	7	15	6.4	S	3.6
Annual	89	158	5.9	S	5.5

[1] I/61–XII/67, except I/62–XII/62.
[2] II/55–XII/67, except I/57; I/62–XII/62.
[3] II/55–XII/67, except I, II/57; I/62–XII/62; I/64–XII/64.
[4] I/56–XII/67, except I, II/57; XI/61; I/62–XII/62.
[5] I/56–XII/67, except I/57; I/62–XII/62.
[6] I/56–XII/67, except I/57; XI/61; I/62–XII/62.
[7] I/57–XII/67, except I/62–XII/62.

TABLE XXXVI

CLIMATIC TABLE FOR MCMURDO

Latitude 77°53′S, longitude 166°44′E, elevation 24 m

Month	Mean sta. press.[1] (mbar)	Temperature (°C)				Mean net rad.[5] (Ly/h)	Mean cloudiness[6] (tenths)
		daily mean[2]	daily range[3]	extremes[4]			
				max.	min.		
Jan.	990.5	−3.4	5.0	5.6	−15.6	—	6.4
Feb.	991.5	−8.3	5.0	4.4	−21.8	—	7.4
Mar.	987.9	−18.9	6.1	−3.3	−43.3	—	7.2
Apr.	989.5	−21.2	7.2	−5.0	−39.4	−1.17	6.6
May	991.5	−23.9	6.7	−7.2	−44.4	−1.25	5.2
June	989.8	−23.4	7.2	−7.8	−40.0	−1.17	5.6
July	989.2	−25.5	10.0	−4.4	−50.6	−1.19	4.8
Aug.	987.9	−27.8	10.0	−7.8	−49.4	−1.20	5.2
Sept.	982.1	−24.1	8.9	−7.8	−41.1	−1.25	6.4
Oct.	981.1	−19.9	7.2	−4.5	−38.3	—	6.4
Nov.	984.3	−8.8	7.2	2.8	−27.8	—	5.8
Dec.	988.8	−3.8	3.9	5.1	−16.6	—	6.7
Annual	987.8	−17.4	7.0	5.6	−50.6	—	6.1

Month	Wind		
	preval. direct.[7]	mean speed[8] (m/sec)	max. wind gust[7] (m/sec)
Jan.	E	5.3	24
Feb.	E	7.0	29
Mar.	E	7.3	27
Apr.	E	6.1	28
May	E	6.9	43
June	E	7.2	43
July	E	6.5	36
Aug.	ENE-E	6.4	38
Sept.	E	6.9	41
Oct.	E	6.2	37
Nov.	E	5.4	35
Dec.	E	6.5	24
Annual	E	6.5	43

[1] I/57–XII/67.
[2] III/56–II/68. For the McMurdo Sound stations (1902–04, 1908–09, 1911–12) the five year annual average also is −17.4. See VOWINCKEL (1957, table 7).
[3] I/56–XII/61, from PHILLPOT (1967).
[4] III/56–II/66.
[5] IV/65–IX/65 only, from C.S.I.R.O. Net-radiometer.
[6] III/56–VIII/65, except XI/59; I, II/60; II/63; VII/65.
[7] III/56–VII/61, from PHILLPOT (1967).
[8] III/56–VIII/65, except VII/65.

TABLE XXXVII

CLIMATIC TABLE FOR HALLEY BAY
Latitude 75°30′S, longitude 26°39′W, elevation 30 m

Month	Mean sealevel press.[1] (mbar)	Temperature[1] (°C)				Mean vap. press.[1] (mbar)	Number of days with[1]	
		daily mean	daily range	extremes			drifting or blowing snow	snow-fall
				max.	min.			
Jan.	991.2	−5.0	5.9	2.8	−22.2	3.5	11	16
Feb.	993.1	−10.0	7.5	4.0	−29.6	2.4	12	15
Mar.	987.1	−16.0	8.9	−2.0	−37.6	1.5	18	15
Apr.	988.5	−21.1	7.8	−2.1	−46.4	1.0	17	16
May	992.3	−25.0	8.8	−2.3	−49.0	0.8	16	12
June	992.3	−26.8	9.1	−6.2	−51.0	0.6	16	10
July	989.7	−27.5	9.0	−5.0	−50.6	0.6	19	12
Aug.	989.4	−28.5	9.1	−4.4	−52.5	0.6	16	12
Sept.	987.3	−26.7	9.0	−5.0	−50.3	0.6	19	13
Oct.	983.1	−19.8	8.8	−1.8	−45.3	1.2	17	15
Nov.	986.4	−12.3	7.5	−1.5	−32.3	2.0	16	17
Dec.	991.4	−5.7	6.0	2.2	−14.6	3.3	11	14
Annual	989.3	−18.7	8.1	4.0	−52.5	1.5	188	167

Month	Mean cloudi-ness[1] (tenths)	Mean sunshine hours[2]	Wind		
			preval. direct.[3]	mean speed[4] (m/sec)	peak gust[4] (m/sec)
Jan.	7.5	256	E	4.5	31
Feb.	7.3	167	E	4.6	29
Mar.	7.0	114	E	4.9	38
Apr.	7.0	48	E	5.0	35
May	6.0	—	E	4.8	35
June	4.9	—	E	4.6	40
July	5.8	—	E	5.1	39
Aug.	6.1	15	E	4.7	36
Sept.	6.3	75	E	5.4	35
Oct.	6.9	201	E	5.5	42
Nov.	7.0	225	E	4.4	34
Dec.	7.3	258	E	4.3	38
Annual	6.6	1,359	E	4.8	42

[1] III/56–XII/67.
[2] I/57–XII/59.
[3] I/62–XII/67.
[4] VII/56–XII/61, from PHILLPOT (1967).

TABLE XXXVIII

CLIMATIC TABLE FOR EIGHTS
Latitude 75°14′S, longitude 77°10′W, elevation 421 m

Month	Mean sta. press.[1] (mbar)	Temperature[1] (°C)				Days with snowfall[2]	Snow-fall[2] (cm)
		daily mean	daily range	extremes			
				max.	min.		
Jan.	942.3	−10.0	5.6	1.1	−24.4	17	38
Feb.	947.0	−18.0	5.4	−1.7	−32.8	19	34
Mar.	939.0	−25.0	6.4	−2.2	−41.7	20	34
Apr.	945.7	−31.1	7.0	−6.7	−46.7	18	15
May	942.0	−33.1	8.1	−9.4	−49.4	22	39
June	946.0	−33.5	9.5	−1.1	−50.6	14	22
July	949.0	−33.5	7.8	−9.4	−52.2	20	19
Aug.	946.7	−37.0	8.2	−13.3	−60.0	14	7
Sept.	937.0	−34.3	8.3	−8.3	−51.7	21	22
Oct.	937.0	−28.3	6.8	−3.3	−48.3	18	17
Nov.	941.5	−17.3	6.4	−4.4	−31.1	17	20
Dec.	937.0	−11.3	6.8	2.2	−28.3	19	19
Annual	942.5	−26.0	7.2	2.2	−60.0	219	286

Month	Mean cloudiness[1] (%)			Wind[1]		
	0–3 tenths	4–7 tenths	8–10 tenths	preval. direct.	mean speed (m/sec)	peak gust (m/sec)
Jan.	15	9	76	S, NW	4.9	25
Feb.	12	11	77	S, SSE	4.4	18
Mar.	13	7	80	S	5.5	26
Apr.	24	7	69	S, NNW	5.0	37
May	17	7	76	S, NNW	5.0	32
June	33	6	61	S, NNW	5.0	35
July	17	8	75	S, NNW	5.4	34
Aug.	28	9	63	NNW, S	5.7	34
Sept.	16	4	80	S, NNW	5.9	32
Oct.	13	5	82	S, NNW	5.5	32
Nov.	13	9	78	S, NNW	6.2	32
Dec.	23	12	65	S	6.0	26
Annual	19	8	73		5.4	37

[1] XII/61–I/62; XII/62–X/65.
[2] XII/61–I/62; XII/62–XII/63; I/65–X/65.

TABLE XXXIX

CLIMATIC TABLE FOR HALLETT
Latitude 72°18′S, longitude 170°19′E, elevation 5 m

Month	Mean sta. press.[1] (mbar)	Temperature (°C)				Blowing snow[3] (%)	Total number of weather observ.[3]
		mean daily[1]	daily range[2]	extremes[2]			
				max.	min.		
Jan.	990.0	−1.1	4.6	5.6	−9.4	3	1,735
Feb.	988.9	−3.2	3.2	4.4	−8.9	5	1,804
Mar.	986.7	−10.5	2.9	−1.1	−22.8	9	1,734
Apr.	990.1	−17.8	4.3	−3.9	−31.1	9	1,677
May	991.3	−22.6	5.0	−6.7	−35.0	10	1,736
June	991.9	−23.0	5.9	−3.9	−36.7	14	1,679
July	987.4	−26.4	6.3	−6.1	−40.6	9	1,735
Aug.	991.6	−26.6	6.3	−7.8	−47.8	10	1,736
Sept.	984.9	−24.5	7.3	−7.2	−40.0	8	1,680
Oct.	982.0	−18.3	7.2	−4.4	−37.2	8	1,950
Nov.	984.3	−8.0	6.7	1.7	−24.4	1	1,919
Dec.	989.8	−1.7	4.9	5.0	−11.7	3	1,982
Annual	988.2	−15.3	5.4	5.6	−47.8	7	21,367

Month	Days with snowfall[2]	Mean cloudiness[2] (%)			Wind[2]		
		0–3 tenths	4–7 tenths	8–10 tenths	mean speed (m/sec)	peak gust (m/sec)	preval. direct.
Jan.	14	23	15	62	3.1	32	SW,SSW
Feb.	18	17	12	71	4.4	41	SSW
Mar.	23	16	8	76	4.9	40	SSW
Apr.	18	35	10	55	3.7	39	SW,SSW
May	15	44	10	46	3.7	41	SSW,SW
June	14	45	11	44	4.5	41	SSW,SW
July	15	50	11	39	3.4	41	SSW
Aug.	12	45	10	45	3.3	40	SSW,SW
Sept.	12	47	12	41	2.8	46	SSW
Oct.	16	33	12	55	3.3	51	SSW,SW
Nov.	11	38	14	48	2.9	43	SSW
Dec.	13	29	16	55	3.0	28	SSW,SW
Annual	181	35	12	53	3.6	51	SSW,SW

[1] II/57–II/64; X/64–I/65; X/65–XII/65.
[2] From record II/57–I/64.
[3] Relative frequency of occurrence, % of all weather observations.

TABLE XL

CLIMATIC TABLE FOR NOVOLAZAREVSKAYA
Latitude 70°46′S, longitude 11°49′E, elevation 87 m

Month	Mean sta. press.[1] (mbar)	Temperature (°C) daily mean[2]	daily range[8]	extremes[3] max.	min.	Mean relat. humid.[4] (%)	Global rad.[9]	Albedo[9] (%)
Jan.	978.4	−1.2	5.5	6.6	−11.7	58	20.6	22
Feb.	978.2	−3.9	4.9	4.0	−15.1	51	11.5	22
Mar.	976.4	−8.7	5.6	2.9	−22.3	52	5.5	24
Apr.	976.3	−13.4	7.0	−1.1	−32.2	46	1.4	25
May	977.0	−13.5	5.9	−1.3	−30.8	49	0.1	27
June	979.6	−14.1	6.1	−2.1	−36.9	50	—	—
July	978.4	−17.9	7.6	−1.0	−35.5	51	0.0	—
Aug.	973.3	−18.0	6.9	−1.3	−41.0	53	0.7	34
Sept.	973.8	−17.6	7.3	−3.5	−40.6	54	4.1	28
Oct.	970.6	−13.7	6.6	0.8	−30.3	51	11.1	30
Nov.	973.1	−6.4	6.4	5.2	−24.2	53	17.9	26
Dec.	977.5	−0.8	5.2	5.3	−10.9	60	20.2	26
Annual	976.1	−10.8	6.2	6.6	−41.0	52	93.1	26

Month	Mean precip.[8] (mm)	Number of days with drifting or blowing snow[8]	Mean cloudiness[5] (tenths)	Wind preval. direct.[8]	mean speed[6] (m/sec)	max. speed[7] (m/sec)
Jan.	13	2	5.6	SE	7.4	28.6
Feb.	4	4	6.4	SE	9.6	26.5
Mar.	4	5	6.3	SE	10.5	30.8
Apr.	7	8	5.3	SE	10.5	29.5
May	13	13	6.0	SE	12.1	28.3
June	75	14	6.0	SE	13.0	32.2
July	29	6	5.5	SE	10.7	29.2
Aug.	32	10	5.8	SE	11.4	33.2
Sept.	44	9	5.5	SE	9.8	31.4
Oct.	64	10	5.2	SE	10.9	33.0
Nov.	13	6	6.4	SE	10.1	29.3
Dec	5	4	6.7	SE	7.5	23.8
Annual	303	91	5.9	SE	10.3	29.7

[1] VII/61–III/68.
[2] II/61–III/68, except XII/66.
[3] II/61–VIII/67, except XII/66.
[4] VII/61–XI/67, except X/61; III, IV,V/62; XII/63.
[5] VII/61–VIII/67, except XII/61; IV, VI, X/62; XII/63.
[6] II/61–VIII/67, except III/66.
[7] II/61–XII/64; X/65–VIII/67.
[8] I/61–XII/65.
[9] Monthly and annual global radiation (in 10^3 Ly) and albedo. I/63–XII/65.

TABLE XLI

CLIMATIC TABLE FOR NORWAY STATION AND SANAE

Latitude 70°30′S, 70°19′S[1], longitude 2°32′W, 2°21′W[1], elevation 56 m, 52 m[1]

Month	Mean sealevel press.[2] (mbar)	Temperature (°C)				Mean relat. humid. (%)	
		daily mean[2]	daily range[2]	extremes			
				max.[3]	min.[2]	([4])	([5])
Jan.	990.0	−4.4	8.6	7.3	−25.3	90	78
Feb.	992.0	−8.9	8.7	2.8	−31.3	85	79
Mar.	986.7	−14.6	9.4	−0.8	−35.5	86	81
Apr.	987.1	−20.2	9.9	−2.8	−43.9	86	78
May	990.1	−22.7	9.8	−2.0	−51.0	89	79
June	990.7	−22.3	10.3	−5.2	−47.1	90	76
July	988.7	−26.1	10.2	−5.5	−47.1	88	73
Aug.	987.7	−26.8	10.4	−3.8	−50.0	88	74
Sept.	985.9	−26.0	10.7	−1.9	−48.3	87	74
Oct.	982.8	−18.3	10.2	−0.8	−42.9	85	78
Nov.	985.2	−10.7	9.6	3.1	−33.0	83	84
Dec.	989.6	−5.2	8.8	5.2	−21.4	85	79
Annual	988.0	−17.2	9.7	7.3	−51.0	87	78

Month	Mean cloudiness[2] (tenths)	Mean sunshine hours[6]	Sunshine duration[6] (% of poss.)	Mean wind speed[7] (m/sec)
Jan.	7.1	238	33	6.1
Feb.	7.3	148	29	6.9
Mar.	6.6	106	27	7.3
Apr.	6.1	72	29	7.6
May	5.7	19	23	7.7
June	6.4	0	—	8.8
July	6.1	0.3	2	8.1
Aug.	5.7	43	25	7.6
Sept.	5.9	100	30	7.2
Oct.	6.7	188	38	7.9
Nov.	7.0	228	34	7.3
Dec.	7.0	247	33	6.1
Annual	6.5	1,389		7.4

[1] Second figure applies from April 1962 (see [2]).
[2] IV/57–I/62; IV/62–III/68, pressure refers to sea level.
[3] IV/57–I/62; IV/62–III/68. See text, p. 273 (VINJE, 1965).
[4] IV/57–XII/59.
[5] VI/62–XII/67.
[6] VII/62–III/68.
[7] IV/57–III/68, except II, III/64, I/68. I/60–XII/64 from PHILLPOT (1967).

TABLE XLII

CLIMATIC TABLE FOR PIONERSKAYA
Latitude 69°44′S, longitude 95°30′E, elevation 2,740 m

Month	Mean sta. press.[1] (mbar)	Temperature[1] (°C)			Global rad.[2]	Albedo[2] (%)
		daily mean	extremes			
			max.	min.		
Jan.	698.5	−23.4	−15.0	−35.0	22.8	83
Feb.	693.0	−32.4	−20.6	−47.4	15.2	78
Mar.	690.6	−38.4	−24.0	−51.2	8.1	81
Apr.	690.9	−39.2	−25.6	−51.0	1.6	88
May	696.9	−42.9	−23.7	−55.4	0.0	92
June	692.6	−46.0	−30.8	−57.7	—	—
July	687.4	−47.3	−28.8	−62.1	—	—
Aug.	683.9	−48.3	−30.7	−62.4	0.7	84
Sept.	681.2	−44.5	−26.7	−60.5	4.1	85
Oct.	685.0	−39.3	−26.2	−55.5	11.5	84
Nov.	681.4	−31.4	−20.0	−43.1	19.7	83
Dec.	698.0	−23.1	−16.3	−32.7	25.1	85
Annual	690.0	−38.0	−15.0	−62.4	108.8	84

Month	Number of days with[1]			Mean cloudiness[1] (tenths)	Wind[1]	
	drifting or blowing snow	$\bar{T}<$ −20°C	$\bar{T}<$ −40°C		preval. direct.	mean speed (m/sec)
Jan.	20	28	0	6.4	SE	10.6
Feb.	18	26	2	5.4	SE	11.0
Mar.	22	31	11	6.2	SE	11.0
Apr.	26	30	14	6.8	SE	12.6
May	24	31	20	6.0	SE	9.6
June	21	30	24	5.1	SE	9.6
July	22	31	26	6.2	SE	10.5
Aug.	24	31	27	5.5	SE	11.0
Sept.	23	30	22	6.0	SE	11.5
Oct.	23	31	14	6.3	SE	10.4
Nov.	20	30	0	5.4	SE	9.9
Dec.	17	25	0	6.6	SE	9.4
Annual	260	354	160	6.0	SE	10.6

[1] V/56–XII/58.
[2] Monthly and annual global radiation (in 10^3 Ly) and albedo; V/56–XII/58.

TABLE XLIII

CLIMATIC TABLE FOR SYOWA BASE

Latitude 69°00′S, longitude 39°35′E, elevation 15 m

Month	Mean sta. press.[1] (mbar)	Temperature[1] (°C) daily mean	daily range	extremes max.	min.	Mean vap. press.[1] (mbar)	Mean relat. humid.[2] (%)
Jan.	987.0	−1.0	5.7	7.8	−11.6	4.0	71
Feb.	989.4	−3.5	6.3	4.5	−17.0	3.0	65
Mar.	982.9	−6.5	5.4	3.6	−22.1	2.8	75
Apr.	984.7	−10.2	5.8	0.4	−29.1	2.2	77
May	989.3	−14.1	7.0	−2.4	−36.2	1.6	72
June	988.1	−15.3	6.2	−0.7	−35.0	1.4	72
July	985.1	−18.2	7.4	−3.6	−42.7	1.2	72
Aug.	983.8	−19.2	7.3	−3.9	−39.6	1.1	74
Sept.	982.2	−19.4	8.4	−3.9	−42.1	1.1	74
Oct.	982.4	−13.0	7.9	−1.6	−29.4	1.7	75
Nov.	984.6	−6.8	7.0	3.5	−23.9	2.7	76
Dec.	987.0	−1.7	6.4	8.1	−12.2	3.7	70
Annual	985.5	−10.7	6.7	8.1	−42.7	2.2	73

Month	Number of days with[1] winds ≥15m/sec	snow-fall	clear	cloudy	Mean cloudiness[1] (tenths)	Mean sunshine[3] (h)	(%)	Wind preval. direct.[1]	mean speed[4] (m/sec)	max. speed[4] (m/sec)	max. gust[5] (m/sec)
Jan.	4	9	5	15	6.5	322 (3)	46	NE	4.7	32	34
Feb.	3	9	6	14	6.4	237 (4)	49	ENE	4.4	26	29
Mar.	11	19	3	21	7.7	110 (5)	27	NE	6.9	35	40
Apr.	11	18	3	20	7.7	57 (5)	22	ENE	7.3	33	37
May	10	11	7	16	6.1	28 (5)	22	ENE	6.2	35	45
June	11	14	7	16	6.3	— (5)	—	NE	7.0	35	48
July	12	13	8	15	6.0	9 (5)	17	NE	6.6	36	49
Aug.	9	16	6	17	6.6	57 (5)	26	ENE	5.8	43	53
Sept.	7	13	7	14	5.9	136 (5)	41	NE	5.0	38	49
Oct.	6	14	5	18	7.0	199 (5)	42	NE	5.3	36	48
Nov.	9	12	5	17	7.0	269 (5)	44	NE	6.6	35	41
Dec.	4	7	7	14	5.8	425 (5)	57	NE	5.0	28	35
Annual	97	155	69	197	6.6	1,849		NE	5.9	43	53

[1] III/57–I/58; II/59–I/62; II/66–I/68.
[2] III/57–I/58; II/59–I/62; II/66–II/67, see text.
[3] Number of months (between brackets) used for hours and percentage of possible sunshine.
[4] From three-cup anemometer, 5 m above ground; maximum speed is 10-min mean.
[5] From aerovane-type instrument, 6 m above ground. No record for 1957–1958.

TABLE XLIV

CLIMATIC TABLE FOR DAVIS

Latitude 68°35′S, longitude 77°58′E, elevation 12 m

Month	Mean sta. press.[1] (mbar)	Temperature[1] (°C)				Dew point temp.[1] (°C)
		daily mean	daily range	extremes		
				max.	min.	
Jan.	987.4	−0.2	3.8	7.1	−8.6	−7.1
Feb.	986.2	−2.6	4.2	5.6	−15.2	−10.2
Mar.	983.8	−8.2	5.1	2.7	−26.7	−13.9
Apr.	986.0	−12.9	4.9	−0.2	−33.2	−17.9
May	988.1	−15.3	5.7	0.0	−38.3	−20.3
June	992.6	−15.6	5.8	−0.4	−32.2	−20.4
July	986.9	−17.4	5.7	−1.3	−37.9	−21.8
Aug.	985.2	−16.7	5.9	−1.1	−36.9	−21.4
Sept.	981.7	−17.0	6.2	−1.8	−34.3	−22.4
Oct.	981.7	−12.0	5.9	0.3	−27.0	−17.3
Nov.	983.5	−5.4	4.8	3.2	−15.0	−12.7
Dec.	986.4	−0.5	4.2	9.5	− 9.9	− 7.5
Annual	985.5	−10.3	5.2	9.5	−38.3	−16.2

Month	Mean cloudiness[1] (tenths)	Mean sunshine hours[1]	Wind[1]		
			preval. direct.	mean speed (m/sec)	max. gust (m/sec)
Jan.	6.2	251	NNE	4.7	30
Feb.	6.9	178	NNE,ENE	5.4	43
Mar.	7.2	77	ENE,NNE	5.2	45
Apr.	6.9	60	ESE,ENE	4.2	52
May	6.1	25	E,ESE	4.7	45
June	5.9	0	ESE,ENE	4.9	51
July	6.5	3	ENE,E	4.9	43
Aug.	6.5	50	ENE,E	5.0	48
Sept.	6.1	114	E,ENE	4.0	44
Oct.	6.8	146	ENE	4.9	49
Nov.	6.3	204	NNE,ENE	5.5	47
Dec.	6.6	257	NNE,ENE	5.7	47
Annual	6.5	1,365	ENE,NNE	4.9	52

[1] III/57–X/64.

TABLE XLV

CLIMATIC TABLE FOR MOLODEZHNAYA

Latitude 67°40′S, longitude 45°50′E, elevation 42 m

Month	Mean sta. press.[1] (mbar)	Temperature (°C) daily mean[1]	daily range[5]	extremes[2] max.	extremes[2] min.	Mean relat. humid.[3] (%)
Jan.	988.8	−0.6	5.6	8.3	− 8.9	65
Feb.	987.8	−4.4	7.2	3.9	−16.8	60
Mar.	981.5	−8.3	7.4	2.4	−22.4	65
Apr.	980.3	−11.2	6.2	−1.5	−26.7	73
May	983.6	−13.2	6.2	−2.6	−29.5	69
June	986.4	−15.4	6.5	−2.3	−29.9	68
July	986.5	−18.6	7.3	−4.2	−41.5	66
Aug.	981.4	−18.1	6.7	−3.5	−37.1	68
Sept.	979.3	−17.4	7.3	−2.6	−37.5	65
Oct.	975.6	−14.5	7.7	−4.5	−28.6	63
Nov.	980.2	−6.9	7.8	3.1	−23.2	64
Dec.	985.5	−1.3	6.9	8.3	−13.7	63
Annual	983.1	−10.8	6.9	8.3	−41.5	66

Month	Mean precip.[5] (mm)	Number of days with[5] drifting or blowing snow	$\overline{T}$ < −20°C	Mean cloudiness[4] (tenths)	Wind preval. direct.[5]	mean speed[2] (m/sec)	max. speed[2] (m/sec)
Jan.	31	2	0	7.2	E	5.4	19.0
Feb.	7	4	0	6.3	SE	7.4	22.0
Mar.	57	23	0	7.4	SE	12.0	27.6
Apr.	114	24	1	8.1	SE	14.6	29.4
May	90	24	4	7.0	SE	13.7	27.8
June	99	25	8	6.3	SE	13.5	29.8
July	75	22	14	6.0	SE	11.1	29.0
Aug.	97	21	10	6.8	SE	11.2	30.6
Sept.	70	19	10	7.1	SE	9.8	28.0
Oct.	50	17	2	6.4	SE	9.9	31.5
Nov.	21	7	0	6.5	E	8.0	23.5
Dec.	1	2	0	5.6	E	6.5	21.3
Annual	712	190	49	6.7	SE	10.3	26.6

[1] II/63–IV/68.
[2] II/63–VIII/67.
[3] III/63–XI/67, except XII/61.
[4] II/63–VIII/67, except XII/61.
[5] I/63–XII/65.

TABLE XLVI

CLIMATIC TABLE FOR MAWSON

Latitude 67°36′S, longitude 62°53′E, elevation 8 m

Month	Mean sta. press.[1] (mbar)	Temperature (°C)				Mean dew point temp.[3] (°C)
		daily mean[2]	daily range[3]	extremes[3]		
				max.	min.	
Jan.	989.2	−0.2	5.1	7.7	− 8.4	− 6.9
Feb.	988.5	−4.2	5.7	6.3	−15.0	−10.9
Mar.	986.8	−9.9	5.6	0.6	−25.0	−16.5
Apr.	988.7	−14.7	4.9	−0.3	−28.7	−19.1
May	990.6	−15.5	5.1	−2.2	−34.5	−21.9
June	991.7	−16.4	5.2	−1.4	−30.6	−22.5
July	989.1	−17.6	5.1	1.7	−34.6	−24.1
Aug.	987.4	−18.5	5.0	−2.7	−35.4	−24.3
Sept.	982.9	−17.7	6.2	−0.7	−33.4	−23.6
Oct.	983.9	−13.2	6.3	−0.6	−25.4	−20.2
Nov.	985.7	−5.5	6.5	4.4	−20.0	−13.0
Dec.	988.0	−0.4	5.4	8.8	−16.7	−6.8
Annual	987.7	−11.2	5.5	8.8	−35.4	−17.5

Month	Mean cloudiness[4] (tenths)	Mean sunshine hours[4]	Wind		
			preval. direct.[7]	mean speed[5] (m/sec)	max. gust[6] (m/sec)
Jan.	8.0	271	ESE	8.8	41
Feb.	7.6	215	ESE	11.0	53
Mar.	7.7	144	SSE,ESE	12.0	52
Apr.	6.6	122	ESE	10.6	44
May	6.4	41	ESE	11.9	54
June	6.4	0	ESE	11.3	61
July	7.0	15	ESE	11.7	56
Aug.	7.0	87	ESE	11.6	60
Sept.	6.3	151	ESE	11.1	62
Oct.	7.1	228	ESE	10.8	54
Nov.	8.0	261	ESE	10.8	57
Dec.	8.1	260	ESE	9.2	53
Annual	7.2	1,795	ESE	10.9	62

[1] II/54–I/68.
[2] III/54–II/68.
[3] III/54–XII/67.
[4] I/57–XII/67.
[5] II/54–XII/67.
[6] I/57–XII/66.
[7] I/57–XII/65.

TABLE XLVII

CLIMATIC TABLE FOR DUMONT D'URVILLE
Latitude 66°42'S, longitude 140°00'E, elevation 41 m

Month	Mean sealevel press.[1] (mbar)	Temperature[1] (°C)			
		daily mean	daily range	extremes	
				max.	min.
Jan.	989.3	−1.1	4.5	6.0	−9.0
Feb.	988.8	−4.2	4.2	4.5	−14.9
Mar.	985.0	−8.5	3.2	2.7	−22.3
Apr.	986.5	−12.0	5.1	1.1	−25.1
May	990.8	−15.3	5.3	0.2	−32.3
June	991.7	−17.6	5.5	−2.3	−33.4
July	992.6	−17.1	5.9	−0.2	−33.4
Aug.	988.6	−16.9	5.5	−0.7	−33.2
Sept.	984.9	−16.8	5.6	−1.1	−36.5
Oct.	982.5	−13.2	6.3	0.1	−28.5
Nov.	985.8	−7.1	6.2	3.5	−22.4
Dec.	989.9	−2.4	6.6	6.1	−10.6
Annual	988.0	−11.0	5.3	6.1	−36.5

Month	Mean sunshine hours[2]	Direct. strongest wind[1](°)	Mean wind speed[1] (m/sec)
Jan.	268	140	10.2
Feb.	182	140	12.1
Mar.	145	140	12.8
Apr.	93	160, 140	12.5
May	42	140, 160	11.6
June	10	140	10.6
July	12	140, 120	9.7
Aug.	63	160, 140	11.3
Sept.	147	140, 120	10.2
Oct.	230	140	9.9
Nov.	316	140, 160	10.2
Dec.	328	140	9.3
Annual	1,836	140	10.9

[1] I/56–XII/67; mean pressure values refer to sea level
[2] I/58–XII/67.

TABLE XLVIII

CLIMATIC TABLE FOR MIRNY

Latitude 66°33′S, longitude 93°01′E, elevation 30 m

Month	Mean sta. press.[1] (mbar)	Temperature (°C) daily mean[1]	daily range[2]	extremes[3] max.	min.	Mean relat. humid.[1] (%)	Inversion[4] strength (°C)	number of soundings
Jan.	986.2 (9)	−1.8 (12)	6.3	8.0	−14.1	69 (9)	−2.2	360
Feb.	987.0 (10)	−5.1 (12)	6.7	5.2	−18.6	67 (10)	−1.7	341
Mar.	982.8 (10)	−10.0 (12)	6.6	−0.1	−29.0	68 (10)	−0.8	391
Apr.	986.7 (10)	−13.8 (12)	6.1	−1.0	−31.3	71 (10)	1.8	381
May	986.6 (10)	−15.5 (12)	6.2	0.0	−40.0	72 (10)	0.1	415
June	988.6 (10)	−16.4 (12)	6.3	0.0	−32.8	71 (10)	0.4	394
July	984.3 (11)	−16.8 (12)	6.6	5.2	−36.9	73 (11)	−0.1	401
Aug.	981.7 (11)	−17.3 (12)	6.5	−2.9	−40.3	73 (11)	0.2	355
Sept.	979.3 (10)	−17.1 (12)	6.9	−2.0	−37.3	72 (11)	3.4	376
Oct.	977.6 (10)	−13.8 (12)	7.2	−0.8	−32.0	68 (11)	−1.3	426
Nov.	981.1 (10)	−7.3 (12)	7.2	5.0	−21.6	68 (11)	−1.8	407
Dec.	986.3 (9)	−2.7 (12)	6.2	8.0	−16.2	71 (9)	−1.9	427
Annual	984.0	−11.5	6.6	8.0	−40.3	70	−0.3	4,674

Month	Mean precip.[2] (mm)	Number of days with drifting or blowing snow[2]	$\overline{T} <$ −20°C[2]	winds[5] >15 m/sec	Mean cloudiness[1] (tenths)	Wind preval. direct.[2]	mean speed[3] (m/sec)	Global rad.[6]	Albedo[6] (%)
Jan.	13	6	0	7	6.6 (9)	E	7.9	20.2	76
Feb.	19	10	0	10	6.1 (10)	SE	9.5	12.8	77
Mar.	51	13	0	17	6.3 (10)	SE	11.2	7.3	82
Apr.	44	21	4	21	6.5 (9)	SE	12.8	2.6	87
May	92	25	8	25	6.3 (9)	SE	14.1	0.5	86
June	67	24	9	22	5.8 (10)	SE	13.5	0.0	86
July	77	25	10	23	6.8 (11)	SE	13.1	0.2	83
Aug.	95	25	8	23	6.6 (11)	SE	13.5	1.5	86
Sept.	52	21	10	23	6.0 (10)	SE	12.8	5.5	84
Oct.	43	16	2	18	6.7 (10)	SE	10.4	12.2	82
Nov.	46	12	0	13	6.0 (10)	SE	10.1	18.6	83
Dec.	26	7	0	8	7.2 (9)	E	8.8	22.3	82
Annual	625	205	51	210	6.4	SE	11.5	103.7	83

[1] Number of months used between brackets; I/56–XII/67.
[2] I/56–XII/65.
[3] I/56–VIII/67.
[4] Average temperature increase from surface to the 500 m level, 470 m above ground; I/56–XII/61.
[5] I/56–XII/63.
[6] Monthly and annual global radiation (in 10^3 Ly) and albedo. I/56–XII/59 and I/63–XII/65.

TABLE XLIX

CLIMATIC TABLE FOR WILKES

Latitude 66°15′S, longitude 110°35′E, elevation 12 m

Month	Mean sta. press.[1] (mbar)	Temperature[1] (°C)				Mean dew point temp.[2] (°C)
		daily mean	daily range	extremes		
				max.	min.	
Jan.	985.7	−0.3	4.4	7.8	−8.3	−4.3
Feb.	985.5	−2.3	4.9	6.1	−13.9	−6.3
Mar.	981.2	−6.2	6.2	3.6	−20.7	−10.2
Apr.	982.6	−11.5	6.6	3.3	−32.2	−16.1
May	985.6	−13.9	7.1	3.9	−33.9	−17.2
June	988.1	−16.3	7.4	5.0	−33.3	−19.3
July	984.8	−15.3	7.7	1.7	−37.2	−18.5
Aug.	981.3	−15.0	8.3	0.6	−36.7	−18.4
Sept.	977.4	−14.9	7.3	1.1	−35.6	−18.1
Oct.	977.6	−11.4	7.2	0.8	−26.7	−14.9
Nov.	981.4	−5.4	6.9	5.6	−27.2	−9.6
Dec.	986.8	−1.3	5.5	6.8	−15.6	−5.7
Annual	983.2	−9.4	6.3	7.8	−37.2	−13.2

Month	Days with precip.[3]	Mean cloudiness[1] (tenths)	Mean wind speed[1] (m/sec)	Max. gust[4] (m/sec)
Jan.	6	7.8	5.1	40
Feb.	7	7.3	6.1	51
Mar.	10	7.0	6.5	45
Apr.	7	6.5	6.5	51
May	8	6.6	7.3	46
June	10	5.8	7.2	59
July	9	6.4	8.0	51
Aug.	12	7.0	7.8	50
Sept.	11	7.3	7.5	50
Oct.	11	7.3	7.5	62
Nov.	7	7.3	6.7	52
Dec.	6	7.0	5.3	42
Annual	104	6.9	6.9	62

[1] I/57–XII/67.
[2] I/59–XII/67.
[3] I/61–XII/61; I/63–XII/66.
[4] I/57–XII/66.

TABLE L

CLIMATIC TABLE FOR ARGENTINE ISLANDS
Latitude 65°15′S, longitude 64°15′W, elevation 11 m

Month	Mean sea level press.[1] (mbar)	Temperature (°C) daily mean[2]	daily range[3]	extremes[3] max.	min.	Mean relat. humid.[4] (%)
Jan.	990.3	0.2	3.9	10.0	−10.6	84
Feb.	988.6	−0.2	3.7	11.7	−12.2	85
Mar.	987.5	−1.3	3.9	7.8	−16.1	87
Apr.	988.3	−4.7	4.9	7.2	−34.4	85
May	991.6	−6.6	5.8	6.7	−34.4	83
June	992.1	−9.6	6.6	6.1	−34.4	86
July	992.2	−11.2	8.0	4.4	−40.6	86
Aug.	992.1	−11.9	8.4	7.2	−43.3	87
Sept.	989.2	−9.0	7.9	5.0	−38.9	87
Oct.	985.6	−5.2	6.7	6.1	−28.9	88
Nov.	985.8	−2.7	5.8	6.7	−22.2	85
Dec.	991.5	−0.4	4.4	6.7	−11.7	87
Annual	989.6	−5.2	5.8	11.7	−43.3	86

Month	Mean cloudiness[3] (tenths)	Sunshine[4] (% of possible)	Mean wind speed[4] (m/sec)
Jan.	8.4	20	3.1
Feb.	8.4	22	3.4
Mar.	8.4	15	3.6
Apr.	7.9	14	4.4
May	7.4	13	4.0
June	7.3	5	4.0
July	7.1	6	4.2
Aug.	7.6	14	3.8
Sept.	8.2	17	4.6
Oct.	8.6	15	5.3
Nov.	8.5	19	3.7
Dec.	8.3	18	2.8
Annual	8.0		3.9

[1] Mean pressure at sea level, I/47–XII/67.
[2] I/47–XII/67.
[3] I/47–XII/60, from PHILLPOT (1967).
[4] I/47–XII/55, from BURDECKI (1957).

TABLE LI

CLIMATIC TABLE FOR MELCHIOR

Latitude 64°20′S, longitude 62°59′W, elevation 8 m

Month	Mean sea level press.[1] (mbar)	Temperature (°C) daily mean[2]	daily range[1]	extremes[1] max.	min.	Relat. humid.[1] (%)
Jan.	991.0	1.1	4.5	9.2	−5.0	83
Feb.	986.3	0.5	5.8	7.7	−6.6	85
Mar.	989.9	−0.7	3.5	6.4	−11.4	86
Apr.	988.3	−2.6	3.3	3.0	−11.0	86
May	994.8	−4.8	3.6	2.7	−20.5	85
June	991.2	−7.3	4.1	2.4	−22.0	87
July	992.0	−9.2	5.2	2.2	−26.5	88
Aug.	992.1	−8.7	6.6	3.2	−29.6	87
Sept.	991.0	−6.2	5.9	4.1	−25.3	86
Oct.	986.9	−3.2	5.0	4.0	−21.6	86
Nov.	986.0	−1.5	2.9	7.3	−12.1	83
Dec.	993.3	0.2	4.0	7.0	−4.9	83
Annual	990.2	−3.7	4.5	9.2	−29.6	85

Month	Winds >12 m/sec[3] (%)	Number of wind observ.[3]	Precip.[4] (mm)	Mean cloudiness[5] (tenths)	Wind preval. direct.[1]	mean speed[3] (m/sec)	relat. frequ. of calms[1] (%)
Jan.	2.4	1,030	43	6.0	S	2.0	45
Feb.	4.6	935	98	6.6	S	3.3	33
Mar.	8.8	974	124	6.3	S	3.7	28
Apr.	7.6	1,079	143	5.5	S	4.0	27
May	4.7	1,173	85	5.9	S	3.3	37
June	7.3	1,136	113	5.5	S	3.7	41
July	6.4	1,147	109	5.0	S	3.1	42
Aug.	8.1	1,147	104	5.8	SSW	3.6	40
Sept.	9.7	1,110	96	5.4	SSW	3.9	32
Oct.	9.9	1,144	130	5.0	SSW	4.5	25
Nov.	6.8	1,110	84	5.9	S	3.4	32
Dec.	2.2	1,155	60	6.0	S	1.9	47
Annual	6.6	13,140	1,189	5.8	S	3.4	36

[1] IV/47–XII/56; mean pressure refers to sea level.
[2] IV/47–X/61, except I, II, XII/60.
[3] IV/47–XII/56, except I, II, III/48.
[4] IV/47–XII/56; water equivalent in mm.
[5] I/52–XII/56.

TABLE LII

CLIMATIC TABLE FOR BAHÍA ESPERANZA, HOPE BAY

Latitude 63°23′S, 63°24′S[1], longitude 57°00′W, 56°59′W[1], elevation 7 m, 11 m[1]

Month	Mean sea level press.[2] (mbar)	Temperature[3] (°C) daily mean	daily range	extremes max.	extremes min.	Relat. humid.[3] (%)
Jan.	991.1	0.4	5.4	14.2	−8.5	84
Feb.	989.2	−1.0	5.4	7.6	−11.8	86
Mar.	990.1	−5.1	7.3	5.4	−17.6	85
Apr.	990.2	−6.8	7.5	14.2	−23.2	80
May	992.2	−9.8	7.1	8.4	−27.6	78
June	993.3	−9.8	8.3	10.5	−28.0	79
July	994.6	−9.9	8.7	8.9	−28.2	80
Aug.	993.9	−9.1	6.9	6.9	−32.1	81
Sept.	990.5	−6.8	7.8	9.5	−24.4	83
Oct.	985.0	−4.0	7.0	14.6	−25.4	82
Nov.	986.5	−1.3	5.9	12.4	−11.0	80
Dec.	992.1	−0.3	5.1	13.0	−10.9	85
Annual	990.7	−5.3	6.9	14.6	−32.1	82

Month	Winds[4] >12 m/sec (%)	Number of wind observ.[4]	Mean cloudiness[3] (tenths)	Wind mean speed[4] (m/sec)	Wind preval. direct.[5]	Wind mean speed[6] (m/sec)	Wind relat. frequ. of calms[6] (%)
Jan.	15.6	248	8.1	4.4	SSW,WSW	5.7	14
Feb.	19.0	227	8.0	5.7	SSW	9.7	13
Mar.	18.8	247	8.5	6.5	SSW	8.5	21
Apr.	20.0	240	7.8	6.7	SSW	12.4	20
May	26.5	244	7.5	6.1	SSW,WSW	6.8	42
June	27.8	240	6.5	6.6	SSW,WSW	8.4	29
July	34.0	247	6.6	6.4	SSW	9.5	25
Aug.	45.8	248	6.8	6.7	WSW	11.5	22
Sept.	23.8	240	6.6	6.5	WSW	10.2	20
Oct.	53.5	248	7.6	7.4	WSW	13.1	11
Nov.	37.8	240	7.8	7.0	SSW	9.6	26
Dec.	15.6	248	8.1	6.6	SSW	5.2	21
Annual	28.2	2,917	7.5	6.4	SSW	9.2	22

[1] Second position refers to Hope Bay.
[2] III/45–XI/61, except XI/48–IV/52. Mean pressure refers to sea level.
[3] IV/52–XII/56.
[4] III/45–XII/48 for Hope Bay.
[5] XII/45–I/49 for Hope Bay.
[6] I/55–XII/56 for Bahía Esperanza.

TABLE LIII

CLIMATIC TABLE FOR DECEPCIÓN

Latitude 62°59′S, longitude 60°43′W, elevation 8 m

Month	Mean sea level press.[1] (mbar)	Temperature (°C)				Mean relat. humid.[3] (%)
		daily mean[2]	daily range[3]	extremes[3]		
				max.	min.	
Jan.	990.0	1.4	3.6	10.0	−5.6	89
Feb.	989.2	1.1	3.6	6.5	−9.0	89
Mar.	989.1	0.1	3.3	6.1	−11.0	90
Apr.	990.1	−2.1	3.7	6.6	−18.0	90
May	992.8	−4.3	3.8	5.5	−20.0	91
June	992.5	−6.3	4.9	3.6	−21.9	90
July	994.1	−8.0	4.8	4.6	−26.9	90
Aug.	993.6	−7.7	6.2	3.2	−30.0	91
Sept.	991.0	−4.8	6.5	5.3	−22.6	91
Oct.	987.2	−2.4	4.6	4.3	−16.0	91
Nov.	987.8	−1.0	3.9	6.3	−13.8	90
Dec.	992.8	0.5	3.1	7.0	−8.0	87
Annual	990.9	−2.8	4.3	10.0	−30.0	90

Month	Winds[3] >12 m/sec (%)	Number of wind observ.[3]	Precip.[4] (mm)	Mean cloudiness[3] (tenths)	Wind[3]		
					preval. direct.	mean speed (m/sec)	relat. frequ. of calms (%)
Jan.	9.7	868	45	8.9	WSW	4.9	14
Feb.	13.3	791	52	8.9	WSW	5.6	10
Mar.	12.8	961	42	8.7	WSW	5.8	14
Apr.	19.1	929	32	8.6	WSW	7.2	14
May	16.3	961	23	8.5	WSW	6.1	22
June	20.2	930	27	8.0	WSW	6.6	17
July	18.0	960	27	7.9	WSW	6.7	17
Aug.	19.8	868	22	8.4	WSW	6.3	19
Sept.	21.3	930	20	8.1	WSW	7.0	16
Oct.	16.8	961	46	8.6	WSW	6.7	15
Nov.	12.0	930	30	9.0	WSW	5.7	16
Dec.	7.5	651	32	8.9	WSW	4.7	17
Annual	15.8	10,740	398	8.5	WSW	6.1	16

[1] II/44–IX/66, except VI/44; I/46–I/47.
[2] III/48–XI/67.
[3] III/48–XII/56.
[4] III/48–XII/56; water equivalent in mm.

TABLE LIV

CLIMATIC TABLE FOR ORCADAS (LAURIE ISLAND)
Latitude 60°44′S, longitude 44°44′W, elevation 4 m

Month	Mean sea level press.[1] (mbar)	Temperature (°C)				Mean vap. press.[4] (mbar)	Mean precip.[5] (mm)	Mean sunshine[6]	
		daily mean[2]	daily range[3]	extremes[3]				(h)	% of possible
				max.	min.				
Jan.	990.9	0.2	3.0	12.2	−7.0	5.3	34.6	47.6	9.3
Feb.	990.9	0.4	3.3	9.0	−9.8	5.3	40.1	38.3	9.8
Mar.	990.8	−0.4	3.9	10.8	−15.1	5.1	49.3	34.0	9.9
Apr.	990.9	−3.0	5.1	8.2	−31.5	4.3	41.2	25.2	10.1
May	992.6	−6.8	6.9	9.2	−31.9	3.2	33.5	16.9	10.8
June	994.7	−9.8	8.4	6.4	−38.3	2.4	26.9	9.7	15.2
July	994.6	−10.6	9.2	7.8	−36.9	2.3	31.6	18.3	13.2
Aug.	994.7	−9.9	8.9	8.2	−40.1	2.4	32.3	44.1	22.5
Sept.	993.2	−6.5	7.6	6.5	−32.6	2.9	29.6	65.8	21.8
Oct.	991.4	−3.7	5.8	8.7	−31.2	4.1	28.6	69.1	17.6
Nov.	988.4	−2.1	4.1	8.8	−20.4	4.4	31.5	54.8	11.2
Dec.	992.1	−0.6	3.3	9.6	−13.2	4.9	25.6	62.7	12.0
Annual	992.1	−4.4	5.8	12.2	−40.1	3.9	404.8	486.5	13.5

Month	Number of days					Mean cloudiness[6] (tenths)	Mean low cloud.[10] (tenths)	Mean wind speed[11] (m/sec)	Low cloud motion[12] (%)	Mean daily water temp.[13] (°C)
	clear[7]	cloudy[8]	precip.[9]							
			≥0.3 mm	>1.0 mm	snowfall					
Jan.	0.0	28.9	14.3	8.2	20.0	9.4	9.2	4.1	43.8	−0.4
Feb.	0.0	25.6	14.5	9.0	17.7	9.3	9.1	4.8	56.5	−0.1
Mar.	0.1	27.8	16.4	10.7	19.9	9.2	9.0	5.0	66.3	−0.3
Apr.	0.1	25.3	17.4	11.2	23.1	9.0	8.8	5.4	62.2	−0.9
May	0.6	22.5	16.0	9.1	22.9	8.4	8.1	5.3	64.9	−1.5
June	0.7	19.0	14.5	7.8	22.2	8.0	7.7	4.8	67.0	−1.7
July	1.0	18.2	15.2	8.8	22.7	7.8	7.4	5.1	74.6	−1.8
Aug.	1.0	19.0	16.0	9.0	22.7	7.9	7.5	5.3	73.6	−1.8
Sept.	0.6	21.0	14.9	8.9	23.3	8.2	8.0	5.5	76.7	−1.7
Oct.	0.2	25.3	15.3	8.4	24.3	8.8	8.5	5.6	77.2	−1.7
Nov.	0.0	27.2	15.3	8.8	23.0	9.2	8.9	4.8	69.0	−1.4
Dec.	0.0	28.0	12.8	6.9	20.6	9.3	9.0	4.0	65.0	−0.9
Annual	4.3	287.8	182.6	106.8	262.4	8.7	8.4	5.0	66.4	−1.2

[1] Pressure at sea level, 63 year mean, April 1903 – March 1966.
[2] 65 year averages, April 1903–March 1968.
[3] April 1904–November 1956.
[4] January 1906–December 1950.
[5] 1904, 1906, March 1908–November 1956.
[6] May 1903–December 1950.
[7] Days with ⩽2/10 average cloud cover, May 1903–December 1950.
[8] Days with ⩾8/10 average cloud cover, April 1903–December 1950.
[9] 1906, March 1908–December 1950.
[10] 1907–1950.
[11] June 1903–December 1950.
[12] Relative frequency (%) of low-cloud motion from SW, W, or NW; (all cases of observation of the motion of low clouds = 100%). Record: 1925–1949.
[13] Mean daily water temperature in Bahía Escocia, about 300 m southeast from the observatory. Record: 1904–1950.

References Index

Geographical Index

Subject Index